Springer Advanced Text in Life Sciences

David E. Reichle, Editor

Springer Advanced Texts in Life Sciences
Series Editor: David E. Reichle

Simon A. Levin   Mark A. Harwell
John R. Kelly   Kenneth D. Kimball
Editors

Ecotoxicology:
Problems and Approaches

With 91 Figures

Springer-Verlag
New York Berlin Heidelberg
London Paris Tokyo

Simon A. Levin
Center for Environmental Research
and Ecosystems Research Center
Cornell University
Ithaca, NY 14853, USA

Mark A. Harwell
Center for Environmental Research
and Ecosystems Research Center
Cornell University
Ithaca, NY 14853, USA

John R. Kelly
Ecosystems Research Center
Cornell University
Ithaca, NY 14853, USA

Kenneth D. Kimball
Research Department
Appalachian Mountain Club
Gorham, NH 03581, USA

Series Editor:
David E. Reichle
Environmental Sciences Division
Oak Ridge National Laboratory
Oak Ridge, TN 37830, USA

Library of Congress Cataloging-in-Publication Data
Ecotoxicology: problems and approaches/Simon A. Levin . . . [et al.],
  editors.
      p. cm.—(Springer advanced texts in life sciences)
    Bibliography: p.
    Includes index.
    ISBN 0-387-96762-1
    1. Pollution—Environmental aspects.   I. Levin, Simon A.
  II. Series.
  QH545.A1E293   1988
  574.5′222—dc19   67517   88-16051

Printed on acid-free paper

Typeset by Bi-Comp, Inc., York, Pennsylvania.
Printed and bound by R.R. Donnelley & Sons, Harrisonburg, Virginia.
Printed in the United States of America.

9 8 7 6 5 4 3 2 1

ISBN 0-387-96762-1 Springer-Verlag New York Berlin Heidelberg
ISBN 3-540-96762-1 Springer-Verlag Berlin Heidelberg New York

Preface

Ecotoxicology is the science that seeks to predict the impacts of chemicals upon ecosystems. This involves describing and predicting ecological changes ensuing from a variety of human activities that involve release of xenobiotic and other chemicals to the environment.

A fundamental principle of ecotoxicology is embodied in the notion of change. Ecosystems themselves are constantly changing due to natural processes, and it is a challenge to distinguish the effects of anthropogenic activities against this background of fluctuations in the natural world. With the frustratingly large, diverse, and ever-emerging sphere of environmental problems that ecotoxicology must address, the approaches to individual problems also must vary. In part, as a consequence, there is no established protocol for application of the science to environmental problem-solving.

The conceptual and methodological bases for ecotoxicology are, however, in their infancy, and thus still growing with new experiences. Indeed, the only robust generalization for research on different ecosystems and different chemical stresses seems to be a recognition of the necessity of an ecosystem perspective as focus for assessment. This ecosystem basis for ecotoxicology was the major theme of a previous publication by the Ecosystems Research Center at Cornell University, a special issue of *Environmental Management* (Levin et al. 1984). With that effort, we also recognized an additional necessity: there should be a continued development of methods and expanded recognition of issues for ecotoxicology and for the associated endeavor of environmental management. Thus, this volume addresses the growth of the science of ecotoxicology by examining a number of problems and approaches in depth. We envisioned a volume with a broad scope that would serve as assessment of the state and promise of the science—useful for scientists in the field, as well as for the environmental managers and regulators who would draw upon the scientific findings, and for graduate students who will be the next generation of environmental problem-solvers.

Ecotoxicology will continue to evolve. We hope this volume, drawing on a variety of tools and experiences, adequately reflects some of the progress and maturation of the field; and that it will stimulate further development.

The editors have many people to thank for guiding this long process through to its completion. The authors of the chapters were responsive to our many queries and reviews and were patient with our criticisms and suggested revisions. We also thank the reviewers of the individual chapters, especially including our colleagues at the Ecosystems Research Center.

Colleen Martin was instrumental in coordinating the initial stages of the project. Kathy May assisted in the typing of many manuscripts, and the conversion among a number of word processing files speaking languages of their own. The final typing, formatting, and preparation of the book was done by Carin Rundle, with her typical speed and efficiency. She patiently endured endless versions and last minute additions.

The greatest debt is owed to Cynthia Berger, who acted as book coordinator for the final phases of its completion. Juggling many tasks and coping with multiple editors' and authors' complaints and quirks, she effectively held us on course. Without her able help, this volume could well exist only as an intermediate draft—bound in spiral notebooks.

This publication is ERC-154 of the Ecosystems Research Center (ERC), Cornell University, and was supported by the U.S. Environmental Protection Agency Cooperative Agreement Number CR812685-01. Additional funding was provided by Cornell University. The work and conclusions herein represent the views of the authors or editors and do not necessarily represent the opinions, policies, or recommendations of the Environmental Protection Agency.

The Editors

Contents

Contributors

Birk, E.M.   181
Connolly, J.P.   221
Farrington, J.W.   279
Ford, J.   99
Garland, E.J.   221
Gearing, J.N.   411
Geiger, R.J.   315
Gillett, J.W.   367
Harwell, C.C.   497, 517
Harwell, M.A.   3, 9, 517
Herricks, E.E.   351
Howarth, R.W.   69
Kelly, J.R.   3, 9, 473

Kimball, K.D.   3
Klerks, P.L.   41
Lerman, A.   315
Levin, S.A.   3, 213
Levine, S.N.   145
Levinton, J.S.   41
MacKenzie, F.T.   315
O'Connor, D.J.   221
Perry, J.A.   351
Schaeffer, D.J.   351
Thomann, R.V.   245
Weinstein, D.A.   181

Contributors' addresses can be found at the beginning of each chapter.

Part I
Ecotoxicology: Problems and Approaches

Chapter 1
Ecotoxicology: Problems and Approaches

Simon A. Levin,[1,2] Mark A. Harwell,[1,2] John R. Kelly,[2] and Kenneth D. Kimball[3]

Virtually any discussion of the the risk assessment of releasing chemicals into the environment begins by organizing the fundamental considerations into three categories: evaluation of fate, transport, and effects. Each of these is contingent on the ecosystem into which the chemical is introduced; therefore none can be resolved adequately without a perspective that considers the ecosystem and its interactions with the introduced chemicals. Yet, as discussed further in Chapter 2 (Kelly and Harwell), it is a long step from this recognition to the development and implementation of methods that address these needs.

For example, determination of effects relies heavily on standardized laboratory bioassays, and on extrapolation from these to prediction of effects under field conditions. The inadequacy of laboratory bioassays for this purpose has been apparent for some time (e.g., Levin and Kimball 1984; Kimball and Levin 1985; Cairns 1986); but this inadequacy has been, to some extent, counterbalanced by the ease and economy of application of such tests, and by the body of experience that has developed concerning them. Furthermore, laboratory bioassays are valuable screening tools, and the development of standardized laboratory tests has made it possible to evaluate the effects of a large numbers of chemicals in relation to a substantial number of species. Thus, despite the limitations of such tests, their availability has led to dramatic improvements in water quality and the protection of ecosystems (Mount 1982).

[1] Center for Environmental Research, Cornell University, Ithaca, New York 14853
[2] Ecosystems Research Center, Cornell University, Ithaca, New York 14853
[3] Appalachian Mountain Club, Gorham, New Hampshire 03581

The case has been made repeatedly that assessment of effects on ecosystems requires the development of approaches that go beyond laboratory tests. It has long been recognized that we need to develop more sophisticated microcosm and modeling methodologies. Furthermore, we need to couple these with the results of experimental manipulations of whole systems to develop cross-system comparisons of the responses of ecosystems to stress, and to lay the foundations for the beginnings of a predictive ecotoxicology. This book is dedicated to these objectives, and to an analysis of what is known and what still needs to be known.

Kimball and Levin (1985) review the basic arguments for an ecosystem perspective. Faced with the need to evaluate the proliferation of chemicals being released into the environment, the Environmental Protection Agency in the United States, and other organizations in other countries, have been charged with the responsibility to establish criteria to protect human health and the environment, including the structure and function of ecological communities. The Clean Water Act and earlier legislation recognized that organic pollutants reduce dissolved oxygen levels in rivers. Early researchers (Forbes 1887; Forbes and Richardson 1913; Kolkwitz and Marrson 1908, 1909; Richardson 1928) documented that the ability of organic pollutants to deplete dissolved oxygen levels affected species composition in lentic habitats. These studies predated the development of the ecosystem concept, and thus it is not surprising that laboratory toxicity bioassays provided the framework upon which are based many of our current legislative acts that relate to the evaluation and regulation of chemicals.

However, using the laboratory toxicity bioassay to assess the impacts of pollutants does not provide an adequate basis for extrapolating to effects on populations, communities, or ecosystems, or for anticipating transformation processes of xenobiotic chemicals in the environment. Howarth (Chapter 4, this volume) elucidates some of the problems associated with laboratory bioassays in his case study based on the effects of oil pollution in the marine environment. Under some conditions, as in Howarth's study, LD_{50} data may be used to underestimate the potential for harm. In numerous other studies, such data would cause researchers to overestimate possible harm, for example because of behavioral responses of individuals, or compensation by functional groups of organisms. Other complications, confounding extrapolations drawn from laboratory data, can occur in field settings. These include cascading indirect effects, for example as critical species are affected, or the evolution of resistance to heavy metals, as discussed by Klerks and Levinton (Chapter 3, this volume). Such effects cannot be predicted without explicit consideration of the linkages among species, and of the connections between the biota and the biogeochemical processes that maintain the ecosystem.

In earlier publications (Levin and Kimball 1984; Kimball and Levin 1985), we have argued for the importance of developing a hierarchy of test systems, ranging from laboratory tests through microcosms to controlled field experimentation. We have argued further for the need to develop a theoretical framework based on comparisons, across systems and across stresses, of the ways systems respond to stresses. Such a theoretical framework must build upon syntheses derived from studies carried out in a wide variety of systems. There is a tendency to regard each new experience—each system and each stress—as unique because it is easy to see the special features of each situation. However, such a point of view is a prescription for defeat, because it implies that one cannot know the fate, transport, and effects of a chemical to be released into a particular ecosystem without actually carrying out the release. Thus, while recognizing explicitly that there will be uncertainty associated with any prediction, we must proceed on two fronts to develop a theoretical framework. We must develop functional classifications based on experiences from different systems, and we must simultaneously develop understanding of the mechanisms underlying the patterns we observe. Tools such as mathematical modeling can help to make the connections between levels of organization, and statistical analyses can help to detect and organize patterns; but the first necessary steps are insightful syntheses of existing information. Such syntheses are carried out in this volume by Weinstein and Birk (Chapter 7) for terrestrial systems, and by Ford and Levine (Chapters 5 and 6) for aquatic systems.

Prediction builds on the body of experiences from case studies, and on the understanding of mechanisms and correlations and incorporation into models that provide the means of extrapolation. For the fate and transport of chemicals in the environment, such models can be derived, to large extent, from first principles; although typically they still incorporate parameters that must be fitted to particular situations rather than estimated independently. The applications of such models are discussed in separate chapters by O'Connor et al., Thomann, and Farrington in Part III of this volume.

Physical models, such as the microcosms and mesocosms discussed in Chapter 14 (Gillett), in Chapter 15 (Gearing), and in the case study in Chapter 16 (Kelly), provide another critical component in the hierarchy. These models serve as a bridge between laboratory bioassays and whole system perturbations. There remains, of course, an extrapolation problem, albeit less severe, and the need for the development of a theory to serve as the basis for extrapolation.

Science can go a long way toward reducing the unknown, but uncertainty will remain in any evaluation. There is a tendency on the part of regulators to expect deterministic predictions from scientists, and to regard any uncertainty as resulting from inferior methodology. But there is an ineluctable core of uncertainty associated with predictions concerning the dynamics of any natural system, and the degree of uncertainty is dependent on the scale of interest. Good management must be based on explicit recognition of that uncertainty, and good science must characterize the uncertainty in any set of predictions. In the face of uncertainty, management schemes must be designed to respect that uncertainty. Typically, predictions should be expected to decline in reliability as the extrapolation distance increases; and this implies that long-term prediction should be regarded, at best, as a crude procedure. This implies that prediction should be coupled with monitoring, in a mutual feedback system that incorporates iterative prediction and adaptive management (Holling 1978; Walters 1986). The problem of decision making under uncertainty, and the complementation of risk assessment with risk management, is one of the central problems in ecotoxicology, and is the subject of the last section of this volume, particularly Chapter 18 by Harwell and Harwell.

This is not the first book to appear on the subject of ecotoxicology. The area is a rapidly expanding one, with a broadening disciplinary basis. Over the past decade, the need for improved ecological approaches to risk assessment and risk management has been increasingly recognized. The challenge now is to develop the methods to address those concerns. We hope that this volume is a step in that direction.

Acknowledgments. This publication is ERC-166 of the Ecosystems Research Center, Cornell University, and was supported by the U.S. Environmental Protection Agency Cooperative Agreement Number CR812685–01. Additional funding was provided by Cornell University.

The work and conclusions published herein represent the views of the authors, and do not necessarily represent the opinions, policies, or recommendations of the Environmental Protection Agency.

References

Cairns J Jr (1986) The myth of the most sensitive species. Bioscience 36: 670–672
Forbes SA (1887) The lake as a microcosm. Bull Peoria Sci Assn (reprinted 1925). Bull Ill State Nat Hist Surv 15: 537–550
Forbes SA, Richardson RE (1913) Studies on the biology of the upper Illinois River. Bull Ill State Lab Nat Hist 9: 481–574
Holling CS (ed) (1978) *Adaptive Environmental Assessment and Management.* Sponsored by the United Nations Environment Program. IIASA International Series on Applied Systems Analysis. New York: Wiley, 377 pp.
Kimball KD, Levin SA (1985) Limitations of laboratory bioassays: The need for ecosystem-level testing. Bioscience 35(3): 165–171
Kolkwitz R, Marrson M (1908) Ökologie der pflanzlichen Saprobien. Berichte der Deutschen Botanischen Gesellschaft 26a: 505–519. (Transl. 1967. Ecology of plant saprobia. In: Keup LE, Ingram WM, Mackenthun KM (eds) *Biology of Water Pollution.* Federal Water Pollution Control Admin., Washington DC: U.S. Dept. of the Interior, pp. 47–52)
Kolkwitz R, Marrson M (1909) Ökologie der tierischen Saprobien. Beiträge zur Lehre von der biologischen Gewässerbeurteilung. Int. Rev. der Gesamten Hy-

drobiol. und Hydrogeographie 2: 126–152. (Transl. 1967. Ecology of animal saprobia. In: Keup LE, Ingram WM, Mackenthun KM (eds) *Biology of Water Pollution*. Federal Water Pollution Control Admin., Washington DC: U.S. Dept. of the Interior, pp. 85–95)

Levin SA, Kimball KD (eds) (1984) New perspectives in ecotoxicology. Environ Manage 8: 375–442

Mount DI (1982) Progress in research on ecotoxicity: Single species tests (Part One). In: Mason WT Jr (ed.), Research on Fish and Wildlife Habitat. Office of Research and Development, U.S. Environmental Protection Agency, EPA-600–8/82–022, pp. 143–149.

Richardson RE (1928) The bottom fauna of the middle Illinois River, 1913–1925: its distribution, abundance, valuation, and index value in the study of stream pollution. Bull Ill State Nat Hist Surv 17: 387–475

Walters C (1986) *Adaptive Management of Renewable Resources*. New York: Macmillan, 429 pp.

Chapter 2
Indicators of Ecosystem Response and Recovery

John R. Kelly[1] and Mark A. Harwell[1,2]

To facilitate effective protection of environmental systems subjected to anthropogenic activities, there must be basic understanding of three areas: how the variety of biological components of ecosystems are exposed to stress; how the ecosystems respond to that disturbance; and how they recover or adapt. Given a solid understanding of ecosystem exposure–response–recovery relationships and their uncertainties, we might reasonably balance risks to ecological systems with risks and benefits to other systems of human concern, such as economic or societal systems. While the approach to the problem seems straightforward, the simple fact is that we presently lack sufficient ecological understanding in all three areas for most environmental stresses. With limited ability to make reliable stress–response predictions, we are greatly constrained in making appropriate environmental decisions. Consequently, instances of unexpected, adverse effects on the environment from a particular human activity continue to intermingle with instances of expensive over-protection from other activities. In principle, ecological risk assessment would minimize these problems.

The task for ecological risk assessment is to illustrate and accommodate uncertainties in stress-response relationships and, furthermore, to develop necessary informational bases for balancing environmental issues with economic and societal issues through the associated process of risk management. While human health effects analyses have usable methodologies for risk assessment from chemical exposures, *ecological* risk as-

[1] Ecosystems Research Center, Cornell University, Ithaca, New York 14853
[2] Center for Environmental Research, Cornell University, Ithaca, New York 14853

sessment is in its nascency, and the current reality is quite remote from the ideal paradigm of a well-developed methodology with tightly coupled risk assessment-risk management processes. This is so for a number of important reasons.

For example, we do not have a satisfactory basis for making cross-comparisons of the value of effects. How much money is a single human life worth? How much for an endangered species? How much does the value differ for an endangered whale species versus an endangered liver-wort species? Ecological and societal system responses typically cannot be reduced to simple economic terms; consequently, economic aspects tend to dominate, in part because they can be couched in terms that have obvious, quantifiable importance.

Even if there were a common currency available, we still have real difficulty in establishing a common level of acceptability of risk across different problems. This occurs, in part, because emotional and subjective factors play a major role in defining acceptability. For example, society pays great attention to airplane disasters, and we have an extensive system of imposing safety on the air transportation system. Yet over the five decades of the airline industry in the U.S., the total human fatalities from commercial airplane accidents have been of the same magnitude as predictably occurs in only a few weeks on U.S. highways from automobile accidents. Many factors create this disparity. One is that there is a very different perception of voluntary vs. involuntary risks; another is our inability objectively to treat infrequent, catastrophic events on the same basis as more constant and predictable, but relatively low-level consequential events.

Apart from subjective factors, for many problems it is the scientific, ecological uncertainties that complicate questions of comparable risk. Foremost are examples where the trade-off between the spatial or temporal scale of impact and the level of the ecological disruption is not clear. For example, it is not clear whether, over the long term, destruction of some given fraction of the viable bottom spawning area of a fishery would be equivalent to a quantitatively similar change, either as a reduction in the number of eggs produced or as an increase in larval mortality for a population of demersal fish; this is true because density-dependent and density-independent factors affecting different species' population levels vary, but are not often well-described. Moreover, this is complicated further because the relationship between spatial reduction of habitat and changes in fecundity or early life history survival rates is probably non-linear—perhaps having thresholds or critical points—and thus may vary with the magnitude of area or portion of population affected. Such relations are not characterized for populations interacting in complex natural ecosystems.

A second example in which scientific uncertainties obscure comparisons of ecological risk is exemplified by more global questions, such as:

how might the essential abolishment of agricultural productivity of 1% of the Earth's area by a process like desertification compare to a 1% reduction in global terrestrial productivity resulting from increased levels of biologically damaging UV-B radiation because of the chemical destruction of stratospheric ozone? With this type of problem there exists the issue of comparing the reduction in quantity and quality of biomass yield predicted for either stress condition. But beyond these seemingly straightforward calculations, which have their own sets of uncertainties, there are many other aspects involved. For example, we must consider also the possibility of global biogeochemical and climatological alterations, where the potential ecological ramifications could extend for periods longer than the original stress. The issue of scale, including the heterogeneity or patchiness of disruption, is thus central. Yet this is unresolved for problems involving assessments of ecological risks crossing from local to regional or global scales. Clearly, a variety of environmental decisions on dispersion vs. containment of chemical wastes hinges on these issues. Recognition of both scaling issues and global environmental problems is increasing swiftly, yet progress will be made only through attention to the problem by researchers from many fields, including ecotoxicology.

There is an immediate and related, if seemingly more parochial, problem for ecotoxicology. The efficacy of ecological risk assessment currently is limited seriously by our inability accurately to predict responses to stress at the scale of ecosystems. There are a host of reasons for this limitation, including the tremendous variety of ecosystems and potential types of anthropogenic perturbations to those ecosystems; the extreme range of spatial, temporal, and organizational scales that are inherent even within a single ecosystem; the lack of adequate baseline data for comparison of disturbed ecosystems to undisturbed ecosystems; fundamental limitations in ecological theory and understanding, creating uncertainties that presumably could be reduced by further research; and the pervasiveness of environmental stochasticity and other irreducible forms of uncertainty associated with stress–response–recovery predictions. Related to these items, later chapters in this book describe some of the variety of ecosystems and the degree to which their structural or functional responses to some types of stress can be described or predicted (Part II). Additionally, some of the empirical and theoretical models and approaches holding the promise of reducing some uncertainties of prediction are discussed in Part III. Finally, some of the scientific needs, regulatory concerns, and approaches to decision-making for ecotoxicology in the face of uncertainties are examined in Part IV.

While hindered by these factors, there also has been far too little attention given to the problem of *what should or could be measured to indicate the response of ecosystems when they are exposed to chemical stresses.* Certainly this is surprising, because it is such an obvious and essential aspect. This oversight may follow from the strong historical emphasis on laboratory assays for toxicology, the results of which help establish at what level of a stress the effects on some species may become demonstrable, but which results have little or no application toward the principal problem of assessing change in ecosystems. The complexity and variability in ecosystems can be overwhelming and confounding; in part, for these reasons, success in understanding relations between stresses and ecosystems has been uneven and incomplete. The process would benefit from development of a soundly based theoretical and conceptual approach.

Our approach here is to focus on the very fundamental issue of how to characterize an ecosystem's response to disturbance and its subsequent recovery upon removal of stress. Thus, the role of this chapter is partially introductory for the book as a whole, and to the problem of chemicals and ecological change—we present here the basic skeletal aspects on which the body of an adequate ecological risk assessment would rest for any given ecological disturbance. Initially, principal characterization of ecosystem exposure, response, and recovery (Table 2.1) are discussed in depth; the scope of characterization examined here contrasts with that actually conducted for ecosystem and/or stress-specific problems addressed in succeeding book chapters, in which the reality is that characterization of these elementary aspects rarely has been achieved to a de-

Table 2.1. Exposure, response, and recovery characterization, and uncertainties

---

EXPOSURE OF ECOSYSTEMS TO ANTHROPOGENIC STRESS
- *Fate, transport,* and environmental *modification*
- *Duration, frequency, intensity,* and *novelty* of exposure
- *Differential exposure regime* within an ecosystem

RESPONSE OF ECOSYSTEMS TO ANTHROPOGENIC STRESS
- Effects on *components* of ecosystems
- Effects on *processes* of ecosystems
- *Relevant scales and characterization* of response
- *Endpoints* for response characterization
- *Relevant indicators* for endpoints

RECOVERY OF ECOSYSTEMS FROM ANTHROPOGENIC STRESS
- Indicators for *components* and *processes*
- *Irreparable harm* and (or) the *ability to adapt*
- *Resilience and homeorhesis*
- *Scales of physical and biotic renewal processes*

UNCERTAINTIES
- *Variability in exposure*
- *Variability in ecosystems*
- *Extrapolation across types of stresses*
- *Extrapolation across types of ecosystems*

---

sired level for a risk assessment. Yet, critically, these three basic concerns fundamentally affect choices of how to measure ecological change. From consideration of these, the final focus on qualities of indicators of ecological effects from anthropogenic and chemical stress follows naturally.

2.1 Stress, Ecosystem Response, and Recovery

A goal of environmental protection is, essentially, to ensure the "health" of ecosystems. Yet unlike indicators of adverse human health effects, there are no comparable integrative, simple measures or indices that show the effects of disturbances on ecosystems. Attempts at an analogy of ecological health to human health (e.g., Rapport et al. 1985) have been unsatisfying, in part because the exposure of ecosystems to stress is very complex, with differential exposure to different—and unlike an organism, loosely connected—component parts within the ecosystem; in part because ecosystems are more diverse and more complex than the human metabolic system; and in part because ecosystems are much less internally coordinated and less able to respond to stress by controlled compensatory mechanisms that engender homeostasis of the system. If ecosystems truly could be seen as superorganisms, where a few key components or processes reflected the state of being of the ecosystem–superorganism, then ecological response and recovery predictions in principle could be as reliable as human health check-ups and prognoses. But the reality is that predictive ecology lags far behind because ecological systems are less robustly defined, their dynamics thus being inherently less tractable, and their state not so easily fully characterized.

This situation is reflected by the semantic dialogue in stress ecology, which it is not our intention to enter. Indeed, the term *stress* itself may be subject to greatest contention; note that the authors of the chapters in this book use a variety of descriptors. For our purposes, the terms *stress, disturbance,* and *perturbation* are all used essentially synonymously to refer to some agent or action, especially human-induced, that affects an ecosystem beyond its normal condition or dynamics. The *response* of an ecosystem is its change of state or dynamics as a consequence of the stress, up to the point of maximum deviation from its unstressed condition. *Recovery* is the rate and manner in which the ecosystem subsequently returns to its unstressed condition or, for example, later follows a chronological sequence of development (often termed "trajectory") that would coincide with an unstressed reference condition, if it indeed does either. Further characterization of these concepts (Table 2.1), in relation to ecosystems, is helpful to set the stage for a discussion of ecological indicators.

2.1.1 The Nature of Stress and Its Influences on Ecosystem Response

Whether one primarily is concerned with anthropogenic stresses in the environment or non-anthropogenic disturbances, such as natural climatic changes, storm events, or invasions by exotic organisms, some of the same general issues apply to ecosystem responses: e.g., what are the important response variables; how does one separate ecosystem stress response from background? An ecosystem response to perturbation of any kind must be characterized in relation to the particular stress, where consideration typically includes the *intensity, frequency, and duration* of disturbance. For *chemical* stress, similar consideration extends to characterizing the exposure regime for multiple components over time, accounting for environmental modifications that enhance or reduce toxicity. Additionally, the *novelty* of the stress is an important issue, since the ability of organisms and ecosystems to accommodate, or buffer against, a disturbance can depend on adaptation.

Ecosystems often are adapted to cope with many types of natural disturbances, especially those that are predictable (e.g., intertidal ecosystems adapted to diurnal and monthly cycles) or are periodic (e.g., grassland ecosystems adapted to fire). When these stresses are small relative to the scale of the ecosystem, and when they have occurred commonly during the historical development of the ecosystem, often the disturbance will be absorbed within the system structure, adding more heterogeneity but not changing the basic ecosystem functioning. For example, in response to wind-induced treefalls, a shifting mosaic pattern forms in northern temperate forests (Bormann and Likens 1979) where patches of early successional, mature, and late successional vegetation composition can be found throughout the landscape. Wind damage provides the mechanism for the replacement of established patches with early successional patches that undergo subsequent dynamical processes of maturing. Chronic disturbances that do not mimic the frequency of natural disturbances can alter the ability of a forest to absorb damage, and in turn alter basic properties of the ecosystem (Reiners 1983). An extreme case, where long-term exposure to chronic stress may be consequential, is exemplified where prolonged flooding results in death to a riparian forest that might well have accommodated frequent, but brief, periods of intermittent flooding.

In contrast, for acute stresses, although timing still can be key in terms of critical periods for some populations and seasonal cycles of growth and metabolism for the whole ecosystem, eliciting a significant response may relate more to the intensity of the perturbation. Research experiences with acute disturbances thus often have centered on abrupt, large magnitude changes in some characteristic of an ecosystem, exemplified by substantial removal of live biomass (Grime 1979) or by the removal of total biotic material (Reiners 1983). These large magnitude changes typically involve alterations of species composition, as in the example of the conversion of Vietnamese forest to grass and bamboo systems following the spraying of defoliants (Tschirley 1969). Elimination of sensitive species, reduction in pools of organic matter, and decreases in diversity have been observed by numerous researchers to occur simultaneously following acute disturbance (Weinstein and Bunce 1981; Freedman and Hutchinson 1980; Woodwell 1970; Gordon and Gorham 1963). Unfortunately, these knock-out blows to ecosystems, while excellent and informative studies, may yield little insight into the majority of environmental concerns, which are with insidiously low-level chronic stresses and gradual ecological changes.

In addition to acute or chronic duration of the disturbance, qualitatively different responses by ecosystems will occur depending on the frequency and novelty of the disturbance in the evolutionary history of the ecosystem. Thus, the same disturbance can have dramatically different consequences on different ecosystems, e.g., fire affecting grassland ecosystems versus tropical rain forests. In the former case, fire is a natural part of the long-term biogeochemical cycles of the ecosystem, necessary to rejuvenate a biotic community that is adapted to survival or redevelopment after fire. In the case of a tropical rain forest, a fire would lead to extreme disruption of the physical habitat and nutrient reservoirs, and reestablishment of the biotic community would take a very long time, if ever. Similarly, a particular ecosystem will likely respond differently to different disturbances; for example, the grassland may do well in the presence of fire, but be devastated by overgrazing. The responses of ecosystems vary widely across systems and across stresses. Insofar as an anthropogenic stress mimics a natural stress, then stress-response relationships are likely to be set by the ecosystem's adaptations to that type of stress. But for novel stresses, stress-response relationships can be quite unpredictable.

Not all chemical stresses are novel in the sense that ecosystems may already experience concentrations and circulation of them through biogeochemical cycles. For these, the novelty relates to concentrations that have not been commonplace.

For a fraction of the many chemicals anthropogenically produced that are not present in nature, ecosystems may be protected to some degree by internal processes. Processes can operate such that a particular component of the system is not highly exposed to the stress for more than short periods; for example, some highly particle-reactive chemicals may be quickly, essentially permanently removed by coagulation and sedimentation from the surface waters of some planktonic-driven aquatic ecosystems. Other internal processes may buffer changes in response to chemical stress. For example, microorganisms may be capable of biochemical degradation, converting toxics to non-toxic substances (Smith 1980). In some ecosystems, toxic materials are removed from sites of biological activity by the organisms themselves, which uptake compounds and sequester them in storage tissues (Amundson and Weinstein 1980). Thus, low doses of air pollutants can be absorbed by forests with no immediate changes, although the possibility exists that delayed damage could occur when accumulations reach toxic levels (Smith 1980). In general, for chemical stresses that are truly novel, i.e., xenobiotics, tolerance may be low and the limit to environmental mollification of effects may be achieved quickly; therefore dose levels are a key interacting factor.

Finally, the intensity of the stress, independent of novelty and other factors, often can be the primary determinant of response of the ecosystem. Sudden onset of an intense stress causes quite different effects on an ecosystem than a gradual, chronic stress of the same type. A hurricane will have far more devastating effects than a rain shower containing low velocity winds. Sensitivity to intense stress may depend on the phenological state of the system and the degree of preadaptation. For example, sudden exposure to freezing temperatures in the middle of summer can severely affect a temperate forest ecosystem, whereas the same temperature imposed gradually, over normal seasonal cycles, may cause no effect whatsoever (Harwell and Hutchinson 1985).

For chemical stress, differences in the intensity of loading rates and resultant environmental concentrations will determine whether different components are or are not exposed to damaging levels. Along a gradient of concentrations, as different organisms become significantly exposed, qualitatively different sets of functional ecological interactions therefore can arise. For example, when a critical predator is exposed and affected, but not its prey (or vice versa), the biotic network of the system can change substantially. Changes in biological structure and interactions, related to, but still unpredicted from, the intensity of most stresses, also can alter the capacity for ecosystem response to continued stresses of similar or different kinds.

Given the variety of factors of stress influencing response, simple generalities are not realistic in predicting effects of chemicals across different ecosystems. However, it may be possible to categorize the nature of different natural and anthropogenic stresses, and we should strive for such classification that would aggregate both classes of stress and types of ecosystems to develop a complex paradigm of ecosystem exposure-response relationships.

2.1.2 Characterizing Ecosystem Response

On what scale should response to stress be measured? No single scale can be selected exclusively; this is so for ecosystems, populations, and even different tissues of key individual organisms. For example, photosynthesis functions at the cellular level on a time scale of from minutes to hours, yet this process controls a life cycle for the tissue in leaves or needles as measured in months or years, and produces the structural tissue in bole, which may stand for decades to centuries. Furthermore, effects on photosynthesis, leaf litterfall, and standing biomass can be interdependent within forested ecosystems, but each effect can have different consequences in duration and in terms of the the sequence of ecological changes to follow.

At a population level, the dynamics of soil bacteria are extremely rapid and fluctuate wildly in time and space; by contrast, the population dynamics of forest bears are long-term and cover a large spatial extent. Responses of these populations must be measured on very different time scales; otherwise, critical characteristics of population dynamics, occurring asynchronously with measurement intervals, may be missed and the response of the population misinterpreted.

Changes at the ecosystem level also may be affected by slowly changing external factors such as climate, which can operate over centuries to millenia. Whereas at a human time-space scale, a mature forest may be perceived as essentially a steady-state climax ecosystem, identifiable by characteristic structure and processes, at another scale the same system is seen as only transitory, responding primarily to climatic alterations over geological time. For example, a present temperate forest might evolve toward the diversity of a tropical deciduous forest or slowly become the near monoculture of an arctic coniferous forest as the global climate changes.

Thus, a great continuum of scales is involved for ecosystem dynamics. Yet the scale of measurement itself gives bias to the perception of the ecosystem, and one must recognize that pattern seen as characteristic of any system is "neither a property of the system alone nor of the observer, but of an interaction between them" (Levin 1987; see also Levin, Chapter 8). Ecological examples of this dilemma of inseparability have been recognized acutely in oceanographic studies on zooplankton and phytoplankton, where patchiness and variability occur on virtually every scale of investigation and, of course, are inextricably linked to a vertically and horizontally turbulent physical environment that makes space and time inseparable for any practical purposes (e.g., Stommel 1963; Steele 1978; Harris 1980). Indeed, definition of the open ocean planktonic "ecosystem" is problematic compared to a physically defined area like a lake or watershed (Kelly and Levin 1986); but like anywhere, vastly different scales are emphasized by physiologists, ecologists, and climatologists who study the euphotic zone of the oceans.

These characteristics indicate that an ecosystem can be perceived and defined only in an operational context (Levin et al. 1984; O'Neill et al. 1986), as no unique characterization of an ecosystem exists and no single measure of its status will suffice. The consequence is that effects on ecosystems also must be examined operationally, with several major aspects being considered (Table 2.1). For chemical stress, this need seems especially clear, because the stress itself can be viewed as differentially applied—different parts of the system are exposed at different scales (i.e., concentrations, timing and duration of exposure)—and the problem of characterization as single, integrated response would be compounded even further. Thus, in practice, in attempting to characterize the response of an ecosystem to a stress, one must select from a suite of indicators, where these indicators are measures that reflect only some facet of the ecosystem—biotic or abiotic, structural or functional—at some spatio-temporal scale of observation.

2.1.3 Issues for Response Indicators

We can begin to appreciate the need to establish response indicators for practical use by considering, as an example, the case of a forested ecosystem subjected to an input of a toxic chemical. By what measure does one decide if the ecosystem has been altered by this exposure? And how long does one wait for a response? If the composition and productivity of the forests' trees appear unchanged, does that demonstrate no effect? What about changes in deer populations; or increased incidences of tree disease and pest outbreaks; or changes in the relative abundance of soil invertebrate species; or alterations in the rates of nitrogen release from decomposing leaf litter; or replacement of one bacteria species by another that performs the same functions; and what about mortality from the toxic chemical to a breeding pair of raptors or a breeding pair of protozoans? Which of these responses would constitute a serious ecological effect from the toxic input?

Although there is a wealth of direct and indirect ecological responses that can occur, it is obvious that each is not of equal importance. For example, loss of a species that plays a vital role in the ecosystem's structure or function would be more important than loss of a species of, say, fungi for which there is a redundant species capable of replacing it. Ecosystems are dynamic entities, even without human interference. Consequently, the various changes, such as those listed above for a forest, which could occur following input of the toxic chemical must be set against a backdrop of constant fluctuations, as species replacements and in processes. The signal of the response to the chemical stress must be discernable from the noise of natural variations across time or space; otherwise, the response would be undetectable and, hence, for all practical purposes, non-existent. In part, such recognition has provided the basis for intact ecosystem experiments and, more recently, precipitated pleas for long-term monitoring of natural ecosystems (e.g., Likens 1983; Schindler 1987).

Detection of response is part of the problem, but every detectable change cannot be considered necessarily to constitute changes in the health of the ecosystem; otherwise, every human activity would be found ultimately to result in ecosystem health changes. Killing a bacterium simply should not be construed as constituting an ecosystem effect; on the other hand, killing all members of the dominant tree species does constitute an ecosystem effect, irrespective of the context. Eliminating one species from the ecosystem may or may not constitute a change of importance to the ecosystem; loss of all species performing a common ecosystem function certainly would. Of course, evaluating changes at other than the extremes of conditions presents a more difficult challenge. Thus, we turn to indicators.

We have argued that the health of an ecosystem must be examined contextually, yet some general concepts are germane to defining response issues by which to identify useful qualities of indicators. First, one measure of the ecosystem's health is the displacement of the selected indicators of the ecosystem compared to the baseline state. This change in indicators may be considered in an absolute or a relative sense. There might be a focus on *changes* from a base state or a focus on the *state* of the system itself; e.g., it may be more relevant to know that primary productivity changed by 25% than to know what the actual productivity level is. This particular need especially extends to ecosystem manipulations, as illustrated by the long-term studies of Schindler et al. (1985). They recognized that it is imperative to compare the productivity of an acidification-stressed lake by reference to control lakes, because the latter had identifiable temporal trends; their critical concern was to determine if experimental values exceed the bounds of reference values, and thus the focus is on change from a (varying) base state.

A second general concept also may be valuable. How readily ecosystem indicators change in response to a given disturbance is an inverse measure of the *resistance* of the ecosystem to that type of disturbance (following definitions in Webster et al. 1975 and Harwell et al. 1978). Thus, a highly resistant ecosystem would change only slightly in response to the same stress that would cause major displacement in a low-resistant ecosystem. Note the reference to a specific type of stress. All evaluations of an ecosystem's resistance (or other measures of response) must be with respect to a particular stress, or combination of stresses, because, as we have discussed, the nature (particularly as affecting exposure) of different ecosystems, or individual components of a single ecosystem, may evoke different responses. In this light, one cannot accurately characterize a type of ecosystem as being intrinsically resistant; further, resistance may be seen in one indicator of the ecosystem, but not in another.

A similar concept to resistance is the idea of *sensitivity*. A sensitive ecosystem is one that responds readily to a particular stress; an insensitive ecosystem may be oblivious to the stress. Sensitivity is not identical to resistance, although both measure how much an ecosystem is affected by a disturbance. Sensitivity also has a temporal component, and a system that responds more rapidly than another, or to lower levels of disturbance, is considered to be more sensitive. By and large, most ecological research on response indicators to date has focused on sensitivity issues (Kelly, Chapter 16).

2.1.4 Recovery Issues

The *stability* of an ecosystem in the face of stress includes the aspect of how readily it recovers, in addition to how it initially responds. The separate concept of *recovery*, i.e., how the ecosystem may recuperate following removal of the stress, is often as critical to assess as the response. In particular, irreparable harm is a major environmental concern, and there are serious efforts that focus on techniques to foster restoration of ecological conditions where impacts have occurred and where the natural recovery processes may be slow or diminished because of the stress. There are two aspects of recovery, one related to how rapidly the ecosystem recovers, the other to how effectively the ecosystem recovers. The temporal aspect is characterized as the ecosystem's *resilience,* which is defined as the inverse of the length of time required for an ecosystem to return to near-normal. Note that one cannot reasonably define this as a complete return to a pre-perturbed state, because natural heterogeneity might preclude ever attaining that precise state. For example, the time required for a return of an ecosystem to within 5% of its pre-stress state could be a measure of resilience (Harwell et al. 1981).

One complicating factor sometimes considered is that the non-perturbed ecosystem may well not be at steady-state, even in the absence of human interference. Properties of the ecosystem may change over time. For example, diversity of a forest ecosystem will increase during the early stages of ecosystem development, decline in the middle stages of succession, and increase again during the later stages (Woodwell 1970). In this case, comparison is made not to a single set of steady-state values for the ecosystem, but to a moving set of values describing the spatio-temporal sequences expected for the undisturbed ecosystem. How resilient a non-steady-state ecosystem is to disturbance reflects the mechanism of *homeorhesis* of the ecosystem, in which feedbacks tend to direct the ecosystem's state along a specific time sequence. The analogue for the steady-state ecosystem responding to disturbance is *homeostasis,* a more commonly known term because of its applicability to physiological control in humans, such as maintaining body temperature or blood pH. The human analogue to homeorhesis is the developmental sequence and timing associated with ontogeny during gestation, following through maturation of the individual into an adult.

Beyond the resilience issues are questions as to whether or not the ecosystem effectively ever will return to its pre-perturbation state or trajectory. It is possible that complex systems, like ecosystems, when subjected to particular disturbances will become irreversibly transformed into another system, having different components, steady-states, and dynamics. This is a well-known characteristic of many ecosystems. For example, deforestation in the coastal hills of Venezuela has changed soil structure, seed sources, and the local physical environment sufficiently so that forests cannot return, even after the areas are abandoned by humans. This phenomenon repeats the irreversible loss of the great forests in Britain during neolithic times, as humans cleared land for agricultural production and energy resources. Perhaps the above examples merely reflect an exceedingly long time period of recovery, and the ecosystem will eventually recover. Yet for practical purposes of human interest, these examples of ecosystem change are permanent.

Non-anthropogenic stresses can also be irreversible, such as when the biotic community of a shallow lake gradually is replaced by other species as the lake fills in with sediments from surrounding terrestrial systems. Here, the time scale of examination is important, as what seems irreversible change in one time frame may not be the case over a much longer period. For example, the once lush coniferous forest covering Mount. St. Helens is now dramatically changed from its pre-eruption state. The initial images of a wasteland and of total devastation are fading slowly, as already initial signs of recovery are being documented. More significantly, the presence of those types of forests on the many volcanic peaks of the Cascade Mountains is testament to an eventual redevelopment of the ecosystem on areas that at one time also were exposed to lava and ash.

As in the case for response, recovery of ecosystems is dependent in part upon the characteristics of the stress and in part upon the history of the ecosystem. Related to the latter, an ecosystem that has been subjected to repeated disturbances may tend to deteriorate over time because of loss of nutrient reserves or substrate. Examples of such deterioration can be found in the forests of the San Bernadino Mountains of California following periodic ozone exposures (Miller 1973), and in salt marshes exposed to a series of oil spills (Baker 1973).

Recovery from repeated stress may be rapid if most of the important species within the ecosystem complete their life cycles within the interim between disturbance events (Noble and Slatyer 1980). Alternatively, for single disturbances of longer duration, recovery will be promoted if the important organisms within the ecosystem are capable of outlasting the toxicant by remaining in a latent or resting stage. For example, poor recovery has been noted in grassland systems exposed to oil because oil degradation proceeds slowly, and the actively growing portions of the grasses become directly exposed to the toxicant during their growth periods (Hutchinson and Freedman 1978).

Characterizing recovery of ecosystems has the same problems as characterizing the ecosystem response to stress, specifically which *indicator* to examine. Is an ecosystem recovered when its pools of nutrients are back to the pre-stressed state, or when a specific species has reestablished its population at a particular density, or when the residues of a toxic chemical in sediments or in biological tissues have decreased to below some threshold? Just as an ecosystem *functions* and *responds* to stress at widely differing rates, hierarchical levels, and spatial extents, it also *recovers* differentially. Selecting the appropriate suite of indicators is not a trivial task. And, as before, there are substantial difficulties added in establishing an appropriate baseline for comparing the stressed ecosystem, especially because when evaluating homeorhesis, not only an adequate existing baseline is required, but also a representation of what the ecosystem dynamics would have been had the ecosystem not been disturbed. Also, natural heterogeneity and fluctuations again raise the issue of detecting signals from the noise of natural variations.

2.2 A Focus on Useful Ecological Endpoints

The necessity of operational definitions of response and recovery, as following from the multifaceted aspects of stress, and the hierarchy of scales in ecosystems, leads one to some unfortunate realities. Simple schemes to overcome the intrinsic complexity are inadequate and untrustworthy. To some, this reality means each new problem must be addressed as unique at each new site. Yet an additional reality is that we do not have resources to deal with all problems in all places as if they were completely new, and therefore we must develop generic guidance on the response and recovery assessment. Clearly, to start, we must have multilayered, and adaptive, schemes of measuring and judging ecosystem response and recovery. From what essential threads could we weave this fabric of many colors?

The singlemost useful criterion to apply in order to reduce the measures of ecosystem health down to a manageable level is the requirement of *relevance to issues of concern to humans*. That is, a change in an ecosystem is only considered relevant if it relates directly or indirectly to something of concern to humans. By focusing on such *ecological endpoints,* a structured way of evaluating ecological effects can be developed, and a framework created for incorporating non-ecological issues into environmental decision making. Further guidance in some cases can be provided where society had to delve into a problem to agree on *regulatory endpoints,* which prescribe, by environmental legislation and the regulatory process, the particular aspects of the ecosystem deemed valuable and of human concern. Ecological and regulatory endpoints have been examined in previous reports (Limburg et al. 1986a; Harwell et al. 1986; Kelly et al. 1987; Harwell et al. 1987) and are covered in depth in the last chapters in this book. Ecological endpoints, in particular, are touched on here, as indicators may be chosen to relate to these valued concepts.

Ecological endpoints are categorized vis-à-vis issues of human concern (Table 2.2). The first item, human health, not environmental health, actually dominates environmental regulation and protection in the United States. Our concern here is limited, and applies only where ecosystems are vectors for human exposure to potentially harmful substances. Because the ecosystem-as-vector endpoint is so closely related to human health, however, it is easily identified as an important ecological endpoint. As an example, having radioactive cesium in fallout deposited on tundra ecosystems and subsequently being biomagnified to dangerous levels for human consumption of reindeer constitutes a measurable ecological endpoint, even if ecologically no adverse reactions occur. In this case, the radioactive cesium is unlikely to affect the biota population levels, or productivity, or nutrient cycling rates, or any other ecological indicator, in the tundra ecosystem, yet its presence in potential human food-chain pathways is *prima facie* a condition of great importance, as is dramatically demonstrated in Scandinavia in the aftermath of Chernobyl.

All of the other categories of ecological endpoints are not directly related to human health issues; consequently these endpoints are much more associated with ecological responses and recovery. These can be

Table 2.2. Ecological endpoints—issues of concern to humans

HUMAN HEALTH
  • *vector for exposure to humans*
SPECIES-LEVEL ENDPOINTS
  • *direct interest*
    • economic, aesthetic, recreational, nuisance, endangered species
  • *indirect interest*
    • bi-species effects (predation, competition, pollination)
    • habitat role
  • *ecological role*
    • trophic relationship
    • functional relationship
    • critical species
COMMUNITY-LEVEL ENDPOINTS
  • *food-web structure*
  • *species diversity*
  • *biotic diversity*
ECOSYSTEM-LEVEL ENDPOINTS
  • *ecologically important process*
  • *economically important process*
  • *water quality*
  • *habitat quality*

From Harwell et al. 1987.

separated into a traditional hierarchy of levels (Table 2.2). More extensive examples than given below are offered by Harwell et al. (1987).

The simplest ecological endpoints concern *direct effects* on particular species that have a direct interest to humans. Obviously, this includes species valued for economic, food, recreational, or aesthetic reasons; but other qualities may be involved. For example, many species on Earth are endangered to the point of the threat of extinction; unprecedented losses of species are underway, especially in tropical biomes. A select few of those species also hold some particular place of importance to humans, usually because of aesthetic values rather than because of a particular ecological value (after all, an endangered species typically is too rare to play a major ecological role). Additionally, many species are of direct concern for their negative value. Such nuisance species include disease vectors (e.g., certain species of mosquitoes), exotic plants that out compete native vegetation (e.g., kudzu, *Casurina*), and other noxious species, such as blue-green algal blooms.

Directly important species do not exist in vacuo, and *indirect effects* on these species, mediated by effects on other components of the ecosystem, must also be considered (e.g., Levin et al. 1984). If the other component is another species that itself is not of importance to humans, it may be considered a secondary ecological endpoint. Bi-specific interactions, such as predation, parasitism, and many types of symbioses, can be important here. Note that if one knows much about the chains and networks of species interactions of an ecosystem, then species that are many steps removed from the one of direct concern may be valuable as higher-order endpoints. The concept of ecological indicators follows naturally from such understanding.

Other such secondary endpoints may be found in effects mediated by habitat alterations, that is, in changes to the physical environment that supports biological populations. One important mechanism for indirect effects is through changes in the physical structure of the environment, such as in changing the vertical or horizontal heterogeneity of the ecosystem. This is particularly important when biota themselves constitute a significant part of the environmental structure for other species in the community, such as ecosystems that are near monoculture or have very dominant species, particularly of autotrophs. For example, one critical role played by mangrove tree species is that of providing a physical substrate for other plant and animal species, concomitantly allowing niche differentiation and the development of a diverse ecological community. Loss of the mangroves would consequently result in loss of habitat for a very large number of other species, which might not be directly affected by the stress that destroys the mangroves, but which might have a particular importance for humans.

Another important mechanism for habitat-mediated indirect effects is the amelioration of physicochemical conditions performed by biota or other components of the ecosystem. For example, the presence of tree and shrub species in semi-arid environments can induce evapotranspiration, which in turn results in increased precipitation; hence, the biota act as an enhancing pump to the hydrological cycle upon which virtually all terrestrial ecosystems depend. Adverse impacts on this role can indirectly result in other ecological effects, as starkly demonstrated by the current positive-feedback paradigm of desertification underway in sub-Saharan Africa; indeed, a major component of the droughts and famines recently experienced is the anthropogenic impacts on the environment through biomass harvesting for energy and food. Thus, the ecological endpoint of concern (tree and shrub productivity) is of both *direct* importance (with respect to the economic value of this resource) and of indirect, habitat-mediated importance (with respect to inducing local- to synoptic-scale reductions in precipitation).

A third component of indirect interest at the species level relates to the ecological role the affected species plays in the maintenance of the biotic community. For example, critical species have been identified in several ecosystems; a well-known example is provided in Paine's work (1974, 1980) on keystone species, particular predatory invertebrates whose presence or absence is the determinant of the presence or absence of other species in certain intertidal ecosystems.

Considerations of species roles are similar to, but distinct from, another type of ecological endpoint that involves higher-level attributes that are valued in their own right. For communities, this includes the overall trophic structure per se, not as a mechanism for supporting a particular species of concern but as a separate characteristic worth concern. Some such ecological endpoints have been refined by regulatory endpoints, including those which specify "maintenance of species diversity" as an endpoint and a measure of overall ecological health (see Harwell, Chapter 18). Humans have come to value the diversity of ecosystems as having intrinsic worth, and consequently any change in species diversity constitutes an ecological change of concern.

The issue of species diversity applies to ecosystem-scale biological units; but the idea also extends across landscape and regional units incorporating many different ecosystems and ecosystem types. Anthropogenic and other disturbances that extend across this spatial scale can result in loss of species, or at least regional-scale extirpation of species. Consequently, a new community-level concern arises, often termed a concern for *biotic diversity*. Thus, the rampant elimination of large numbers of species in the tropics, primarily through anthropogenic clearing of forests for biomass and for agricultural uses, is of immense importance because of the overall reduction in biotic diversity that is occurring on essentially a global scale. And, of course, extinction is of heightened concern compared to extirpation because the former involves the permanent loss of the species. Again, the biotic diversity itself is an ecological endpoint, as it constitutes an ecological change that is of direct (and indirect) importance to humans.

This brings us to the final level of ecological endpoints, i.e., those which cut across the biotic–abiotic boundary. Again there are direct and indirect components of the issue of importance ascribed to ecosystem-level changes. A direct interest is for maintenance of processes that are of particular economic importance, such as maintenance of the biogeochemical cycles in wetlands as a mechanism for reduction of anthropogenically elevated levels of nutrients in wastewater, or the processes by which a standing forest provides tremendous influence on water retention and flood control from major watersheds. Often the roles that ecosystems perform in ameliorating environmental extremes that would be of real concern to humans are greatly under-appreciated, but clearly many instances exist of ecosystem processes having significant economic or other societal benefit to humans.

Another set of ecosystem process-level endpoints relates to how changes in ecosystem processes translate into other changes of concern to humans; i.e., in the importance of the ecological role of the process. As an example, changes in primary productivity typically translate into ecological responses of concern, even if the production itself is not a major resource for humans. A general case can be made that changes in the biotic populations of ecological systems may or may not be directly or indirectly of importance, and redundancy or other compensatory mechanisms may mitigate against adverse effects on ecosystem processes. But the converse is not true, and changes in ecosystem processes almost invariably result in changes in biological constituents. As Odum (1985) puts it, functional disruption may "signal a loss of homeostasis."

Finally, other ecosystem process-level endpoints are specified directly by the regulatory framework, such as endpoints relating to quality of surface waters in the U.S., and endpoints relating to habitat quality, especially for the remaining wetland ecosystems. We have argued that there are no simple measures of such concepts of ecosystems, and this applies equally to water or habitat quality. But often physical (e.g., soil structure), chemical (e.g., dissolved oxygen levels), or biological parameters (e.g., available habitat for waterfowl, or forage base for deer populations) offer tested surrogates for these ecosystem-level concerns.

2.3 Ecosystem Indicators

2.3.1 Categories of Purposes

Given that we decide for a particular circumstance what ecological endpoints are applicable, the next step is to identify the indicators that should be measured to detect potential changes in the ecological endpoints.

There are innumerable components of the ecosystem that could be evaluated, but some careful thought can reduce these to a manageable set of indicators selected to optimize the detection of potential or actual changes in the selected ecological endpoints (Table 2.3).

One first purpose is for an indicator to have *intrinsic importance;* that is, the *indicator is the endpoint*. This is exemplified by many instances of monitoring for the populations of direct concern in particular systems, such as the striped bass population of the Hudson River and of other estuaries of the east coast of the United States. This species has developed, through the regulatory and especially the litigation framework, into

Table 2.3. Indicators of ecological effects

PURPOSES FOR INDICATORS
- *intrinsic importance*-key: indicator is endpoint
  - economic species
- *early warning indicator*-key: rapid indication of potential effect
  - use when endpoint is slow or delayed in response
  - minimal time lag in response to stress; rapid response rate
  - signal-to-noise ratio low; discrimination low
  - screening tool; accept false positives
- *sensitive indicator*-key: reliability in predicting actual response
  - use when endpoint is relatively insensitive
  - stress specificity
  - signal-to-noise ratio high
  - minimize false positives
- *process/functional indicator*-key: endpoint is process
  - monitoring other than biota; e.g., decomposition rates
  - complement structural indicators

CRITERIA FOR SELECTING INDICATORS
- *signal-to-noise ratio*
  - sensitivity to stress
  - intrinsic stochasticity
- *rapid response*
  - early exposure; e.g., low trophic level
  - quick dynamics; e.g., short life span, short life cycle phase
- *reliability/specificity of response*
- *ease/economy of monitoring*
  - field sampling
  - lab identification
  - pre-existing data base; e.g., fisheries catch data
  - easy process test; e.g., decomposition, chlorophyll
- *relevance to endpoint*
  - addresses "so what" question
- *monitoring feedback to regulation*
  - adaptive management

From Harwell et al. 1987.

the central concern for major anthropogenic disturbances to the Hudson. Concerns were expressed as to the location of thermal power plants, the management of the PCB-laden sediments in the river, and the recently resolved Westway Project (see Limburg et al. 1986a, b). As the dominant ecological endpoint, it also became a primary indicator of ecological effects, especially through evaluations of population levels, age structures, recruitment rates, mortality rates, migratory patterns, and so on. Demonstrated changes in striped bass as an indicator would constitute demonstrated impacts on the striped bass as an ecological endpoint. Many other examples of the endpoint as the indicator exist; e.g., deer population levels, breeding success in bald eagles, productivity of Douglas fir stands, and harvest yields of shrimp. The common theme for this intrinsic importance criterion is some direct, usually economic value of the species or processes.

Too much reliance can be placed on just monitoring indicators that are the ecological endpoints of concern; in particular, it is often the case that by the time an effect shows up in the indicator, it is too late for effective management or mitigation. Thus, another category of indicators is warranted, i.e., *early warning indicators*.

The key characteristic of an early warning indicator is for it to respond rapidly to the stress. This involves having minimal delay in response once the stress is imposed. Usually then, a good early warning indicator needs to be exposed to the stress early in its introduction into the ecosystem; note therefore that the specific indicator in a certain ecosystem often will vary with the different chemicals that may be partitioned differently throughout the ecosystem. Further, the indicator needs to respond rapidly once exposed. Thus, there is both a time-lag and a rate of response factor involved.

Because the key issue is the rapid indication of a potential effect, the early warning indicator is a red flag hoisted to alert ecologists that closer attention be paid to a potential problem. Consequently, the discrimination of the indicator can be rather low, that is, it need not provide all the information needed to evaluate effects on the ecological endpoints of concern, and tight, causal relationships between the stress and the triggering of the early warning indicator are not required. Hence, this functions as a screening tool, where false positives are acceptable at a relatively high rate (i.e., having the flag go up even though further evaluation demonstrates no ecological effects of concern). Conversely, a primary need is to minimize false negatives; thus, early warning indicators need to avoid missing a warning for a problem which is real. One way of enhancing this protective aspect is to incorporate more than one early warning indicator in an environmental protection scheme.

Another purpose for indicators is one specifically to minimize false positives, thus having one opposing quality to the early warning signal. Thus, a *sensitive and reliable* indicator has very high fidelity in characterizing an adverse effect on an ecological endpoint of concern. Note that this category of indicators is focused on *actual* ecological effects rather than on *potential* ecological effects. Thus, the key issue is not the rapidity of response, but rather is the reliability for indicating changes in ecological endpoints. This type of indicator does require strong evidence of causal relationships with the stress, and the response should be relevant to the state of the ecosystem.

One would use this type of indicator when the ecological endpoint itself is relatively insensitive to the disturbance or when it is difficult to separate actual stress-induced changes in the ecological endpoint from the normal variation that occurs in that endpoint over time and/or space. Stress specificity is of major importance here if the indicator is to provide the inferences of causality often needed to bridge the gap between demonstrated effects on an ecological endpoint and specific management or protection practices necessary to alleviate the adverse impact (a problem, for example, in developing a politically acceptable regulatory approach to acid precipitation in North America and Western Europe). Also, a criterion for this category of indicators is to minimize false positives, since incorrectly predicting unacceptable adverse impacts could lead to uneconomical over-regulation.

A variant on this sensitive and reliable indicator theme is for an indicator specifically attuned to long-term damage or significant change in ecological endpoints of concern. This relates to an issue discussed previously, of differences between concern for a *change of state* in the ecosystem versus concern for the *absolute level* in the ecosystem. Thus, long term indicators might be necessary to reflect alterations at large spatial-temporal scales, alterations that might not become evident by only examining short-term indicators. As an example, remote sensing of land use patterns can illustrate loss of estuarine habitats in coastal regions of southwest Florida, alterations that might not be as clearly apparent by monitoring, say, the population levels of tarpon.

Another category of purposes for indicators is to represent alterations in ecological functions and processes. It may well be that such *process indicators* are the ecological endpoints themselves, but it is more likely that they would represent the potential for changes in other ecological endpoints of more immediate concern to humans. Note that process indicators are not excluded from also being early-warning indicators, or sensitive indicators of change or of state. The category is specified here to highlight the fact that indicators need not be limited to species; rather the indicator is of some physicochemical condition or some process rate.

Much has been made of the relative value of structural indicators (i.e., biotic indicators of population and community structures) as opposed to functional indicators (i.e., of ecosystem processes). These issues are discussed more in Kelly et al. (1987) and in Chapter 16; we will just recapitulate some salient issues here.

Often it is stated that structural indicators, involving effects on biotic populations in aquatic ecosystems in particular, are more sensitive and better early warning indicators than are functional indicators (cf., Schindler et al. 1985). Although terrestrial scientists would not necessarily agree with this generalization, a few reasons are offered for it: a) Ecological effects are first manifested as effects on individual organisms and subsequently on populations; thus, functional responses would imply prior associated changes in biotic populations performing those functions. b) There is often a functional redundancy in ecosystems, so that effects on specific biota may not translate into ecosystem function effects because of replacements by other species performing a similar function role. c) Recovery of biotic structure of an ecosystem often lags behind recovery of functional attributes.

However, there are instances where functional indicators are equally or even more rapid to respond and more sensitive to toxic stress than most of their structural counterparts (Kelly et al. 1987). Intrinsically, preference for functional over structural indicators or vice versa should not be argued; rather, the point is that carefully selected functional indicators can significantly enhance our ability to evaluate ecological responses to stress, and their increased role as complementary to far more prevalent biotic indicators is warranted.

2.3.2 Criteria for Selection

The second theme concerning indicators is the identification of criteria by which to select a particular indicator to perform a particular role relative to particular ecological endpoints (Table 2.3). Different categories of indicators will require different criteria for selection; however, it is not our intent to specify those criteria for each indicator type, but, rather, to discuss selection criteria in general.

One criterion is the sensitivity of the indicator to perturbation, that is, how extensive is the response to a unit stress. This measure of indicator resistance is important in particular with respect to the normal variation that the indicator experiences over time and space in the absence of stress. These two factors combine to form the major issue of the signal-to-noise ratio for the indicator. A high signal-to-noise ratio is required for sensitive, stress-specific indicators; a low signal-to-noise ratio is acceptable for screening indicators, especially involving inexpensive or easily measured variables. As an example of this issue, consider the purported impacts on the striped bass population in the Hudson River ecosystem exposed to thermal power plants in comparison to the natural stochasticity of that species. Having density-independent mechanisms as the primary control for these populations means a poor signal-to-noise relationship, and experts were able to argue effectively on both sides of the associated controversy concerning the presence or absence of demonstrated effects, compensatory mechanisms, and other issues. By contrast, consider the data on CO_2 concentrations in the atmosphere monitored at the Mauna Loa observatory in Hawaii. The annual cycle in CO_2 levels, related to seasonal turning on and off the tremendous primary production potential of the Northern Hemisphere, is clearly discernable. It is superimposed over a constant, inexorable rise in the annually averaged CO_2 levels, reflecting effects of anthropogenic inputs of CO_2 and anthropogenic destruction of primary production. Here the signal-to-noise ratios of both the long-term trend and the annual cycle are quite good, and a convincing indicator is illustrated.

A second criterion relates to the rapidity of response of the indicators, involving the lack of a time lag and the high rate of signal processing, as discussed previously. Early exposure is important to maintaining early responses; consequently, for some stresses, especially transmitted through food-webs, the rapidly responding indicator is likely to occur at lower trophic levels. Quick response also implies quick population dynamics, such as having a short life span or at least a short-duration for one phase of the life cycle; for example, changes in phytoplankton are likely to occur much more rapidly than changes in cetacean populations.

Another criterion is the specificity of the response indicator. High specificity may be critical to establish causal relationships and, hence, appropriate management decisions. Conversely, broader response characteristics (i.e., low specificity) may be much more appropriate for screening indicators.

The criterion of the ease and/or economy of monitoring historically has been of special importance. In one sense, seeking the most economical indicator is of major benefit, in that a greater data base is likely to ensue. For example, historical records such as fisheries catch data or board feet of lumber harvested provide a major mechanism for comparing current environmental conditions with those in existence prior to the anthropogenic stresses of industrial society. On the other hand, we seem to become too enamored with the ease or historical precedent for indicators, and we may find ourselves focusing efforts on lots of data with great precision, but with poor accuracy or poor response rate, and little relationship to all of ecological endpoints of concern. This problem applies to laboratory testing (e.g., bioassays on easily maintained but ecologically insignificant species), field sampling (e.g., counting the 95th species in benthic samples even though looking at the top dozen or so will provide virtually all the relevant information), and pre-existing data bases (e.g., fishery catches, for which although the endpoint is the indicator, the endpoint is also a poor indicator of anthropogenic stress on the environment because of poor signal-to-noise ratios).

A final criterion here is the direct relevance of the indicator to the ecological endpoint of concern. Clearly, if the indicator itself is identical to the endpoint (e.g., the population levels of an endangered species), the relevancy is maximal. Otherwise, the closer linked the indicator is to the ecological endpoint(s) of concern, the less difficult it is to answer the "so what" question that often haunts demonstrations of environmental change. Such a question relates to how to characterize an ecosystem, as discussed previously; for example, where a certain isopod species in a forest appears as a sensitive indicator, the key to its relevance may lie in an ability to ecologically link its abundance or activity long-term to something like tree production or deer populations. Process indicators tend automatically to be considered more relevant than, for example, sensitive species indicators, because loss of a species sensitive to stress but otherwise not of particular note for humans or ecosystems raises the "so what" question (cf., Kelly et al. 1987; Kelly, Chapter 16).

2.4 Conclusion

Ecosystems are complex and varied, multiscaled and multitiered, and subject to continual change and adaptation. In the absence of a super organism at the ecosystem level, we are left to define ecosystems, their exposures to stress, and their subsequent responses and recovery, operationally. Consequently, we argue that a more sophisticated approach is needed to characterize ecological effects from anthropogenic activities, relying on a suite of ecological response–recovery indicators that reflect the status of the variety of facets about the ecosystem, or endpoints, of concern to humans. Focusing on these suites of indicators and endpoints can provide a systematic framework for optimal incorporation of scientific knowledge and understanding into a broader process of ecological risk assessment. It is unreasonable and futile to expect that a simplistic, generically applicable, single measure of ecosystem health can ever be realized. However, such a scheme is not required, and an ecological risk assessment paradigm for enhancing environmental decision making is a feasible goal for development in the ecological sciences.

Acknowledgments. The chapter is based in part on a previous ERC report, "Anthropogenic Stresses of Ecosystems: Issues and Indicators of Response and Recovery," authored by Mark Harwell, Christine Harwell, David Weinstein, and John R. Kelly (ERC-153, Ecosystems Research Center, Cornell University, Ithaca, New York 14853). Many of the ideas were developed over a series of discussions among the staff at the Ecosystems Research Center, including Barbara Bedford, Elaine Birk, Linda Buttel, Jesse Ford, Jim Gillett, Mark Harwell, Bob Howarth, Jack Kelly, Simon Levin, Suzanne Levine, Karin Limburg, Roxanne Marino, David Weinstein, and including Dr. David Yount, a visiting scientist at the ERC from the U.S. EPA Environmental Research Laboratory at Duluth. However, this report represents the views of the authors.

This is publication ERC-165 of the Ecosystems Research Center (ERC), Cornell University, and was supported by the U.S. Environmental Protection Agency Cooperative Agreement Number CR812685–01. Additional funding was provided by Cornell University.

The work and conclusions herein represent the views of the authors and do not necessarily represent the opinions, policies, or recommendations of the Environmental Protection Agency.

References

Amundson RG, Weinstein LH (1980) Effects of airborne fluoride on forest ecosystems. In: *Proceedings of Symposium on Effects of Air Pollutants on Mediterranean and Temperate Forest Ecosystems.* U.S.D.A. Forest Service Pacific SW Forest Experiment Station General Technical Report PSW-43, Berkeley, California. pp. 63–78

Baker JM (1973) Recovery of salt marsh vegetation from successive oil spillages. Environ Pollut 4: 223–230

Bormann FH, Likens GE (1979) *Pattern and Process in a Forested Ecosystem.* New York: Springer-Verlag

Freedman B, Hutchinson TC (1980) Long-term effects of smelter pollution at Sudbury, Ontario, on forest community composition. Can J Bot 58: 2123–2140

Gordon AG, Gorham E (1963) Ecological effects of air pollution from an iron-sintering plant at Wawa, Ontario. Can J Bot 41: 1063–1078

Grime JP (1979) *Plant Strategies and Vegetation Processes.* Chichester, UK: Wiley

Harris GP (1980) Temporal and spatial scales in phytoplankton ecology. Mechanisms, methods, models, and management. Can J Fish Aquat Sci 37:877–900

Harwell MA, Cropper WP, Ragsdale HL (1978) Nutrient cycling and stability: A reevaluation. Ecology 58:660–666

Harwell MA, Cropper WP Jr, Ragsdale HL (1981) Analyses of transient characteristics of a nutrient cycling model. Ecol Modelling 12:105–131

Harwell MA, Hutchinson TC, with Cropper WP Jr, Harwell CC, Grover HD (1985) *Environmental Consequences of Nuclear War. Volume II. Ecological and Agricultural Effects.* Chichester, UK: Wiley

Harwell MA, Harwell CC, Weinstein DA, Kelly JR (1987) *Anthropogenic Stresses on Ecosystems: Issues and Indicators of Response and Recovery.* ERC-153. Ithaca, NY: Ecosystems Research Center, 32 pp.

Harwell MA, Harwell CC, JR Kelly (1986) Regulatory endpoints, ecological uncertainties, and environmental decision-making. Oceans 86 Proceedings 3:993–998. Washington DC: Marine Technology Society

Hutchinson TC, Freedman W (1978) Effects of experimental spills on subarctic boreal forest vegetation near Norman Wells, N.W.T., Canada. Can J Bot 56: 2424–2433

Kelly JR, Duke TW, Harwell MA, Harwell CC (1987) An ecosystem perspective on potential impacts of drilling fluid discharges on seagrasses. Environ Manage 11(4):537–562

Kelly JR, Levin SA (1986) A comparison of aquatic and terrestrial nutrient cycling and production processes in natural ecosystems, with reference to ecological concepts of relevance to some waste disposal issues. In: Kullenberg G (ed) *The Role of the Oceans as a Waste Disposal Option*. NATO Advanced Research Workshop Series. Dodrecht Holland: D. Reidel Publishing Co

Levin SA (1987) Scale and predictability in ecological modeling. In: Vincent TL, Cohen Y, Grantham WJ, Kirkwood GP, Skowronski JM (eds) *Modeling and Management of Resources Under Uncertainty*. Lecture Notes in Biomathematics 72. Berlin: Springer-Verlag

Levin SA, Kimball KD, McDowell WH, Kimball SH (eds) (1984) New perspectives in ecotoxicology. Environ Manage 8(5): 375–442

Likens GE (1983) A priority for ecological research. Bull Ecol Soc Amer 64:234–243

Limburg KE, Levin SA, Harwell CC (1986a) Ecology and estuarine impact assessment: Lessons learned from the Hudson River and other estuarine experiences. J Environ Manage 22:255–280

Limburg KE, Moran MA, McDowell WH (1986b) *The Hudson River Ecosystem*. New York: Springer-Verlag

Miller PL (1973) Oxidant-induced community change in a mixed conifer forest. In: Naegle JA (ed) *Air Pollution Damage to Vegetation*. Adv Chem Series No 122, Washington DC: American Chemical Society, pp. 101–117

Noble IR, Slatyer RO (1980) The use of vital attributes to predict successional changes in plant communities subject to recurrent disturbances. Vegetatio 43:5–21

Odum EP (1985) Trends expected in stressed ecosystems. Bioscience 35:419–422

O'Neill RV, DeAngelis DL, Waide JB, Allen TFH (1986) *A Hierarchical Concept of Ecosystems*. Monographs in Population Biology 22. Princeton, NJ: Princeton University Press

Paine RT (1974) Intertidal community structure: experimental studies on the relationship between a dominant competitor and its principal predator. Oecologia 15:93–120

Paine RT (1980) Food webs: linkage, interaction strength, and community infrastructure. J Animal Ecol 49:667–685

Rapport DJ, Regier HA, Hutchinson TC (1985) Ecosystem behavior under stress. Amer Nat 125(5): 617–640

Reiners WA (1983) Disturbance and basic properties of ecosystem energetics. In: Mooney HA, Godron M (eds) *Disturbance and Ecosystems*. New York: Springer-Verlag, pp. 83–98

Schindler DW (1987) Detecting ecosystem responses to anthropogenic stress. Can J Fish Aquat Sci 44, Supplement No. 1:6–25

Schindler DW, Mills KH, Malley DF, Findlay DL, Shearer JA, Davies IJ, Turner MA, Linsey GA, Cruikshank DR (1985) Long-term ecosystem stress: The effects of years of experimental acidification on a small lake. Science 228: 1395–1401

Smith WH (1980) Air pollution - a 20th century allogenic influence on forest ecosystems. In: *Proceedings of Symposium on Effects of Air Pollutants on Mediterranean and Temperate Forest Ecosystems*. U.S.D.A. Forest Service Pacific SW Forest Experiment Station General Technical Report PSW-43, Berkeley, CA, pp. 79–87

Steele JH (ed) (1978) *Spatial Pattern in Plankton Communities*. NATO Conference Series IV: Marine Sciences, Vol. 3. New York: Plenum Press

Stommel H (1963) Varieties of oceanographic experience. Science 139:572–576

Tschirley FH (1969) Defoliation in Vietnam. Science 167: 779–786

Webster JR, Waide JB, Patten BC (1975) Nutrient recycling and the stability of ecosystems. In: Howell FG, Gentry JB, Smith MH (eds) *Mineral Cycling in Southeastern Ecosystems*. ERDA CONF-740513, Springfield, VA: NTIS

Weinstein LW, Bunce HWF (1981) Impact of emissions from an alumina reduction smelter on the forests at Kitimat, B.C.: A synoptic review. In: *Proceedings of the 74th Annual Meeting of the Air Pollution Control Association,* Philadelphia, PA, June 21–26, 1981, pp. 2–16

Woodwell GM (1970) Effects of pollution on the structure and physiology of ecosystems. Science 168: 429–433

Part II
Responses of Ecosystems to Chemical Stress

During the last several decades, steady progress has been made in the gathering of data on the responses of ecosystems to stress. As examined in the last chapter, there are many possible ways to characterize a response to stress, and accordingly, a number of ecological properties have been monitored in the face of anthropogenic activities that may directly or indirectly affect natural ecosystems. These properties range from species to ecosystem levels, where many recent efforts explicitly have grappled with understanding the roles of different species within the context of the ecosystem. For example, there has been focus on changes in major fish species as modified by indirect effects or by compensatory changes, either of which occur as a consequence of existence within an ecosystem, i.e., through interactions with other organisms and with the geochemical environment. Additionally, effects on a variety of integrated ecological properties have been studied in separate, site-specific observations of areas known or suspected to be changing in response to an anthropogenic perturbation. Included are functional measures such as primary productivity, decomposition and biogeochemical cycling, and structural characteristics, including levels of biomass and species richness.

With studies of ecosystems, as opposed to controlled laboratory studies of individual organisms, the issue of establishing causality between stress and an ecological response often has arisen. Related to this issue, recent efforts have made strides in understanding scales of variability in ecosystems; such information would aid in separating the ecosystem response to anthropogenic stress from the background environmental variability, i.e., the common problem of distinguishing "signal" from "noise." In particular, long-term monitoring of specific ecosystems and their properties, such as supported by NSF's program for Long Term

Ecological Research, will make possible some statistical analyses of the effects of stress on ecosystems. Yet it is also well recognized that observational data on structure and function of disturbed ecosystems need to be supplemented with other approaches.

One major advance has been provided by experimental ecosystem studies, including *in situ* manipulation of field ecosystems, as well as experimental mesocosm and microcosm techniques. These have allowed direct observation of ecological responses, across a hierarchy of ecological organizational levels, for a given stress and a defined ecosystem. The coupling of experiments, where effects are separable by use of unmanipulated controls, with long-term observational data is required; together these approaches facilitate a clearer understanding of the response of ecosystems to stress.

A variety of these observational and experimental approaches have been used to study responses to *chemical* stress, including ecological changes resulting from release of xenobiotics to the environment. The focus of these studies ranges from the site-specific to generic levels. For example, many individual sites and ecosystems subjected to an accidental spill or to a chronic exposure have been studied intensively; these published studies of specific cases provide the basis for many of the discussions in this section. Additionally, a number of syntheses that examine a given chemical stress upon a type of ecosystem (e.g., ozone on forests) have been published. By these, much insight can be gained, particularly in identifying those functional characteristics of traditional "classes" of ecosystems (e.g., coniferous forests, riparian wetlands, temperate lakes) that are critical in determining the nature of an individual ecosystem's response to a certain type of chemical stress. However, a fascinating, but more difficult, next step of synthesis, i.e., the comparison of behavior and responses of different types of ecosystems to a wide variety of different chemical stress, is virtually unexplored. Attempts at more general empirical synthesis are, in part, limited by lack of sufficiently detailed data for very many ecosystems or chemical stresses. As data gathering continues this situation will improve, but theoretical insights are also essential to the more general problem of prediction of ecosystem responses to stress.

The chapters of this section were solicited to reflect some of the variety of experiences, problems, and approaches taken to examine responses of ecosystems exposed to chemical stresses. In organization, the chapters flow from the specific to the more general, encompassing a large part of the range and level of synthesis just described. Thus, an example of ecotoxicological issues addressed in a site-specific study (Klerks and Levinton), is followed by the synthesis of one type of chemical stress on a class of marine ecosystems (Howarth). These, in turn, are followed by chapters that examine structural (Ford) and functional (Levine) responses for aquatic ecosystems of many types. Where possible, encumbered by the limits of empirical data, these chapters consider comparative responses across a selection of different chemical stresses. Similarly, the final chapter of this section (Weinstein and Birk) discusses structural properties of terrestrial ecosystems, in general examining the basis of ecosystem response to chemical stress in relation to physical disturbance response paradigms.

Chapter 3

Effects of Heavy Metals in a Polluted Aquatic Ecosystem

Paul L. Klerks[1] and Jeffrey S. Levinton[2]

The field of ecotoxicology includes the study of (1) the effects of toxic substances on single species and (2) the consequences for the resulting composition of species assemblages, which we shall call communities. In the study of single species we seek to understand how toxic substances affect the physiological function of individuals, and how toxic substances might affect the demography of the population. In essence, we wish to predict whether or not the species will decline as the result of the influence of the toxic substance. On the community level, minimally, we wish to know how the presence of a toxic substance alters the relative and absolute abundance of the species present. If there were no interactions among the species, this objective would involve merely a bookkeeping process whereby the toxic effects on all of the individual species would be summed up to predict the effect on the assemblage. Given the presence of species interactions such as predation and competition, such additivity may not occur. The effect of a toxic substance might vary among species. Some toxic substances are, for instance, transferred through a food web without harming lower trophic levels, but, owing to peculiarities of physiology, may have substantial effects at higher levels. Mechanisms of detoxification and uptake, furthermore, may profoundly affect such transfer to higher trophic levels, influencing the effect on the overall community. It is therefore desirable to understand the mechanisms of the physiological stress encountered, as well as detoxification mechanisms.

[1] Chesapeake Biological Laboratory, University of Maryland, Solomons, Maryland 20688
[2] Department of Ecology and Evolution, State University of New York at Stony Brook, Stony Brook, New York 11794

The processes described above result from the underlying phenomena associated with toxic effects on physiology, exchanges with the environment, and movement through the food web. To assess the first of these three components, investigators typically perform one of a variety of "bioassays," and then pursue the mechanisms of toxicity and detoxification. It is the purpose of this chapter to consider the additional complexity generated by the evolution of resistance to toxic substances. It is our contention that this factor strongly alters our perception of the type of answers typical toxicology studies can give, especially with regard to prediction of single species and community changes in natural habitats (see Luoma 1977; Levinton 1980). At worst, evolutionary response changes the predictions so completely that standard bioassays done on laboratory stocks are wholly inadequate, as they imply incorrect conclusions as to the fate of individual species and even the potential for transfer of toxic substances through the food web.

We present evidence for the evolution of resistance to heavy metals and will focus on our own research, which demonstrates the consequences of the evolution of resistance of invertebrates to cadmium in a freshwater tidal cove in the Hudson River (New York, USA). Our results demonstrate that evolutionary factors can alter the potential transfer of toxic substances through the food chain and the potential species composition of the cove.

3.1 Approaches

3.1.1 Ecotoxicology and Community Composition

A common approach in ecotoxicology is to look at the effects of pollutants on the species composition. The notion that severe pollution will lead to a reduction in the number of species is certainly not new. For example, Blegvad (1932) showed the absence of benthic animals in the outfall zone of the Copenhagen domestic waste discharge. Similarly, Reish (1959) showed a decline in the number of benthic species with an increase in pollution level in Los Angeles–Long Beach Harbors, with a total absence of animals in the most polluted bottoms. Since then, many researchers have reported a reduction in the taxonomic diversity or the number of species in a metal-polluted environment (McLean 1975; Beers et al. 1977; Thomas and Seibert 1977; Sanders et al. 1981; Foster 1982a; Sanders and Osman 1985). Davies (1976) demonstrated the presence of variation in mercury sensitivity among algal species, which could lead to a change in the species composition upon mercury pollution. According to Gray (1979), a departure from a log-normal distribution of individuals among species (due to an increase in abundance of several neither rare nor common species) would be the earliest detectable change caused by pollution in a community. Gray (1982) hypothesized that this change is due to interspecific differences in life-history traits, rather than interspecific differences in their tolerance to stress.

3.1.2 Ecotoxicology and Evolution

Our main approach is to look at the evolution of resistance to pollutants. Evidence that such adaptations might occur frequently comes from human experience with pesticides; pest species are known to have developed resistance to many pesticides. The same processes can be expected to lead to resistance to pollutants in populations inhabiting polluted environments, if the pollution is severe enough to have a detrimental effect on these organisms and genetic variation in sensitivity is present in the population. The evolution of resistance to environmental pollutants has been documented by Antonovics, Bradshaw and their co-workers on plants at spoils of abandoned mines in the British Isles (e.g., Antonovics et al. 1971; Bradshaw and McNeilly 1981). An increased resistance in organisms from polluted environments has also been reported for metal-polluted aquatic environments. Arima and Beppu (1964), Nelson and Colwell (1975), Timoney et al. (1978), Devanas et al. (1980), and Timoney and Port (1982) report the presence of metal-resistant strains of bacteria in metal-polluted areas. Similar results are reported for algal populations by Russell and Morris (1970), Stokes et al. (1973), Jensen et al. (1974), Foster (1977), Say et al. (1977), Fisher and Frood (1980), Murphy et al. (1982), and Foster (1982b). The studies by Bryan and co-workers (Bryan 1974; Bryan and Hummerstone 1973 a,b) demonstrated an increased resistance in the polychaete, *Nereis diversicolor,* from a metal-polluted estuary in southwest England. An increased resistance in animals from polluted environments has also been reported for bivalves (Nevo et al. 1984), crustaceans (Brown, 1976, 1977, 1978; Fraser et al. 1978; Moraitou-Apostolopoulou et al. 1979; Callahan and Weis 1983), insects (Wentsel et al. 1978), and fishes (Weis et al. 1981; Bailey and Saltes 1982; Smith 1983; Benson and Birge 1985). There is some evidence that not all species from a polluted environment exhibit an increased resistance. Antonovics et al. (1971) report that many of the plant species around a mine were not present on the mine spoils. And Rahel (1981) reported the absence of increased resistance in common shiners, *Notropis cornutus,* from a zinc-polluted stream relative to these fish from a nearby unpolluted stream. Furthermore it is hard to evaluate how many negative results were never published.

There are two phenomena that can account for an increased resistance in organisms from a polluted area. Exposure to a pollutant can lead to physiological acclimation or behavioral changes within the life-span of the exposed animal. But, as indicated above, exposure to a pollutant can also lead to natural selection for an increased resistance, resulting in genetic adaptation to the pollutant. If physiological acclimation is the cause of the increased resistance, it will not be passed on to offspring of the exposed animals. But in the case of genetic adaptation, this resistance will be maintained in offspring. To assess long-term changes as a result of pollution it is important to distinguish if acclimation rather than adaptation is the cause of the increased resistance.

Most of the above-cited studies, especially the ones on animals, present very little evidence that the observed resistance is based on natural selection for resistance, rather than just physiological acclimation (Klerks and Weis 1987). Moreover, it is in most situations impossible to distinguish between the two possibilities for an increased resistance, unless it is possible to obtain offspring of resistant animals that were born and raised in a clean environment. It is even preferable to use second generation offspring, to diminish the influence of maternal effects (Levinton 1980).

The physiological mechanism that enables a resistant organism to withstand higher pollution levels is of interest from a scientific point of view. But insight in such mechanisms is also important when attempting to predict how the presence of resistant organisms in a polluted environment will affect organisms at other trophic levels. For example, a reduced metal accumulation in resistant animals will result in lower pollutant levels in its predators. Such a reduced metal accumulation in metal-resistant organisms has been reported repeatedly (e.g., Beppu and Arima 1964; Bryan and Hummerstone 1973a; Hall et al. 1979; Weis et al. 1981). But an increased metal accumulation in resistant organisms has been found as well (e.g., Bryan and Hummerstone 1971). If resistant organisms accumulate higher levels of a pollutant, they must have some mechanism that prevents the pollutant from exerting a toxic effect. The exact nature of such a mechanism might have implications for effects at other trophic levels. If increased sequestering in a non-toxic form is the mechanism for the increased resistance, the pollutant might remain in this non-toxic form upon transfer to a predator.

One mechanism for the detoxification of several heavy metals, including cadmium, is binding to certain proteins. These metallothioneins were first isolated from horse kidney by Margoshes and Vallee (1957) and characterized by Kagi and Vallee (1960). Metallothionein-like proteins have since been found in fish (e.g., Bouquegneau et al. 1975), invertebrates (e.g., Olafson et al. 1979; Suzuki et al. 1980; Thompson et al. 1982), and even fungi (Lerch 1980). Metallothioneins are low-molecular-weight cytosolic proteins, with a high affinity for cadmium, mercury, zinc, copper, silver, and possibly other heavy metals. Their synthesis is induced by these metals. These proteins probably play a role in the regulation of essential metals (Brouwer et al. 1986). An induction of these proteins coincides with an increased survival when exposed to the metals (Bouquegneau 1979; Benson and Birge 1985). No increased presence of these proteins in genetically adapted organisms has yet been demonstrated, though an increased production has been demonstrated for metal-resistant mammalian cell lines (Griffith et al. 1981; Rugstad and Norseth 1975; Beach and Palmiter 1981). In yeast, metal resistance seems to be associated with amplification of the metallothionein genes (C. Paquin, pers. comm.).

Another detoxification mechanism that has been reported is the sequestering of heavy metals in vesicles or granules (Brown 1977; Coombs and George 1978; George and Pirie 1979; Lowe and Moore 1979; Mason et al. 1984). In oligochaetes, uptake of heavy metals in chloragosomes has been reported (Ireland and Richards 1977; Prent 1979). It is possible that the two detoxification mechanisms are related; e.g., heavy metals could first become bound to metallothionein-like proteins and then become enclosed in a vesicle.

3.2 Some Background on Metal-Polluted Foundry Cove

Foundry Cove is a tidal freshwater bay on the east side of the Hudson River. Figure 3.1 shows its location near Cold Spring, New York, 87 km (54 miles) upriver from Battery Park (New York City). The cove connects with the Hudson River through an opening in a railroad trestle separating the cove from the main part of the river. Foundry Cove is adjacent to the Marathon Battery Site in Cold Spring. The facility on this site, constructed in 1952, was used for the production of nickel-cadmium batteries from 1953 through 1979. Several companies used the facility; in the earlier years of its existence batteries were produced under military contracts, later the plant was owned by private companies that produced batteries for commercial use (Resource Engineering 1983).

Concentrated cadmium and nickel nitrate solutions were used in the manufacturing process. At one time, cobalt was used as an additive (Kneip and Hazen 1979). Wastewater from the manufacturing process had a pH range from 12 to 14 and contained cadmium as metallic hydroxides, mostly as suspended solids (Resource Engineering 1983). The waste was initially discharged into the Hudson River through the Cold Spring sewer system, but about 10% was discharged in a bypass system emptying into Foundry Cove. The manufacturing company on the site in 1965 was ordered to disconnect from the sewer system in that year, after which all the waste was discharged directly into Foundry Cove. From 1971 to 1979, wastewaters for which maximum discharge limits had then been set were again discharged via the sewer system into the Hudson River (Resource Engineering 1983). Resource Engineering (1983) has estimated that during the manufacturing operations a total of 179,105 kilograms (394,860 lbs) of

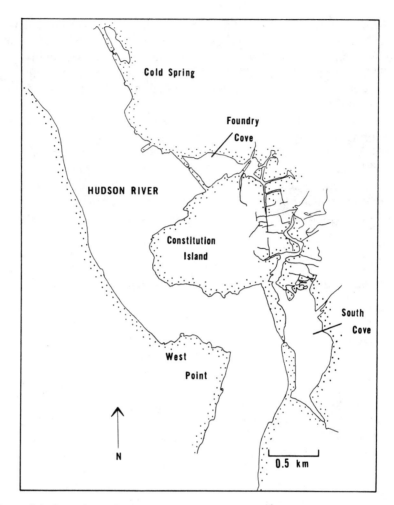

Figure 3.1. Location of metal-polluted Foundry Cove and our control area, South Cove. (From Knutson et al. 1987. With permission from Elsevier Science Publishers Ltd., England.)

cadmium were discharged. Of this, 51,004 kilograms (112,444 lbs) of particulate cadmium and 1,569 kilograms (3,460 lbs) of soluble cadmium were discharged into Foundry Cove. The remainder was discharged into the Hudson River at the Cold Spring Pier.

A "Final Consent Judgement" was issued in 1972 resulting from a civil suit filed against Marathon Battery Company in 1970. This agreement required the removal of all sediment exceeding 900 ppm cadmium (based on wet weight). Dredging to meet this requirement was done in 1972 and 1973. Only 10% of the cadmium released in the cove was removed in this process (Resource Engineering 1983). It is therefore not surprising that extremely high cadmium and nickel concentrations were encountered in subsequent years; up to 50,000 and 11,000 ppm (based on dry sediment), respectively (Kneip 1978; Kneip and Hazen 1979; Hazen and Kneip 1976; Occhiogrosso et al. 1979). The main effect of the dredging seems to have been a change in the the horizontal distribution pattern (Kneip and Hazen 1979). In 1975, about 30% of the cove still had surface sediment levels in excess of 1000 mg cadmium per kg dry sediment (Knutson et al. 1987).

In 1983 we determined cadmium concentrations in the surface sediments (0–5 cm) of Foundry Cove. Our results showed that cadmium concentrations were in general much lower than those found in 1974 and 1975 (Hazen and Kneip 1976). We determined that in 1983 only 8% of the total area of the cove contained surface sediment with cadmium levels in excess of 1000 ppm (based on dry weight), relative to 30% in 1974 and 1975 (Knutson et al. 1987). A contour map of the surficial cadmium concentrations in 1983 is shown in Figure 3.2.

We determined depth profiles of the cadmium concentrations in Foundry Cove, as we hypothesized that the polluted sediment is possibly being covered through the deposition of new sediment. One profile that is representative for the more polluted parts of Foundry Cove, is shown in

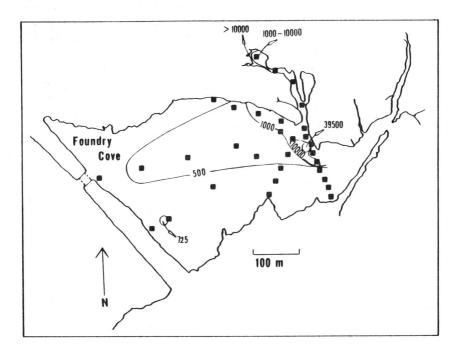

Figure 3.2. Contour map of cadmium concentrations in surficial sediment (0- to 5-cm depth) of Foundry Cove, in 1983. Values are in micrograms of Cd per gram dry sediment. (From Knutson et al. 1987. With permission from Elsevier Science Publishers Ltd., England.)

Figure 3.3. It is taken where the channel coming from the outfall enters into the main part of the cove. Additional profiles are presented by Knutson et al. (1987). The subsurface peak at a depth of several centimeters suggests that following cessation of the cadmium discharge, the highly polluted sediment is becoming covered by new sediment. Alternatively, cadmium in the sediment near the surface has been lost to the overlying water and thereby lost from the cove in the tidal water exchange. Loss of metals from polluted sediments to the water column has been reported (Hunt and Smith 1983). Various estimates of the total amount of cadmium present in the cove have been reported (Bower et al. 1978; Kneip and Hazen 1979; Resource Engineering 1983; Knutson et al. 1987) and do not indicate a significant loss of cadmium from the cove. In addition, Resource Engineering (1983) estimated that about 0.8% of the deposited cadmium is lost from the cove yearly through transport of suspended solids and dissolved metal. The decrease in the cadmium content of the surface sediment is about an order of magnitude greater than that figure for the yearly loss, also indicating that the decreased cadmium content of the surface sediment is more likely due to burial than to loss of the cadmium. If one assumes that the surface sediment has been recently deposited, background cadmium levels could be expected in this sediment. The surface sediment still has fairly high cadmium levels nevertheless, which could be due to either biological mixing, diffusion, or sedimentation of resuspended polluted sediment.

   The above measurements illustrate that organisms in Foundry Cove have been exposed to elevated levels of cadmium and nickel for about 30 years (although metal levels at the sediment surface have decreased,

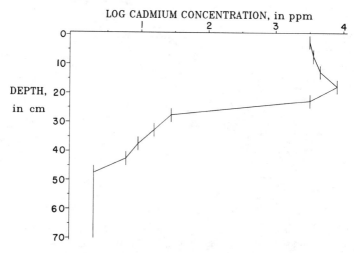

Figure 3.3. Representative profile for Foundry Cove, showing cadmium concentration (based on sediment dry weight) versus depth.

probably since 1971). Both short- and long-term processes have therefore had ample time to affect the macrofauna in the polluted sediments of this ecosystem.

3.3 Effects of Heavy Metals on the Composition of the Macrobenthos

Occhiogrosso et al. (1979) showed a reduction in the density of the benthic macrofauna in 1973 (post-dredging) and 1974 at a highly polluted site in Foundry Cove. Our survey of the macrobenthos was aimed at determining the current impacts of the pollution on macrobenthos abundance and comparing macrofauna compositions in more detail. We determined the composition of the macrobenthos in Foundry Cove (at several sites with a different level of pollution) and in our closely situated control area (South Cove). We later included a site with extremely high metal levels that we encountered in the outfall in Foundry Cove. Replicate cores (surface area 35 cm^2) were taken at these sites from the top 5 cm of the sediment, sieved with a 500-μm mesh-opening sieve and preserved.

There are significant temporal differences in the total macrobenthos density, but no significant differences between the sites (Figure 3.4). This evidence suggests that the heavy metal pollution has not reduced the macrofaunal abundance. Of the two major groups comprising the macrofauna, oligochaetes are slightly (though not significantly) more abundant at the most polluted site (Figure 3.5), whereas the chironomids are less abundant (Figure 3.6). The reduction in chironomid density was also found in 1973 and 1974 (Occhiogrosso et al. 1979). But contrary to that study, no effect on overall density is evident in our data (based on 1984 samples). In our study, the oligochaete *Limnodrilus hoffmeisteri* and the chironomid *Tanypus neopunctipennis* are more abundant at the site with the highest cadmium levels (Figures 3.7 and 3.8). The difference in the composition of the macrofauna between the sites is clearly reflected in the number of taxa present (Figure 3.9). The most polluted site has an impoverished benthic macrofauna. This, however, could result from factors other than the metal pollution, as this site is a small cul-de-sac on a tidal creek while the other sites are situated in a more open body of water.

3.4 The Evolution of Resistance to Heavy Metals

For an evaluation of differences in resistance between populations from Foundry Cove and the control area, we concentrated our efforts on the oligochaete *L. hoffmeisteri,* the most abundant species in these areas. To

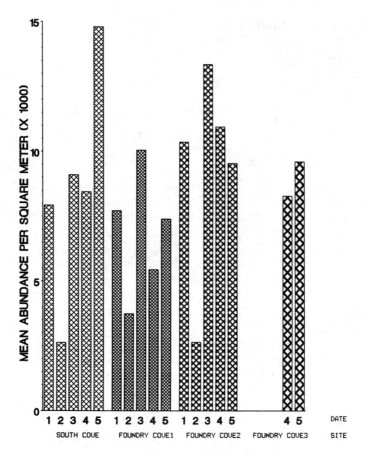

Figure 3.4. Mean density of all macrobenthos species combined at a site in South Cove (*CS*) and at three different sediment pollution levels in Foundry Cove, (*FC*) sampled at different times of the year (dates 1–5 are respectively, 4/11, 6/4, 8/1, 10/11, and 12/12/84). Sediment cadmium concentrations at the sites are respectively 19, 533, 7,050, and 52,020 ppm (dry weight). Foundry Cove 3 was sampled in October and December only. General Linear Models: no significant difference between sites, a significant difference between dates (p < 0.001) and no significant interaction.

determine if *L. hoffmeisteri* from the polluted area has an altered heavy metal resistance, we exposed worms from both areas to metal-rich Foundry Cove sediment as well as to sediment from South Cove. Survival was determined after 28 days. The results of this experiment, shown in Figure 3.10, demonstrate that *L. hoffmeisteri* from the polluted area has an increased resistance to the metal-polluted sediment of Foundry Cove. We performed an additional toxicity experiment, in which we exposed worms to a solution of cadmium, nickel, and cobalt in water (in the same ratios as present in Foundry Cove sediment). This experiment was in-

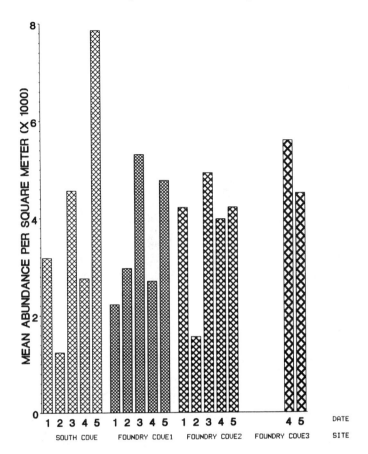

Figure 3.5. Mean density of oligochaetes at a site in South Cove and at three different sediment pollution levels in Foundry Cove. See legend to Fig. 3.4 for additional details. General Linear Models: no significant difference between sites, a significant difference between dates (0.01 < p < 0.05) and no significant interaction.

tended to confirm that the above-mentioned resistance differences are indeed differences in resistance to cadmium, nickel, and cobalt, and not resistance differences to another toxic factor of the Foundry Cove sediment. The worms were checked every hour for survival. The results (Table 3.1) confirm that the two groups of worms differ in their resistance to a mixture of cadmium, nickel, and cobalt.

To distinguish between acclimation and adaptation, we performed experiments with the second generation offspring of Foundry Cove worms. Both first and second generation offspring were born and raised in the lab in clean sediment. Heavy metal resistance of second generation offspring was compared to the resistance of worms collected from Foundry Cove

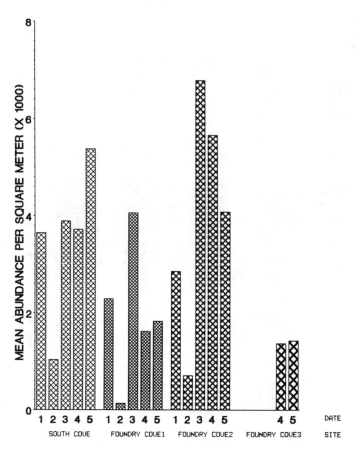

Figure 3.6. Mean density of chironomids at a site in South Cove and at three different sediment pollution levels in Foundry Cove. See legend to Fig. 3.4 for additional details. General Linear Models: a significant difference between sites $(0.01 < p < 0.05)$ and between dates $(0.001 < p < 0.01)$, but no significant interaction.

Table 3.1. Survival Time of *L. hoffmeisteri*

Origin	Survival time
South Cove	13.67 ± 0.47 hr
Foundry Cove	23.55 ± 1.40 hr

Survival times (mean $\pm$ S.E., $N=42$) of *L. hoffmeisteri* from Foundry Cove and South Cove, in a solution of cadmium, nickel, and cobalt (respectively, 1, 0.6, and 0.03 ppm) in reconstituted fresh water (soft).
The difference between the two groups is highly significant ($p < 0.001$, ANOVA on log-transformed data).

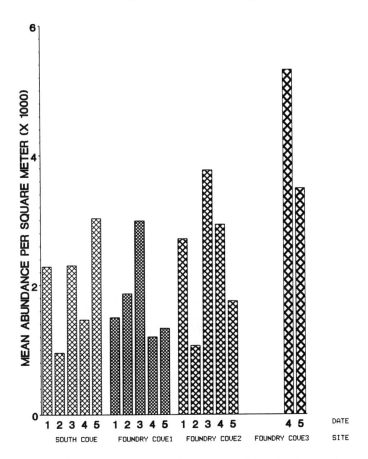

Figure 3.7. Mean density of the oligochaete *Limnodrilus hoffmeisteri* at a site in South Cove and at three different sediment pollution levels in Foundry Cove. See legend to Fig. 3.4 for additional details. General Linear Models: a significant difference between sites, as well as between dates (both: $0.01 < p < 0.05$), but no significant interaction.

and from the control area by exposing the three groups of worms to Foundry Cove sediments with different heavy metal contents. Survival was again determined after 28 days. These results (Figure 3.11) show that much of the increased resistance of the Foundry Cove worms has a genetic basis. The second generation offspring of Foundry Cove worms were slightly less resistant than the worms from Foundry Cove, but the difference was not statistically significant. This possible reduction indicates the presence of an environmental component in the resistance of Foundry Cove worms, or could be due to a relaxation of the selection pressure when culturing these worms for two generations in clean sedi-

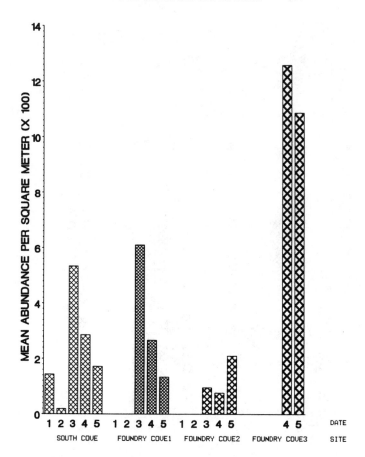

Figure 3.8. Mean density of the chironomid *Tanypus neopunctipennis* at a site in South Cove and at three different sediment pollution levels in Foundry Cove. See legend to Fig. 3.4 for additional details. General Linear Models: a significant difference between sites and between dates (both: $p < 0.001$), but no significant interaction.

ment. The latter implies a cost associated with the increased resistance: a reduced fitness in clean sediment.

3.5 Heavy Metal Accumulation and Detoxification in Resistant Biota

L. hoffmeisteri from Foundry Cove and the control area were kept in metal-rich Foundry Cove sediment. Cadmium concentrations were then determined, after acid digestion, by flameless atomic absorption spectro-

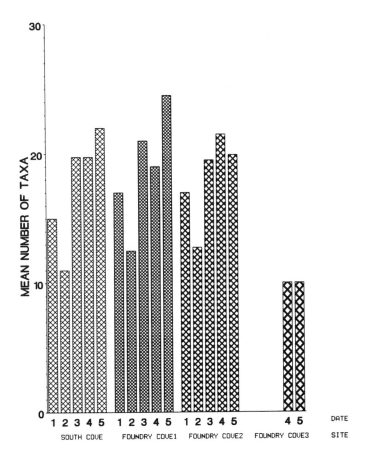

Figure 3.9. Mean number of macrobenthos taxa at a site in South Cove and at three different sediment pollution levels in Foundry Cove. See legend to Fig. 3.4 for additional details. General Linear Models: a significant difference between sites ($0.001 < p < 0.01$), between dates ($p < 0.001$), but no significant interaction.

photometry. Foundry Cove worms have a significantly higher net uptake of cadmium when kept in metal-rich sediment (Figure 3.12). We also exposed worms collected from Foundry Cove and South Cove, as well as the aforementioned second generation offspring of Foundry Cove worms born and raised in clean sediment, to the radioactive isotope [109]Cd in water for 6 days. The worms were exposed in petri dishes with reconstituted hard water (ASTM 1980), containing a total concentration of 1 ppm cadmium and a radioactivity of about 0.6 μCi/ml. After exposure, the worms were homogenized and their radioactivity determined in a gamma counter. This radioactivity is a direct measure of the net cadmium uptake. These results (Table 3.2) confirm the increased cadmium uptake in the

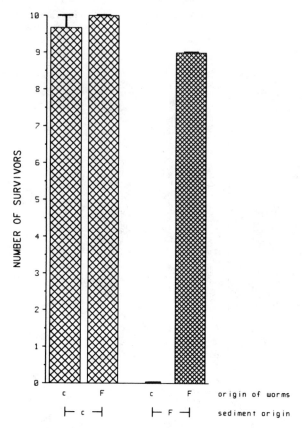

Figure 3.10. Long-term survival (28 days) of *L. hoffmeisteri* from Foundry Cove (*F*) and South Cove (*c*) in sediments of each cove. Values are the means for three replicates, while lines represent the standard errors of the means. The experiment was started with 10 worms per replicate. ANOVA: significant differences between origins, between sediments, and a significant interaction (all: $p < 0.001$).

Table 3.2. Cadmium Uptake of *L. hoffmeisteri*

Origin	Cadmium uptake
South Cove	1010 ± 157 nmoles/g
Foundry Cove*	2040 ± 46 nmoles/g
Foundry Cove	2232 ± 182 nmoles/g

Cadmium uptake of *L. hoffmeisteri* from Foundry Cove and South Cove, and second generation offspring of Foundry Cove worms born and raised in clean sediment (Foundry Cove*). The worms were exposed to 1 ppm Cd (incl. [109]Cd) in water for 6 days. Values are mean ± S.E. (*N*=3), in nmoles per g animal wet weight (1 nmol/g = 0.1124 ppm).
The groups differ significantly (ANOVA: $0.01 < P < 0.001$).
Comparison of means, using Least Significant Difference: $S < F^* = F$ ($p < 0.05$).

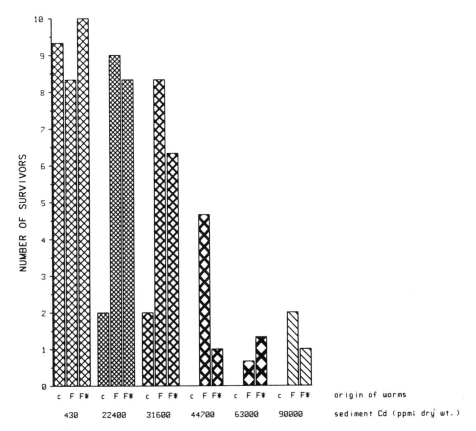

Figure 3.11. Long-term survival (28 days) of *L. hoffmeisteri* from Foundry Cove (*F*) (both recently collected and second generation offspring [F*]) and from South Cove (*c*), exposed to metal-rich sediments. Values are the means for three replicates, the experiment was started with 10 worms per replicate. ANOVA: significant differences in survival between different cadmium levels, between origins, and a significant interaction (all: p < 0.001). Least Significant Difference: c < F = F* (at α = 0.05).

more resistant (Foundry Cove) worms. They also show that, like the resistance differences, these differences in cadmium uptake have a large genetic component. We mentioned earlier that in several instances of an increased metal resistance, this was shown to be due to a reduced accumulation. Our results on the cadmium accumulation do not explain the increased resistance to metals in Foundry Cove worms. The increased cadmium accumulation in Foundry Cove worms would be expected to result in a decreased rather than an increased resistance. So mechanisms other than an increased metal exclusion must be responsible for the increased resistance.

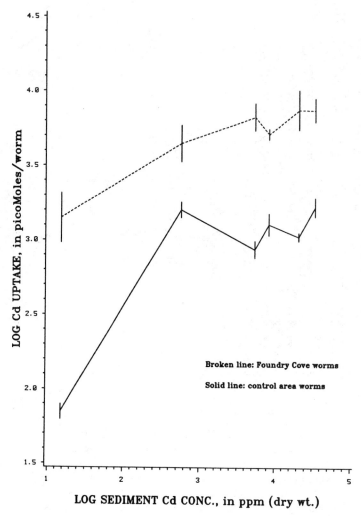

Figure 3.12. Cadmium uptake (in 28 days) of *L. hoffmeisteri* from Foundry Cove and South Cove, in sediments differing in metal content. The experiment was started with 10 individuals per replicate, three replicates per group. Values are based on Cd concentrations of surviving animals (6 to 10 per replicate) minus concentrations prior to exposure (three replicates of 10 animals from each location). Lines connect the means of the replicates, and vertical lines represent the standard errors of the means. ANOVA: significant differences in uptake between the origins, between the cadmium levels (both $p < 0.001$), and a significant interaction ($0.001 < p < 0.01$).

The same animals that were exposed to the ^{109}Cd were used to determine the subcellular and cytosolic cadmium distributions. The homogenate was fractionated by differential centrifugation (Nash et al. 1981) to yield a nuclear fraction, a mitochondrial fraction, a microsomal fraction, and the cytosol. The subcellular distribution was then obtained by gamma counting of these fractions. Table 3.3 gives the cadmium concentrations in the subcellular fractions. There are differences among some groups in all fractions. The reason for the difference within the mitochondrial fraction is not readily apparent. Possibly the different biomasses of the groups resulted in differences in adsorption of the cadmium. The differences in the nuclear and microsomal fractions follow the same pattern as the differences in resistance, suggesting that sequestering of cadmium in structures of these fractions could be partly responsible for the differences in resistance. Additional research on the presence of cadmium in these fractions using electron probe microanalysis is ongoing.

The three groups differ significantly in the amount of cadmium in the cytosol, which constitutes the largest cadmium pool for every group. Those differences were further analyzed by determining the cytosolic cadmium distribution. To achieve this, the cytosol was presieved with a 0.4-μm pore size filter and up to 700 μl was injected on a TSK-3000) high performance liquid chromatography steric exclusion column. The 1 ml fractions (separated in principle on the basis of molecular weight) were analyzed by gamma counting. The cadmium concentrations of the cytosol fractions are shown in Figure 3.13. This figure shows the presence of a low-molecular-weight protein with a high affinity for cadmium. It has an apparent molecular weight of about 16,000 daltons, as determined with molecular weight markers on the same HPLC column. Awaiting further

Table 3.3. Cadmium Concentrations in Subcellular Fractions of *L. hoffmeisteri*

	Fraction			
Origin	Nuclear	Mitochondrial	Microsomal	Cytosol
South Cove	113 ± 22	131 ± 32	78 ± 8	218 ± 10
Foundry Cove*	309 ± 22	327 ± 22	199 ± 27	522 ± 50
Foundry Cove	218 ± 22	109 ± 2	214 ± 10	1161 ± 238

Cadmium concentrations (mean $\pm$ S.E.; $N=3$) in subcellular fractions of *L. hoffmeisteri* from Foundry and South Coves, and second generation offspring of Foundry Cove worms born and raised in clean sediment (Foundry Cove*). The worms were exposed to 1 ppm Cd (incl. ^{109}Cd) in water for 6 days. Values are expressed in nmoles per g animal wet weight (1 nmol/g = 0.1124 ppm).

ANOVA on log-transformed data: significant differences between origins and between fractions, as well as significant interaction (all: $p < 0.001$). Within each fraction, significant differences between the origins. Comparisons of means within each fraction using Least Significant Difference, respectively: S < F = F*; F = S < F*; S < F* = F; S < F* < F ($p < 0.05$).

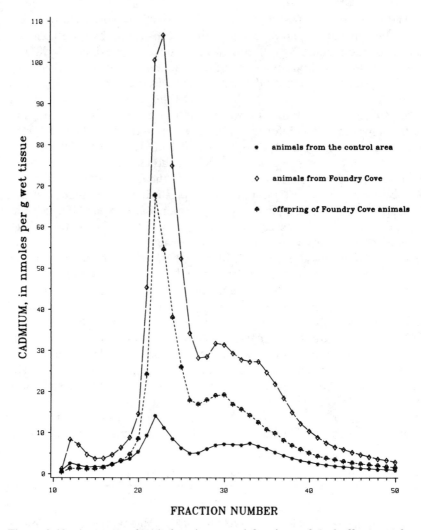

Figure 3.13. Amounts of cadmium in cytosol fractions of *L. hoffmeisteri* from Foundry and South Cove, after a 6-day exposure to 1 ppm Cd (including ^{109}Cd) in water. Values are adjusted for differences in sample wet weight, and are means of three replicates

characterization, we termed it a metallothionein-like protein. The molecular weight fractions were pooled in three groups: a high-molecular-weight group, a metallothionein-like group, and a very low-molecular-weight group. Table 3.4 shows that worms from Foundry Cove have much more cadmium in the metallothionein pool than worms from South Cove. We conclude that worms from Foundry Cove detoxify more cadmium by binding it to the metallothionein-like protein. But Foundry Cove worms

Table 3.4. Cadmium Concentration in Pooled Cytosol Fractions of *L. hoffmeisteri*

	Pooled fractions		
Origin	High mol. wt. (mw > 28000) fract. 11–20	Metallothionein (mw 22000–6000) fractions 21–26	Very low mol. wt. (mw < 5000) fractions 27–50
South Cove	26 ± 5	55 ± 4	96 ± 14
Foundry Cove*	26 ± 3	229 ± 13	208 ± 26
Foundry Cove	64 ± 7	415 ± 76	388 ± 59

Cadmium concentrations in pooled cytosol fractions of *L. hoffmeisteri* from Foundry and South Coves, and second generation offspring of Foundry Cove worms born and raised in clean sediment (Foundry Cove*). The worms were exposed to 1 ppm Cd (incl. [109]Cd) in water for 6 days. Values are mean ± S.E. (N=3), in nmoles per g animal wet weight (1 nmol/g = 0.1124 ppm).
ANOVA on log-transformed data: significant differences between origins and between fractions, as well as significant interaction (all: $p < 0.001$). Within each fraction, significant differences between the origins. Comparisons of means within each fraction using Least Significant Difference, respectively: S = F* < F; S < F* < F; S < F* < F ($p < 0.05$).

take up more cadmium, so they have to cope with higher internal cadmium concentrations to begin with. Whether the increased binding to the metallothionein-like protein in these worms is actually enough to compensate for the increased cadmium uptake, and to be responsible for the reduced toxicity of the accumulated cadmium as well, depends on the importance of the metallothionein binding relative to the amount of cadmium that is actually exerting a toxic effect. We are unable at present to pinpoint precisely the most critical toxic effect of cadmium. Binding of cadmium to enzymes of the high-molecular-weight pool has been mentioned as a likely candidate for the critical injury. Our results did not reveal the presence of more cadmium in the more sensitive worms (from South Cove) in this pool as a whole. There is a tendency for more cadmium in the highest molecular weight fractions of this pool, which could be responsible for the earlier onset of toxic effects. At this point it is not possible to conclude that an increase in the amount of metallothionein in the resistant population is the only mechanism responsible for this resistance; it may be just one of several mechanisms.

3.6 Conclusion

Our results show that the heavy metals in Foundry Cove have affected the taxonomic composition of the macrobenthos only at a small and extremely polluted site in this cove. Yet *L. hoffmeisteri* from the main part of Foundry Cove is distinctly more resistant to the metals than its conspecifics from the control area. Our survey for differences in resistance to pollutants has proven more sensitive in detecting effects of pollution than a taxonomic survey, in so far as these very different effects are comparable.

Changes in the taxonomic composition of polluted ecosystems and changes in resistance within a species are not necessarily unrelated. Not all species in a polluted environment that undergo detrimental effects from this pollution might be equally predisposed to become more resistant, resulting in an altered taxonomic composition. Thus the increased abundance of *L. hoffmeisteri* at the most polluted site could be due to a predisposition for evolving an increased resistance. But changes in the taxonomic composition upon pollution can also be due to interspecific differences in their intrinsic sensitivity to pollutants rather than in their ability to adapt.

Which phenomenon we expect to be more pronounced—an increased resistance in species or a change in the taxonomic make-up of a community—will depend on the nature of the pollution process. A sudden release of a pollutant at a concentration that is so toxic that most organisms succumb is likely to have an immediate drastic effect on the density and species-distribution in the ecosystem. A slow build-up of pollutants in an ecosystem is more likely to allow for a slow increase in resistance to build up over time through natural selection. In such a situation, changes in the taxonomic composition can be expected to be less severe. The increased resistance in *L. hoffmeisteri* has nevertheless evolved fairly rapidly (in less than 30 years).

Demonstrated resistance to a specific toxic substance is indirect evidence for a negative influence of the substance on the ecosystem. When dealing with areas polluted by a complex of toxicants, a survey for resistance could possibly be used as a tool to decide which toxic compounds have affected the ecosystem. The evidence for a negative effect on an ecosystem holds especially if it can be demonstrated that the increased resistance has a genetic basis, as this entails that the pollutant has either killed the more sensitive individuals or reduced their number of offspring. But if a population is genetically resistant to a given toxic substance, it is not necessarily valid that an introduction of this very substance was the cause of the evolution of resistance. It has been shown that organisms resistant to toxic substances may also be resistant to others, sometimes even those that the organism has never confronted before (Fisher 1977). The simplest explanation for this phenomenon is that resistance mechanisms for one compound may be biochemically and physiologically sufficient to enable resistance to other (most likely similar) compounds. For instance, a previous exposure to cadmium reduces the toxicity of mercury, probably via the induction of metallothioneins (Bouquegneau 1979; Roesijadi and Fellingham 1987).

Changes in the taxonomic composition of a polluted area relative to a control area, can never be definitively attributed to the pollution factor. Many environmental factors, as well as random events, influence the composition of the biota. From that point of view a demonstration of a genetically-based increased resistance to a pollutant is more convincing, given our caveat, as it constitutes evidence that this pollutant has indeed exerted an effect.

In conclusion, we have demonstrated that strong selection has resulted in the evolution of resistance to metals in an aquatic oligochaete. This phenomenon is largely ignored in ecotoxicology and may lead to severe errors in the understanding of the response of species to toxic substances and the fate of toxic substances in food webs. As an example, a bioassay experiment on naive oligochaetes might lead to the incorrect conclusion that the introduction of the substance would result in the demise of the population. With the evolution of resistance, the population may not become extinct, but may rather maintain population densities comparable to those of pristine areas. The mechanism of resistance, moreover, may result in enhanced uptake of the toxic substance, with the resultant danger that it may be transferred through the food web. Studies of the evolution of resistance therefore increase our understanding of ecotoxicology.

Acknowledgments. Much appreciation goes to Amy Knutson for considerable technical assistance. We thank Kenneth Kimball for comments on the manuscript, and Brenda Sanders and Ken Jenkins of the Molecular Ecology Institute of the California State University at Long Beach for making their lab and expertise available for P.L.K.'s work on metallothioneins. This is contribution number 600 in Ecology and Evolution at the State University of New York at Stony Brook.

This research has been financed through a research grant from the Hudson River Foundation for Science and Environmental Research Inc. The views expressed herein do not necessarily reflect the views of the Hudson River Foundation or of the Hudson River Panel. The Hudson River Foundation and the Hudson River Panel assume no liability for the contents or use of the information herein.

References

Antonovics J, Bradshaw AD, Turner RG (1971) Heavy metal tolerance in plants. Adv Ecol Res 7:1–85

Arima K, Beppu M (1964) Induction and mechanisms of arsenite resistance in *Pseudomonas pseudomallei.* J Bacteriol 88:143–150

ASTM (1980) Standard practice for conducting acute toxicity tests with fishes, macroinvertebrates, and amphibians. ASTM, E 729–80

Bailey GC, Saltes JG (1982) The development of some metal criteria for the protection of Spokane River rainbow trout. Project Completion Report to Washington State Department of Ecology. Pullman, WA: Department of Civil and Environmental Engineering

Beach LR, Palmiter RD (1981) Amplification of the metallothionein-I gene in cadmium-resistant mouse cells. Proc Natl Acad Sci USA 78:2110–2114

Beers JR, Stewart GL, Hoskins KD (1977) Dynamics of micro-zooplankton populations treated with copper: controlled ecosystem pollution experiment. Bull Mar Sci 27:66–79

Benson WH, Birge WJ (1985) Heavy metal tolerance and metallothionein induction in fathead minnows: results from field and laboratory investigations. Env Toxicol Chem 4:209–217

Beppu M, Arima K (1964) Decreased permeability as the mechanism of arsenite resistance in *Pseudomonas pseudomallei*. J Bacteriol 88:151–157

Blegvad H (1932) Investigations of the bottom fauna at the outfalls of drains in the sound. Dan Biol Stn 37:1–20

Bouquegneau JM (1979) Evidence for the protective effect of metallothioneins against inorganic mercury injuries to fish. Bull Environm Contam Toxicol 23:218–219

Bouquegneau JM, Gerday C, Disteche A (1975) Fish mercury-binding thionein related to adaptation mechanisms. FEBS Letters 55:173–177

Bower PM, Simpson HJ, Williams SC, Li YH (1978) Heavy metals in the sediments of Foundry Cove, Cold Spring, New York. Environ Sci Technol 12:683–687

Bradshaw AD, McNeilly T (1981) *Evolution and Pollution*. London: Edward Arnold Ltd

Brouwer M, Whaling P, Engel DW (1986) Copper-metallothioneins in the american lobster, *Homarus americanus:* Potential role as Cu(I) donors to apohemocyanin. Environ Health Persp 65:93–100

Brown BE (1976) Observations on the tolerance of the isopod *Asellus meridianus* Rac. to copper and lead. Water Research 10:555–559

Brown BE (1977) Uptake of copper and lead by a metal tolerant isopod *Asellus meridianus* Rac. Freshwater Biol 7:235–244

Brown BE (1978) Lead detoxification by a copper-tolerant isopod. Nature 276:388–390

Bryan GW (1974) Adaptation of an estuarine polychaete to sediments containing high concentrations of heavy metals. In: Vernberg FJ, Vernberg WB (eds) *Pollution and Physiology of Marine Organisms*. New York: Academic Press, pp. 123–135

Bryan GW, Hummerstone LG (1971) Adaptation of the polychaete *Nereis diversicolor* to sediments containing high concentrations of heavy metals. I. General observations and adaptation to copper. J Mar Biol Assn UK 51:845–863

Bryan GW, Hummerstone LG (1973a) Adaptation of the polychaete *Nereis diversicolor* to estuarine sediments containing high concentrations of zinc and cadmium. J Mar Biol Assn UK 53:839–857

Bryan GW, Hummerstone LG (1973b) Adaptation of the polychaete *Nereis diversicolor* to manganese in estuarine sediments. J Mar Biol Assn UK 53:859–872

Callahan P, Weis JS (1983) Methylmercury effects on regeneration and ecdysis in fiddler crabs (*Uca pugilator, U. pugnax*) after short-term and chronic pre-exposure. Arch Environ Contam Toxicol 12:707–714

Coombs TL, George SG (1978) Mechanisms of immobilization and detoxication of metals in marine organisms. In: McLusky DS, Berry AJ (eds) *Physiology and Behaviour of Marine Organisms*. Oxford: Pergamon Press, pp. 179–187

Davies AG (1976) An assessment of the basis of mercury tolerance in *Dunaliella tertiolecta*. J Mar Biol Assn UK 56:39–57

Devanas MA, Litchfield CD, McLean C, Gianni J (1980) Coincidence of cadmium and antibiotic resistance in New York Bight Apex benthic organisms. Mar Poll Bull 11:264–269

Fisher NS (1977) On the differential sensitivity of estuarine and open-ocean diatoms to exotic chemical stress. Amer Nat 111:871–895

Fisher NS, Frood D (1980) Heavy metals and marine diatoms: influence of dissolved organic compounds on toxicity and selection for metal tolerance among four species. Mar Biol 59:85–93

Foster PL (1977) Copper exclusion as a mechanism of heavy metal tolerance in a green alga. Nature 269: 322–323

Foster PL (1982a) Species associations and metal contents of algae from rivers polluted by heavy metals. Freshwater Biol 12:17–39

Foster PL (1982b) Metal resistances of Chlorophyta from rivers polluted by heavy metals. Freshwater Biol 12:41–61

Fraser J, Parkin DT, Verspoor E (1978) Tolerance to lead in the freshwater isopod, *Asellus aquaticus*. Water Research 12:637–641

George SG, Pirie BJS (1979) The occurrence of cadmium in sub-cellular particles in the kidney of the marine mussel, *Mytilus edulis,* exposed to cadmium. The use of electron microprobe analysis. Biochim Biophys Acta 580:234–244

Gray JS (1979) Pollution-induced changes in populations. Philos Trans R Soc Lond Biol Sci 286:545–561

Gray JS (1982) Why do ecological monitoring? Mar Pollut Bull 11:62–65

Griffith JK, Enger MD, Hildebrand CE, Walters RA (1981) Differential induction by cadmium of a low-complexity ribonucleic acid class in cadmium-resistant and cadmium-sensitive mammalian cells. Biochemistry 20:4755–4761

Hall A, Fielding AH, Butler M (1979) Mechanisms of copper tolerance in the marine fouling alga *Ectocarpus siliculosis*. - Evidence for an exclusion mechanism. Mar Biol 54:195–199

Hazen RE, Kneip TJ (1976) The distribution of cadmium in the sediments of Foundry Cove. Fourth Symposium on Hudson River Ecology, Bear Mountain, New York

Hunt CD, Smith DL (1983) Remobilization of metals from polluted marine sediments. Can J Fish Aquat Sci 40 (Suppl. 2):132–142

Ireland MP, Richards KS (1977) The occurrence and localisation of heavy metals and glycogen in the earthworms *Lumbricus rubellus* and *Dendrobaena rubida* from a heavy metal site. Histochemistry 51:153–166

Jensen A, Rystad B, Melsom A (1974) Heavy metal tolerance of marine phytoplankton. I. The tolerance of three algal species to zinc in coastal sea water. J Exp Mar Biol Ecol 15:145–157

Kagi JHR, Vallee BL (1960) Metallothionein: a cadmium- and zinc-containing protein from equine renal cortex. J Biol Chem 235:3460–3465

Klerks PL, Weis JS (1987) Genetic adaptation to heavy metals in aquatic organisms: a review. Environ Pollut 45:173–205

Kneip TJ (1978) Effects of cadmium in an aquatic environment. In: *Cadmium 77:* Edited Proceedings of the First International Cadmium Conference, San Francisco. London: Metal Bulletin Ltd., pp. 120–124

Kneip TJ, Hazen RE (1979) Deposit and mobility of cadmium in a marsh-cove ecosystem and the relation to cadmium concentration in biota. Environ Health Persp 28:67–73

Knutson AB, Klerks PL, Levinton JS (1987) The fate of heavy metal contaminated sediments in Foundry Cove, New York. Environ Pollut 45:291–304

Lerch K (1980) Copper metallothionein, a copper-binding protein from *Neurospora crassa*. Nature 284:368–370

Levinton JS (1980) Genetic divergence in estuaries. In: Kennedy VS (ed) *Estuarine Perspectives*. New York: Academic Press, pp. 509–520

Lowe DM, Moore MN (1979) The cytochemical distribution of zinc (ZnII) and iron (FeIII) in the common mussel, *Mytilus edulis,* and their relationship with lysosomes. J Mar Biol Assn UK 59:851–858

Luoma SN (1977) Detection of trace contaminant effects in aquatic ecosystems. J Fish Res Bd Can 34:436–439

Margoshes M, Vallee BL (1957) A cadmium protein from equine kidney cortex. J Am Chem Soc 79:4813–4814

Mason AZ, Simkiss K, Ryan KP (1984) The ultrastructural localization of metals in specimens of *Littorina littorea* collected from clean and polluted sites. J Mar Biol Assn UK 64:699–720

McLean RO (1975) Zinc tolerances of *Hormidium rivulare* Kutz. Br Phycol J 10:313

Moraitou-Apostolopoulou M, Verriopoulos G, Palla P (1979) Temperature and adaptation to pollution as factors influencing the acute toxicity of Cd to the planktonic copepod *Acartia clausi.* Tethys 9:97–101

Murphy LS, Guillard RRL, Gavis J (1982) Evolution of resistant phytoplankton strains through exposure to marine pollutants. In: Mayer GF (ed) *Ecological Stress and the New York Bight: Science and Management.* Columbia, SC: Estuarine Research Federation, pp. 401–412

Nash WW, Poor BW, Jenkins KD (1981) The uptake and subcellular distribution of lead in developing sea urchin embryos. Comp Biochem Physiol 69C:205–211

Nelson JD, Colwell RR (1975) The ecology of mercury-resistant bacteria in Chesapeake Bay. Microbial Ecol 1:191–218

Nevo E, Ben-Shlomo R, Lavie B (1984) Mercury selection of allozymes in marine organisms: prediction and verification in nature. Proc Natl Acad Sci USA 81:1258–1259

Occhiogrosso TJ, Waller WT, Lauer GJ (1979) Effects of heavy metals on benthic macroinvertebrate densities in Foundry Cove on the Hudson River. Bull Environm Contam Toxicol 22:230–237

Olafson RW, Sim RG, Boto KG (1979) Isolation and chemical characterization of the heavy metal-binding protein metallothionein from marine invertebrates. Comp Biochem Physiol 62B:407–416

Prent P (1979) Metals and phosphate in the chloragosomes of *Lumbricus terrestris* and their possible physiological significance. Cell Tissue Research 196:123–134

Rahel F (1981) Selection for zinc tolerance in fish: results from laboratory and wild populations. Trans Amer Fish Soc 110:19–28

Reish DJ (1959) An ecological study of pollution in Los Angeles - Long Beach Harbors, California. Occas. Pap., Allan Hancock Found. No. 22, pp. 1–119

Resource Engineering (1983) Preliminary site background data analysis of Foundry Cove, prepared for Vinson & Elkins. Houston, TX: Resource Engineering

Roesijadi G, Fellingham GW (1987) Influence of Cu, Cd, and Zn preexposure on Hg toxicity in the mussel *Mytilus edulis*. Can J Fish Aquat Sci 44:680–684

Rugstad HE, Norseth T (1975) Cadmium resistance and content of cadmium-binding protein in cultured human cells. Nature 257:136–137

Russell G, Morris OP (1970) Copper tolerance in the marine fouling alga *Ectocarpus siliculosus*. Nature 228:288–289

Sanders JG, Batchelder JH, Ryther JH (1981) Dominance of a stressed marine phytoplankton assemblage by a copper-tolerant pennate diatom. Botanica Mar 24:39–41

Sanders JG, Osman RW (1985) Arsenic incorporation in a salt marsh ecosystem. Estuar Coast Shelf Sci 20:387–392

Say PJ, Diaz BM, Whitton BA (1977) Influence of zinc on lotic plants. I. Tolerance of *Hormidium* species to zinc. Freshwater Biol 7:357–376

Smith GEA (1983) Comparative zinc toxicity and tolerance of indigenous fish populations. M.S. thesis, Pullman, WA: Washington State University

Stokes PM, Hutchinson TC, Krauter K (1973) Heavy metal tolerance in algae isolated from contaminated lakes near Sudbury, Ontario. Can J Bot 51:2155–2168

Suzuki KT, Yamamura M, Mori T (1980) Cadmium-binding proteins induced in the earthworm. Arch Environ Contam Toxicol 9:415–424

Thomas WH, Seibert DLR (1977) Effects of copper on the dominance and the diversity of algae: controlled ecosystem pollution experiment. Bull Mar Sci 27:23–33

Thompson KA, Brown DA, Chapman PM, Brinkhurst RO (1982) Histopathological effects and cadmium-binding protein synthesis in the marine oligochaete *Monopylephorus cuticulatus* following cadmium exposure. Trans Am Microsc Soc 101:10–26

Timoney JF, Port JG (1982) Heavy metal and antibiotic resistance in *Bacillus* and *Vibrio* from sediments of New York Bight. In: Mayer GF (ed) *Ecological Stress and the New York Bight: Science and Management*. Columbia, SC: Estuarine Research Federation, pp. 235–248

Timoney JF, Port J, Giles J, Spainer J (1978) Heavy metal and antibiotic resistance in the bacterial flora of sediments of New York Bight. Appl Environ Microbiol 36:465–472

Weis JS, Weis P, Heber M, Vaidya S (1981) Methylmercury tolerance of killifish (*Fundulus heteroclitus*) embryos from a polluted vs nonpolluted environment. Mar Biol 65:283–287

Wentsel R, McIntosh A, Atchinson G (1978) Evidence of resistance to metals in larvae of the midge *Chironomus tentans* in a metal contaminated lake. Bull Environ Contam Toxicol 20:451–455

Chapter 4
Determining the Ecological Effects of Oil Pollution in Marine Ecosystems

Robert W. Howarth[1]

Massive amounts of oil hydrocarbons enter coastal oceans from both accidental spills and routine, chronic discharges. Although it is known that oil is toxic to many organisms, current knowledge does not allow precise estimation of ecological damage resulting from oil pollution. At present, most governmental agencies that regulate chronic sources of oil pollution rely on an excessively simplistic approach to estimate damage: determination of lethal toxicity in short-term bioassays. This approach probably underestimates ecological damage, perhaps greatly so. This chapter is a synthesis of what is known about the ecological effects of oil pollution in marine ecosystems, with an emphasis on more sophisticated approaches to estimating effects. Public attention following oil spills often centers on the death of sea birds and contamination of shellfish; these indeed can be serious problems (Teal and Howarth 1984; National Academy of Sciences 1985). However, the potential for ecological damage is much greater than this, and it is the less obvious but potentially more far-reaching problems that are emphasized here. Portions of this manuscript have been previously published as part of a site-specific assessment of the potential effects of oil pollution on Georges Bank (Howarth 1987).

A recent report of the National Academy of Sciences (1985) estimated that oil input to the world's oceans is between 1.7 and 8.8 million metric tons in an average year, most of this occurring in coastal areas. A small amount of this estimated input is from natural sources; most input is anthropogenic. Of the anthropogenic inputs, roughly half is estimated to come from tankering and transportation sources, and the rest is from a

[1] Section of Ecology & Systematics, Corson Hall, Cornell University, Ithaca, New York 14853

wide range of sources including municipal and industrial wastes. Offshore oil production is thought to contribute only 1 to 2 % of anthropogenic inputs in most years (National Academy of Sciences 1985). Nonetheless, the most disagreement over regulation of oil pollution has revolved around offshore oil development. In part, this is probably because off-shore oil fields frequently are developed in relatively pristine areas within the territorial waters of developed countries, whereas other sources of industrial and municipal wastes typically occur in industrialized regions or in third-world countries. Also, average statistics can severely under-state the pollution potential. Thus, a single accident, the blowout of an exploratory offshore well in the Bay of Campheche (Mexico) in 1979, spilled some 0.5 to 1.4 million tons of oil (Teal and Howarth 1984), obvi-ously a very sizeable percentage of all oil inputs to the world's oceans in that year. Further, routine discharges of oil from offshore oil operations are thought to be rapidly rising over time as new fields are developed and as existing fields age (Johnston 1980; Read 1980).

Ability to estimate the damage from oil discharges and oil spills is necessary for environmental managers to set regulatory guidelines and procedures. Several approaches to estimating these effects of oil pollution have been attempted. However, the dominant approach used by environ-mental management agencies, at least in the United States (Department of Interior 1985), is the determination of concentrations of oil that cause acute toxicity, that is, the LD_{50} approach (i.e., the dose found to be lethal to 50% of exposed organisms in a laboratory experiment). This is surpris-ing because the current scientific consensus both in the United States (National Academy of Sciences 1985) and in the United Kingdom (Clark 1987) is that the LD_{50} approach is a crude and naive one with little if any ability to predict real-world responses.

4.1 Acute Toxicity, the LD_{50} Approach

In the LD_{50} approach, test organisms are exposed to different concentra-tions of dissolved oil in the laboratory for a period of 96 hours; the con-centration that causes the death of 50% of the test organisms, the LD_{50} value, is calculated. Typically reported values for dissolved oil are 1 to 3 mg l^{-1} (Department of Interior 1985). When using such data for estimating effects of offshore oil development, the Minerals Management Service of the U.S. Department of the Interior applies a 10-fold "safety factor." That is, they assume that oil concentrations less than those 10-fold lower than LD_{50} values, or approximately 0.1 to 0.3 mg l^{-1}, will not cause ecological harm (Department of Interior 1985). There is absolutely no scientific basis for this procedure, and as this review suggests, it often underestimates the potential for harm from oil pollution.

The problems with using the LD_{50} results in this manner are many; some of these problems are general ones that would apply to any pollutant; other problems are specific to using this approach to assess harm from oil pollution. A general problem with using the LD_{50} approach to estimate environmental harm is that it ignores sublethal effects and ecological interactions, which often may dominate the effects of pollution in ecosystems. Sublethal effects of oil pollution on marine organisms include alterations in behavior (Jacobson and Boylan 1973; Pearson et al. 1981; Linden 1977), in growth rate (Gilfillan and Vandermeulen 1978), and in reproductive success (Kuhnhold et al. 1978; Steele 1977). These sublethal effects of oil and oil hydrocarbons can occur at concentrations that are at least 1,000-fold lower than typically reported LD_{50} values, that is concentrations as low as a few micrograms per liter or even less. For instance, sexual reproduction in macroalgae was found to be completely inhibited by 0.2 micrograms per liter of #2 fuel oil (Steele 1977). Concentrations of one to a few micrograms per liter of the water-soluble fraction of kerosene were found to disrupt the normal feeding behavior of snails and crabs (Jacobson and Boylan 1973; Johnson 1977).

Such sublethal effects may have major ramifications on real-world populations and ecosystems. After a spill from the barge *Florida* in West Falmouth, Massachusetts, Krebs and Burns (1977, 1978) found behavioral changes in salt-marsh fiddler crabs not directly killed by the oil spill. The crabs dug abnormally shallow burrows and moved more slowly when threatened. Because they moved more slowly, they were more vulnerable to predation; and because their burrows were abnormally shallow, they were more likely to succumb to freezing in winter. In another study following a spill from the tanker *Tsesis* in the Baltic Sea, Linden (1980) found that clams of the species *Macoma balthica* collected from the spill area one week after the spill burrowed into clean sand in the laboratory significantly more slowly than clams from a control area. The slower burrowing presumably would subject the clams to greater predation. Pearson et al. (1981) demonstrated that when littleneck clams burrow more slowly because of oiling, they are subjected to greater predation by Dungeness crabs. Sublethal effects are best studied in the context of the ecosystem in which the organisms live in order to understand the real-world significance of the effects and to avoid biases inherent in laboratory experiments (see discussion below).

Another problem with using LD_{50} data to extrapolate to the field situation also applies to using laboratory sublethal effect data: species vary greatly in their tolerance to pollution, and most toxicity studies use fairly pollution-tolerant species. Pollution-tolerant species also tend to be resistant to other forms of stress and are called "opportunistic" because they readily colonize environments where stress has eliminated most other species. LD_{50} studies and other toxicity studies typically use these opportunistic species because they are easier to maintain under the stressful conditions of laboratory culture; more sensitive species die even in laboratory control treatments not subject to pollutants. By relying on the more resilient species, the lab studies clearly underestimate the potential for pollutant damage to more sensitive species. Yet another problem related to the differential sensitivity of species to pollution is that organisms are often more sensitive at one point in their life cycle than at another. In general, larvae are much more sensitive than adults (Wells and Sprague 1976; Caldwell et al. 1977; Rosenthal and Alderdice 1976; Kuhnhold et al. 1978), yet LD_{50} studies typically use adults.

Some problems specific to using the LD_{50} approach to assess harm from oil pollution are caused by the volatility and reactivity of oil hydrocarbons and also apply to lab studies of sublethal effects. In most lab studies, dissolved oil concentrations are calculated from the amount of oil added and are not actually measured. Consequently, actual concentrations of oil are lower than assumed, often greatly so, since much of the added oil can be lost through evaporation, biodegradation, and adsorption to container walls (Anderson et al. 1974; Neff et al. 1976). The most readily extrapolatable studies are those that maintain dissolved oil concentrations at a known level by continuously adding oil in a flow-through chamber and that measure the actual concentrations, but such studies are relatively rare (National Academy of Sciences 1985).

Even most flow-through laboratory experiments probably underestimate the potential toxicity of oil under field conditions, since the degradation of oil under field conditions probably increases its toxicity. Photo-oxidation in particular can increase the toxicity of oil components, with ultraviolet radiation at the sea surface rapidly oxidizing some components and producing compounds such as peroxides and phenols that are more toxic than the parent compound (Larson et al. 1977; Lacaze and Villedon de Naide 1976). Laboratory toxicity studies are usually conducted under conditions of relatively low ultraviolet illumination that minimizes this photo-oxidation. Larson et al. (1977) showed that exposure to ultraviolet light approximating the level found in nature increased by four fold the toxicity of an oil to yeasts. Light was also found to increase the toxicity of a crude oil to the alga *Phaeodactylum* sp. (Lacaze and Villedon de Naide 1976). Microbial degradation also can result in the formation of hydrocarbons more toxic than the parent compound (Hinga et al. 1980; Varanasi and Gmur 1981), and such degradation is minimized in laboratory flow-through experiments.

4.2 Ecosystem-Level Approaches

All of the problems raised above illustrate the need to obtain information on oil pollution in intact ecosystems, where environmental conditions mimic the conditions of interest and where major ecological interactions are intact. The regular occurrence of unexpected ecosystem-level effects from pollution (Teal and Howarth 1984) further necessitates ecosystem-level experiments. These can take the form of controlled experiments, in which ecosystems are oiled and compared to control ecosystems. Or they can be "experiments" of opportunity, either actual oil spills or chronic sources of oil pollution. This section explores the problems and advantages with each of these approaches and data sources.

A great deal of effort has gone into investigating the ecological effects of accidental oil spills, and some valuable insights and data have resulted (Teal and Howarth 1984; National Academy of Sciences 1985). However, despite the effort, one must conclude that most oil spills, and particularly spills that occur offshore, are quite poorly studied, in part because it is difficult to mount a timely research response. Frequently, it is difficult to identify suitable control sites because little is known about the ecological functioning of areas in the spill vicinity. This problem is compounded by a high degree of natural variability. The best ecological studies of oil spills (such as those for the *Florida, Tsesis,* and *Amoco Cadiz* spills) have occurred in areas where there were ongoing research programs (Teal and Howarth 1984).

Some of the problems inherent in studying the effects of oil pollution from accidental spills theoretically can be overcome in studying chronic pollution in offshore oil fields. Offshore oil fields in the Gulf of Mexico (where extensive development began in the early 1950's) and in the North Sea (where development commenced in 1969) have received the most research attention. Most of the North Sea studies on chronic oil pollution have focused on benthic effects, and these have yielded some valuable data (Addy et al. 1978). In contrast, rather little effort has gone into examining the effects on plankton or fish from chronic oil pollution in the North Sea because most British scientists considered such effects unlikely (Royal Society 1980; Royal Commission on Environmental Pollution 1981). Also, surprisingly little effort has gone into measuring concentrations of oil in either the sediments or the water column of North Sea oil fields, and as discussed later, this makes it difficult to estimate effects.

In the Gulf of Mexico, three large projects have investigated the effects of chronic oil pollution: the Offshore Ecology Investigation (O.E.I.) conducted by the Gulf Universities Research Consortium (GURC) from 1972–1974 (GURC 1974); the Central Gulf Platform project from 1978 to 1979 (Bedinger et al. 1980); and the Buccaneer Field project from 1976 to 1980 (Middleditch 1982). These studies concluded that chronic oil pollution had caused no damage or that damage was limited to the vicinity of individual rigs, a conclusion that has been widely disseminated to policymakers and the public. However, all three studies were critically flawed, and in no way do their data support this conclusion (Sanders 1981; Sanders and Jones 1981; Carney 1985). The concentrations of oil hydrocarbons were not measured in any of the three studies, and so the exposure of organisms to pollutants is not known. It seems likely, in fact, that "control" sites may have been as polluted by oil as were the sites near the rigs, since oil development had proceeded for so long prior to the studies (Bender et al. 1979; Sanders 1981; Sanders and Jones 1981; Carney 1985). Sanders (1981) concluded that the benthic communities in both the near-rig and control sites were typical of those to be expected in a highly polluted environment. And none of the three studies considered how to determine oil-induced changes in the face of "considerable environmental variation near the mouth of the Mississippi" (Carney 1985). Any final conclusion on the ecological health of the area of oil and gas production in the Gulf of Mexico will have to await better designed studies that incorporate measurements of oil hydrocarbons. The choice of suitable control sites will be difficult and critical, and the lack of pre-oil-production data may permanently limit our understanding. Existing data from the Gulf tell us little or nothing about the potential impacts of oil pollution.

Probably the best information on the ecological effects of oil has come from controlled experiments in mesocosms. Mesocosms are experimental ecosystems consisting of enclosures generally larger than 10 cubic meters (Grice and Reeve 1982). They are designed to mimic natural communities as much as possible, but they allow experimental manipulations with replication and under controlled conditions that are seldom possible in nature. The best known mesocosm experiments on oil pollution are those of the Controlled Ecosystem Pollution Experiment (CEPEX) project in Saanich Inlet, British Columbia, and the Marine Ecosystem Research Laboratory (MERL) project along Narragansett Bay, Rhode Island. The CEPEX mesocosms were large bags designed to study planktonic ecosystems, whereas the MERL mesocosms are large tanks containing both seawater and sediments and are designed to study the coupling between benthic and planktonic ecosystems. Other successful mesocosm experiments on oil pollution have been the planktonic bags at Loch Ewe, Scotland (Davies et al. 1980) and the hard-bottom benthic mesocosms in Oslo, Norway (Gray 1987). The mesocosm approach has proven powerful, and much of the information in the chapter is based on mesocosm research. When comparisons between the effects of accidental oil spills, chronic oil pollution, and mesocosm experiments have proven possible, the mesocosms have done a good job of representing the real-world ecosystems (Elmgren and Frithsen 1982; Teal and Howarth 1984; National Academy of Sciences 1985; Gray 1987).

The next two sections of this paper discuss what is known about the effects of oil pollution on benthic and planktonic ecosystems.

4.3 Effects of Oil Pollution on Benthic Communities

There is now widespread agreement that oil pollution deleteriously affects benthic communities (National Academy of Sciences 1985). Sedimentation of oil seems to be a general process that occurs from oil spills and chronic oil discharges, whether nearshore or offshore (Teal and Howarth 1984). While it may appear paradoxical that oil can end up in sediments since oil is less dense than water and therefore floats in slicks, there are in fact a variety of mechanisms for oil sedimentation. One mechanism for the sedimentation of oil offshore is by adsorption onto clays and other sediment particles suspended in the water (Meyers and Quinn 1973). Another is uptake by phytoplankton, followed by the sedimentation of the phytoplankton (Lee et al. 1978a). Yet another mechanism is sedimentation of oil in zooplankton fecal pellets (Conover 1971). Zooplankton have been found to feed directly on particles of suspended oil following several spills (Conover 1971; Grose and Mattson 1977; Johansson 1980). Not all of these mechanisms are important after all oil spills; for instance, sedimentation in zooplankton fecal pellets has not been observed after all spills and is dependent upon, among other factors, the abundance of zooplankton at the time of the spills (Johansson 1980).

Chronic oil releases also can result in elevated levels of hydrocarbons in sediments, as clearly demonstrated in the MERL mesocosm studies (Gearing et al. 1980; Grassle et al. 1981; Elmgren and Frithsen 1982; Farrington et al. 1982). A dispersion of #2 fuel oil in water was added for 25 weeks to the inlet seawater of flow-through tanks containing seawater, sediments, and benthic animals and designed to mimic a coastal ecosystem. The oil-in-water dispersion consisted of 190 μg oil per liter of water. By 20 weeks the upper 2 cm of sediment contained over 100 μg of oil hydrocarbons per gram of sediment. Contamination of the benthic fauna had increased by at least an order of magnitude (Farrington et al. 1982). Mass budget calculations showed that roughly half of the oil added to the water had become incorporated into the sediment (Gearing et al. 1980). There is general agreement that oil in sufficient concentration has deleterious effects on benthic communities (Sanders et al. 1980; Cabioch et al. 1981; d'Ozouville et al. 1979; Glemarec and Hussenot 1981; Addy et al. 1978; Elmgren and Frithsen 1982; Grassle et al. 1981; Linden et al. 1980; Elmgren et al. 1980a,b; Boucher 1981; Teal and Howarth 1984). Massive kills of the benthic fauna have occurred when sufficiently large quantities of oil have reached the bottom following spills (Sanders et al. 1980; Cabioch et al. 1981; Teal and Howarth 1984). Oil in sediment, even at relatively low concentration (a few grams per square meter or 100 μg per gram, Elmgren and Frithsen 1982; see Table 4.1), can change the structure of the benthic community whether the oil comes from a spill (Sanders et al. 1980; Cabioch et al. 1981; d'Ozouville et al. 1979; Glemarec and Hussenot 1981; Elmgren and Frithsen 1982; Linden et al. 1980; Elmgren et al. 1980a,b; Boucher 1981) or from chronic pollution such as that associated with offshore oil production (Addy et al. 1978; Elmgren et al. 1980b; Elmgren and Frithsen 1982; Grassle et al. 1981). Sensitive species die or emigrate and are replaced by oil-tolerant, opportunistic species. The total number of species in the community decreases, and generally

Table 4.1. Concentrations of Oil Hydrocarbons Typically Found in Surface Sediments Compared with Concentration in MERL Experiment Found to Have Deleterious Effects.

	Micrograms oil hydrocarbons per gram of sediment
Georges Bank (offshore New England)[a]	1–20
Buzzards Bay, MA[b]	10–400
Off New Jersey coast[c]	200–1,300
New York Bight[d]	10–2,800
Narragansett Bay, RI[e]	25–5,700
Concentration in MERL experiment that resulted in altered benthic community structure and lowered biomass[f]	100

[a] Boehm et al. 1979
[b] Farrington et al. 1977, as reported in Olsen et al. 1982
[c] Gibson et al. 1979, as reported in Olsen et al. 1982
[d] Farrington and Tripp 1977, as reported in Olsen et al. 1982
[e] VanVleet and Quinn 1978 and Olsen and Lee 1979, as reported in Olsen et al. 1982
[f] Elmgren and Frithsen 1982

the biomass decreases. A partial exception to this generality were the findings of Elmgren et al. (1980a) after the *Tsesis* spill. There the clam *Macoma balthica* was the major contributor to the benthic biomass before the spill. Since this clam appears to be a relatively pollution-tolerant species (Elmgren et al. 1980a), its numbers were unaffected by the oil, and community biomass did not decrease measurably, although species diversity did. Gilfillan and Vandermeulen (1978) have shown that oil can decrease production in commercially important benthic species such as *Mya arenaria,* a soft-shelled clam.

The effects of a chronic oiling experiment on benthic organisms in the MERL mesocosms were similar to those seen in a variety of oil spills (Elmgren and Frithsen 1982). In the MERL experiment, decreases in abundance and biomass of both the macrofauna and meiofauna were seen at concentrations of oil of approximately 100 μg per gram of sediment in the upper 2 cm (Grassle et al. 1981; Elmgren and Frithsen 1982). Ampeliscid amphipods and ostracods were found to be particularly sensitive to the oil. The mesocosm experiments allow a discrimination of cause and effect that is seldom possible after oil spills and help to estimate the concentrations of oil in sediment necessary to cause significant effects on the benthos.

Natural oil seeps provide another opportunity for examining the effects of chronic oil input on benthic ecosystems. Benthic communities in areas of intense and moderate seepage were compared to the benthos in nearby areas without seepage in the Santa Barbara Channel, California. The concentration of hydrocarbons dissolved in the interstitial waters of the sediment in the areas of intense seepage were 1.3 mg l^{-1}; in the areas of moderate seepage they ranged from 45 to 110 μg l^{-1} (Stuermer et al. 1982); in the areas without seepage, concentrations ranged from 0.2 to 5.0 μg l^{-1}. The areas of moderate seepage had a community structure similar to that of the non-seepage areas (Spies and Davis 1979; Davis and Spies 1980; Stuermer et al. 1982). The same species of organisms were present in the two kinds of places, but the deposit feeders were more abundant and certain amphipods were less so at the sites with seepage. Animal densities were higher in the moderate seeps, perhaps because of organic enrichment resulting from production by chemosynthetic bacteria supported by sulfides in the seepage (Spies et al. 1980). In the areas of intense seepage the community structure was that of a highly stressed system. Typically, nematodes and capitellid polychaetes, well-known opportunists (Grassle and Grassle 1974), were the only infaunal organisms present in any abundance (Spies and Davis 1979).

Why the MERL mesocosm experiment resulted in a benthic community so different from the one at the moderate-seepage site in Santa Barbara, which had a similar level of oiling, is unknown (National Academy of Sciences 1985). The difference may be the result of adapted resistance to oil in the seep systems, but this has not yet been demonstrated, and there is at least some evidence against this hypothesis (National Academy of Sciences 1985). If there is such adapted resistance to oil by communities that have been exposed for thousands of years, it would have little relevance in predicting the response of unadapted organisms in more pristine environments. Also, the community at the site of intense seepage shows clearly that there are limits to such adaptation. Another explanation for differences between the MERL experiment and the composition of seep communities is that seepage oil may be low in toxicity because of weathering as it passes through crustal rocks and sediments, but the evidence on this is limited (National Academy of Sciences 1985).

Some of the effects of oil in benthic ecosystems would not have been predicted from knowledge of effects on populations of individual species. For instance, populations of benthic ciliates and foraminifera increased significantly during the MERL experimental oilings, apparently because of the reduced bioturbation and predation that resulted from smaller populations of macroscopic animals (Elmgren et al. 1980b; Grassle et al. 1981). Another unanticipated ecosystem-level effect occurred after the *Tsesis* spill. The hatching success of herring eggs, which develop on the sediment surface, spawned the spring after the spill was only about half as great in oil-contaminated sediments as in control areas (Nellbring et al. 1980). Apparently this was not a direct effect of the oil on the herring eggs, but rather was primarily a result of increased fungal infection of the eggs. Normally, gammarid amphipods graze the fungi, keeping the infection of the herring eggs low. It is presumed that the spilled oil killed the amphipods and that the resulting uncontrolled infection caused the low hatching success (Nellbring et al. 1980).

Most of the clearly documented effects of spilled oil on benthic ecosystems are associated with nearshore spills. Offshore spills have proven very difficult to examine, and there are very few studies of them and their effects. However, the small amount of evidence available suggests that oil concentrations in sediments offshore following individual spills remain low enough so as to cause little harm. Thus, after the *Argo Merchant* spill very little oil could be found in the bottom sediments or benthic fauna of Georges Bank and Nantucket Shoals except in the immediate vicinity of the wreck (MacLeod et al. 1978). Similarly, there was little evidence of much accumulation of oil in the sediments from the Bravo blowout in the North Sea in the few weeks following the spill (Audunson 1977), even though sediment-trap studies a month after the spill clearly showed sedimentation of oil (Mackie et al. 1978).

In contrast to the lack of evidence for the introduction of much oil into offshore sediments following spills, there is convincing evidence that chronic oil release from oil production contaminates sediments. Ward et al. (1980) found evidence of sediment contamination with aromatic hydrocarbons within 30 km of oil rigs in several fields in the North Sea. Audunson (1977) also reported low-level contamination (maximum concentration of 5 μg per gram of sediment) of sediments near the Ekofisk field in the North Sea. He concluded that this hydrocarbon contamination probably came from small random spills and chronic discharges. That such low-level contamination can have biological effects is indicated by the results of Addy et al. (1978). These authors found decreases in both the number of individual animals and the number of animal species in sediments within about 5 km of a storage platform in the Ekofisk Field after four years of operation. Although several factors associated with the platform, such as domestic waste or physical disturbance of the sediment, may have caused this, the most likely factor is the accumulation of oil hydrocarbons in the sediments as a result of the routine discharge of the platform's displacement water as new oil was pumped into the storage tanks. Oil released in formation waters may also have been important. The release of oil in formation waters is estimated to have made up more than half of the chronic oil discharge in the North Sea in 1980 (Read 1980).

The persistence of benthic damage from oil pollution can be quite long-lasting but depends in part upon the physical environment. Oil preferentially adsorbs to finer-grained, organic-rich sediments (Blumer et al. 1971; Beslier et al. 1980). Thus, the coarse sediments found in high-energy environments tend to be cleaned of most oil within a year or two following a spill as the finer-grained, oil-contaminated particles are winnowed away. This process was well documented following the *Amoco Cadiz* spill in France (Beslier et al. 1980). Of course, this oil does not just disappear but rather ends up in the fine-grained sediments of low-energy environments. Here, the oil can persist because the anoxic conditions typically found in such sediments favor the preservation of oil hydrocarbons, particularly the more toxic aromatic fraction. Oil hydrocarbons have been observed to persist in anoxic sediments for at least 12 years after a spill, the limits of observation at present (Thomas 1977; Teal et al. 1978; Gilfillan and Vandermeulen 1978; Thomas 1978; Krebs and Burns 1978; Sanders et al. 1980; Gundlach et al. 1982; Teal and Howarth 1984). Some of the effects of the oil persist for as long as the oil remains (Sanders et al. 1980; Krebs and Burns 1978; Gilfillan and Vandermeulen 1978; Teal and Howarth 1984). In fact, effects can last even longer than the oil persists because of lack of recruitment and because pollution can cause community structure to shift into alternative, stable configurations (Thomas 1978; Southward and Southward 1978).

4.4 Effects of Oil Pollution on Planktonic Communities

No clear consensus exists that oil pollution poses serious threats to planktonic communities, in sharp contrast to the situation for benthic communities. In fact, numerous papers and reports have concluded that there is unlikely to be much damage to open-sea planktonic organisms or communities from oil pollution (Lee et al. 1978b; Royal Society 1980; Royal Commission on Environmental Pollution 1981; Davenport 1982; Gray 1982; Department of Interior 1977; 1983; 1985). However, clear evidence exists that dissolved oil can harm planktonic organisms and communities if it persists in sufficiently high concentrations for sufficiently long (National Academy of Sciences 1985). Why then do so many reports and papers conclude that serious damage to planktonic communities is unlikely? This conclusion is based largely on the assumption that dissolved oil does not persist in sufficiently high concentrations in open waters. For instance, the Royal Commission on Environmental Pollution (1981) assumed that oil is not dispersed in harmful concentrations to depths greater than 1 or 2 meters. Another common assumption is that evaporation rapidly removes the more toxic constituents of an oil slick, so that the toxicity of a spill becomes negligible in a matter of hours (Royal Society 1980) or days (Department of Interior 1983; 1985).

Surprisingly few studies have tested these assumptions, but available data show them to be incorrect (Table 4.2).

Perhaps the best set is that collected following the spill of the tanker *Argo Merchant* off the Massachusetts coast in December, 1976. These data were collected as part of the baseline monitoring program for offshore oil exploration on Georges Bank, not because anyone expected the

Table 4.2. Concentrations of Dissolved Oil Hydrocarbons in the Water Column Typically Found in Unpolluted Oceanic Waters, Urban Harbors, Offshore Oil Fields, and Following Oil Spills.

	Micrograms per liter
Unpolluted oceanic waters[a]	less than 1
Urban harbors	
Providence River[b]	30–500
New York Harbor[c]	15–270
Boston Harbor[d]	100–820
Tokyo Harbor[e]	30
Bedford Basin, N.S.[f]	5–70
North Sea oil fields[g]	0.4–88
	(mean = 5)
Following oil spills	
Bravo[h]	up to 6
Tsesis, after several days[i]	50–60
Amoco Cadiz	
after 2 weeks[j]	26–340
after 1 month[k]	2–200
Argo Merchant	
immediately[l]	up to 210
after 2 months[a]	10–100
	(mean = 44)
after 5 months[a]	1–50
	(mean = 11)
after 8 months[a]	less than 1
("background")	

Concentrations higher than a few mg per liter may have deleterious effects (see Table 4.3).

[a] Boehm et al. 1978
[b] VanVleet and Quinn 1978, Farrington and Quinn 1973, and Hurtt 1979, as reported in Olsen et al. 1982
[c] Searl et al. 1977, as reported in Olsen et al. 1982
[d] Ahmed et al. 1974, as reported in Olsen et al. 1982
[e] Brown et al. 1976, as reported in Olsen et al. 1982
[f] Keizer and Gordon 1973, as reported in Olsen et al. 1982
[g] Ward et al. 1980, Gunkel et al. 1980, and Massie et al. 1985
[h] Mackie et al. 1978
[i] Elmgren and Frithsen 1982
[j] Calder et al. 1978
[k] Law 1978, as reported in Olsen et al. 1982
[l] Vandermeulen 1982

Argo Merchant oil to persist. But the spilled oil was rapidly mixed throughout the entire water column over a vast area; the dissolved oil, including toxic aromatic hydrocarbons, persisted throughout the water column of Georges Bank at elevated concentrations for at least 5 months following the spill (Boehm et al. 1978; Farrington and Boehm 1987; Howarth 1987). Immediately following the spill, concentrations of oil of up to 210 μg l^{-1} were detected at a depth of 20 m on Nantucket Shoals (Vandermeulen 1982). By February, concentrations were lower, but still averaged 44 μg l^{-1} over all of Georges Bank, and oil was uniformly distributed throughout the water column (Boehm et al. 1978). By May, five months after the spill, concentrations had fallen further, but were still elevated over background levels and averaged 11 μg l^{-1} (Boehm et al. 1978). Concentrations had fallen to background levels by the time of the next measurements, in August (Boehm et al. 1978). The quality of these data are due in part to the method of analysis, gas chromatography/mass spectrometry. For further discussion of these data, see Farrington and Boehm (1987) and Howarth (1987).

A near-uniform contamination of the water column also occurred following the *Amoco Cadiz* spill off the coast of France, with concentrations of up to 100 μg l^{-1} found at a depth of 100 m (Marchand 1978). Surface concentrations ranged up to 340 μg l^{-1} (Calder et al. 1978). Once oil is dispersed in the water column, the toxic components are much more slowly lost than they are from a surface slick (Vandermeulen 1982).

Studies of dissolved oil concentrations resulting from chronic discharges in offshore oil fields are also few and have tended to use relatively unsophisticated, fluorescence measurements. I know of no such measurements made in any offshore oil field by gas chromatography/mass spectrometry. Reported values for dissolved aromatic hydrocarbons (the more toxic component of oil) in the North Sea since the start of oil production range from 0.35 to 88 μg l^{-1} (Ward et al. 1980; Gunkel et al. 1980; Massie et al. 1985), with average concentrations for the entire North Sea since 1976 seeming to be roughly 5 μg l^{-1} (Gunkel et al. 1980; Massie et al. 1985). Total dissolved oil hydrocarbons tend to be much higher (Ward et al. 1980; Gunkel et al. 1980), but the analytical method used to determine this tends to be poor. Concentrations of dissolved aromatic hydrocarbons are variable but tend to be higher near rigs (Ward et al. 1980) and are higher in the vicinity of oil fields than in more pristine areas of North Sea (Gunkel et al. 1980). One should be somewhat skeptical of the concentration data determined by fluorometry, but taken at face value, existing data suggest that the potential for harm may be great. Monitoring of oil hydrocarbons in the waters and sediments of existing offshore oil fields such as those in the North Sea and in the Gulf of Mexico using state-of-the-art methodology should be accorded high priority.

Planktonic ecosystems are characterized by large natural variations, both spatial and temporal, making it extremely difficult to determine ef-

fects from oil pollution even in controlled mesocosm experiments, particularly if the effects are fairly subtle. Of course, field studies are even more
difficult. This difficulty, combined with the wide-spread (although probably incorrect) perception that toxic concentrations of oil never persist in
pelagic environments, have resulted in rather few investigations. Our
knowledge of planktonic effects is quite limited as a result (Teal and
Howarth 1984).

The only data on planktonic ecosystem processes are for primary production, and it appears that production is sometimes increased and sometimes decreased by oil pollution (Teal and Howarth 1984; National Academy of Sciences 1985). The mechanisms responsible for this are not
entirely clear, but there is some evidence showing that oil stimulates
photosynthesis in pure cultures of some phytoplankton species (Parsons
et al. 1976) while it inhibits it in others (Pulich et al. 1974; Lee et al. 1977).
However, ecological interactions can also be important, and in at least
some instances a stimulation of phytoplankton production has apparently
resulted from reduced grazing pressure by zooplankton (Johansson 1980).
Ecological interactions also can reduce primary production; bacterioplankton frequently bloom after oil spills or experimental oilings, perhaps
because of increases in hydrocarbon-degrading component of the bacterioplankton (Lee et al. 1978b; Johansson 1980; Davis et al. 1979). The
bacterial uptake of nutrients can be enough potentially to limit phytoplankton nutrient uptake and production where nutrients are low (Lee et
al. 1978b).

Some evidence suggests that low-level oil pollution can cause a change
in the species composition of the phytoplankton community, with larger
species being replaced by smaller species. Lee et al. (1977) found that a
single addition of oil (initially 40 μg l^{-1}, decreasing to zero by two weeks)
to CEPEX microcosms resulted in the replacement of the large diatom
Ceratualina bergonii by the small nanoflagellate *Chrysochromulina
kappa*. A similar bloom of *Chrysochromulina kappa* was caused in an
earlier mesocosm experiment by the addition of 20 μg l^{-1} of oil (Parsons et
al. 1976). These results are consistent with the pure-culture experiments
showing an inhibition by oil of growth in *Ceratualina* (Lee et al. 1977) and
a stimulation of growth in *Chrysochromulina* (Parsons et al. 1976). Some
other CEPEX mesocosm experiments found a replacement of a large-
celled (20 to 50 μ) *Chaetoceros* species (a diatom) by a small-celled (less
than 5 $^{-1}$) *Chaetoceros* species (Lee et al. 1978b). In two other experimental oilings in which *Chaetoceros* species were initially dominant, no major
changes in phytoplankton species composition took place, apparently because these are fairly resistant species (Lee et al. 1978b). No major
changes occurred in phytoplankton species composition in the oil-addi-
tion experiments in the MERL mesocosms, either (Elmgren et al. 1980b).
But in those experiments nanoflagellates and small centric diatoms such

as *Leptocylindrus minimus* dominated both oiled and control systems, species that are apparently quite resistant to oiling (Elmgren et al. 1980b).

To date, very little work has been done on the abundance and composition of phytoplankton in offshore oil fields (Royal Society 1980). Similarly, the response of phytoplankton species composition to actual oil spills has received little study, in part because this has proven very difficult under spill conditions (Teal and Howarth 1984). The best data on effects of a spill on phytoplankton probably are those from the *Tsesis* spill. There, phytoplankton species composition was unchanged, but as in the MERL experiments, nanoflagellates, which may be relatively insensitive to oil (Johansson 1980; Parsons et al. 1976; Lee et al. 1978b), were already dominant at the time of the spill (Johansson 1980). Following the *Torrey Canyon* spill, the nanoflagellate *Chrysochromulina kappa* was found to dominate (Smith 1968), as in the CEPEX oiling experiments. However, there were no pre-spill data. After the *Bravo* spill, diatoms dominated the phytoplankton both before and after the spill (Rey et al. 1977); however, the *Bravo* spill is somewhat unusual in that only low levels of aromatic hydrocarbons were detected in the water (less than 1 μg l^{-1}; Rey et al. 1977; also, see Table 4.2), probably because the oil shot 50 m into the air at a temperature of 75 to 90° C before falling to the water, greatly facilitating evaporation (Audunson 1977). Following most other oil spills, including *IXTOC*, *Amoco Cadiz*, and *Argo Merchant*, there were few if any observations on phytoplankton species composition (Teal and Howarth 1984).

Zooplankton community structure can also be altered by oil pollution (National Academy of Sciences 1985). Again, the best evidence for this comes from mesocosm experiments (Lee and Takahashi 1977; Giesy 1980; Davies et al. 1980). Such alterations have been demonstrated at surprisingly low concentrations, as low as 5 to 15 μg l^{-1} (Davies et al. 1980). However, there are no unequivocal data indicating that this has occurred in the field as a result of oil pollution. Reid (1987) documented that major changes in the zooplankton and phytoplankton communities and an almost 10-fold decrease in planktonic biomass in the North Sea (ICES statistical division IV) have occurred over the period from 1948 to 1982. But these trends were occurring even before the advent of oil and gas development in the North Sea. Therefore, the changes are generally ascribed to natural variability, although one cannot rule out the effects of overfishing or of other sources of pollution, including the rather large amount of oil pollution from land-based sources surrounding the North Sea.

Many fish species have fish eggs and larvae that are planktonic, and numerous studies and reviews have now demonstrated that the eggs and larvae of several fish species are particularly sensitive to the toxic effects of dissolved oil (National Academy of Sciences 1985; Miranov 1968; Johannessen 1976; Kuhnhold et al. 1978; Wiseman et al. 1982; Vandermeulen and Cappuzo 1982; Teal and Howarth 1984; Howarth 1987). However, I am unaware of any studies that have examined the impacts of chronic oil pollution in offshore oil fields on fish egg and larval viability, and surprising few studies have even investigated the effects of large oil spills on planktonic fish eggs and larvae (Teal and Howarth 1984).

One of the few such studies was conducted after the *Argo Merchant* spill off of Massachusetts. Although limited, the evidence suggests a major effect. An average of 20% of cod eggs and 46% of pollock eggs were dead or damaged at all stations sampled near the oil slick (Longwell 1978). (This result has been discounted by Grose and Mattson 1977, but is supported by the analysis of Teal and Howarth 1984.) Evidence for high pilchard-egg mortality was reported after the *Torrey Canyon* spill, although there the use of oil dispersants may have caused or contributed to the mortality (Smith 1968). However, that oil spills kill fish eggs and larvae should not be surprising given the conclusion of the review by Vandermeulen and Capuzzo (1983) that dissolved oil concentrations in the range of from 2 to 10 μg l^{-1} can decrease larval fish viability; as we have seen, concentrations much higher than this can persist in the water, even in offshore environments, for periods of months following a spill (see also Wiseman et al. 1982; Howarth 1987). Existing data, although poor (as discussed above), suggest that concentrations higher than this may occur regularly throughout offshore oil fields as well as a result of routine pollution discharges. It is therefore surprising that little or no effort has been made directly to examine the effects of such chronic pollution on fish eggs and larvae.

4.5 Significance of the Observed Ecosystem Effects

As the previous two sections have shown, oil pollution from both accidental spills and routine discharges in offshore oil operations is sufficient to cause alterations in both benthic and planktonic communities. However, the public and environmental decision makers are generally not concerned directly with changes of this sort; moreover, only an expert eye would detect many of these changes. The question discussed in this section is, do such alterations in ecosystem structure potentially change the availability of resources of more general interest, such as commercial fishery resources?

Two approaches to this question can be taken. One is directly to examine whether fishery resources are lowered following oil spills or the advent of offshore oil operation. The other approach is to attempt to infer what effects ecosystem alterations might have on fishery resources. Unfortunately, the direct approach is much more difficult and thus less pow-

erful than it might at first appear. The natural variability in fishery recruitment is great (Hennemuth et al. 1980). Consequently, it is unlikely that
the reduction of a year class from an oil spill could be detected by stock
assessment methods unless the reduction was very large (Sinclair 1982;
Wiseman et al. 1982; Longhurst 1982; Ware 1982). An analysis by the
Canadian Department of Fisheries and Oceans stressed the need to distinguish between "no effect" and "no statistically detectable effect" on
fishery recruitment; the analysis concluded that even with well-studied
fisheries, oil-induced mortalities of juvenile and pre-recruit fish of less
than an order of magnitude would be undetectable (Longhurst 1982; Wiseman et al. 1982).

Some evidence exists that oil pollution can slow the growth of adult
fish (Conan and Friha 1980) and perhaps decrease fish recruitment when
pollution occurs in localized nearshore waters of small bays (National
Academy of Sciences 1975; Laubier 1980; Desaunay 1981). However,
after most major oil spills offshore, there is no conclusive evidence that
recruitment to stocks either has or has not been affected (Royal Society
1980; Royal Commission on Environmental Pollution 1981; McIntyre
1982; Teal and Howarth 1984). Although there was a significant mortality
of pilchard eggs following the *Torrey Canyon* spill, there was no detectable effect on the population of adult pilchards in following years (Royal
Commission on Environmental Pollution 1981). This also appears to be
the case for cod and pollock stocks following the *Argo Merchant* spill
(unpublished stock assessment data of the National Marine Fisheries Service, Woods Hole, MA).

As for oil spills, there is little evidence to show that chronic oil pollution associated with offshore oil production either has or has not affected
fishery recruitment in such areas as the Gulf of Mexico and the North Sea
(National Academy of Sciences 1975; Manners 1982). However, the
Royal Society (1980) report notes that monitoring in offshore oil fields has
largely not investigated effects on the species composition and abundance
of the fish fauna. What data are available indicate problems with the
commercial fisheries both in the Gulf of Mexico and in the North Sea.
Catches per unit effort of commercially valuable fisheries have fallen in
the Gulf since the start of oil development (National Academy of Sciences
1975). And in the North Sea, not only have catches per unit effort fallen
since 1969 (Manners 1982), but actual total fish catches have been falling
since 1974 when oil development expanded substantially (Reid 1987). But
for both the North Sea and the Gulf of Mexico, it is impossible to distinguish the impact of oil pollution from the effects of other pollutants,
overfishing, other alterations of the environment, and natural fluctuations
(National Academy of Sciences 1975, 1985; Manners 1982; Reid 1987).

Although there are no conclusive data showing an effect of most oil
spills or offshore oil operations on fishery resources, one can infer that
such effects are quite possible from the observed alterations in ecosys-

tems discussed above. For instance, the observed changes in benthic community structure if these occurred on a sufficiently large scale could adversely affect commercial fisheries because so many commercially important fish feed on benthic invertebrates; a decreased rate of benthic production clearly could decrease demersal fish production. Amphipods, particularly ampeliscids, are both among the most important foods for some fish (Langton 1983) and are quite sensitive to oil pollution, disappearing as a result of spills aud chronic discharge (Elmgren and Frithsen 1982; Teal and Howarth 1984; Cabioch et al. 1981; d'Ozouville et al. 1979; Sanders et al. 1980; Elmgren et al. 1980a). Their loss from the community might slow the growth of adult fish or result in decreased spawning.

Changes in the species composition of phytoplankton communities could also adversely affect fishery resources. Greve and Parsons (1977) have speculated that a change from an ecosystem dominated by large phytoplankton to one dominated by small phytoplankters such as nanoflagellates might alter food chains to the detriment of the production of commercially valuable fish species. Their hypothesis is that communities dominated by large diatoms result in food chains leading to large zooplankton, then to young fish, and that communities dominated by nanoflagellates have food chains leading to small zooplankton, then to ctenophores and medusae. Thus, Greve and Parsons (1977) conclude that "the release of very small amounts of hydrocarbons in ocean areas where oil exploration is in progress could cause a decrease in the harvestable fisheries of those areas through indirect interference in the natural food web." Their hypothesis is as yet untested. Another possible result of the change in phytoplankton toward smaller species might be to increase the number of steps in the food chain, thereby decreasing the efficiency with which the energy of primary production is transferred to fish (Ryther 1969). This too would decrease fish production.

Perhaps the most obvious way that oil pollution might lower fishery resources, other than just by killing adult fish, is by reducing spawning or survival of fish eggs and larvae, thereby decreasing recruitment. And although the factors that control fishery recruitment are very poorly understood (Hennemuth et al. 1980), numerous lines of evidence suggest ways in which oil may reduce recruitment. For instance, oil pollution may lead to decreased spawning (Nellbring et al. 1980), kill fish eggs and larvae (Johannessen 1976; Longwell 1978; Grose and Mattson 1977; Kuhnhold et al. 1978; Nellbring et al. 1980; Miranov 1968; Teal and Howarth 1984), or delay the development of fish eggs (Kuhnhold et al. 1978).

If the sediments of spawning grounds become contaminated with oil, decreased spawning may result (Nellbring et al. 1980). No one knows how much of a decrease in spawning is necessary to cause a decrease in recruitment, but clearly a large enough decrease in spawning will decrease recruitment. Contaminated spawning grounds may have another adverse effect on recruitment. Exposure of adult fish during gonad maturation to an average concentration of 10 μg l^{-1} of dissolved oil for 10 days can cause reduced survival and growth of the larvae coming from the eggs laid by such fish (Kuhnhold et al. 1978).

Although it might seem obvious that the death of fish eggs and larvae would lower fishery recruitment, this point is in fact hotly argued. Two British studies conclude that oil induced mortality of fish eggs and larvae would have little effect on recruitment because of density compensation; these studies argue that the proportional survival of the remaining eggs and larvae would increase, compensating for the mortality (Royal Society 1980; Royal Commission on Environmental Pollution 1981). A model developed by Spaulding et al. (1983) to predict the effect of an oil spill on Georges Bank assumes some density compensation, but less than assumed by the British studies. A typical run of their model predicted that a 40% oil-induced mortality among cod larvae resulted in a 25% loss of adult fish from the resulting age class.

Such assumptions about density-dependent compensation are based on observations in unpolluted environments. The effects of pollutant-induced mortality on density-dependent mortality have never been tested. It is at least arguable that density-dependent compensating mechanisms are not likely to be important following an oil spill, if for instance the spill kills entire patches of larvae while leaving other patches unaffected (Howarth 1987). Density-dependent compensating mechanisms probably exist only within patches rather than between patches if cannibalism among larvae is the major mechanism of compensation (Ellertsen et al. 1981; Reed et al. 1981). Models of the effects of oil spills on fish recruitment are very sensitive to assumptions about density-dependent compensation among fish larvae (Reed et al. 1981).

Low concentrations of oil have sublethal effects on fish eggs and larvae that also may be important to recruitment. For instance, oil can delay the hatching of eggs and slow the growth of larvae at concentrations lower than those which directly cause death (Kuhnhold et al. 1978). Since eggs and larvae are very susceptible to predation (Oiestad 1982; Ellertsen et al. 1981), a slower development rate could greatly increase mortality by predation. Furthermore, predation on fish eggs and larvae may be a major factor controlling recruitment (Cushing 1976; Laurence et al. 1979; Oiestad 1982). Indeed, Cushing (1976) has hypothesized that larval growth rate, by changing the time of exposure to predation, may be the most important factor controlling recruitment. Thus, sublethal concentrations of oil might decrease recruitment by slowing development.

4.6 Conclusions

Oil pollution can damage both benthic and planktonic communities and ecosystems. The best documented effects are changes in species composition, with sensitive species being replaced by more pollution-tolerant species. Effects on ecosystems processes such as primary production have proven more variable and more difficult to study. The benthic effects of oil are better studied, more long lasting, and more widely accepted by scientists than are planktonic effects. Nonetheless, planktonic effects occur and can potentially be quite damaging. The best data on the effects of oil pollution on marine ecosystems come from controlled mesocosm experiments and clearly show effects of dissolved oil at concentrations as low as 20 μg l^{-1} (Lee and Takahashi 1977) to 29 μg l^{-1} (Gray 1987). Virtually no attempt has yet been made to determine if even lower concentrations also have serious deleterious effects in mesocosm experiments, but data on sublethal effects of oil on individual organisms show effects at concentrations as low as 0.2 μg l^{-1} (Steele 1977). We need more mesocosm experiments to explore the effects of these low concentrations of oil on ecosystems and to determine better the dose-response curves of ecosystems. Nonetheless, the current reliance on acute lethal toxicity data, such as from LD$_{50}$ tests, for environmental management clearly is permissive. Typically used LD$_{50}$ values are 1 to 3 mg l^{-1}, or some 10- to 100-fold greater than concentrations demonstrated to cause effects in mesocosms (Table 4.3).

Table 4.3. Concentrations of Dissolved Oil Hydrocarbons Used by U.S. Department of Interior to Estimate Effects of Offshore Oil Development Compared with Those Actually Found to Have Deleterious Effects on Organisms or Ecosystems.

	Micrograms per liter
LD$_{50}$ values used by U.S. Dept. of Interior[a]	1,000–3,000
"Safe" level estimated by U.S. Dept. of Interior from applying 10-fold safety margin to LD$_{50}$ data[a]	100–300
Inhibition of sexual reproduction in macroalgae[b]	0.2
Disruption of snail feeding[c]	1–3
Decreased viability of larval fish[d]	2–10
Alteration in zooplankton community composition[e]	5–15
Alteration in benthic community composition (Solbergstrand mesocosms)[f]	29
Alteration in phytoplankton community composition (CEPEX microcosms)[g]	less than 40

[a] Department of Interior 1985
[b] Steele 1977
[c] Jacobson and Boylan 1973
[d] Vandermeulen and Capuzzo 1983
[e] Davies et al. 1980
[f] Gray 1987
[g] Lee et al. 1977

Many biologists and most environmental decision-makers assume that toxic levels of oil are dispersed within a few days following oil spills and that concentrations of dissolved oil in offshore oil fields remain low despite chronic pollution. Actual data relating to these assumptions are rare, but existing data show that the assumptions are wrong (Table 4.2). Rather, following one offshore tanker spill, dissolved oil concentrations averaged 44 μg l^{-1} 2 months later and 11 μg l^{-1} 5 months later over a wide area (Boehm et al. 1978). Concentrations of dissolved aromatic hydrocarbons from oil chronic pollution in the North Sea average roughly 5 μg l^{-1}, with concentrations higher in the offshore oil fields, and with concentrations ranging up to 88 μg l^{-1} (Ward et al. 1980; Gunkel et al. 1980; Massie et al. 1985). We need more and better chemical data in offshore oil fields and for longer periods of time after oil spills if we are to improve our prediction of the effects of oil spills and oil development.

Numerous lines of evidence suggest mechanism whereby low levels of oil pollution may harm important resources such as commercial fisheries. However, the extreme variations in fish populations and fishery recruitment make it extremely difficult to determine directly whether oil spills and chronic oil pollution actually have had major deleterious effects on fisheries. A rather poor understanding of the factors that affect recruitment exacerbate this situation. For instance, one cannot determine if the recent decline in the fishery in the North Sea (Reid 1987) is or is not related to oil pollution.

Clark (1987), in summarizing a recent discussion meeting of the British Royal Society, stated "although marine pollution research has yielded valuable insights into the responses of individuals, populations, and communities to perturbations, natural as well as man-made, it is not likely that future problems associated with oil extraction from the sea will be as stimulating to fundamental research." He argues that marine scientists should be directed to study other questions. But on the contrary, the remaining questions concerning the effects of oil pollution are central to understanding the relationship between structure and function of marine ecosystems and to understanding fishery recruitment. Research directed toward these questions is not only critical to better management of offshore oil development but also promises to add materially to our basic understanding of marine ecosystems.

Acknowledgments. I thank Donald F. Boesch, Michael Connor, John W. Farrington, Ragnar Elmgren, Richard C. Hennemuth, John E. Hobbie, Cindy Lee, Richard F. Lee, Mark Reed, Saul B. Saila, Howard Sanders, John M. Teal, and John H. Vandermeulen for their comments, advice, and information used in preparing the background for this paper. Portions of this paper have appeared previously as part of a site-specific assessment of the potential effects of oil pollution on Georges Bank (Howarth 1987).

This publication is ERC-164 of the Ecosystems Research Center, Cornell University, and was supported by the U.S. Environmental Protection Agency Cooperative Agreement Number CR812685–01. Additional funding was provided by Cornell University.

The conclusions published here represent the views of the author, and do not necessarily represent the opinions, policies or recommendations of Cornell University or of the Environmental Protection Agency.

References

Addy JM, Levell D, Hartley JP (1978) Biological monitoring of sediments in Ekofish oilfield. In: Proceedings of the Conference *Assessment of Ecological Impacts of Oil Spills,* A.I.B.S. pp. 515–539

Anderson JW, Neff JM, Cox BA, Tatem HE, Hightower GM (1974) Characteristics of dispersions and water-soluble extracts of crude and refined oils and their toxicity to estuarine crustacea and fish. Marine Biology 27: 75–88

Audunson T (1977) *The Bravo Blow-Out. Field Observation, Results of Analyses and Calculations Regarding Oil on the Surface.* Trondheim, Norway: Report of the Institutt for Kontinentalsokkelundersokelser. 287 pp.

Bedinger CA, Childers RE, Cooper JW, Kimball KT, Kwok A (1980) *Central Gulf Platform Study.* Vol. 1. *Pollutant Fate and Effects Studies.* Part 1. Background, program organization and study plan. Southwest Research Inst., Houston, Texas, 54 pp.

Bender ME, Reish DJ, Ward CH (1979) Independent appraisal. Reexamination of the offshore ecology investigation. Rice University Studies 65:35–118

Beslier A, Birrien JL, Cabioch L, Larsonneur C, Le Borgne L (1980) La pollution des Baies de Morlaix et de Lannion par les hydrocarbures de l'Amoco Cadiz: Repartition sur les fonds et évolution. Helgol Wiss Meeresunters 32:209–224

Blumer M, Sanders HL, Grassle JF, Hampson GR (1971) A small oil spill. Environment 13: 2–12

Boehm PD, Perry G, Fiest DL (1978) Hydrocarbon chemistry of the water column of Georges Bank and Nantucket Shoals, February-November 1977. In: *"In the Wake of the Argo Merchant."* Symposium, Center for Ocean Management Studies, U.R.I. pp. 58–64

Boehm PD, Steinhauer WG, Fiest DL, Mosesman N, Barak JE, Perry GH (1979) A chemical assessment of the present levels and sources of hydrocarbon pollutants in the Georges Bank region. In: *Proceedings of the 1979 Oil Spill Conference.* Am. Petr. Inst. Publ. #4308. pp. 333–341

Boucher G (1981) Effets à long terme des hydrocarbons de l'Amoco Cadiz sur la structure des communaut–s de n–matodes libres des sables fins sublittoraux. In: *"Amoco Cadiz, Fates and Effects of the Oil Spill".* Paris: CNEXO, pp. 539–549

Cabioch L, Dauvin JC, Gentil F, Retiere C, Rivain V (1981) Perturbations induites dans la composition et le fonctionnement des peuplements benthiques sublittoraux sous l'effet des hydrocarbures de l'Amoco Cadiz. In: *"Amoco Cadiz, Fates and Effects of the Oil Spill".* Paris: CNEXO, pp. 513–526

Calder JA, Lake J, Laseter J (1978) Chemical composition of selected environmental and petroleum samples from the Amoco Cadiz spill. In: *Oil Spill.* NOAA/EPA Special Report.

Caldwell RS, Calderone EM, Mallon MH (1977) Effects of a seawater-soluble fraction of Cook Inlet crude oil and its major aromatic components on larval stages of the Dungeness crab, *Cancer Magister* Dana. In: *Fate and Effects of Petroleum Hydrocarbons in Marine Ecosystems and Organisms.* Proceedings Symposium, Seattle, WA, Nov., 1976. New York: Pergamon Press, pp. 210–220

Carney RS (1985) A review of study designs of field for the detection of long-term environmental effects of offshore petroleum activities. In: Boesch DF, Rabalais N (eds) *Long-term Effects of Offshore Oil and Gas Development: An Assessment and Research Strategy.* Final Report to the National Marine Pollution Program, Office of N.O.A.A. for the Interagency Committee on Ocean Pollution Research, Development, and Monitoring. Rockville, MD, pp. 14–1 to 14–48

Clark RB (1987) Summary and conclusions: environmental effects of North Sea oil and gas developments. Phil Trans R Soc Lond B 316: 669–677

Conan G, Friha M (1981) Effets des pollutions par les hydrocarbures du pétrolier Amoco Cadiz sur la croissance des soles et des plies dans l'estuaire de l'Aber Benoit. In: *"Amoco Cadiz, Fates and Effects of the Oil Spill".* Paris: CNEXO, pp. 749–773

Conover RJ (1971) Some relations between zooplankton and bunker C oil in Chedabucto Bay following the wreck of the tanker Arrow. J Fish Res Board Can 28:1327–1330

Cushing DH (1976) Biology of fishes in the pelagic community. In: Cushing DH, Walsh JJ (eds) *The Ecology of the Seas.* Philadelphia: WB Saunders, pp. 317–340

d'Ozouville L, Hayes MO, Gundlach ER, Sexton WJ, Michel J (1979) Occurrence of oil in offshore bottom sediments at the *Amoco Cadiz* oil spill site. pp. 187–192 in Proceedings of the 1979 Oil Spill Conference. Am. Pet. Inst. Publ. #4308

Davenport J (1982) Oil and planktonic ecosystems. Phil Trans R Soc Lond B 297:369–384

Davies JM, Baird IE, Massie LC, Hay SJ, Ward AP (1980) Some effects of oil-derived hydrocarbons in a pelagic food web from observations in an enclosed ecosystem and a consideration of their implications for monitoring. Rapp PV Reun Cons Int Explor Mer 179:201–211

Davis PG, Hefferman RF, Sieburth JM (1979) Heterotrophic microbial populations in estuarine microcosms: Influence of season and water accommodated hydrocarbons. Trans Am Micros Soc 98:152

Davis PH, Spies RB (1980) Infaunal benthos of a natural petroleum seep: study of community structure. Mar Biol 59:31–41

Department of Interior (1977) Final Environmental Statement, Proposed 1977 Outer Continental Shelf Oil and Gas Lease Sale Offshore the North Atlantic States, OCS Sale #42 Bureau of Land Management

Department of Interior (1983) Final Environmental Impact Statement, Proposed April 1984 North Atlantic Outer Continental Shelf Oil and Gas Lease Offering, OCS. Oil and Gas Sale #52. Bureau of Land Management

Department of Interior (1985) Final Environmental Impact Statement, North Aleutian Basin Lease Sale #92, Bureau of Land Management

Desaunay Y (1981) Evolution des stocks de poissons plats dans la zone contaminée par l'Amoco Cadiz. In: *Amoc Cadiz, Fates and Effects of the Oil Spill.* Paris: CNEXO, pp. 727–735

Ellertsen B, Moksness E, Solemdal P, Tilseth S, Westgaard T, Øiestad V (1981) Growth and survival of cod larvae in an enclosure, experiments and a mathematical model. Rapp PV Réun, Cons Int Explor Mer 178: 45–57

Elmgren R, Hansson S, Larsson U, Sundelin B (1980a) Impact of oil on deep soft bottoms. In: Kineman JJ, Elmgren R, Hansson S (eds) *The Tsesis Oil Spill,* N.O.A.A., Boulder, CO, pp. 97–126

Elmgren R, Vargo GA, Grassle VF, Grassle JP, Heinle DR, Ianglois G, Vargo SL (1980b) Trophic interactions in experimental marine ecosystems perturbed by oil. In: Giesy JP (ed) *Microcosms in Ecological Research.* U. S. Dept. of Energy, Symp. Series 52 (CONF-781101), pp. 779–800

Elmgren R, Frithsen JB (1982) The use of experimental ecosystems for evaluating the environmental impacts of pollutants: A comparison of an oil spill in the Baltic Sea and two long-term, low-level oil addition experiments in mesocosms. In: Grice GD, Reeve MR (eds) *Marine Mesocosms: Biological and Chemical Research in Experimental Ecosystems.* New York: Springer-Verlag, pp. 153–165

Farrington JW, Boehm PD (1987) Natural and pollutant organic compounds. In: Backus R (ed) *An Atlas of Georges Bank.* Cambridge, MA: M.I.T. Press, pp. 195–207

Farrington JW, Tripp BW, Teal JM, Mille G, Tjessem K, Davis AC, Livramento JB, Hayward NA, Frew NM (1982) Biogeochemistry of aromatic hydrocarbons in the benthos of microcosms. Toxicol Environ Chem 5:331–346

Gearing PJ, Gearing JN, Pruell RJ, Wade TL, Quinn JG (1980) Partitioning of #2 fuel oil in controlled estuarine ecosystems: Sediments and suspended particulate matter. Env Sci Tech 14: 1129–1136

Giesy JP, Jr (1980) Microcosms in ecological research. U. S. TIC Conference 781101

Gilfillan ES, Vandermeulen JW (1978) Alternations in growth and physiology of soft-shelled clams, *Mya arenaria,* chronically oiled with Bunker C from Chedabucto Bay, Nova Scotia, 1970–1976. J Fish Res Bd Canada 35: 630–636

Glemarec K, Hussenot E (1981) Définition d'une succession écologique en milieu meublé anormalement enrichi en matières organiques à la suite de la catastrophe de L'Amoco Cadiz. In: *Amoco Cadiz, Fates and Effects of the Oil Spill.* Paris: CNEXO, pp. 499–512

Grassle JF, Grassle JP (1974) Opportunistic life histories and genetic systems in marine benthic polychaetes. Marine Research 32:253–284

Grassle JF, Elmgran R Grassle JP (1981) Response of benthic communities in MERL experimental ecosystems to low level chronic additions of No. 2 fuel oil. Mar Environ Res 4: 279–297

Gray JS (1982) Effects of pollutants on marine ecosystems. Neth J Sea Res 16:424–443

Gray JS (1987) Oil pollution studies of the Solbergstrand mesocosms. Phil Trans R Soc Lond B 316:641–654

Greve W, Parsons TR (1977) Photosynthesis and fish production: Hypothetical effects of climatic change and pollution. Helgol Wiss Meeresunters 30:666–672

Grice GD, Reeve MR (1982) Introduction and description of experimental ecosystems. In: Grice GR, Reeve MR (eds) *Marine Mesocosms: Biological and Chemical Research in Experimental Ecosystems.* New York: Springer-Verlag, pp. 1–9

Grose PL, Mattson JS (1977) *The Argo Merchant Oil Spill: A Preliminary Scientific Report.* N.O.A.A., Boulder, CO, 133 pp.

Gundlach ER, Domeracki DD, Thebeau LC (1982) Persistence of Metula oil in the Straits of Magellan six and one-half years after the incident. Oil and Petroleum Pollution 1: 37–48

Gunkel W, Gassmann G, Oppenheimer CH, Dundas I (1980) Preliminary results of baseline studies of hydrocarbons and bacteria in the North Sea: 1975, 1976, 1977. In: Ponencias del Simposio Internacional en: *Resistencia a Los Antibioticos y Microbiologia Marina.* VI Congreso Nacional Microbiologia, 6–9 Julio, 1977, Santiago de Compostela, Spain, pp. 223–247

G.U.R.C. (1974) Final Project Planning Council Concensus Report. Gulf Universities Research Cosortium Report no. 138

Hennemuth RC, Palmer JE, Brown BE (1980) A statistical description of recruitment in eighteen selected fish stocks. Northwest Atlantic Fish Sci 1:101–111

Hinga KR, Pilson MEQ, Lee RF, Farrington JW, Tjessem K, Davis AC (1980) Biogeochemistry of benzanthracene in an enclosed marine ecosystem. Environ Sci Tech 14:1136–1143

Howarth RW (1987) The potential effects of petroleum on marine organisms on Georges Bank. In: Backus R (ed) *An Atlas of Georges Bank.* Cambridge, MA: M.I.T. Press, pp. 540–551

Jacobson SM, Boylan DB (1973) Effect of seawater soluble fraction of kerosene on chemotaxis in a marine snail, *Nassarius obsoletus.* Nature 241: 213–215

Johannessen KI (1976) Effects of seawater extract of Ekofisk oil on hatching success of Barents Sea capelin. Int. Council for the Exploration of the Sea publ. C.M. 1976/E:29

Johansson S (1980) Impact of oil on the pelagic ecosystem. In: Kineman JJ, Elmgren R, Hansson S (eds) *The Tsesis Oil Spill,* U.S. Dept. of Commerce, NOAA, Washington, D.C.,m pp. 61–80

Johnson FG (1977) Sublethal biological effects of petroleum hydrocarbon exposure: Bacteria, algae, and invertebrates. In: Malins DC (ed) *Effects of Petroleum on Arctic and Subarctic Marine Environments and Organisms,* Vol. 2. Academic Press, pp. 271–318

Johnston CS (1980) Sources of hydrocarbons in the marine environment. In: Johnston CS, Morris RJ (eds) *Oily Water Discharges.* London: Applied Science Publishers, Ltd., pp. 41–62

Krebs CT, Burns KA (1977) Long-term effects of an oil spill on populations of the salt marsh crab *Uca Pugnax.* Science 197: 484–487

Krebs CT, Burns KA (1978) Long-term effects of an oil spill on populations of the salt marsh crab *Uca Pugnax.* J Fish Res Bd Can 35:648–649

Kuhnhold WW, Everich D, Stegeman JJ, Lake J, Wolke RE (1978) Effects of low levels of hydrocarbons on embryonic, larval and adult winter flounder (*Pseudopleuronectes americanus*). In: *Proceedings of Conference on Assessment of Ecological Impacts of Oil Spills.* A.I.B.S. pp. 677–711

Lacaze JC, Villedon de Naide O (1976) Influence of illumination on phytotoxicity of crude oil. Mar Poll Bull 7:73

Langton RW (1983) Food habits of yellowtail flounder, *Limanda ferruginea* (Storer) from off the northeastern United States. Fisheries Bull 81:15–22

Larson RA, Hunt LL, Blankenship DW (1977) Formation of toxic products from a #2 fuel oil by photo-oxidation. Env Sci Tech 11:492–496

Laubier L (1980) The Amoco Cadiz oil spill: An ecological impact study. Ambio 9:268–276

Laurence GC, Halavik TA, Burns BR, Smigielski AS (1979) An environmental chamber for monitoring "in situ" growth and survival of larval fishes. Trans Am Fish Soc 108:197–203

Lee RF, Takahashi M (1977) The fate and effect of petroleum in controlled ecosystem enclosures. Rapp PV Reun Cons Int Explor Mer 171:150–156

Lee RF, Takahashi M, Beers JR, Thomas WH, Siebert DLR, Koeller P, Green DR (1977) Controlled ecosystems: their use in the study of the effects of petroleum hydrocarbons on plankton. In: Vernberg FJ, Calabrese A, Thurberg FP, Vernberg WB (eds) *Physiological Responses of Marine Biota to Pollutants.* Academic Press, New York, pp. 323–342

Lee RF, Gardner WS, Anderson JW, Blaylock JW, Barwell-Clarke J (1978a) Fate of polycyclic aromatic hydrocarbons in controlled ecosystem enclosures. Env Sci Technol 12:832–838

Lee RF, Takahashi M, Beers JR (1978b) Short term effects of oil on plankton in controlled ecosystems. In: *Proceedings of Conference on Assessment of Ecological Impacts of Oil Spills.* A.I.B.S., pp. 635–650

Linden O (1977) Sublethal effects of oil on mollusc species from the Baltic Sea. Water Air Soil Pollution 8:305–313

Linden O, Elmgren R, Westin L, Kineman J (1980) Scientific summary and general discussion. In: Kineman JJ, Elmgren R, Hansson S (eds) *The Tsesis Oil Spill.* N.O.A.A., Boulder, CO, pp. 43–58

Linden O (1980) Burrowing behaviour in the clam Macoma balthica. In: Kineman JJ, Elmgren R, Hansson S (eds) *The Tsesis Oil Spill.* N.O.A.A., Boulder, CO, pp. 97–126

Longhurst A (1982) Executive summary. In: Longhurst A (ed) *Consultation on the Consequences of Offshore Oil Production on Offshore Fish Stocks and Fishing Operations.* Canadian Technical Report of Fisheries and Aquatic Sciences #1096, pp. 7–11

Longwell AC (1978) Field and laboratory measurements of stress responses at the chromosome and cell levels in planktonic fish eggs and the oil problem. In: *In the Wake of the Argo Merchant.* Symposium, Center for Ocean Management Studies, Univ. of Rhode Island, Kingston, pp. 116–125

Mackie PR, Hardy R, Whittle KJ (1978) Preliminary assessment of the presence of oil in the ecosystem at Ekofisk after the blowout, April 22–30, 1977. J Fish Res Board Can 35:544–551

MacLeod WD, Uyeda MY, Thomas LC, Brown DW (1978) Hydrocarbon patterns in some marine biota and sediments following the Argo Merchant spill. In: *In the Wake of the Argo Merchant.* Symposium, Center for Ocean Management Studies, Univ. of Rhode Island, Kingston, pp. 72–79

Manners IR (1982) *North Sea Oil and Environmental Planning: The United Kingdom Experience.* Austin: Univ. of Texas Press, 332 pp.

Marchand M (1978) Estimation par spectrofluorométrie des concentrations d'hydrocarbures dans l'eau de mer en Manche Occidentale à la suite du naufrage de l'Amoco Cadiz, du 30 Mars au 18 Avril 1978. In: *Amoco Cadiz: Actes de Colloques.* Paris: CNEXO, pp. 27–38

Massie LC, Ward AP, Davies JM, Mackie PR (1985) The effects of oil exploration and production in the northern North Sea: Part 1—The levels of hydrocarbons in water and sediments in selected areas, 1978–1981. Mar Environ Res 15: 165–213

McIntyre AD (1982) Oil pollution and fisheries. Phil Trans R Soc Lond B 297:401–411

Meyers PA, Quinn JG (1973) Associations of hydrocarbons and mineral particles in saline solution. Nature 244: 23–24

Middleditch BS (1982) *Environmental Effects of Offshore Oil Production. The Buccaneer Gas and Oil Field Study.* New York: Plenum Press

Miranov OG (1968) Hydrocarbon pollution of the sea and its influences on marine organisms. Helgol Wiss Meeresunters 17:335–339

National Academy of Sciences (1975) *Petroleum in the Marine Environment.* Washington DC

National Academy of Sciences (1985) *Oil in the Sea, Inputs, Fates, and Effects.* Washington DC

Neff JM, Anderson JW, Cox BA, Laughlin RB, Rossi SS, Tatem HE (1976) Effects of petroleum on survival, respiration, and growth of marine animals. In: *Sources, Effects, and Sinks of Hydrocarbons in the Aquatic Environment.* Washington DC: A.I.B.S., pp. 516–539

Nellbring S, Hansson S, Aneer G, Westin L (1980) Impact of oil on local fish fauna. In: Kineman JJ, Elmgren R, Hanson S (eds) *The Tsesis Oil Spill.* U. S. Department of Commerce, N.O.A.A. pp. 193–201

Oiestad V (1982) Application of enclosures to studies on the early life history of fishes. In: Grice GD, Reeve MR (eds) *Marine Mesocosms.* New York: Springer-Verlag, pp. 49–62

Olsen S, Pilson MEQ, Oviatt C, and Gearing JN, (1982) Ecological consequences of low, sustained concentrations of petroleum hydrocarbons in temperate estuaries. Marine Ecosystems Research Laboratory, Graduate School of Oceanography, Univ. of Rhode Island, Narragansett, technical report.

Parsons TR, Li WKW, Waters R (1976) Some preliminary observations on the enhancement of phytoplankton growth by low levels of mineral hydrocarbons. Hydrobiology 51:85–89

Pearson WH, Woodruff DL, Sugarman PC, Olla BL (1981) Effects of oiled sediment on the littleneck clam, *Protothaca staminea,* by the Dungeness crab, *Cancer magister.* Est Coast Shelf Sci 13:445–454

Pulich WM, Winters K, Van Baalen C (1974) The effects of a No. 2 fuel oil and two crude oils on the growth and photosynthesis of microalgae. Mar Biol 28: 87–94

Read AD (1980) Treatment of oily water at North Sea installations - A progress report. In: Johnston CS, Morris RJ (eds) *Oil Water Discharges.* London: Applied Science Publishers, Ltd., pp. 127–136

Reed M, Spaulding ML, Cornillion P (1981) A fishery-oilspill interaction model: simulated consequences of a blowout. In: Haley KB (ed) *Applied Operations Research in Fishing.* New York: Plenum Press, pp. 99–114

Reid PC (1987) The importance of the planktonic ecosystem of the North Sea in the context of oil and gas development. Phil Trans R Soc Lond B 316:587–602

Rey F, Seglem K, Johannessen M (1977) The Ekofisk Bravo blow out. Phytoplankton and primary production investigations. I.C.E.S. Rept. CM 1977/E:55, pp. 8.1–8.18

Rosenthal H, Alderdice DF (1976) Sublethal effects of environmental stressors, natural and pollutional, on marine fish eggs and larvae. J Fish Res Bd Can 33:2047–2065

Royal Commission on Environmental Pollution (1981) Eighth report. *Oil Pollution of the Sea*. H. M. Stationery Office, London. 307 pp.

Royal Society (1980) *The effects of oil pollution: some research needs*. London, 103 pp.

Ryther JH (1969) Photosynthesis and fish production in the sea. Science 166:72–76

Sanders HL, Jones C (1981) Oil, science, and public policy. In: Jackson TC, Reische D (eds) *Coast Alert: Scientists Speak Out*. San Francisco: Friends of the Earth Publishers

Sanders HL (1981) Environmental effects of oil in the marine environment. In: *Safety and Offshore Oil: Background Papers of the Committee on Assessment of Safety of OCS Activities*. National Research Council, Washington DC: National Academy Press

Sanders HL, Grassle JF, Hampson GR, Morse LS, Garner-Price S, Jones CC (1980) Anatomy of an oil spill: long-term effects from the grounding of the barge Florida off West Falmouth, Massachusetts. J Mar Res 38:265–380

Sinclair M (1982) What kind of observational programs would be required to detect the effects on biota? In: Longhurst A (ed) *Consultation on the Consequences of Offshore Oil Production on Offshore Fish Stocks and Fishing Operations*. Canadian Technical Report of Fisheries and Aquatic Sciences #1096, pp. 54–63

Smith JE (ed) (1968) *Torrey Canyon: Pollution and Marine Life*. Mar Biol Assoc U.K. Plymouth Laboratory Rep. XIV. Cambridge, Cambridge Univ. Press. 196 pp.

Southward AJ, Southward EC (1978) Recolonization of rocky shores in Cornwall after use of toxic dispersants to clean up the Torrey Canyon spill. J Fish Res Bd Can 35:682–706

Spaulding ML, Saila SB, Anderson E (1983) Oil spill fishery impact assessment model: Application to selected Georges Bank fish species. Est Coast Shelf Sci 16:511–541

Spies RB, Davis PH (1979) The infaunal benthos of a natural oil seep in the Santa Barbara Channel. Mar Biol 50:227–238

Spies RB, Davis PH, Stuermer DH (1980) Ecology of a submarine petroleum seep off the California coast. In: Geyer R (ed) *Environmental Pollution*. I. Hydrocarbons. Amsterdam: Elsevier, pp. 208–263

Steele RL (1977) Effects of certain petroleum products on reproduction and growth of zygotes and juvenile stages of the alga *Fucus edentatus* De la Pyl (Phaeophyceae: Fucales). In: Wolfe D (ed) *Fate and Effects of Petroleum Hydrocarbons in Marine Ecosystems and Organisms*. New York: Pergamon, pp. 115–128

Stuermer DH, Spies RB, Davis PH, Ng DJ, Morris CJ, Neal S (1982) The hydrocarbons in the Isla Vista marine seep environment. Marine Chemistry 11:413–426

Teal JM, Howarth RW (1984) Oil spill studies: A review of ecological effects. Environmental Management 8:27–44

Teal JM, Burns K, Farrington J (1978) Analyses of aromatic hydrocarbons in intertidal sediments resulting from two spills of No. 2 fuel oil in Buzzards Bay, Massachusetts. J Fish Res Board Can 35:510–520

Thomas MLH (1977) Long term biological effects of bunker C oil in the intertidal zone. In: Wolfe DA (ed) *Fate and Effects of Petroleum Hydrocarbons in Marine Organisms and Ecosystems*. New York: Pergamon Press, pp. 238–245

Thomas MLH (1978) Comparison of oiled and unoiled intertidal communities in Chedabucto Bay, Nova Scotia. J Fish Res Board Can 35:707–716

Vandermeulen J (1982) What levels of oil contamination may be expected in water, sediments, and what would be the physiological consequences for biota? In: Longhurst A (ed) *Consultation on the Consequences of Offshore Oil Production on Offshore Fish Stocks and Fishing Operations*. Canadian Technical Report of Fisheries and Aquatic Sciences #1096, pp. 22–53

Vandermeulen JH, Capuzzo JM (1983) Understanding sublethal pollutant effects in the marine environment. In: Champ MA, Trainor M (ed) *Ocean Waste Management Policy and Strategies*. International Ocean Disposal Symposium Series. Melbourne, FL: Center for Academic Publication, 11 pp.

Varanasi U, Gmur DJ (1981) Hydrocarbons and metabolites in English sole (*Paraphrys vetulus*) exposed simultaneously to (3H)benzo(a)pyrene and (14C)naphthalene in oil contaminated sediment. Aquatic Toxicol 1:49–67

Ward AP, Massie LC, Davies JM (1980) A survey of the levels of hydrocarbons in the water and sediments in the areas of the North Sea. I.C.E.S. report CM 1980/E:48

Ware D (1982) What is the likelihood that such effects would impair recruitment, and that such impaired recruitment would be separable from natural variation? In: Longhurst A (ed) *Consultation on the Consequences of Offshore Oil Production on Offshore Fish Stocks and Fishing Operations*. Canadian Technical Report of Fisheries and Aquatic Sciences #1096, pp. 64–73

Wells PG, Sprague JP (1976) Effects of crude oil on American lobster (*Homarus americanus*) larvae in the laboratory. J Fish Res Bd Can 33:1604–1614

Wiseman RJ, Payne J, Akenhead S (1982) Rapporteurs' comments. In: Longhurst A (ed) *Consultation on the Consequences of Offshore Oil Production on Offshore Fish Stocks and Fishing Operations*. Canadian Technical Report of Fisheries and Aquatic Sciences #1096, pp. 85–94

Chapter 5

The Effects of Chemical Stress on Aquatic Species Composition and Community Structure

Jesse Ford[1]

Biological communities are sensitive to their chemical environment, although the degree of sensitivity varies among species and communities. Aquatic ecosystems are particularly responsive to chemical stress, because pollutants tend to be well-distributed throughout zones of active mixing, and because communities tend to be dominated by motile, short-lived species with high reproductive rates. Response to chemical stress often involves rapid changes in species composition of aquatic ecosystems that can translate into changes in various aspects of community structure, such as species richness.

This chapter is an introduction to the effects of chemical stress on several aspects of aquatic ecosystem structure, including indicator species, community composition, biotic indices, biomass, abundance, species richness, and species diversity. It treats a wide range of information, and is not exhaustive in any particular subject area. Its purpose is to stimulate discussion on the identification of biological early warning signals and parameters that can track the probable course of ecosystem deterioration for various pairs of stressors and ecosystem types.

[1] Ecosystems Research Center, Cornell University, Ithaca, New York 14853 (Current address: Environmental Research Laboratory, United States Environmental Protection Agency, Corvallis, Oregon 97333)

5.1 Information Required for Effective Resource Management

One of the challenges of modern aquatic environmental protection and resource management is to evaluate potential stress effects on ecosystems before actual, unacceptable ecosystem degradation occurs. The type of information needed to meet this challenge is usually perceived to be a matrix of the effects of treatments X_1, X_2, . . . X_n on ecosystems Y_1, Y_2, . . . Y_m. Typical treatments might include, for example, deliberate applications of substances into particular ecosystems (e.g., pesticides, $CuSO_4$), accidental spills (e.g., oil, industrial wastes), or chronic pollution by substances perceived to be necessary if unfortunate consequences of economic development (e.g., mining wastes, road salt, acid precipitation). Given the proliferation of chemical stressors and the diversity of aquatic systems, however, collection of the vast array of substance- and system-specific information necessary to create all matrices of interest is impractical. Further, even well-defined ecosystem types (e.g., dimictic temperate lakes) can have a variety of trophic structures and redundancy characteristics as well as different key elements and processes in which the major actors vary widely in sensitivity to any given chemical stressor. Because of this diversity the stressor-system pair approach may often be inadequate to deal with management questions concerning any particular ecosystem of interest.

Effective ecosystem management requires three basic types of information. First, we need to know something about the baseline condition of the ecosystem and its natural range of variation. Second, we need to identify the point at which we can say confidently that the system has begun to deviate from its normal condition. Third, given that a system has begun to change, we need to be able to describe the range of probable trajectories, and the point(s) (if any) at which change can be stopped or redirected. Once these three kinds of information are known, decisions can be made about whether the anticipated changes are acceptable.

The first issue, that of baseline studies and natural variability, is important and difficult. For the purposes of this chapter it is assumed that the reference condition of any ecosystem of interest can be adequately characterized. There are several ways of doing this including: 1) accessing previous work on that or similar ecosystems (e.g., Edmondson et al. 1956; Brooker and Edwards 1973; Myren and Pella 1977; Johannson et al. 1980; Schindler et al. 1985); 2) using simultaneous studies of upstream, adjacent, or nearby ecosystems as controls (e.g., Schindler 1974; Winner et al. 1980; Chadwick and Canton 1983); or 3) using the stratigraphic record to reconstruct site-specific baseline conditions (e.g., Bradbury and Megard 1972; Edmondson 1974; Brugam and Speziale 1983; Charles 1984; Ford 1986). The importance of addressing the issues of baseline conditions and natural variability cannot be overemphasized. Failure to have this reference data will cast doubt on the causal linkage between stress and effect, because observed "effects" may fall within the range of natural variability of the ecosystem.

The second issue, identifying the point at which a system has begun to deviate from baseline conditions, is essentially the issue of early warning signals of ecosystem deterioration. For any given stressor and ecosystem type, there may or may not exist a threshold below which there is no effect. However, whether or not there is a threshold of action, there is certainly a perceptual threshold below which we will not detect effects. Making this perceptual threshold as low as possible is an important challenge, because the earlier we can detect the onset of ecosystem dysfunction, the more options are available for avoiding undesirable consequences.

A major challenge in early detection studies is to identify diagnostic changes against a complex background rich in spatial and temporal heterogeneity. Heterogeneity is a particularly severe problem in aquatic ecosystems because we cannot directly observe the variability we are trying to describe. Aquatic community structure is often cryptic, and stochastic factors are important in determining ecosystem structure and dynamics. Even the normal course of seasonal succession can exhibit a much higher degree of plasticity than in terrestrial ecosystems. Because we cannot directly observe many of the characteristics that we are trying to describe, our understanding of the structure of aquatic ecosystem structure depends critically on sampling design. For planktonic studies, sampling designs must deal with instantaneous three-dimensional spatial patchiness superimposed on a temporally dynamic physical structure. The benthos, although no less cryptic, is more easily sampled. However, spatial patchiness must still be addressed in order to have a statistically valid sampling design, and temporal sampling issues may still be very significant. In either case the importance of sampling design in descriptions of aquatic community composition and structure cannot be overemphasized.

The third issue is the art and science of defining potential trajectories. The accuracy of these efforts, which entirely depends on the depth of our understanding of ecosystem dynamics, must draw heavily on descriptive, comparative, and experimental field studies, complemented by appropriate laboratory investigations. Even so, the identification of potential trajectories is a complex and difficult task because of the diversity of aquatic ecosystems and the array of stochastic factors that can be involved.

5.2 Methodologies Used in the Study of Chemical Stress Effects

Aquatic ecosystems vary in both spatial and temporal characteristics of structural organization. Large lakes and open oceans may be well-mixed, temporally stable physicochemical environments that nonetheless can exhibit surprisingly finely scaled spatial patchiness in community structure (e.g., Fiedler 1983; Levine and Lewis 1985). In these ecosystems, the sampling intensity necessary to characterize the range of natural variability in species composition and community structure can be formidable. Further, both lakes and open oceans have a dynamic physical structure associated with stratification and mixing processes, and can therefore exhibit extreme patchiness in both time and space, with physicochemical habitats emerging, changing, and disappearing in a year, a month, or even a single day. This cryptic spatial heterogeneity must be dealt with in making assessments of likely stress effects. At the other extreme there are temporally variable lotic ecosystems, such as streams and rivers, which can be affected heavily by the nature and timing of both upland processes and regional climatic factors. The monitoring period necessary to adequately characterize the range of natural variability in communities that inhabit lotic ecosystems may be significantly longer than that for more temporally constant environments.

There are six major approaches used singly or in combination to study stress effects on aquatic ecosystems: microcosm studies, mesocosm studies, whole ecosystem manipulations, regional or gradient studies, paleoecological studies, and modeling studies. The most direct way to study biological stress effects on the ecosystem level is, of course, by whole-system manipulation. These studies can be either deliberate experimental manipulations (e.g., Hall et al. 1980; Winner et al. 1980; Schindler et al. 1985) or opportunistic studies that take advantage of planned or unplanned pollutant releases in areas where substantial background data already exist (e.g., Kineman et al. 1980; Hargis and Shannon 1984). Stream ecosystems are particularly favorable environments for experimental whole-system manipulations because upstream stations can serve as controls, avoiding the need for extensive long-term background data. On the other hand, many marine whole-ecosystem studies of stress effects are initiated following accidental acute pollution events. In these cases adequate reference data characterizing the unstressed system may not exist, making it difficult to evaluate the significance of the observations (e.g., Foster and Holmes 1977).

Experimental manipulations of whole ecosystems are the most useful way to study ecosystem-level stress effects, but, with the exception of stream studies, they tend to be expensive when compared to funding levels for other types of ecological research. The investment, however, usually pays off in terms of a significantly deepened understanding of ecosystem structure and function. Proper site selection is critical, partly because of the magnitude of the investment required by this single experiment, but also to ensure that results can be extrapolated from the single, intensively studied ecosystem to other sites of interest. For most kinds of stress-response analyses, experimental whole ecosystem studies are unavailable, and other, complementary approaches have been used.

Microcosm studies attempt to approximate real-world systems well enough to assess the response of specific biological components to particular stresses. Microcosms are small, laboratory scale ecosystems created either by combining selected components of the real system of interest, or by isolating a small portion of the parent system (e.g., core top studies in which the sediment–water interface is preserved). Microcosms are generally used for experimental purposes, and have been applied successfully to a number of areas, including toxicity bioassays (e.g., Geckler et al. 1976; Giesy 1980; Giddings 1983; Taub 1983; Kelly in Chapter 16, this volume). Under certain conditions, microcosms are able to reveal important characteristics of novel compounds (e.g., whether they bioaccumulate), but they cannot give reliable information about the behavior of larger-scale, multi-component systems with emergent and often unpredictable characteristics (Kimball and Levin 1985).

Mesocosms are larger-scale units that have been defined as being "larger than benchtop containers but smaller than and isolated from any subunit of the natural environment" (Grice and Reeve 1982). Mesocosms can either be synthetic laboratory tanks (e.g., the MERL experiments) or in-situ enclosures (e.g., the CEPEX experiments), and are commonly used for experimental studies of pollutant fate and transport and associated biological effects, as well as for simple observations of the dynamic behavior of natural systems. As with microcosm studies, there are issues of scale (how simple can the experimental system be and still be a useful fascimile of the original system) and of enclosure effects (how does the increased ratio of substrate to open water affect the extrapolation of observations and experimental findings in mesocosms to actual field situations; e.g., Kuiper et al. 1983). Most microcosm and mesocosm studies run for shorter time periods than whole ecosystem studies. Furthermore, neither are typically capable of incorporating large animals (e.g., fish), which may distort long-term projections concerning stress effects (Harrass 1983). Despite these drawbacks, however, both types of scaled-down studies have generated useful information about stress-induced ecosystem-level effects in marine and freshwater environments (e.g., Reeve et al. 1976; Yan and Stokes 1978; Elmgren and Frithsen 1982; Gearing in Chapter 15, this volume), although field validation is generally necessary for predictions to be made with high confidence.

Regional or gradient approaches to studies of stress effects have often been used to link pollutants with stress effects. This approach has been successfully applied, for instance, in the estimation of aquatic effects of acid precipitation. For example, regional inventories have documented that the loss of fish and certain aquatic macroinvertebrates is highly correlated with regional patterns of hydrogen-ion and sulfate loading (e.g., J. Økland 1969; K. Økland 1969, 1980; Jensen and Snekvik 1972; Langlois et al. 1983, 1984; Schofield 1976). Similarly, gradient studies extending outward from oil storage tanks in the North Sea have documented changes in benthic community composition and reductions in species richness with increasing proximity to point sources (Addy et al. 1978). Gradient studies downstream of pollutant inputs to streams and rivers are also commonly used to assess stress effects.

Paleoecological studies offer the opportunity to examine changes in community structure at sites that have already experienced chemical stress. These studies deal with well preserved organisms, many of which, such as diatoms, are known to be quite sensitive to physicochemical conditions. They offer a direct and efficient way to analyze aspects of lentic ecosystem response to anthropogenically imposed stresses, where the nature of the stress is known or strongly inferred (e.g., Bradbury and Megard 1972; Brugam 1978; Charles 1984; Charles et al. 1987), and can be helpful in identifying useful indicator species. Furthermore, they can establish the context within which recent effects should be assessed. By choosing sites carefully to achieve both replication and control, paleoecological studies of lake ecosystems can be valuable tools for testing hypotheses about ecosystem response to stress (e.g., Smol and Dickman 1981; Battarbee et al. 1985; Wright et al. 1986).

Modeling is becoming increasingly popular as a tool to predict stress effects (e.g., Turgeon 1983; Rhea and Malanchik 1985; Wright et al. 1986). Several different types of models are used, of which the most realistic are the process-oriented, mechanistic models. The advantage of this technique is that, once developed and critically verified (e.g., Mankin et al. 1975), a good mechanistic model permits behavior of a system to be studied under a variety of scenarios. The primary drawback of this technique is the difficulty of creating and validating ecologically realistic models (Patten et al. 1983; Limburg et al. 1986) which depend on detailed knowledge about system processes and interactions that are often poorly understood. Because of the uncertainties, sensitivity analysis is an essential part of the development of ecologically realistic models, as it analyzes model behavior over a range of values for particularly important or uncertain parameters. One drawback is that even when the information required for sensitivity analyses exists, it is often based on laboratory dose-response toxicity data, which can be a poor predictor of effects at the ecosystem level (e.g., Bartell et al. 1986). This underscores the necessity for rigorous validation before process-oriented models can be used to give accurate real-world results.

It is important to recognize that these six methodologies are complementary rather than mutually exclusive approaches to the study of stress effects. Each method has its characteristic strengths and weaknesses, and differences in predictions based on different techniques reveal where a deeper understanding of ecosystem dynamics is needed. In the following review, several generalizations are made about the usefulness of various structural parameters in assessing the status of aquatic ecosystems. Because these generalizations are based on many individual stress-effects studies, the study methodologies (e.g., whole-system experiment, microcosm experiment, paleoecological study) and types of study system (e.g., near-shore marine, temperate stream) are given to provide the reader with a context within which to evaluate the generalizations.

5.3 Early Studies of Community Composition and Structure as Indicators of Chemical Stress: The Historical Context

The use of structural parameters of aquatic ecosystem organization for water quality assessment and monitoring has a long history. Kolkwitz and Marsson (1908) proposed the "saprobien" spectrum, which used community composition to assess the effects of organic pollution on aquatic ecosystems. The saprobien spectrum was followed by the development of other spectra (e.g., the halobion spectrum (Kolbe 1927), and the pH spectrum (Hustedt 1937–39)). The intent of these early semi-quantitative classifications was to assess water quality using the biological community as a bioassay of water quality conditions.

In 1949, Patrick published what is now a classic paper showing how various aspects of community structure could be used to monitor the condition of streams affected by sewage and industrial waste. Six streams in the Conestoga basin of Pennsylvania were studied, and assemblages at several different trophic levels were used to create profiles of community structure associated with different levels of pollution. Patrick demonstrated that: 1) quantitative changes in the distribution of species numbers and biomass occurred in stream ecosystems subjected to chemical stress; 2) the diverse stressor/ecosystem changes could be aggregated into a small set of classes of changes; 3) these classes could be replicated; and 4) changes in community structure preceded changes in then-standard integrative chemical parameters such as biological oxygen demand (BOD).

Patrick (1949) concluded from the Conestoga basin study that assessments of water quality could best be made using biological, rather than physicochemical, monitoring techniques. There were three major parts to her argument. First, it is difficult if not impossible to reduce the complex physicochemical environment to a few simple parameters for monitoring purposes, and it is impossible to test for all possible toxic or hazardous agents in every waterway. Second, bioassays are temporally integrated measures of physical and chemical streamwater characteristics over relatively long time periods (days or weeks), whereas physicochemical measures reflect only instantaneous conditions. Third, standard integrative physicochemical measurements cannot always detect ecosystem impact. For example, significant changes in community structure often occurred at sites that would qualitatively be described as "polluted" but that had biological oxygen demand (BOD) measures within normal limits.

With the passage of the National Environmental Protection Act (NEPA) in 1969, the need arose for simple descriptors of stress effects for the purposes of impact assessment. The 1972 amendments to the Federal Water Pollution Control Act and the 1977 Clean Water Act Amendment created a mandate for the maintenance or improvement of water quality. Taken together, these legislative measures created the necessity among state and federal agencies for developing realistic criteria to assess water quality and for monitoring the integrity of ecosystems under their jurisdictions. Increased attention was given to the problem of describing existing ecosystems and predicting probable responses to stress. For a variety of reasons, including the magnitude of the task and the resources needed to accomplish it, the scientific response to this new set of needs was often inadequate (see Schindler 1976; Kibby and Glass 1980; Rosenberg et al. 1981). Nevertheless, the perceived need for simple descriptors of stress effects by responsible managers, planners, and decision makers became intense, particularly for aquatic ecosystems, which are the traditional receptors for municipal and industrial wastes. These challenges were initially addressed by sampling or monitoring the physicochemical environment, a technique that became more meaningful as technological advances began to allow direct measurement of low levels of pollutants with good precision and accuracy. However, the proliferation of papers by Patrick, Cairns, and their co-workers stressing the inadequacy of physicochemical data for assessment of water quality (e.g., Patrick 1950, 1955, 1968, 1972; Patrick et al. 1968a, 1968b; Cairns 1974; Cairns et al. 1972; Cairns and Dickson 1971; Herricks and Cairns 1982; Matthews et al. 1982) had an impact. Much basic research was initiated to define the relationship between alterations of physicochemical conditions and community composition and structure. The use of biological measures to complement physicochemical measures in monitoring programs began to increase. By 1984, biological measures were incorporated into water pollution monitoring programs in 33 of 50 states (Perry et al. 1984). Actual applications of these measures often suffered a host of inadequacies, especially as applied to routine monitoring studies. However, despite the flaws in individual applications, species composition and community structure have begun to regain their status as valuable tools in assessing pollutant effects on aquatic ecosystems. Various aspects of this information have been applied to particular situations, including indicator species, community composition, synthetic "biotic indices," biomass, abundance, species richness, and species diversity.

5.4 Structural Changes

5.4.1 Community Composition

Indicator Species

The use of indicator species in assessing environmental conditions is based on the idea that the distributions of at least some species or groups of species are constrained to a narrow range of environmental conditions. Changes in chemical conditions will therefore affect the occurrence of those species, eliminating some and encouraging colonization by others. In this context, the task is to identify species whose appearance or disappearance is associated with low-level (statistically or analytically undetectable) chemical stress, so that we may be forewarned of more widespread changes should chemical stress continue to accumulate or escalate.

During the earliest phases of chemical stress, species that are sensitive to particular stressors often begin to decline in large numbers and to experience reproductive failures, leading to the rapid disappearance of sensitive species from the affected ecosystems. This was the case when opposum shrimp (*Mysis relicta*), a predaceous copepod (*Epischura lacustris*), and fathead minnows (*Pimephales promelas*) were lost from the experimentally acidified Lake 223 in the Experimental Lakes Area (ELA), southern Ontario, in the earliest stages of lake acidification (Malley et al. 1981; Nero and Schindler 1983; Schindler et al. 1985). In a regional study of 47 acidified lakes near Sudbury, Ontario, *E. lacustris* was also the first of several species to disappear under the combined stresses of acid and heavy metal deposition (Sprules 1975).

Several benthic and shallow-water crustaceans are particularly sensitive to chemical stress. A regional study of 79 Ontario lakes (Stephenson and Mackie 1986) indicates that the freshwater amphipod *Hyalella azteca* is absent from lakes with a negative epilimnetic summertime alkalinity, although it is well-distributed in other lakes of the region. This species was found to be the most acid-sensitive of several benthic invertebrates in an experimental stream mesocosm study (Zischke et al. 1983). Regional studies indicate that the "fresh-water shrimp" *Gammarus lacustris* (an amphipod) and the tadpole shrimp *Lepidurus arcticus* (a notostrachan) are among the earliest species lost from anthropogenically acidified Norwegian lakes (K. Økland 1969, 1980; Økland and Økland 1986). Loss of *Gammarus fossarum* and *G. pulex* in the upper reaches of streams in forested German drainage basins appears to be a good early warning signal of forest damage associated with acidification (Økland and Økland 1986). The sowbug *Asellus aquaticus* (an isopod) appears to be lost later in the acidification process (K. Økland 1980). Ampeliscid amphipods as a group tend to be lost at the lowest concentrations of oil pollution in the marine environment (Teal and Howarth 1984). Fairy shrimp (anostracans) were lost within two weeks by experimental oiling of freshwater ponds on Alaska's North Slope, and were the only organisms to be eliminated completely from these systems over the seven-week sampling period (Abraham 1975).

Aquatic insects can also be useful as early warning indicators. Mayflies (Ephemeroptera), especially those in the genus *Baetis,* are particularly sensitive to a wide variety of chemical stresses and are among the first species lost in copper-contaminated streams (Winner et al. 1975, 1980), acidified streams (Sutcliffe and Carrick 1973; Hendrey et al. 1976; Hall et al. 1980; Arnold et al. 1981; Canton and Ward 1981), experimentally oiled freshwater ponds (Giddings et al. 1984), arsenic-contaminated lakes (Wagemann et al. 1978), streams receiving DDT for control of dipterans or from spraying of adjacent forests (Adams et al. 1949; Corbet 1958; Ide 1967; Keenleyside 1967), and experimental ponds receiving the organophosphate insecticide Dursban (Hurlburt et al. 1972) and the carbamate insecticide Larvin (Ali and Stanley 1982). Beckett (1978) found that mayflies are eliminated under the complex chemical stresses typical of industrialized river systems, as are rheotanytarsid chironomids (Diptera). Several chironomids, including *Nilotanypus fimbriatus,* are early casualties of oil-contaminated freshwater systems (Rosenberg and Wiens 1976). Small caddisflies (Trichoptera) such as *Glossoma ventrale* are lost in western streams affected by coal-mine drainage (Chadwick and Canton 1983), and following DDT applications to control blackflies (Diptera) (Corbet 1958; Hatfield 1969).

The indicator species approach depends precisely on the use of species, rather than genera, families, or orders. For example, a study of acidified Ontario lakes found mayflies in low pH lakes (Collins et al. 1981), although this group of organisms is generally lost during lake acidification. The surviving mayflies were burrowers in the genus *Hexagenia,* rather than the more sensitive species in the genera *Baetis* and *Ephemera.* However, within the relatively sensitive genus *Baetis,* a study of 48 streams in the Swedish highlands indicated a wide variation in the apparent pH tolerance limits of six species, with *B. lapponicus* being very acid sensitive (not found in streams of pH < 6.0) and *B. rhodani* being acid tolerant (found in streams with pH as low as 4.6) (Engblom and Lingdell 1984). Collins et al. (1981) also found crayfish at low pH, although the surviving crayfish were *Cambarus bartoni,* rather than *Orconectes virilis,* the species lost from experimentally acidified Lake 223 (Schindler et al. 1985). Mierle et al. (1986) point out that three congeners (*O. virilis, O. propinquus,* and *O. rusticus*) of the five common Canadian species of crayfish are all sensitive to low pH, whereas the two crayfish in the genus *Cambarus* (*C. bartoni* and *C. robustus*) are considered acid tolerant.

Changes in chemical conditions can also result in the appearance of characteristic taxa, although frequently these appearances represent dramatic increases in abundance of previously inconspicuous taxa rather than colonization *de novo*. For instance, the early stages of acidification in north temperate lakes often are characterized by the rapid expansion of zygnematacean algae in the genus *Mougeotia* (Grahn et al. 1974; Stokes 1981, 1984a; Schindler et al. 1985). *Mougeotia* and other chlorophycean algae also have been found at unusually high abundances in moderately oil stressed ponds in northern Alaska (Barsdate et al. 1972). The appearance of the chrysophyte species *Mallomonas hindonii, M. hamata,* and *Synura macracantha* is also closely associated with recent lake acidification (Smol et al. 1984; Christie and Smol 1986; Smol 1986), and *M. hindonii* may be a specific indicator of the onset of recent, anthropogenic, changes in snowmelt chemistry (Gibson et al. 1987). Similarly, the proliferation of the filamentous bacterium *Sphaerotilus* signals the beginning of significant stream degradation from organic enrichment (e.g., Patrick 1950; Mackenthun 1969).

Shifts in dominance caused by colonization by opportunists also occur at other trophic levels. In freshwater streams, dipterans are often tolerant opportunists. For example, chironomids increase in importance in the macroinvertebrate fauna of stream reaches with increasing concentrations of the pollutant copper (Butcher 1946; Winner et al. 1975, 1980) and with increasing acidity (Økland and Økland 1986). Capitellids often become dominant following oil spills, as occurred following the "Amoco Cadiz" spill off the Brittany coast and the "Florida" spill off the coast of Massachusetts (Teal and Howarth 1984). They are also associated with pollution of marine sediments by organic wastes (Rosenberg 1972; Grassle and Grassle 1974; Pearson 1975; Young and Young 1982). Oil spills off the coasts of both France and Nova Scotia also reportedly resulted in increases in dominance of the lugworm *Arenicola* (Teal and Howarth 1984). In sludge-polluted waters, tolerant opportunistic annelids such as the freshwater oligochaetes *Tubifex tubifex* (Brinkhurst 1965) and *Limnodrilus* species (Goodnight 1973) or the complex (Grassle and Grassle 1974) of marine polychaetes referred to as *Capitella capitata* (Pearson and Rosenberg 1978) may dominate the bottom fauna. The literature on characteristic opportunistic species associated with organic pollution in the marine environment has been thoroughly reviewed by Pearson and Rosenberg (1978).

Chemical stress can also result in individual species replacements when stressor-tolerant species replace stressor-sensitive competitors. For example, in experimentally acidified Lake 223 the acid-tolerant cladoceran, *Daphnia catawba* replaced acid-sensitive *D. galeata mendotae,* and pearl dace, *Semotilus margarita,* a relatively acid-tolerant minnow, replaced the more sensitive fathead minnow (*Pimephales promelas*) (Mills 1984; Schindler et al. 1985).

With increasing levels or duration of chemical stress, secondary effects begin to appear. Blooms of opportunistic species normally excluded or controlled by competition or predation may appear. For example, pesticide applications remove herbivorous invertebrates, resulting in blooms of blue-green and benthic filamentous algae (Hurlburt 1975). Herbicide applications can have the same effect, presumably by decreasing competition from aquatic macrophytes (Hurlburt 1975). Because these treatments create new food supplies for decomposer populations, they also can produce temporary increases in abundance of organisms that derive their energy from this source, such as oligochaetes, chironomid larvae, and dragonfly nymphs (Hurlburt 1975).

The use of presence–absence data on individual indicator species to assess environmental conditions has been criticized on several grounds (e.g., Gaufin 1973; Cairns 1974; Roback 1974). Although many of the indicator species that disappear in the early stages of chemical stress are common species, others are uncommon or rare in a community, and their presence (or especially, their absence) may be difficult to demonstrate. When post-impact studies reveal the absence of known indicator species, the potential role of other factors such as competition, predation, lack of colonization potential, inadequate sampling intensity, and chance must also be assessed. Similarly, the simple presence of indicator species associated with deteriorated conditions may be misleading, as these cosmopolitan species can be present, although at lower abundances, in undisturbed communities, where they are important components of communities subject to natural disturbances (e.g., periodic accumulations of decaying organic matter; Wurtz 1955; Chandler 1970; Brinkhurst and Cook 1974). The significance of short-term spatial and temporal fluctuations in abundance of indicator species is difficult to evaluate unless there are long term data to indicate ranges of natural variability. Finally, if an ecosystem is subject to more than one chemical stress, the indicator species approach may be difficult to apply, because indicator species often respond differently to different sets of stressors. For example, mesocosm studies at ELA and in Lake Michigan indicate that the cladoceran *Holopedium gibberum* is particularly sensitive to cadmium additions (Marshall and Mellinger 1978; Marshall et al. 1981), although this species is well-represented in the acid- and metal-stressed Sudbury lakes (Sprules 1975), northern Ontario lakes (Roff and Kwiatkowski 1977), and post-acidification Lake 223 (ELA) (Malley and Chang 1986). A related problem occurs in situations where more than one stress is suspected. For example, in Collins et al. (1981) regional study in Ontario, three species of sphaeriid bivalves persisted in low-pH lakes, although these organisms are usually considered extremely sensitive to in low-pH environments. The authors attributed the persistence of these taxa to the low levels of aluminum found at their study sites relative to other North American and Scandinavian studies, suggesting that aluminum, or aluminum–pH relationships (rather than pH per se) is the stress causing the elimination of these species.

The value of the indicator species approach as an assessment technique is low in the absence of other supporting data. Nevertheless, judicious choice of taxa applied to well-focused problems may be of value in detecting regional or site-specific contamination. For example, the indicator species approach was useful in assessing oil impacts on rivers and lakes in the North American arctic (e.g., Rosenberg and Wiens 1976). Similarly, the distribution of a single organism (*Mallomonas hindonii*) in sediment cores has led to new insights concerning the potential timing of acid deposition effects in dilute New England lakes (Gibson et al. 1987). The usefulness of the indicator species approach is enhanced if sets of indicator species are used, particularly if they are chosen from different guilds or trophic levels. If several such taxa all begin to show changes at roughly comparable times or in a systematically related fashion, the likelihood that all changes are being driven by extraneous factors is greatly reduced.

Species Composition and Relative Abundance

Relatively rapid changes in overall community composition (i.e., beyond the simple appearance or disappearance of indicator species) have been found in association with many chemical stresses, including anthropogenically induced acidification (e.g., Hall et al. 1980; Schindler et al. 1985), pesticides (e.g., Hulburt et al. 1972; deNoyelles et al. 1982), heavy metals (e.g., Marshall and Mellinger 1978; Winner et al. 1980), oil (e.g., Johansson et al. 1980), pulp mill effluents (e.g., Pearson 1971; Rosenberg 1972), and organic enrichment (e.g., Pearson and Rosenberg 1978). The literature on lake and stream acidification is particularly comprehensive and is focused on here. More detailed treatments can be found in reviews and compilations by Almer et al. 1974, Braekke 1976, Gorham 1976, Seliga and Dochinger 1976, Almer et al. 1978, Drabløs and Tollen 1980, d'Itri 1981, Haines 1981, Singer 1981, Eilers et al. 1984, Hendrey 1984, Økland and Økland 1986, Smol et al. 1986, and Martin 1987. Økland and Økland's (1986) review of the effects of acid deposition on the benthos is a particularly useful recent contribution, as it summarizes and critiques a large body of work that includes many contributions from the German and Scandinavian literature.

Synoptic surveys in Scandinavia and North America have demonstrated that lake and stream acidification leads to persistent changes in species composition at several levels of biological organization. Some of the species initially lost have been discussed under "Indicator Species"; other groups of organisms that tend generally to decrease in abundance or be lost in acidified lakes include gastropods (J. Økland 1969, 1980; Raddum 1980; Collins et al. 1981; Økland and Økland 1986), sphaeriid bivalves (Sutcliffe and Carrick 1973; Roff and Kwiatkowski 1977; K.

Økland 1980; Økland and Økland 1986), crayfish (especially *Astacus asta-cus* and species of *Orconectes;* Schindler et al. 1985; Økland and Økland 1986; Mierle et al. 1986), leeches (Raddum 1980), daphnid zooplankton (except *D. catawba* and *D. pulicaria;* Almer et al. 1974, 1978; Sprules 1975; Roff and Kwiatkowski 1977; Confer et al. 1983), and several species of mayflies (Ephemeroptera; Sutcliffe and Carrick 1973; Almer et al. 1978; Økland and Økland 1986). Loss of particular groups of organisms is associated with increases in other groups. For instance, in experimentally acidified Lake 223, the decline in abundance of large-bodied zooplankton was accompanied by an increase in the abundance of rotifers and small cladocera (Schindler et al. 1985; Malley and Chang 1986).

Many species of fish disappear from low-pH rivers and lakes, presumably because of the direct toxic effects of high concentrations of hydrogen ions and inorganic aluminum, disruptions to electrolyte balance, or loss of important food chain support, such as benthic macroinvertebrates. Regional surveys have indicated that the most prominent losses include Atlantic salmon (*Salmo salar*), brook trout (*Salvelinus fontinalis*), lake trout (*Salvelinus namaycush*), smallmouth bass (*Micropterus dolomieue*), walleye (*Stizostedion vitreum vitreum*), burbot (*Lota lota*), and various species of minnows and darters (e.g., Beamish and Harvey 1972; Jensen and Snekvik 1972; Hendrey and Wright 1976; Lievestad et al. 1976; Schofield 1976; Harvey 1980; Haines 1981; Wiener et al. 1984; Frenette et al. 1986; Hesthagen 1986; Pauwels and Haines 1986). There is significant regional variability in pH-tolerance thresholds for individual species, as well as significant differences between results from bioassay and regional field studies, which suggests several alternative explanations that are not mutually exclusive. For example there may be considerable regional ecotypic variation with respect to pH sensitivity, or other environmental factors may be involved, or the regional studies on which such tolerances are usually based may be too limited in extent to represent given regions accurately. For example, Pauwels and Haines (1986) found that common shiners (*Notropis cornutus*) become rare in a suite of Maine lakes of pH < 5.9, whereas Frenette et al. (1986) found the threshold for the common shiner to be pH 6.9 in a survey of 37 Quebec lakes, and Smith et al. (1986) found the threshold to be pH 5.27 in a survey of 234 Nova Scotian lakes. Økland and Økland (1986) discussed the issue of absolute lower pH limits with respect to benthic invertebrates, and concluded that there is no absolute lower limit for particular species because: (1) there are differences in pH sensitivity among different stages of the life cycle; (2) there may be ecotypic variations in sensitivity; (3) other physical and chemical factors may modify the effect of low pH; and (4) the population response to decreasing pH is a continuous decline in population density rather than a discrete, all-or-none response. Generally speaking, however, species can be ordered into general sensitivity groups, with cyprinids being the most sensitive and yellow perch among the least sensitive.

Decreases in fish species richness are also strongly correlated with increasing lake and stream acidity. Trends observed in synoptic surveys have been substantiated by smaller-scale comparative studies (e.g., Arnold et al. 1981) and by long-term records from individual lakes (e.g., Beamish 1974; Leivestad and Muniz 1976; Wiederholm and Eriksson 1977; Mills 1984; Schindler et al. 1985).

pH effects on phytoplankton are not as immediately evident in regional surveys, partially because of the inherent difficulty of characterizing spatially variable populations with strong patterns of seasonal succession in a single sampling effort, and partially because many investigators fail to take phytoplankton taxonomy to the necessary level of the species. Kwiatkowski and Roff (1976) sampled six Canadian shield lakes over two field seasons using a weekly sampling schedule of from 7 to 9 samples per lake, and observed a general shift in relative abundance from chlorophytes to cyanophytes in increasingly acidic lakes. They also noted a significant loss of species in all major phytoplankton groups with increasingly acid conditions. Their observation of marked changes in species composition in both diatom and chrysophyte communities with increasingly acid conditions is consistent with regional paleoecological studies that have demonstrated strong relationships between surface water pH and the species composition of diatom and chrysophyte assemblages (e.g., Patrick 1955; Patrick et al. 1986b; Nygaard 1956; Meriläinen 1967; Battarbee 1984; Smol et al. 1984; Charles 1985; Davis and Anderson 1985; Anderson et al. 1986). This relationship is sufficiently strong to permit quantitative estimates of recent lake acidification using microfossil stratigraphy (e.g., Renberg and Hellberg 1982; Davis et al. 1983; Flower and Battarbee 1983; Charles 1984; Arzet et al. 1986; Charles et al. 1986; Ford 1986; Tolonen et al. 1986), and in fact diatom- and chrysophyte-inferred pH reconstructions have become a major tool in acid-precipitation research.

Mesocosm experiments at Carlyle Lake, a low alkalinity Canadian Shield lake, demonstrated changes in the phytoplankton community with increasing acidification. In this case, dinoflagellates (primarily *Peridinium limbatum*) and cryptophytes replaced chrysophytes within 28 days (Yan and Stokes 1978). Mesocosm studies of periphytic algae in other Canadian lakes also showed replacement of chrysophytes by different taxa (Muller 1980; Stokes 1984a), although this result was not found in a regional study of 75 lakes in the Sault Ste. Marie district (Kelso et al. 1982). In the experimentally acidified Canadian Lake 223, dinoflagellates and cyanophytes replaced chrysophytes within six years (Schindler 1980; Findlay 1984; Schindler et al. 1985; Findlay and Kasian 1986). Emerging dinoflagellate dominance in these studies is consistent with the widespread nature of dinoflagellate dominance in acid lakes (e.g., Almer et al. 1974; Hutchinson et al. 1978; Yan and Stokes 1978; Stokes 1984b). Loss of chrysophytes with increasing acidification is also consistent with the complete lack of chrysophytes in arctic ponds made extremely acidic (pH < 3.0) by natural lignite burns (Sheath et al. 1982), even though chrysophytes occur in higher pH ponds in the region. Despite this loss of overall chrysophyte biomass with advancing lake acidification, however, enough diversity is maintained within the chrysophyte assemblage to permit inferences about lakewater acidity down to about pH 4.0 (Smol et al. 1984).

The factors that drive changes in phytoplankton community composition with increasing acidification are as yet unclear, although increased lakewater transparency leading to expanded habitats and increased competitive success of benthic and littoral species are probably involved. This factor may also be involved in the induction of blooms of benthic zygnematacean algae typically associated with acidic lakes (e.g., Grahn et al. 1974; Muller 1980; Stokes 1981), although other factors, such as decreased grazing, decreased competition, and decreased microbial decomposition may be equally important (Stokes 1981).

Changes in diatom species composition have also been observed after low-level oiling experiments in marine mesocosms. In these cases, the large diatom *Ceratulina bergonii* was replaced by smaller diatoms and nanoflagellates, especially *Chrysochromulina kappa* (Lee et al. 1977), an alteration that alters the quality of food chain (Greve and Parsons 1977). When, however, the initial diatom flora consists of smaller-bodied diatoms, little change is observed (Elmgren et al. 1980).

Different stressors can cause opposing effects. For example, unlike the situation with lake acidification in which dinoflagellates replace chrysophytes at low pH, chrysophytes (primarily *Mallomonas pseudocoronata*) replace dinoflagellates (primarily *Peridinium inconspicuum*) within seven days after application of 500 μg of the pesticide Atrazine to experimental hardwater ponds (deNoyelles et al. 1982). Deep-ocean dinoflagellate communities are also very sensitive to pharmaceutical wastes relative to other elements of the phytoplankton (Murphy et al. 1983), although dinoflagellate abundance at the sampling station in this open community recovers within a few hours of acute exposure.

In 1974 a whole-lake acidification experiment was undertaken on Lake 223 in the Experimental Lakes Area (ELA) in southern Canada to study long-term acidification effects (e.g., Schindler 1980; Malley and Chang 1981; Nero and Schindler 1983; Findlay 1984; Mills 1984; Schindler et al. 1985). Changes in species composition at several different trophic levels occurred surprisingly early in the nine-year acidification process, but they were not followed by the changes in functional parameters, such as productivity and decomposition, expected on the basis of laboratory studies. In this case, changes in relative abundance and species replacements provided functional compensation for productivity and decomposition processes affected by acidification in laboratory studies, and laboratory studies significantly underestimated the inherent resilience of the natural ecosystem. In other situations the relationship between laboratory and field studies has been found to be just the reverse, with laboratory studies overestimating the resilience of natural aquatic ecosystems. This occurs when, for example, there are sublethal effects (e.g., changes in growth rate, reproductive success, or behavior) that affect critical species. Such sublethal effects have been well-documented for crabs and clams in coastal oil-spill studies (Howarth 1987).

Nilssen (1980) and Malley et al. (1981) pointed out that the trajectories of lake ecosystem degradation following acidification vary with the initial composition and trophic structure of the ecosystem. For example, the most sensitive species in Scandinavian lakes are typically fish and herbivorous zooplankton. As fish are lost, invertebrate predators increase. The zooplankton community is thus subjected to a change in predation pressure favoring large-bodied over small-bodied zooplankton prey. The typical acid-stressed Scandinavian lake therefore contains invertebrate predators and large-bodied zooplankton prey, or prey with other appropriate adaptations against invertebrate predation. In Canadian Shield lakes, on the other hand, invertebrate predators are among the most sensitive organisms and are lost from the lake. The most sensitive fish species, fathead minnows, is quickly replaced by the ecologically similar pearl dace, a replacement that apparently has few consequences for the overall trophic structure of the lake.

Abiotic environmental characteristics may also influence biological response. For example, anthropogenic acidification affects fish populations at a higher pH in lakes with relatively low ionic strength and low concentrations of dissolved organic carbon (DOC) relative to lakes with higher ionic strength and DOC.

Many of the observed effects of chemical stress on species composition and community structure may be reversed by removing the stressor and/or applying appropriate ameliorative techniques. However, the extent of biological recovery will vary with stressor and ecosystem type, depending on population characteristics, and hysteresis does not appear to be the rule. Unless the physicochemical environment, including cycling rates and processes, is completely restored, and there is a ready supply of propagules, some species may never recover their former role in the community.

In the case of lake acidification, liming appears to reverse some effects, and is currently widely employed as an ameliorative technique. For instance, in Sweden, more than 3000 lakes and 100 streams have been limed (Lessmark and Thornelof 1986). Amelioration appears to affect primarily phytoplankton, zooplankton, and aquatic insects (e.g., Schieder and Dillon 1976; Dillon et al. 1979; Eriksson et al. 1983), although the response of the phytoplankton may not be obvious in the first few years (Raddum et al. 1986) without detailed taxonomic treatment. Recovery of molluscs and benthic crustacea is much slower and may be limited by dispersal problems, although insect biomass and abundance frequently increase, particularly among the Chironomidae (e.g., Raddum et al. 1986). Fish stocking success and recruitment often improve markedly after liming (Booth et al. 1986; Nyberg et al. 1986), especially if food sources recover.

The reversal of changes in species composition by liming complements stream and lake acidification studies in demonstrating the importance of acidification per se in the initiation of changes in species composition. However, liming cannot be regarded as a panacea for reversing the aquatic effects of acid precipitation, because it does not work in some ecosystems (e.g., those high in humic acids) and because the effects do not persist in ecosystems with short turnover times (Eriksson et al. 1983; Scheffe et al. 1986). Furthermore, indirect changes associated with altered inputs from the acid-impacted drainage basins, rather than by direct alteration of the aquatic environment, may not be reversible. Finally, liming itself is a chemical alteration that may produce unexpected side effects.

Even though the effects of chemical stress on species composition can be immediate and obvious to trained observers, there are several problems with using community composition *per se* as an assessment tool. First, the most likely candidates for use as early warning signals for particular combinations of stressors and ecosystem types are often the most taxonomically difficult. For example, both diatoms and aquatic insects appear to be particularly sensitive to a variety of chemical stresses, but are taxonomically very challenging. Under these circumstances, investigators tend either to aggregate species to the level of the genus— losing valuable information and reducing the usefulness of the diagnostic tool (Resh and Unzicker 1975; Collins et al. 1981; Hilsenoff 1982)—or to avoid the technique altogether. Second, implementation of sampling strategies that can provide adequate descriptions of the species composition of spatially and temporally cryptic aquatic communities is often perceived to be prohibitive in terms of cost and logistics, although recent advances in sampling concept and design demonstrate that this is not always the case (Raddum et al., in press). Finally, even when adequate compositional information exists, the ultimate user of the information may face several difficulties. Effects may be subtle, and trajectories following imposition of a particular chemical stress can vary from ecosystem to ecosystem. Cascades of effects can begin with different organisms at different trophic levels, and propagate in a variety of ways depending on the initial species composition and trophic structure, as well as on the nature and mode of action of the stressor. The prior history of, and other loadings to, the ecosystem may be an issue, because various pollutant combinations can act synergistically to either aggravate or ameliorate stress effects of single pollutants (e.g., Klaverkamp et al. 1983).

The inherent complexity of compositional data, combined with all these complicating factors, leads to a literature that quickly becomes voluminous and defies easy generalization. To those interested in actual management applications, information on changes in species composition may seem to be merely "a confounding morass of unusable information" (Kimball and Levin 1985), despite the fact that trained observers familiar with particular target ecosystems are often alerted to the onset of stress-related, ecosystem-level changes by the occurrence of subtle changes in community composition. Because of the difficulties in adequately characterizing overall community composition, together with the infeasibility of using the raw data for rapid and/or regionally extensive assessments of ecosystem status, significant attention has been focused on other, hopefully more robust, descriptors of community structure to both signal the onset of significant ecosystem deterioration and to provide indications of probable ecosystem trajectories. These descriptors include biomass, abundance, biotic indices, species richness, and species diversity (see below).

5.4.2 Community Structure

Biomass and Abundance

Early experience with stressed systems suggested that heavily stressed ecosystems tended to have reduced biomass, abundance, species richness, and species diversity (variously measured) relative to pristine ecosystems. These structural parameters have been examined by many investigators as potential tools for environmental assessment. Biomass and abundance are particularly attractive parameters because they require minimal taxonomic training, and yet often provide useful signals of ecosystem deterioration.

Several studies of the relationships between chemical stress and changes in the biomass and abundance of phytoplankton, zooplankton, and the benthos are summarized in Table 5.1, which shows primarily chronic chemical stresses. Acute stresses such as oil spills or pesticide applications typically result in immediate reductions of biomass and abundance and raise the issue of recovery, rather than impact, trajectories. The table, which is not an exhaustive treatment, is in fact an oversimplification, inasmuch as it shows only two system states—"stressed" and "unstressed"—and does not consider the path between the two. Because not all the studies compare strongly contrasting states, all the entries in Table 5.1 may not be strictly comparable.

Chemically stress-induced changes in biomass and abundance are difficult to demonstrate rigorously for several reasons. Chemical stress can occur intermittently or continuously. Sheehan and Winner (1984) studied two streams receiving continuous additions of copper and found that: 1) mean macroinvertebrate abundance was reduced to <20% of reference immediately below the point of pollutant introduction and 2) good correlations existed between mean macroinvertebrate abundance and copper

Table 5.1. Effects of chemical stress on biomass and abundance

System	Type of study	Stressor	Element of biota	Biomass (B) or abundance (A) in stressed vs. unstressed system	Reference
Temperate lakes	Site comparison ($n=1679$)	Acidity	Fish	Lower (A)	Leivestad et al. 1976
Temperate streams	Site comparison ($n=2$)	Acidity	Aquatic insects	Lower (B)	Arnold et al. 1981
Temperate stream	Whole-stream experiment	Acidity	Aquatic insects	Lower (A)	Hall et al. 1980
Temperate lakes	Site comparison ($n=7$)	Acidity	Benthos	Lower (B,A)	Leivestad et al. 1976
Temperate lakes	Site comparison ($n=34$)	Acidity	Oligochaetes	Similar (A)	Raddum 1980
Temperate lakes	Site comparison ($n=34$)	Acidity	Gastropods and bivalves	Lower (A)	Raddum 1980
Temperate lakes	Site comparison ($n=12$)	Acidity	Gastropods	Lower (B)	Collins et al. 1981
Temperate lakes	Site comparison ($n=34$)	Acidity	Chironomids	Lower (A)	Raddum (1980)
Temperate lakes	Site comparison ($n=12$)	Acidity	Infaunal benthos	Similar or higher (B)	Collins et al. 1981
Temperate lakes	Site comparison ($n=12$)	Acidity	Benthic cladocera	Similar or higher (B)	Collins et al. 1981
Rocky Mountain stream	Gradient study	Coal mine drainage	Aquatic insects	Similar (A,B)	Canton and Ward 1981
Temperate streams	Gradient study	Copper	Aquatic insects	Lower (A)	Sheehan and Winner 1984
Nearshore marine	Whole system	Oil	Benthos	Lower	Jacobs 1980
Nearshore marine	Whole system	Oil	Benthos	Similar (B)	Elmgren et al. 1980a
Temperate lakes	Site comparison	Acidity	Zooplankton	Lower (B)	Raddum 1976 as cited in Hendrey et al. 1976
Temperate lakes	Site comparison ($n=6$)	Acidity	Zooplankton	Lower (B)	Confer et al. 1983
Temperate lake	Whole-lake experiment	Acidity	Crustacean zooplankton	Similar (B)	Malley and Chang 1981
Temperate lake	Mesocosm	Cadmium	Zooplankton	Lower (A)	Marshall and Mellinger 1978
Temperate lake	Mesocosm	Mercury and cadmium	Zooplankton	Lower (A)	Marshall et al. 1981
Arctic lakes	Site comparison ($n=5$)	Acidity	Zooplankton	Lower (A)	Hutchinson et al. 1978

Table 5.1 (continued)

System	Type of study	Stressor	Element of biota	Biomass (B) or abundance (A) in stressed vs. unstressed system	Reference
Marine	Mesocosm	Copper	Zooplankton	Lower (A)	Reeve et al. 1976
Nearshore marine	Whole system	Oil	Zooplankton	Lower	Johansson et al. 1980
Marine	Mesocosm	Oil	Zooplankton	Lower (A)	Elmgren and Frithsen 1982
Temperate lakes	Site comparison ($n=3$)	Acidity	Aquatic macrophytes	Higher (B)	Wile et al. 1985
Temperate lakes	Site comparison ($n=2$)	Acidity	Aquatic macrophytes	Higher (B)	Stokes 1984b
Temperate lakes	Experimental ponds	Atrazine	Phytoplankton	Lower (7-day recovery)(B)	deNoyelles et al. 1982
Temperate stream	Artificial stream	Mercury	Phytoplankton	Lower (B)	Sigmon et al. 1977
Temperate lake	Whole-lake experiment	Acidity	Phytoplankton	Similar or higher (B)	Findlay 1984; Schindler et al. 1985
Temperate lake	Whole-lake study	Acidity	Phytoplankton	Similar (B)	Stokes 1984b
Temperate lake	Mesocosm	Acidity	Phytoplankton	Higher (B)	Yan and Stokes 1978
Temperate lakes	Site comparison ($n=6$)	Acidity	Phytoplankton	Lower (B)	Kwiatkowski and Roff 1976
Temperate stream	Mesocosm	Acidity	Phytoplankton	Lower (B)	Patrick et al. 1968b
Nearshoe marine	Whole-system study	Oil	Phytoplankton	Higher (B)	Kineman and Clark 1980
Marine	Mesocosm	Oil	Phytoplankton	Higher (B)	Elmgren et al. 1980b
Marine	Whole-system study	Oil	Phytoplankton	Higher (B)	Johansson et al. 1980
Marine	Mesocosm	$CuSO_4$	Phytoplankton	Higher (B)	Reeve et al. 1976
Temperate streams	Mesocosm	Acidity	Periphyton	Higher (B)	Leivestad et al. 1976 as presented in Hendrey et al. 1976
Temperate lake	Mesocosm	Acidity	Periphyton	Higher (B)	Muller 1980
Temperate stream	Mesocosm	Acidity (12–72 h)	Periphyton	Similar or lower (B)	Parent et al. 1986
Temperate stream	Mesocosm	Acidity (84 days)	Periphyton	Higher (B)	Parent et al. 1986
Temperate stream	Whole-stream experiment	Acidity	Periphyton	Higher (B)	Hall et al. 1980

Table 5.1 (continued)

System	Type of study	Stressor	Element of biota	Biomass (B) or abundance (A) in stressed vs. unstressed system	Reference
Temperate streams	Site comparison ($n=2$)	Acidity	Periphyton	Similar	Arnold et al. 1981
Temperate lake	Mesocosm transplants	Acidity	Periphyton	Lower (B)	Stokes 1984a
Temperate lake	Whole-lake experiment	CuSO$_4$	Bacteria	Lower (7-day recovery)(B)	Effler et al. 1980
Marine	Mesocosm	Oil	Bacteria	Higher (A)	Elmgren and Frithsen 1982
Marine	Whole-system study	Oil	Bacteria	Higher (A)	Johansson et al. 1980

concentrations in the streamwater. However, similar data for a third stream that received concentrated, aperiodic inputs of copper, chromium, zinc, and cyanide were highly variable, and good correlations between population density and heavy metal concentrations were not found (Sheehan and Winner 1984). The lack of correlation was attributed to extreme variability in the density data, which the authors attributed to invertebrate avoidance of pollutant events by burrowing in this highly unpredictable environment. Burrowing can also affect responses to chronic as well as acute chemical stress. For example, Collins et al. (1981) found that lake acidification affects epifauna (organisms living at the sediment surface) more than infauna (burrowing organisms), presumably because the latter can exploit the higher pH environments found below 1–2 cm of sediment.

Phenological factors are another aspect of temporal variability that can affect stressor impacts on abundance and biomass. For example, Ide (1967) found that the abundance of aquatic insects in streams adjacent to forests sprayed with DDT depended on the stage of the life cycle at the time of spraying. Species that were in active stages (e.g., feeding larvae, nymphs) were almost completely eliminated, whereas species in inactive stages (e.g., eggs, pupae, diapausing larvae) were only slightly to moderately affected.

Systematic changes in biomass with chemical stress often do not occur because of species replacements that result in higher abundance or biomass. For example, Mackay and Kersey (1985) found that total abundance of benthic invertebrates increases with decreasing pH in a suite of

Canadian streams because of increased representation of acid tolerant chironomid species.

Finally, there may be significant variability within the classes of stressors considered in Table 5.1. This may be a particular issue for lake and stream acidification studies, in which pH and alkalinity effects may be confounded with aluminum, lead, zinc, and other metals. For example, although Scandinavian studies of acid lakes typically showed decreases in benthic biomass below pH 5.0, the regional study of Collins et al. (1981) in Ontario did not. These authors suggested that the difference may be related to aluminum levels, which are significantly higher in acidified Swedish and Norwegian lakes than in acidified Ontario lakes, although another factor may be an initial lack of sensitive species in their study lakes. Raddum (1980) pointed out that the higher-pH lakes in parts of southern Norway receiving acid rain have markedly lower bivalve abundances than would be expected based on bivalve abundances in lakes of similar pH in other regions of the country, and believed that this suggests the influence of other pollutant factors in addition to pH in determining bivalve distribution in this region.

When one species dominates the biomass at a certain level of organization, the sensitivity of that species to a stressor controls both the impact and the recovery trajectories. Thus, although oil spills almost always dramatically reduce live benthic biomass, this was not true following the nearshore "Tsesis" spill off the Stockholm archipelago, in which *Macoma* clams dominated the benthic biomass (Elmgren et al. 1983). Similarly, when relatively resistant nematodes dominate the meiofauna, oil spills may not affect total meiofaunal biomass (Teal and Howarth 1984).

Pearson and Rosenberg (1978) reviewed the literature on the effects of organic enrichment on the marine macrobenthos and described generalized trajectories for biomass and abundance along enrichment gradients (Fig. 1 in Pearson and Rosenberg 1978). According to this model, biomass is very low at maximum loadings, and with increasing distance or time it: 1) increases to a secondary peak with colonization by opportunistic stress-tolerant species; then 2) decreases as species composition begins to change in response to more favorable conditions; then 3) increases to its absolute maximum as a high diversity of relatively large-bodied species is achieved; and finally 4) stabilizes at a value lower than for "non-polluted" communities. Total abundance is very low at maximum loadings, increases to an absolute maximum in tolerant opportunistic communities, and stabilizes at lower values for natural conditions.

Although this summary is useful as a conceptual model, comparison with actual data presented in Pearson and Rosenberg (1978) indicates that in only one of five spatial gradients and one of two temporal sequences was the described biomass trajectory actually followed. Therefore, although the biomass model may describe general ecological processes in response to organic enrichment of the marine macrobenthos, it does not have the necessary resolution for application to site-specific problems. Abundance trajectories, on the other hand, are generally simpler and conform better with the generalized model.

Stress-related changes in biomass or abundance may involve several factors. Decreases may be caused by direct or indirect toxicity effects or (over longer time scales) by disruption of reproductive processes, negative effects on food prey items, or changes in species composition. Increases, such as the increase in periphyton and phytoplankton biomass typically found with increasing lake and stream acidification (Table 5.1), may result from enhanced growth and reproduction in the altered environment, decreased grazing or predation pressure, decreased microbial decomposition, or changes in species composition (Hall et al. 1980; Stokes 1981). Because any combination of these contributing factors can occur, early effects may involve increases, decreases, or no change at all in these parameters, making them inappropriate as generalized tools for early warning signals. Furthermore, although it is probably true that severe pollutant stress will eventually result in reduced biomass and abundance in one or more elements of the community (Table 5.1), the route between initial conditions and the endpoint is subject to significant uncertainty. Therefore patterns of fluctuation in biomass and abundance may not be useful in assessing the course of ecosystem deterioration or recovery in individual cases.

Biotic Indices

Biotic indices are designed to collapse complicated compositional data into concise numerical descriptions that both reflect stress effects on particular ecosystems and permit direct comparisons with other ecosystems. They are based exclusively on species abundance and compositional data, which distinguishes them from many other indices used to assess pollution-related degradation of aquatic ecosystems .

Many different biotic indices have been proposed (Kolkwitz and Marsson 1908; Patrick 1949; Beck 1955; Chandler 1970; Gaufin 1973; Goodnight and Whitley 1961; Hilsenoff 1977, 1982; Howmiller and Scott 1977; Winget and Mangum 1979; Karr et al. 1986). Most of these indices were developed to assess the effects of organic enrichment, and rely on benthic macroinvertebrates, primarily oligochaetes, tubificids, and aquatic insects for freshwater ecosystems, and polychaetes and bivalves for marine ecosystems. Reviews comparing the utility of various indices can be found in Goodnight (1973), O'Connor and Swanson (1982), Sheehan (1984), and Washington (1984).

Hilsenoff's (1977, 1982) modification of Chutter's index (1972) was developed for the Wisconsin Department of Natural Resources to assess organic enrichment, and is one of the best verified of the biotic indices.

This index concentrates on insects, amphipods, and isopods, and is suitable for application to stream riffle communities in which these relatively sessile organisms with a life cycle usually greater than one year are abundant and easily collected. After substantial background calibration work, taxa were weighted according to their sensitivity to organic enrichment. The biotic index was then calculated as

$$\text{B.I.} = (\Sigma n_i a_i)/N$$

where

n_i = number of individuals of each species (or genus, if all species have the same tolerance value)
a_i = tolerance value assigned to that species or genus
N = total number of individuals in the sample.

The most time-consuming, labor-intensive part of the work is in developing valid species-specific tolerance values and deciding for which genera species-level identifications are unnecessary and aggregation to the level of genus acceptable. Once this has been done, this simple index can be applied to samples from other stream riffle communities and good assessments of overall ecosystem status can be made. The critical feature of this technique is the development of species weightings, which requires exacting taxonomic proficiency as well as a deep understanding of species ecology. The resulting index, however, can be applied to stream riffle communities in that region by anyone trained to recognize the particular taxa involved, for which a training period of six months or less is sufficient (Hilsenoff 1982). The resulting index, while time-consuming and difficult to create, is quite simple and rapid to apply. For example the time required to collect, sort, identify, and enumerate a sample at the genus level is 63 minutes and at species level is 84 minutes.

This general approach has also been used in paleoecological studies, in which the goal is to use microfossil stratigraphy to reconstruct the history of particular aspects of past environments. For lake ecosystems, the most widespread application of this technique has been the reconstruction of recent lake acidification using diatoms as a means of reconstructing past lakewater pH (e.g., Renberg and Hellberg 1982; Davis et al. 1983; Flower and Battarbee 1983; Charles 1984; Arzet et al. 1986; Charles et al. 1986; Ford 1986; Tolonen et al. 1986). The method depends on the development of statistically rigorous region specific transfer functions that describe the relationships between the community composition of diatom death assemblages in the upper few millimeters of lake sediment and chemical and morphometric parameters of regional lakes (e.g., Charles 1985; Davis and Anderson 1985; Anderson et al. 1986). When pH is found to be an important parameter in the ordination or clustering of lakes (as is usually the case), quantitative functions relating community composition to lakewater pH can be developed. These functions are then applied to diatom

assemblages in sediment cores to reconstruct past lakewater pH measures. When this technique is systematically applied to groups of lakes in
different geographic regions it becomes possible to study the regional
extent of recent lake acidification (Charles et al. 1986).

Biotic indices are concise measures of community composition that
combine pollutant tolerances with the relative abundances of species
within ecologically sensitive groups. Because species identities remain
intact (i.e., taxa are not equivalent, as they are with diversity indices),
biotic indices have a high likelihood of reflecting subtle but ecologically
important changes in species composition that may not be caught by the
indicator species approach (because of its extreme reliance on one or a
very few species) and that are not reflected in diversity indices (which are
not sensitive to species identities). In cases where biotic indices have
been tested against diversity indices (e.g., Howmiller and Scott 1977;
Murphy 1978; Jones et al. 1981), biotic indices have been found to be
more sensitive and less variable in discriminating differences in stream
water quality. Washington (1984) argues that comparisons of biotic indices and diversity indices are inappropriate, because they measure different things. Species diversity is a crude measure of community structure,
which, depending on the initial structure of the community and on the
timing, level, and duration of the stressor, may not respond to chemical
stress. A biotic index, on the other hand, "is a specific measure of pollution . . . based on the reaction of physiological[ly] sensitive organisms"
(Washington 1984).

There are several limitations to the use of existing biotic indices for
evaluation of chemical stress effects. Most existing indices are meant to
refer only to organic enrichment and should not be applied to the analysis
of other types of chemical stress (such as toxins) without further calibration and verification. Studies are needed to develop, calibrate, and verify
indices for other pollutants, other ecosystem types, and other regions. On
the other hand, simpler indices may suffice for toxic pollutants, which
typically benefit few, if any, species at high concentrations (Warren 1971).
For example, Goodnight and Whitley (1961) suggested that abundance of
oligochaetes may be a useful index of both organic enrichment and industrial pollution in streams, and Winner et al. (1980) found that the relative
abundance of chironomids in a benthic sample is a useful index of the
extent of copper pollution in streams.

The potential usefulness of the biotic index approach is high, and the
method deserves renewed scrutiny as a quantitative technique for early
detection of stress-induced changes in ecosystem organization. A well-
calibrated biotic index can be a rapid, sensitive, reliable, and robust
method for evaluating water quality with respect to particular pollutants,
and as such can be a valuable tool for assessment, management, and
regulatory uses. Application of these indices is usually much simpler and
more straightforward than their creation, which is an appropriate weight-

ing of effort, because the purpose of the technique is to reduce complex data sets into simple but biologically meaningful descriptors for use in resource planning. A major advantage of the method is that once the initial groundwork has been done, it can be applied to an array of similar ecosystems. The interpretive power of the technique improves when analyses can be made along spatial (e.g., distance from source) or temporal (e.g., sediment stratigraphy) gradients. Investment in the biotic index technique will be maximized by focusing attention on carefully chosen arrays of stressors and receiving ecosystems.

Species Richness

Species richness is the number of species encountered in a sampling effort, and is the simplest structural parameter that uses but does not retain species identity. Table 5.2, although it is not exhaustive, suggests that an overall reduction in species richness is a reliable indicator of chemical stress in phytoplankton, zooplankton, and benthic communities. This implies that the stress-related rates of loss for sensitive species typically exceed rates of gain of tolerant species. The rate and timing of the overall reduction may vary from ecosystem to ecosystem, and in cases where the stressor is organic enrichment rather than toxics, reductions may be preceded by slight-to-moderate increases in richness (Pearson and Rosenberg 1978). Even in these situations, however, there is less case-to-case variability in stress-related changes in species richness than in either biomass or abundance (See Figs. 1–4 in Pearson and Rosenberg 1978). It must be emphasized that comparisons of species richness can only be made across studies if there is consistency in sampling design and sampling effort, as the number of species reported depends on sampling intensity (e.g., Confer et al. 1983).

Within certain guilds or trophic levels, decreased species richness may be a secondary effect related to factors such as habitat simplification. This is particularly true with substrate-dependent organisms such as chydorid cladocera. To test this possibility for reductions in chydorid species richness with increasing lake acidity, Kenlan et al. (1984) studied the littoral zone of three acidified and non-acidified lakes using artificial substrates placed in beds of the same macrophyte species in each lake. Their finding of lower chydorid species richness on identical substrates in lower-pH versus higher-pH environments supports the hypothesis that pH, rather than substrate, is one of the primary controls over decreased species richness for oligotrophic littoral cladoceran communities.

Winner et al. (1975) compared abundance, species richness, and two diversity indices (Margalef's d and Shannon's H) in an experimental stream study and concluded that species richness was the most sensitive index of copper stress on the macroinvertebrate community. By itself,

Table 5.2. Effects of chemical stress on species richness

System	Type of study	Stressor	Element of biota	Species richness in stressed vs. unstressed system	Reference
Temperate lake	Site comparison (n=12)	Acidity	Fish	Lower	Wiener et al. 1984
Temperate lakes	Site comparison (n=68)	Acidity	Fish	Lower	Harvey 1975
Temperate lakes	Site comparison (n=234)	Acidity	Fish	Lower	Smith et al. 1986
Marine systems	Review paper	Organic enrichment	Macrobenthos	Lower	Pearson and Rosenberg 1978
Nearshore marine	Whole system	Oil	Benthos	Lower	Elmgren et al. 1980a
Nearshore marine	Whole system	Oil	Benthos	Lower	Jacob 1980
Offshore marine	Gradient study	Oil	Benthos	Lower	Addy et al. 1978
Temperate stream	Gradient study	Copper	Aquatic insects	Lower	Winner et al. 1975
Temperate lakes	Site comparison (n=75)	Acidity	Benthic macroinvertebrates	Lower	Leivestad et al. 1976
Temperate lakes	Site comparison (n=34)	Acidity	Chironomids	Lower	Raddum 1980
Temperate stream	Whole-stream experiment	Acidity	Aquatic insects	Lower	Hall et al. 1980
Temperate stream	Whole-stream experiment	Acidity	Aquatic insects	Lower	Pratt and Hall 1981
Temperate streams	Site comparison (n=13)	Acidity	Benthic invertebrates	Lower	Otto and Svensson 1983
Temperate stream	Whole-system study	DDT	Aquatic insects	Lower	Ide 1967
Temperate stream	Gradient study	Copper	Chironomids	Lower	Winner et al. 1980
Rocky Mountain stream	Gradient study	Coal mine drainage	Aquatic insects	Similar	Canton and Ward 1981
Temperate lakes	Site comparison (n=3)	Acidity	Littoral cladocera	Lower	Kenlan et al. 1984
Temperate lakes	Site comparison (n=47)	Acidity	Crustacean zooplankton	Lower	Sprules 1975
Temperate lake	Site comparison (n=2)	Acidity	Crustacean zooplankton	Lower	Stokes 1984b
Temperate lakes	Site comparison (n=37)	Acidity	Chydorid cladocera	Lower	Brakke et al. 1984
Temperate lakes	Site comparison (n=57)	Acidity	Zooplankton	Lower	Leivestad et al. 1976

Table 5.2 (continued)

System	Type of study	Stressor	Element of biota	Species richness in stressed vs. unstressed system	Reference
Temperate lakes	Site comparison ($n=6$)	Acidity	Zooplankton	Lower	Roff and Kwiatkowski 1977
Temperate lake	Whole-lake experiment	Acidity	Zooplankton	Lower	Malley et al. 1981
Temperate lakes	Site comparison ($n=6$)	Acidity	Zooplankton	Lower	Confer et al. 1983
Temperate lakes	Site comparison ($n=3$)	Acidity	Aquatic macrophytes	Lower	Wile et al. 1985
Temperate lakes	Site comparison ($n=17$)	Acidity (SO_4^{2-})	Aquatic macrophytes	Lower	Gorham and Gordon 1963
Temperate lakes	Site comparison ($n=2$)	Acidity	Aquatic macrophytes	Lower	Stokes 1984b
Temperate lakes	Site comparison ($n=6$)	Acidity	Phytoplankton	Lower	Kwiatkowski and Roff 1976
Temperate lake	Mesocosm	Acidity	Phytoplankton	Lower	Yan and Stokes 1978
Temperate lake	Mesocosm	Acidity	Phytoplankton	Lower	Stokes 1984a
Temperate lake	Whole-lake experiment	Acidity	Phytoplankton	Similar	Schindler 1980
Temperate stream	Artificial stream	Acidity	Phytoplankton	Lower	Patrick et al. 1968b
Temperate streams	Site comparison	Acidity	Phytoplankton	Lower	Whitton and Diaz 1980
Temperate stream	Artificial stream	Mercury	Phytoplankton	Lower	Sigmon et al. 1977
Temperate river	Longitudinal study	Complex industrial pollution	Phytoplankton	Lower	Patrick 1977
Temperate lake	Mesocosm	Acidity	Periphyton	Lower	Muller 1980
Temperate stream	Whole-stream experiment	Acidity	Fungi	Lower	Hall et al. 1980
Temperate lakes	Microcosm	$CuSO_4$	Protozoans	Lower	Hart and Cairns 1984

this result cannot be generalized. However, the consistent finding (Table 5.2) of decreased species richness across a large array of organismal groups, ecosystem types, stressor types, timings, exposures, and strengths of contrast between "stressed" and "unstressed" states suggests that decreases in species richness may be a relatively sensitive measure of chemical stress, appearing in the early stages of ecosystem response. On the other hand, the studies of Canton and Ward (1981) on aquatic insects of a Colorado stream affected by strip mining found that species richness, standing crop, and species diversity all failed to discriminate between upstream and downstream sites. Only differences in community composition indicated the presence of unusual conditions downstream of the mine. Apparently, their system was only mildly affected, because the study stream had both a buffer strip of unmined land and a settling pond between the strip mine and the stream. This suggests that subtle but quantifiable changes in community composition are an earlier and more sensitive indicator of chemical stress than decreases in species richness.

Species Diversity

Diversity indices differ from simple species richness in that they typically include information about the partitioning of individuals among species. They have become popular descriptors of community structure because, like biotic indices, they reduce large masses of ecological data into single numerical values. Several diversity indices are in common use, of which the most popular are Brillouin's H (Brillouin 1951) and Shannon's H' or H" (Shannon and Weaver 1949). The merits of the species diversity approach have been extensively debated (e.g., Sanders 1968; Wilhm and Dorris 1968; Hurlburt 1971; Peet 1974; Zand 1976; Hughes 1978; Kaesler et al.1978; Gray 1979; Robinson and Sandgren 1984). Washington (1984) and Sheehan (1984) have both recently provided comprehensive reviews of 18 different diversity indices.

The initial applications of diversity indices to stress effects studies were successful because they were applied to extreme situations in which the imposition of acute stress resulted in reductions in diversity. Problems arose, however, as this approach was more extensively applied. The term ''diversity'' meant different things to different investigators, and its usage in the literature was inconsistent. Several things became increasingly unclear, such as which diversity index to use under what circumstances, what taxocene to apply the indices to (e.g., phytoplankton, benthic invertebrates, zooplankton, fish), and whether the indices ought to be applied to more than taxocene—and if so, to which one(s). Comparative studies have found that different indices rank communities differently (e.g., Robinson and Sandgren 1984) and different diversity indices or the same indices applied to different taxocenes have been found to give conflicting information (e.g., Sigmon et al. 1977). Often diversity measures do not change monotonically across environmental gradients, but change as step functions. In these cases, diversity may remain unchanged even though there have been clear, stress-related changes in community composition. For example, Roff and Kwiatkowski (1977) found that zooplankton com-

munities retained their diversity with increasing lake acidification down to
a pH of about 5.5, after which diversity decreased. Kwiatkowski and Roff
(1976) obtained similar results for phytoplankton diversity. On the other
hand, Marshall and Mellinger (1978) found that zooplankton diversity
decreased with increasing doses of cadmium up to 20 μg/l, after which it
leveled off. Finally, interpreting the significance of changes in species
diversity can be difficult because the highest species diversity can be
associated with intermediate levels of disturbance (e.g., Pearson and Rosenberg 1978). Successional relationships and other aspects of spatial and
temporal variability also cloud the issue.

Some indices appear reasonably correlated with stress effects in particular situations. Both the Sanders rarefaction technique (1968) and a modified log-normal technique appear to be useful indices for marine benthos
(Pearson and Rosenberg 1978; Gray and Pearson 1982; Pearson et al.
1982). On the whole, however, there is general agreement that diversity
indices do not reliably reflect disturbance over time or space (e.g., Cook
1976; Hilsenhoff 1977; Murphy 1978; Gray 1979; Jones et al. 1981; Larson
et al. 1983).

5.5 Conclusions

The imposition of chemical stress on aquatic ecosystems apparently induces a common suite of changes in community composition and structure across a range of stressors and receiving ecosystems. These changes
can serve as early warning signals of ecosystem deterioration, and can
help track the course of stress-related change.

The earliest phase of ecosystem response begins with the loss of species most sensitive to the stressor or stressor complex, possibly accompanied by changes in the relative abundances of native taxa. This stage may
be subtle and difficult to recognize, because it requires detailed knowledge about the stressor-specific response characteristics of the particular
array of species in any given community. The early loss of stressor intol-
erant species may be missed if the initial abundance of sensitive
species is low and spatial and temporal heterogeneity high. Often, however, these changes are readily identifiable by experienced field ecologists
familiar with the target ecosystems even if it is difficult to convincingly
convey these changes. Carefully designed and well calibrated biotic indices can play an important role in this respect, providing quantitative
representations of stressor-specific responses that can be compared
across ecosystems.

Because the identity of the most sensitive species will vary depending
on the nature of the stressor and the array of species present in the
community, standard single-species toxicity testing cannot be relied on to

predict the dosage at which early changes will occur. Laboratory toxicity tests on standard species only rarely test the most sensitive species for particular stressor-system pairs, and they cannot predict the sublethal effects that can dominate the dynamics of real ecosystems.

The consequences of species loss at this early stage will depend entirely on the role played by stressor-sensitive species in the target community and the amount of redundancy that exists in this role. If the species play minor roles or are buffered by redundancy, ecosystem-level consequences may be minimal. Regardless of the eventual consequences of loss of sensitive species, however, the loss itself provides an early warning signal of ecosystem deterioration.

As chemical stress continues to operate, native species respond in individualistic ways, and changes occur in the relative representation of elements of the community. Rapid species rearrangements can occur in the plankton, where generation times are short. Autoecological studies of longer-lived species may uncover changes in condition or reproductive success of particular populations. Changes in relative abundance may be subtle or marked, and must be evaluated within the context of the natural variability of the system. As effects at one level of organization begin to propagate throughout the system, stress-related changes will exceed the range of natural variability or represent a temporally inappropriate condition for the community.

Compositional changes may be difficult to detect without substantial information on pre-stress conditions in the community, but again, tools such as carefully calibrated and verified biotic indices and paleoecological techniques make this information amenable to analysis even in the absence of long-term data. The construction of such diagnostic tools is still in its early phases, and in order for significant progress to be made, both focused attention and serious funding are required. Often it seems that what is wanted are simple, quick, reliable, sensitive, robust measures of ecosystem health that are cheap to develop and easy to apply. This is an unrealistic goal. When the investment necessary for creating and refining compositional indices have been made, the results have justified the investment (e.g., Hilsenoff 1982, Charles et al. 1986).

One of the biggest difficulties to date in using compositional data to reflect ecosystem health has been in the data base management of the large and complex data sets generated by comprehensive studies of species composition. However, as Cairns and Pratt (1986) have emphasized, technological advances in information processing over the last several years undercuts the argument that data sets of this type are too unwieldy to be useful. The key issue here is that programs designed to take advantage of biological parameters in assessing the effects of chemical stress must pay close attention to the management needs for various kinds of data. If early warning signals are wanted, and if there is adequate evidence that compositional data will give such a signal, then it is possible to design a study that can provide this information. But care must be taken to avoid the standard pitfall of lack of definition in addressing the management uses of data to be collected.

Early changes in ecosystem structure are followed by declines in species richness as increased stress is applied to the ecosystem. Depending on the particulars of the situation, these declines may or may not be accompanied by decreases in biomass, abundance, and species diversity. All of these latter measures can follow trajectories that include primary or secondary maxima at intermediate levels of disturbance. These parameters therefore are only useful in situations in which there is detailed knowledge about the behavior of the particular stressor–system pair.

There has been considerable interest in the relationship between stress-related changes in community structure and community function. Levine (this volume) has recently reviewed the literature on the latter for aquatic ecosystems. Generally speaking, except for simple communities composed primarily of opportunistic species, the initial response of most aquatic ecosystems to chemical stress typically involves loss of sensitive species accompanied by changes in relative abundance of rapidly reproducing taxa. The consequences of these structural changes for ecosystem function depend primarily on the structural organization and trophic dynamics of the receiving system. Redundancy buffers the effects of species loss on functional parameters such as primary productivity, and significant changes in species composition can occur without affecting community function. On the other hand, if the most sensitive species are, for example, selective feeders and community dominants, early species losses can initiate a cascade of effects that results in rapid changes in functional parameters. These issues are only of interest, however, insofar as the objective is to identify early warning signals of ecosystem change. This is, in the last analysis, an information issue, and is distinct from the equally important need for conscious identification of acceptable kinds and magnitude of change.

Acknowledgments. I thank my colleagues at the Ecosystems Research Center, Cornell University, for helpful discussions, and M. Fiering, N. Hairston, Jr., M. Harwell, R. Howarth, J. Kelly, K. Kimball, S. Levin, K. Limburg, and P. Spencer for critical reviews of various drafts of the manuscript.

This publication is ERC-084 of the Ecosystems Research Center (ERC), Cornell University, and was supported in part by the U.S. Environmental Protection Agency Cooperative Agreement Number CR811060. The work and conclusions published herein represent the views of the author, and do not necessarily represent the opinions, policies, or recommendations of the Environmental Protection Agency.

References

Abraham KF (1975) Waterbirds and oil-contaminated ponds at Point Storkersen, Alaska. M.S. Thesis. Iowa State University

Adams L, Hanavan M, Hosley N, Johnston D (1949) The effects on fish, birds, and mammals of DDT used in the control of forest insects in Idaho and Wyoming. J Wildlife Manage 13:245–254

Addy JM, Levell D, Hartley JP (1978) Biological monitoring of sediments in Ekofisk oil field. In: Proceedings of the Conference on Assessment of Ecological Impacts of Oil Spills. June 14–17, 1978, Keystone CO: American Institute of Biological Sciences, pp. 515–539

Ali A, Stanley BH (1982) Effects of a new carbamate insecticide, Larvin (VC-51762), on some non-target aquatic invertebrates. Florida Entomol 65:477–483

Almer B, Dickson W, Eckstrom C, Hornstrom E, Miller U (1974) Effects of acidification on Swedish lakes. Ambio 3:30–36

Almer B, Dickson W, Eckstrom C, Hornstrom E (1978) Sulfur pollution and the aquatic ecosystem. In: Nriagu J (ed) *Sulfur in the Environment*. Part II: Ecological Impacts. New York: Wiley, pp. 273–311

Anderson DS, Davis RB, Berge F (1986) Relationships between diatom assemblages in lake surface-sediments and limnological characteristics in southern Norway. In: Smol JP, Battarbee RW, Davis RB, Merilainen J (eds) *Diatoms and Lake Acidification*. Developments in Hydrobiology No. 29. Boston: Dr. W Junk, pp. 97–114

Arnold DE, Bender PM, Hale AB, Light RW (1981) Studies on infertile, acidic Pennsylvania streams and their benthic communities. In: Singer R (ed) *The Effects of Acid Precipitation on the Benthos*. New York: Canterbury Press

Arzet K, Krause-Dellin D, Steinberg C (1986) Acidification of four lakes in the Federal Republic of Germany as reflected by diatom assemblages, cladoceran remains, and sediment chemistry. In: Smol JP, Battarbee RW, Davis RB, Merilainen J (eds) *Diatoms and Lake Acidification*. Developments in Hydrobiology No. 29. Boston: Dr W Junk, pp. 227–250

Barsdate RJ, Alexander V, Benoit RE (1972) Natural oil seeps at Cape Simpson, Alaska: Aquatic effects. In: *Proceedings of the Symposium on the Impact of Oil Resource Development on Northern Plant Communities*. Institute of Arctic Biology, Fairbanks, AK: University of Alaska, pp. 91–95

Bartell SM, Gardner RH, O'Neill RV (1986) The influence of bias and variance in predicting the effects of phenolic compounds in ponds. In: *1986 Eastern Simulation Conference. Supplementary Proceeding*. Society for Computer Simulation, pp. 173–176

Battarbee RW (1984) Diatom analysis and the acidification of lakes. Phil Trans R Soc B 305:451–477

Battarbee RW, Flower RJ, Stevenson AC, Rippey B (1985) Lake acidification in Galloway: a paleoecological test of competing hypotheses. Nature 314:350–352

Beamish RJ (1974) Loss of fish populations from unexploited remote lakes in Ontario, Canada as a consequence of atmospheric fallout of acid. Water Research 8:85–95

Beamish RJ, Harvey HH (1972) Acidification of the LaCloche Mountain Lakes, Ontario and resulting fish mortalities. J Fish Res Bd Can 29:1131–1143

Beck WM Jr (1955) Suggested method for reporting biotic data. Sewage Indust Wastes 27:1193–1197

Beckett DC (1978) Ordination of macroinvertebrate communities in a multi-stressed river system. In: Thorp JH, Gibbons JW (eds) *Energy and Environmental Stress in Aquatic Systems*. U.S. Dept. Energy Technical Information Center, pp. 748–770

Booth GM, Hamilton JG, Molot LA (1986) Liming in Ontario: Short-term biological and chemical changes. Water Air Soil Pollut 31:709–720

Bradbury JP, Megard RO (1972) Stratigraphic record of pollution in Shagawa Lake, northeastern Minnesota. Geol Soc Amer Bull 83:2639–2648

Braekke FH (ed) (1976) *Impact of Acid Precipitation on Forest and Freshwater Ecosystems in Norway*. Research Report 6/76, the Scandinavian Nations Science Foundation project

Brillouin L (1951) Maxwell's demon cannot operate: Information and entropy. I J Appl Phys 22:334–343

Brinkhurst RO (1965) Observations on the recovery of a British river from gross organic pollution. Hydrobiologia 25:9–51

Brinkhurst RO, Cook DG (1974) Aquatic earthworms (Annelida: Oligochaeta). In: Hart CW Jr, Fuller SLJ (eds) *Pollution Ecology of Freshwater Invertebrates*. New York: Academic Press, pp. 143–156

Brooker MP, Edwards RW (1973) Effects of the herbicide paraquat on the ecology of a reservoir. Freshwater Biol 3:157–175

Brugam RB (1978) Human disturbance and the historical development of Linsley Pond. Ecology 59:19–36

Brugam RB, Speziale BJ (1983) Human disturbance and the paleolimnological record of change in the zooplankton community of Lake Harriet, Minnesota. Ecology 64:578–591

Butcher RW (1946) The biological detection of pollution. J Instit Sew Purif 2:92–97

Cairns J (1974) Indicator species vs. the concept of community structure as an index of pollution. Water Resources Bull 10:338–347

Cairns J Jr, Dickson KL (1971) A simple method for the biological assessment of the effects of waste discharges on aquatic bottom-dwelling organisms. J Water Poll Control Fed 43:755–772

Cairns J Jr, Pratt JR (1986) On the relation between structural and functional analyses of ecosystems. Environm Toxicol Chem 5:785–786

Cairns J Jr, Lanza GR, Parker BC (1972) Pollution related structural and functional changes in aquatic communities with emphasis on freshwater algae and protozoa. Proc Acad Nat Sci Philadelphia 124:79–127

Canton SP, Ward JV (1981) The aquatic insects, with emphasis on Tricoptera, of a Colorado stream affected by coal strip-mine drainage. Southwest Nat 25:453–460

Chadwick JW, Canton SP (1983) Coal mine drainage effects on a lotic ecosystem in northwest Colorado, USA. Hydrobiologia 107:25–33

Chandler JR (1970) A biological approach to water quality management. Water Poll Control 4:415–422

Charles DF (1984) Recent pH history of Big Moose Lake (Adirondack Mountains, New York, U.S.A.) inferred from sediment diatom assemblages. Verh Internat Verein Limnol 22:559–566

Charles DF (1985) Relationships between surface sediment diatom assemblages and lakewater characteristics in Adirondack lakes. Ecology 66:994–1011

Charles DF, Whitehead DR, Anderson DS, Bienert R, Camburn KE, Cook RB, Crisman TL, Davis RB, Ford J, Fry B, Hites RA, Kahl JS, Kingston JC, Kreis RG Jr, Mitchell MJ, Norton SA, Roll L, Smol JP, Sweets PR, Uutala A, White JR, Whiting M, Wise R (1986) The PIRLA Project (Paleoecological Investigation of Recent Lake Acidification): Preliminary results for the Adirondacks, New England, N. Great Lakes States, and N. Florida. Water Air Soil Poll 30:355–366

Christie CE, Smol JP (1986) Recent and long-term acidification of Upper Wallface Pond (N.Y.) as indicated by mallomonadacean microfossils. Hydrobiologia 143:355–360

Chutter FM (1972) An empirical biotic index of the quality of water in South African streams and rivers. Water Resources 6:19–30

Collins NC, Zimmerman AP, Knoechel R (1981) Comparisons of benthic infauna and epifauna biomasses in acidified and non-acidified Ontario lakes. In: Singer R (ed) The Effects of Acidic Precipitation on the Benthos. Proceedings of the Symposium of the North American Benthological Society, New York: Canterbury Press, pp. 35–48

Confer JL, Kaaret T, Likens GE (1983) Zooplankton diversity and biomass in recently acidified lakes. Can J Fish Aquat Sciences 40:36–42

Cook SEK (1976) Quest for an index of community structure sensitive to water pollution. Environ Pollut 11:269–288

Corbet PS (1958) Some effects of DDT on the fauna of the Nile. Rev Zool Bot Afr 57:73–95

Davis RB, Anderson DS (1985) Methods of pH calibration of sedimentary diatom remains for reconstructing history of pH in lakes. Hydrobiologia 120:69–87

Davis RB, Norton SA, Hess CT, Brakke DF (1983) Paleolimnological reconstruction of the effects of atmospheric deposition of acids and heavy metals on the chemistry and biology of lakes in New England and Norway. Hydrobiologia 103:113–123

deNoyelles F, Kettle WD, Sinn DE (1982) The responses of plankton communities to Atrazine, the most heavily used pesticide in the United States. Ecology 63:1285–1293

Dillon PJ, Yan ND, Scheider WA, Conroy N (1979) Acidic lakes in Ontario, Canada: characterization, extent and responses to base and nutrient additions. Arch Hydrobiol Bech Ergebn Limnol 13:317–336

d'Itri FM (1982) Acid Precipitation: Effects on Ecological Systems. Ann Arbor, MI: Ann Arbor Science, 506 pp.

Drabløs D, Tollan A (eds) (1980) The Ecological Impact of Acid Precipitation. Oslo, Norway: Proc Conf SNSF Proj, 383 pp.

Edmondson WT (1974) The sedimentary record of the eutrophication of Lake Washington. Proc Natl Acad Science,71:5093–5095

Edmondson WT, Anderson GC, Peterson DR (1956) Artificial eutrophication of Lake Washington. Limnol Oceanogr 1:47–53

Eilers JM, Lein GJ, Bert RG (1984) Aquatic organisms in acidic environments: a literature review. Wisconsin Dept Natural Resources Tech Bull No 150. Madison, Wisconsin, 18 pp.

Elmgren R, Frithsen JB (1982) The use of experimental ecosystems for evaluating the environmental impact of pollutants: a comparison of an oil spill in the Baltic

Sea and two long-term, low level oil addition experiments in mesocosms. In: Grice GD, Reeve MR (eds) *Marine Mesocosms: Biological and Chemical Research in Experimental Ecosystems*. New York: Springer-Verlag, pp. 153–165

Elmgren R, Vargo GA, Grassle JF, Grassle JP, Heinle DR, Langlois G, Vargo SL (1980) Trophic interactions in experimental marine ecosystems perturbed by oil. In: Giesy JP (ed) *Microcosms in Ecological Research*. Washington DC: U.S. Dept of Commerce, N.O.A.A., pp. 779–800

Elmgren R, Hansson S, Larsson U, Sundelin B, Boehm PD (1983) The "Tsesis" oil spill: acute and long-term impact on the benthos. Marine Biol 73:51–65

Engblom E, Lingdell PE (1984) The mapping of short-term acidification with the help of biological pH indicators. In: Institute of Freshwater Research - Drottningholm, Sweden. Report No. 61, pp. 60–68

Eriksson F, Hornstrom E, Mossberg P, Nyberg P (1983) Ecological effects of lime treatment of acidified lakes and rivers in Sweden. Hydrobiologia 101:145–164

Fiedler PC (1983) Fine-scale spatial patterns in the coastal epiplankton off southern California. J Plankton Research 5:865–879

Findlay DL (1984) Effects on phytoplankton biomass, succession and composition in Lake 223 as a result of lowering pH levels from 5.6 to 5.2. Data from 1980 to 1982. Canadian Manuscript Report of Fisheries and Aquatic Sciences No. 1761

Findlay DL, Kasian SEM (1986) Phytoplankton community responses to acidification of Lake 223, Experimental Lakes Area, northwestern Ontario. Water Air Soil Poll 30:719–726

Flower RJ, Battarbee RW (1983) Diatom evidence for recent acidification of two Scottish lochs. Nature 305:130–133

Ford J (1986) The recent history of a naturally acidic lake (Cone Pond, N.H.). In: Smol JP, Battarbee RW, Davis RB, Meriläinen J (eds) *Diatoms and Lake Acidity*. Developments in Hydrobiology, No. 29. Boston, Dr W. Junk, pp. 131–148

Foster MS, Holmes RW (1977) The Santa Barbara oil spill: an ecological disaster? In: Cairns J Jr, Dickson KL, Herricks EE (eds) *Recovery and Restoration of Damaged Ecosystems*. Charlottesville, VA: University Press of Virginia, pp. 166–190

Frenette JJ, Richard Y, Moreau G (1986) Fish responses to acidity in Quebec lakes: A review. Water Air Soil Poll 30:461–475

Gaufin AR (1973) Use of aquatic invertebrates in the assessment of water quality. In: Cairns J Jr, Dickson KL (eds) *Biological Methods for the Assessment of Water Quality*. Amer Soc Test Materials STP 528, pp. 96–116

Geckler JR, Horning WB, Neiheisel TM, Pickering QH, Robinson EL, Stephan CE (1976) Validity of laboratory tests for predicting copper toxicity in streams. EPA-600/3-76-116. Duluth, MN: U.S. Environmental Protection Agency, 191 pp.

Gibson KN, Smol JP, Ford J (1987) Chrysophycean microfossils provide new insight into the recent history of a naturally acidic lake (Cone Pond, N.H.). Can J Fisheries Aquat Res 44:1584–1588

Giddings JM (1983) Microcosms for assessment of chemical effects on the properties of aquatic ecosystems. In: Saxena J (ed) *Hazard Assessment of Chemicals: Current Developments,* vol. 2, New York: Academic Press, pp. 45–94

Giddings JM, Franco PJ, Cushman RM, Hook LA, Southworth GR, Stewart AJ

(1984) Effects of chronic exposure to coal-derived oil on freshwater ecosystems: II. Experimental ponds. Environm Toxicol Chem 3:465–488

Giesy JP (ed) (1980) *Microcosms in Ecological Research*. Washington DC: U.S. Dept. of Commerce, N.O.A.A., 1110 pp

Goodnight CJ (1973) The use of aquatic macroinvertebrates as indicators of stream pollution. Trans Amer Microscopical Soc 92:1–13

Goodnight CJ,Whitley LS (1961) Oligochaetes as indicators of pollution. Proc 15th Indust Waste Conf, Purdue Univ Eng Ext Serv 106:139–142

Gorham E (1976) Acid precipitation and its influence upon aquatic ecosystems – an overview. Water Air Soil Poll 6:457–481

Gorham E, Gordan AG (1963) Some effects of smelter pollution upon aquatic vegetation near Sudbury, Ontario. Can J Bot 41:371–378

Grahn O, Hultberg H, Landner L (1974) Oligotrophication—a self-accelerating process in lakes subject to excessive supply of acid substances. Ambio 3:93–94

Grassle JF, Grassle JP (1974) Opportunistic life histories and genetic systems in marine benthic polychaetes. J Marine Research 32:253–284

Gray JS (1979) Pollution-induced changes in communities using the log-normal distribution of individuals among species. Marine Poll Bull 12:173–176

Gray JS, Pearson TH (1982) Objective selection of sensitive species indicative of pollution-induced change in benthic communities. I. Comparative methodology. Marine Ecol Prog Ser 9:111–119

Greve W, Parsons TR (1977) Photosynthesis and fish production: hypothetical effects of climatic change and pollution. Helgol Wiss Meeresunters 30:666–672

Grice GD, Reeve MR (1982) Introduction and description of experimental ecosystems. In: Grice GD, Reeve MR (eds) *Marine Mesocosms: Biological and Chemical Research in Experimental Ecosystems*. New York: Springer-Verlag, pp. 1–9

Haines TA (1981) Acidic precipitation and its consequences for aquatic ecosystems: A review. Trans Amer Fisheries Soc 110:669–707

Hall R, Likens G, Fiance S, Hendrey G (1980) Experimental acidification of a stream in the Hubbard Brook Experimental Forest, New Hampshire. Ecology 61:976–989

Hargis JR, Shannon LJ (1984) An assessment of historical change in two northern Minnesota lakes. Environ Manage 8:481–488

Harrass MC (1983) Effects of small fish predation on microcosm community responses to selective chemical stress. Ph.D. dissertation. Seattle WA: University of Washington

Harvey H (1980) Widespread and diverse changes in the biota of North American lakes and rivers coincident with acidification. In: Drabløs D, Tollan A(eds) *The Ecological Impact of Acid Precipitation*. Proc Int Conf SNSF Project. Oslo, Norway, pp. 93–98

Hatfield CT (1969) Effects of DDT larviciding on aquatic fauna of Bobby's Brook, Labrador. Can Fish Cult 40:61–72

Hendrey GR (1984) *Early Biotic Responses to Advancing Lake Acidification*. Boston: Butterworth Publishers, 173 pp.

Hendrey GR, Wright RF (1976) Acid precipitation in Norway: effects on aquatic fauna. J Great Lakes Research 2:192–207

Hendrey GR, Baalsrud K, Traaen T, Laake M, Raddum G (1976) Acid precipitation: some hydrobiological changes. Ambio 5:224–227

Herricks EE, Cairns JC Jr (1982) Biological monitoring. Part III - Receiving system methodology based on community structure. Water Research 16:141–153

Hesthagen T (1986) Fish kills of Atlantic salmon (*Salmo salar*) and brown trout (*Salmo trutta*) in an acidified river of SW Norway. Water Air Soil Poll 30:619–628

Hilsenhoff WL (1977) Use of arthropods to evaluate water quality of streams. Wisconsin Dept Natural Resources Tech Bull No 100

Hilsenhoff WL (1982) Using a biotic index to evaluate water quality in streams. Wisconsin Dept Natural Resources Tech Bull No 132.

Howarth RW (1987) The potential effects of petroleum on marine organisms on Georges Bank. In: Backus R (ed) *An Atlas of Georges Bank.* Cambridge MA: M.I.T. Press. pp 540–551

Howmiller RP, Scott MA (1977) An environmental index based on the relative abundance of oligochaete species. J Water Poll Control Fed 49:809–815

Hughes BD (1978) The influence of factors other than pollution on the value of Shannon's diversity index for benthic macroinvertebrates in streams. Water Research 12:359–364

Hurlburt SH (1971) The nonconcept of species diversity: a critique and alternative parameters. Ecology 52:577–586

Hurlburt SH (1975) Secondary effects of pesticides on aquatic ecosystems. Residue Rev 57:81–146

Hurlburt SH, Mulla MS, Willson HR (1972) Effects of an organophosphorus insecticide on the phytoplankton, zooplankton, and insect populations of freshwater ponds. Ecol Monog 42:269–299

Hustedt F (1937–1939) Systematische und ökologische Untersuchungen über die Diatomeenflora von Java, Bali, Sumatra. Arch Hydrobiol (Suppl.) 15 & 16

Hutchinson TC, Gizyn W, Havas M, Zobens V (1978) Effect of long-term lignite burns on Arctic ecosystems at the Smoking Hills, N.W.T. In: Hemphill DD (ed) *Trace Substances in Environmental Health.* Columbus, MO: University of Missouri Press, pp. 317–332

Ide FP (1967) Effects of forest spraying with DDT on aquatic insects of salmon streams in New Brunswick. J Fish Res Board Can 24:769–805

Jensen K, Snekvik E (1972) Low pH levels wipe out salmon and trout populations in southernmost Norway. Ambio 1:223–225

Johannson S, Larsson O, Boehm P (1980) The *Tsesis* oil spill. Mar Poll Bull 11:284–293

Jones JR, Tracyu BH, Sebaugh JL, Hazelwood DH, Smart MM (1981) Biotic index tested for ability to assess water quality of Missouri Ozark streams. Trans Amer Fisheries Soc 110:627–637

Kaesler RL, Herricks EE, Crossman JS (1978) Use of indices of diversity and hierarchical diversity in stream surveys. Amer Soc Testing Materials Spec Tech Publ 652:92–112

Karr JR, Fausch KD, Angermeier PL, Yant PR, Schlosser IJ (1986) Assessing biological integrity in running waters: A method and its rationale. Illinois Natural History Survey Special Pub No 5. Champaign, ILL, 28 pp.

Keenleyside MHA (1967) Effects of forest spraying with DDT in New Brunswick on food of young Atlantic salmon. J Fish Res Bd Can 24:807–822

Kelso JRM, Love RJ, Lipsit JH, Dermott R (1982) Chemical and biological status of headwater lakes in the Sault Ste. Marie District, Ontario. In: d'Itri FM (ed) *Acid Precipitation: Effects on Ecological Systems.* Ann Arbor, MI: Ann Arbor Science, pp. 165–208

Kenlan KH, Jacobson GL Jr, Brakke DF (1984) Aquatic macrophytes and pH as controls of diversity for littoral cladocerans. In: Hendrey GR (ed) *Early Biotic Responses to Advancing Lake Acidification.* Boston: Butterworth Publishers, pp. 63–84

Kibby H, Glass N (1980) Evaluating the evaluations: A review perspective on environmental impact assessment. In: *Biological Evaluation of Environmental Impacts: The Proceedings of a Symposium.* Council on Environmental Quality and U.S. Dept. of Interior, Fish and Wildlife Service. FWS/OBS-80/26, pp. 40–48

Kimball KD, Levin SA (1985) Limitations of laboratory bioassays: the need for ecosystem-level testing. BioScience 35:165–171

Kineman JJ, Elmgren R, Hansson S (eds) (1980) *The Tsesis Oil Spill.* Washington DC: U.S. Dept Commerce, N.O.A.A., 296 pp.

Klaverkamp JF, Hodgins DA, Lutz A (1983) Selenite toxicity and mercury-selenium interactions in juvenile fish. Arch Environ Contam Toxicol 12:405–413

Kolbe RW (1927) Zur Ökologie, Morphologie, und Systematik der Brackwasser-Diatomeen. Pflanzenforschung (Jena) 7:1–146

Kolkwitz R, Marsson M (1908) Ökologie der pflanzlichen Saprobien. Ber Dtsch Bot Ges 26a:505–519

Kuiper J, Brockmann UH, van het Groenewoud H, Hoornsman G, Hammer KD (1983) Influences of bag dimensions on the development of enclosed plankton communities during POSER. Mar Ecol Prog Ser 14:9–17

Kwiatkowski RE, Roff JC (1976) Effects of acidity on the phytoplankton and primary productivity of selected northern Ontario lakes. Can J Bot 54:2546–2561

Langlois G, Vigneault Y, Desilets L, Nadeau A, Lachance M (1983) Evaluation des effects de l'acidification sur la physico-chimie et la biologie des lacs du bouclier canadien (Québec). Can J Fish Aquat Sciences Tech Report No 1233. Ministère des Péches et des Océans, Québec, Canada,129 pp.

Larsen PF, Johnson AC, Doggett LF (1983) Environmental benchmark studies in Casco Bay - Portland Harbor, Maine. NOAA Tech Memorandum NMFS-F/NEC-19.

Lee RF, Takahashi M, Beers JR, Thomas WH, Borbet DIR, Koeller P, Green DR (1977) Controlled ecosystems: their use in the study of the effects of petroleum hydrocarbons on plankton. In: Vernberg FJ, Calabrese A, Thurlberg FP, Vernberg WB (eds) The Physiological Responses of Marine Biota to Pollutants. Proc. of a symposium sponsored by Middle Atlantic Coastal Fisheries Center, National Marine Fisheries Service, and Belle W. Baruch Institute for Marine Biology and Coastal Research, University of South Carolina. New York: Academic Press, pp. 323–342

Leivestad H, Muniz IP (1976) Fish kill at low pH in a Norwegian river. Nature 259:391–392

Leivestad H, Hendrey G, Muniz IP, Snekvik E (1976) Effects of acid precipitation on freshwater organisms. In: Braekke FH (ed) *Impact of Acid Precipitation on Forest and Freshwater Ecosystems in Norway.* Research Report 6/76, the SNSF project, pp. 87–111

Lessmark O, Thornelof E (1986) Liming in Sweden. Water Air Soil Poll 31:809–815

Levine SN, Lewis WM Jr (1985) The horizontal hetergeneity of nitrogen fixation in Lake Valencia, Venezuela. Limnol Ocean 30:1240–1245

Limburg KE, Levin SA, Harwell CC (1986) Ecology and estuarine impact assessment: lessons learned from the Hudson River (U.S.A.) and other estuarine experiences. J Environm Manage 22:255–280

Mackay RJ, Kersey KE (1985) A preliminary study of aquatic insect communities and leaf decomposition in acid streams near Dorset, Ontario. Hydrobiologia 122:3–11

Mackenthun KM (1969) The Practice of Water Pollution Biology. Washington, DC: U.S. Dept. Interior, Fed. Water Poll. Control Admin, 281 pp.

Malley DF, Chang PSS (1981) Response of zooplankton in PreCambrian shield lakes to whole-lake chemical modifications causing pH change. In: Restoration of Lakes and Inland Waters: Proceedings of an International Symposium. Portland, Maine. EPA 440/5–81–010, pp. 108–114

Malley DF, Findlay DL, Chang PSS (1981) Ecological effects of acid precipitation on zooplankton. In: d'Itri FM (ed) Acid Precipitation: Effects on Ecological Systems. Ann Arbor, MI: Ann Arbor Science, pp. 297–327

Malley DF, Chang PSS (1986).Increase in the abundance of cladocera at pH 5.1 in experimentally-acidified Lake 223, Experimental Lakes Area, Ontario. Water Air Soil Poll 30:629–638

Mankin JB, O'Neill RV, Shugart HH, Rust BW (1975) The importance of validation in ecosystem analysis. In: New Directions in the Analysis of Ecological Systems. La Jolla, CA: Simulation Councils Proceedings Series, Simulation Councils Inc, pp. 63–71

Marshall JS, Mellinger DL (1978) An in situ study of cadmium stress in a natural zooplankton community. In: Thorp JH, Gibbons JW (eds) Energy and Environmental Stress in Aquatic Systems. U.S. Dept. Energy Technical Information Center, pp. 316–330

Marshall JS, Parker JJ, Mellinger DL, Lawrence SG (1981) An in-situ study of cadmium and mercury stress in plankton community of Lake 382, Experimental Lakes Area, northwestern Ontario. Can J Fish Aquat Sci 38:1209–1214

Martin HC (ed) (1987) Acid Precipitation. Boston: D. Reidel Publishing Company, Part 1053 pp.; Part 2: 1118 pp.

Matthews RA, Buikema AL Jr, Cairns J Jr, Rodgers JH Jr(1982) Biological monitoring. Part IIA - Receiving system functional methods, relationships and indices. Water Research 16:129–139

Meriläinen J (1967) The diatom flora and the hydrogen-ion concentration of water. Ann Bot Fenn 4:51–58

Mierle G, Clark K, France R(1986) The impact of acidification on aquatic biota in North America: A comparison of field and laboratory results. Water Air Soil Poll 31:593–604

Mills KH (1984) Fish population responses to experimental acidification of a small Ontario lake. In: Hendrey GR (ed) Early Biotic Responses to Advancing Lake Acidification. Boston: Butterworth Publishers, pp. 117–132

Muller P (1980) Effects of artificial acidification on the growth of periphyton. Can J Fish Aquat Sci 37:355–363

Murphy LS, Haugen EM, Brown JF (1983) Phytoplankton: comparison of laboratory bioassay and field measurements. Wastes Ocean 1:219–233

Murphy PM (1978) The temporal variability in biotic indices. Environ Poll 17:227–236

Myren RT, Pella JJ (1977) Natural variability in distribution of an intertidal population of *Macoma balthica* subject to potential oil pollution at Port Valdez, Alaska. Marine Biology 41:371–382

Nero RW, Schindler DW (1983) Decline of *Mysis relicta* during the acidification of Lake 223. Can J Fish Aquat Sci 40:1905–1911

Nilssen JP (1980) Early warning signals of acidification (Abstr). In: Drabløs D, Tollen A (eds) *Ecological Impact of Acid Precipitation.* Oslo, Norway: Proc Int Conf SNSF Proj, p. 344

Nyberg P, Appelberg M, Degerman E (1986) Effects of liming on crayfish and fish in Sweden. Water Air Soil Poll 31:669–687

Nygaard G (1956) Ancient and recent flora of diatoms and chrysophyceae in Lake Gribsø. In: Berg K, Petersen IC (eds) *Studies on the Humic, Acid Lake Gribsø.* Fol Limnol Scand 8:32–94

O'Connor JS, Dewling RT (1986) Indices of marine degradation, their utility. Environ Manage 10:335–343

Økland KA (1969) On the distribution and ecology of *Gammarus lacustris* G.O. Sars in Norway, with notes on its morphology and biology. Norway J Zool 17:111–152

Økland KA (1980) Mussels and crustaceans: studies of 1000 lakes in Norway. In: Drabløs D, Tollen A (eds) *Ecological Impact of Acid Precipitation.* Oslo, Norway: Proc Int Conf SNSF Proj, pp. 324–325

Økland J (1969) Distribution and ecology of the fresh-water snails (Gastropoda) of Norway. Malacologia 9:143–151

Økland J (1980) Environment and snails (Gastropoda): studies of 1000 lakes in Norway. In: Drabløs D, Tollen A (eds), *Ecological Impact of Acid Precipitation.* Oslo, Norway: Proc Int Conf SNSF Proj, pp. 322–323

Økland J , Økland KA (1986) The effects of acid deposition on benthic animals in lakes and streams. Experientia (Basel) 42:471–486

Patrick R (1949) A proposed biological measure of stream conditions based on a survey of the Conestoga Basin, Lancaster County, Pennsylvania. Proc Acad Nat Sci Philadelphia 101:277–341

Patrick R (1950) Biological measure of stream conditions. Sewage Indust Wastes 22:926–938

Patrick R (1955) Diatoms as an indication of river change. Proceedings 9th Industrial Waste Conference, Purdue University Engineering Extension Service 87:325–330

Patrick R (1968) The structure of diatom communities in similar ecological conditions. Amer Nat 102:173–183

Patrick R (1972) Aquatic communities as indices of pollution. In: Thomas WA (ed) *Indicators of Environmental Quality.* New York: Plenum Press, pp. 93–100

Patrick R, Cairns J Jr, Scheier A (1968a) The relative sensitivity of diatoms, snails, and fish to twenty common constituents of industrial wastes. Progressive Fish-Culturist 30:137–140

Patrick R, Roberts W, Davis B (1968b) The effect of changes in pH on the structure of diatom communities. Not Naturalae, Acad Nat Sci, Philadelphia. No. 415, 16 pp.

Patten BC, Webb K, Boynton W, Browder J, Forster W, Grose P, Kitchens W, Koons CB, Kremer J, Scavia D, Schneider ED, Schroeder P, Taft J (1983) Ecosystem modelling as an environmental management tool. In: Turgeon KW (ed) *Marine Ecosystem Modelling*. Proceedings from a workshop held April 6–8, 1982, Frederick, MD: U.S. Dept Commerce, N.O.A.A., pp. 243–255

Pauwels SJ, Haines TA (1986) Fish species distribution in relation to water chemistry in selected Maine lakes. Water Air Soil Poll 30:477–488

Pearson TH (1971) The benthic ecology of Loch Linnle and Loch Fil, a sea–loch system on the west coast of Scotland. III. The effect on the benthic fauna of the introduction of pulp mill effluent. J Exp Mar Biol Ecol 6:211–233

Pearson TH (1975) The benthic ecology of Loch Linnhe and Loch Eil, a sea-loch system on the west coast of Scotland. IV. Changes in the benthic fauna attributable to organic enrichment. J Exp Mar Biol Ecol 20:1–41

Pearson TH, Rosenberg R (1978) Macrobenthic succession in relation to organic enrichment and pollution of the marine environment. Oceanogr Mar Biol Ann Rev 16:229–311

Pearson TH, Gray JS, Johannessen PJ (1982) Objective selection of sensitive species indicative of pollution-induced changes in benthic communities. 2. Data analyses. Mar Ecol Prog Ser 12:237–255

Peet RK (1974) The measurement of species diversity. Ann Rev Ecol Syst 5:285–307

Perry JA, Ward RC, Loftis JC (1984) Survey of state water quality monitoring programs. Environ Manage 8:21–26

Raddum GG (1980) Comparison of benthic invertebrates in lakes with different acidity. In: Drabløs D, Tollen A (eds) *The Ecological Impact of Acid Precipitation*. Oslo, Norway: Proc Int Conf SNSF Proj, pp. 330–331

Raddum GG, Brettum P, Matzow D, Nilssen JP, Skov A, Svealv T, Wright RF (1986) Liming the acid Lake Hovvatn, Norway: A whole-ecosystem study. Water Air Soil Poll 31:721–763

Raddum GG, Fjellheim A, Hesthagen T (In press) Monitoring of acidification through the use of aquatic organisms. Verh Internat Verin Limnol

Reeve MR, Grice GD, Gibson VR, Walter MA, Darcy K, Ikeda T (1976) A controlled environmental pollution experiment (CEPEX) and its usefulness in the study of large marine zooplankton under toxic stress. In: Lockwood APM (ed) *Effects of Pollutants on Aquatic Organisms*. New York: Cambridge University Press, pp. 145–162

Renberg I, Hellberg T (1982) The pH history of lakes in southwestern Sweden, as calculated from the subfossil diatom flora of the sediments. Ambio 11:30–33

Resh VH, Unzicker JD (1975) Water quality monitoring and aquatic organisms: the importance of species identification. J Water Poll Control Fed 47:9–19

Rhea JR, Malanchuk JL (1985) Simulation of toxicant effects on aquatic ecosystem properties. ISEM Journal 7:71–111

Roback SS (1974) Insects (Arthropoda: Insecta). In: Hart CW Jr, Fuller SLH (eds) *Pollution Ecology of Freshwater Invertebrates*. New York: Academic Press, pp. 313–376

Robinson JV, Sandgren CD (1984) An experimental evaluation of diversity indices as environmental discriminators. Hydrobiologia 108:187–196

Roff JC, Kwiatkowski RE (1977) Zooplankton and zoobenthos communities of selected northern Ontario lakes of different acidities. Can J Zool 55:899–911

Rosenberg DM, Wiens AP (1976) Community and species responses of Chironomidae (Diptera) to contamination of fresh waters by crude oil and petroleum products, with special reference to the Trail River, Northwest Territories. J Fish Res Board Can 33:1955–1963

Rosenberg DM, Resh VH, Balling SS, Barnaby MA, Collins JN, Durbin DV, Flynn TS, Hart DD, Lamberti GA, McElravy EP, Wood JR, Blank TE, Schultz DM, Marrin DL, Price DG (1981) Recent trends in environmental impact assessment. Can J Fish Aquat Sci 38:591–624

Rosenberg R (1972) Benthic faunal recovery in a Swedish fjord following closure of a sulphite pulp mill. Oikos 23:92–108

Sanders HL (1968) Marine benthic diversity: a comparative study. Amer Nat 102:243–282

Scheffe RD, Booty WG, DePinto JV (1986) Development of methodology for predicting reacidification of calcium carbonate treated lakes. Water Air Soil Poll 31:857–864

Scheider W, Dillon PJ (1976) Neutralization and fertilization of acidified lakes near Sudbury, Ontario. Proc. 11th Canadian Symposium. Water Poll Research Can 11:93–100

Schindler DW (1974) Eutrophication and recovery in experimental lakes: implications for lake management. Science 194:897–899

Schindler DW (1976) The impact statement boondoggle. Science 192:509

Schindler DW (1980) Experimental acidification of a whole lake: a test of the oligotrophication hypothesis. In: Drabløs D, Tollan A (eds) *The Ecological Impact of Acid Precipitation*. Oslo, Norway: Proc Int Conf SNSF Prog, pp. 370–374

Schindler DW, Mills KH, Malley DF, Findlay DL, Shearer JA, Davies IJ, Turner MA, Linsey GA, Cruikshank DR (1985) Long-term ecosystem stress: the effects of years of experimental acidification on a small lake. Science 228:1395–1401

Schofield C (1976) Acid precipitation: effects on fish. Ambio 5:228–230

Seliga TA, Dochinger LS (1976) *First International Symposium on Acid Precipitation and the Forest Ecosystem*. Water Air Soil Poll 6 (No. 2, 3, 4):135–514

Shannon CE, Weaver W (1949) *The Mathematical Theory of Communication*. Urbana, IL: University of Illinois Press

Sheath RG, Havas M, Hellebust JA, Hutchinson TC (1982) Effects of long-term natural acidification on the algal communities of tundra ponds at the Smoking Hills, N.W.T., Canada. Can J Bot 60:58–72

Sheehan PJ (1984) Effects on community and ecosystem structure and function. In: Sheehan PJ, Miller DR, Butler GC, Bourdeau P (eds) *Effects of Pollutants at the Ecosystem Level. Scope 22*. New York: Wiley, pp. 51–99

Sheehan PJ, Winner RW (1984) Comparison of gradient studies in heavy-metal-polluted streams. In: Sheehan PJ, Miller DR, Butler GC, Bourdeau P (eds) *Effects of Pollutants at the Ecosystem Level. Scope 22*. New York: Wiley, pp. 255–271

Sigmon CF, Kania HJ, Beyers RH (1977) Reductions in biomass and diversity resulting from exposure to mercury in artificial streams. J Fish Res Board Can 34:493–500

Singer R (ed) (1981) *The Effects of Acidic Precipitation on Benthos*. Proc Symp N Amer Benthological Soc, New York: Canterbury Press, 154 pp.

Smith DL, Underwood JK, Ogden JG III, Sabean BC (1986) Fish species distribution and water chemistry in Nova Scotia lakes. Water Air Soil Poll 30:489–496

Smol JP (1986) Chrysophycean microfossils as indicators of lakewater pH. In: Smol JP, Battarbee RW, Davis RB, Meriläinen J (eds) *Diatoms and Lake Acidity.* Developments in Hydrobiology No. 29. Boston: Dr W. Junk, pp. 275–287

Smol JP, Dickman MD (1981) The recent histories of three Canadian Shield lakes: a paleolimnological experiment. Arch Hydrobiol 93:83–108

Smol JP, Charles DF, Whitehead DR (1984) Mallomonadacean microfossils provide evidence of recent lake acidification. Nature 307:628–630

Smol JP, Battarbee RW, Davis RB, Meriläinen J (eds) (1986) *Diatoms and Lake Acidity.* Developments in Hydrobiology No. 29. Boston: Dr W. Junk, 307 pp.

Sprules W (1975) Midsummer crustacean zooplankton communities in acid-stressed lakes. J Fish Res Board Can 32:389–395

Stephenson M, Mackie GL (1986) Lake acidification as a limiting factor in the distribution of the freshwater amphipod *Hyalella azteca.* Can J Fish Aquat Sci 43:288–292

Stokes PM (1981) Benthic algal communities in acidic lakes. In: Singer R (ed) *The Effects of Acidic Precipitation on Benthos.* Proc Symp N Amer Benthological Soc, New York: Canterbury Press, pp. 119–138

Stokes PM (1984a) pH related changes in attached algal communities of softwater lakes. In: Hendrey GR (ed) *Early Biotic Responses to Advancing Lake Acidification.* Boston: Butterworth Publishers, pp. 43–62

Stokes P (1984b) Clearwater Lake: study of an acidified lake ecosystem. In: Sheehan PJ, Miller DR, Butler GC, Bourdeau P (eds) *Effects of Pollutants at the Ecosystem Level.* Scope 22. New York: Wiley, pp. 229–253

Sutcliffe D, Carrick T (1973) Studies on mountain streams in the English Lake District. I. pH, calcium, and the distribution of invertebrates in the River Duddon. Freshwater Biol 3:437–462

Taub FB (1983) Synthetic microcosms as biological models of algal communities. In: Shubert LE (ed) *Algae as Ecological Indicators.* New York: Academic Press, pp. 363–394

Teal JM, Howarth RW (1984) Oil spill studies: a review of ecological effects. J Environ Manage 8:27–44

Tolonen K, Liukkonen M, Harjula R, Patila A (1986) Acidification of small lakes in Finland documented by sedimentary diatom and chrysophycean remains. In: Smol JP, Battarbee RW, Davis RB, Meriläinen J (eds) *Diatoms and Lake Acidification.* Boston: Dr W. Junk, pp. 169–200

Turgeon KW (ed) (1983) *Marine Ecosystem Modelling.* Proceedings from a workshop held April 6–8, 1982, Frederick, Md. U.S. Dept. Commerce, NOAA, 274 pp.

Wagemann R, Snow NB, Rosenberg DM, Lutz A (1978) Arsenic in sediments, water, and aquatic biota from lakes in the vicinity of Yellowknife, Northwest Territories, Canada. Arch Environ Contam Toxicol 7:169–191

Warren CE (1971) *Biology and Water Pollution Control.* Philadelphia: WB Saunders, 434 pp.

Washington HG (1984) Diversity, biotic and similarity indices: A review with special relevance to aquatic ecosystems. Water Res 18:653–694

Wiederholm T, Eriksson L (1977) Benthos of an acid lake. Oikos 29:261–267

Wiener JG, Rago PJ, Eilers JM (1984) Species composition of fish communities in northern Wisconsin lakes: relation to pH. In: Hendrey GR (ed) *Early Biotic Responses to Advancing Lake Acidification,* vol. 6, Acid Precipitation Series. Boston: Butterworth Publishers, pp. 133–146

Wilhm JS, Dorris TC (1968) Biological parameters for water quality criteria. BioScience 18:477–481

Winget RN, Mangum FA (1979) Biotic condition index: integrated biological, physical and chemical stream parameters for management. U.S. Forest Service Intermountain Region, 51 pp.

Winner RW, Dyke JSV, Caris N, Farrel MP (1975) Response of the macroinvertebrate fauna to a copper gradient in an experimentally-polluted stream. Verh Internat Verein Limnol 19:2121–2127

Winner RW, Boesel MW, Farrell MP (1980) Insect community structure as an index of heavy-metal pollution in lotic ecosystems. Can J Fish Aquat Sciences 37:647–655

Wright RF, Cosby BJ, Hornberger GM, Galloway JN (1986) Comparison of paleolimnological with MAGIC model reconstructions of water acidification. Water Air Soil Poll 30:367–380

Wurtz CB (1955) Stream biota and stream pollution. Sewage Indust Wastes 27:1270–1278

Yan ND, Stokes P (1978) Phytoplankton of an acidic lake, and its responses to experimental alterations of pH. Environm Conserv 5:93–100

Young MW, Young DK (1982) Marine macrobenthos as indicators of environmental stress. In: Mayer GF (ed) *Ecological Stress and the New York Bight: Science and Management.* Columbia, SC: Estuarine Research Foundation, pp. 527–539

Zand SM (1976) Indexes associated with information theory in water quality. J Water Poll Control Fed 48:2020–2031

Zischke JA, Arthur JW, Nordlie KJ, Hermanutz RO, Standen DA, Henry TP (1983) Acidification effects on macroinvertebrates and fathead minnows (*Pimephales promelas*) in outdoor experimental channels. Water Res 17:47–63

Chapter 6

Theoretical and Methodological Reasons for Variability in the Responses of Aquatic Ecosystem Processes to Chemical Stresses

Suzanne N. Levine[1]

6.1 The Global Significance of Ecosystem Processes and Chemical Stresses

The interdependency of environment and organism has been a unifying theme of ecology for a long time (Forbes 1887; Tansley 1935; Evans 1956), but for many years research activities necessarily emphasized one half of the interaction, the impact of the abiotic on the biotic elements of an ecosystem. Redfield's seminal papers on the probable role of plankton in regulating seawater composition (Redfield 1934, 1958) provoked serious discussion of the possible influence of life on the inorganic world, but ecologists could not quantify most processes in ecosystems until methods were developed for labeling elements and observing their environmental movement. The advent of radiotracer and stable-isotope tracer technology following World War II was, therefore, a tremendous stimulus for biogeochemistry, and has resulted in the discovery that the cycling of many elements is influenced by organisms. These elements include the twenty or so nutrients required by primary producers (e.g., carbon, nitrogen, phosphorus, sulfur, and iron) and those elements that organisms confuse with nutrients because they have similar chemistries (e.g., arsenic, which resembles phosphorus; and cadmium, which resembles iron).

The oceans, atmosphere, and soils of our planet are all in states of chemical nonequilibrium, probably because organisms transport elements against chemical gradients and with enzymes catalyze reactions that would otherwise be sluggish (Crocker 1952; Broecker 1974; Lovelock

[1] Ecosystems Research Center, Cornell University, Ithaca, New York 14853

1979). Organisms are capable of these feats because many of them have developed the ability to trap solar energy or the chemical energy available in certain inorganic compounds, and store this energy in accessible biochemical bonds. Others obtain energy indirectly through food chains based on autotrophs and their detritus. The geological record indicates that biogeochemical cycles have existed on Earth for at least two billion years (Schopf 1982; Stanley 1986). These cycles have not been consistent in character, however; they have been transformed continually by the evolution of life (Schopf 1982; Stanley 1986). Undoubtedly, natural changes in biogeochemical cycles continue, albeit very slowly.

The potential impact of environmentally destructive human activities on element cycles and energy flow is a current topic of study and debate among ecologists. Of special concern is the release into the environment of synthetic and mined chemicals. The number of chemicals that humans use regularly is large, roughly 63,000, according to Maugh (1983). Chemical exposure is not a new variable for organisms; natural chemicals number about four million (Maugh 1983), and at least some of these compounds (e.g., antibiotics, siderophores, which chelate iron, and phosphatases) modify the suitability of the environment for biological processes. However, anthropogenic activities often introduce chemicals more rapidly, over larger areas, and at greater concentrations than natural processes do. In addition, many synthetic chemicals are xenobiotic (totally foreign to organisms) and may require that organisms develop new biochemical pathways to tolerate their presence. In this chapter, the term *chemical stress* is used to refer to chemical exposures above or below the range of concentrations to which a biological community is adapted. This definition is one of two common in the literature (Barrett and Rosenberg 1981). Sometimes the term is applied when referring to the degraded physiological state of a biological system that may result from its response to a chemical, rather than to the chemical exposure itself. The advantage of the definition chosen here is that it implicates an independent, forcing function rather than a dependent variable. A forcing function may be measured more directly and usually more easily than a syndrome with multiple indicators.

Chemical stresses may affect biogeochemical cycles and energy networks through their influences on the efficiencies and abundances of the organisms performing the various processes, or by altering the availability of process substrates. As indirect consequences, pool sizes and the relative importance of cycling pathways within the networks may be altered. The prospect of changes in important ecosystem processes is alarming, as we can only guess at the ultimate consequences for the structure and composition of animal and plant communities and for ecosystem fertility as a whole. The interlinking of ecosystem processes greatly complicates prediction of these changes. Although chemical stresses are expected to impinge primarily on the local ecosystem or landscape, global consequences such as the accumulation of carbon dioxide, methane, and nitrous oxide in the atmosphere are being documented (Brewer 1978; Weiss 1981). For widespread environmental perturbations, there is a greater concern that the Earth's ability to support life might in some way be degraded.

A stark example of the complex repercussions that may result when anthropogenic chemicals affect ecosystem function has been provided by the eutrophication of countless lakes, rivers, and estuaries in response to domestic wastewater inputs. The presence of nutrients (most importantly, phosphorus and nitrogen) in wastewater often enhances primary production in the receiving water bodies. As a result, secondary production and decomposition rates increase, and changes in species composition may occur at every trophic level. These characteristics of eutrophication are well known (Wetzel 1983). Less frequently discussed are the consequences of increased ecosystem productivity for aquatic biogeochemistry. Greater organic synthesis entails not only more carbon fixation but greater assimilation of all nutrients (Schelske and Stoermer 1971; Lean et al. 1978; Persson et al. 1977; Schindler 1985). Thus, the particulate and the dissolved organic pools of most nutrients increase during eutrophication, while pools of inorganic ions and gases may diminish in size (Schindler 1985). With changes in pool sizes come changes in the relative importances of the different pathways within the cycles. For example, sedimentation becomes an increasingly important sink for nutrients as the ratio of particulate to dissolved nutrient increases (Schindler 1985).

Eutrophication may also influence the cycling of several non-nutritive and marginally bioavailable elements. The elements affected are those whose ionic forms are strongly influenced by redox conditions (i.e., by electron activity, which is estimated as redox potential). During eutrophication, redox potentials in the deeper layers of stratified waters and in sediments often decline dramatically as the detritus load to these regions increases and raises the oxygen demand of decomposers (see Wetzel 1983). Because the various ionic forms of elements differ in their reactivities (Stuum and Morgan 1970), the structures of the cycles of the redox-sensitive elements undergo transformations as oxygen depletion sets in.

Although it is a serious environmental problem, eutrophication may be unrepresentative as an example of the potential effects of chemical stresses on ecosystem functions. It is an unusually strong disturbance because it directly attacks a key ecosystem process, primary productivity, through the manipulation of a limiting factor. Few ecosystem processes have as many links to other processes as does primary productivity and most anthropogenically produced chemicals are not nutrients, much less limiting nutrients.

6.2 The Detection of Ecosystem Responses to Stress

An important managerial issue for ecology has been whether traditional monitoring programs, which emphasize biological species composition and simple metabolic indicators such as dissolved oxygen concentration and pH, can be relied upon to detect major ecosystem responses to chemical stress. Measures related to ecosystem structure (i.e., to the relative abundances of biological species and chemical pools and their distribution in space) are usually less costly and require less technical skill than measurements of process rates. Yet one can justify neglecting functional measures only if notable modifications of ecosystem structure usually precede or at least coincide with alterations of ecosystem functions. If this is not the case, process damage may be incurred before a perturbation is detected by changes in the population densities of indicator species, a decline in species diversity, or other popular structural measures.

Early papers on the expected effects of stress on ecosystems (e.g., O'Neill et al. 1977; Van Voris et al. 1980; Bormann 1983; Rapport et al. 1985; Odum 1985) frequently contended that significant changes in ecosystem functions ought to occur early in a perturbation. The principal argument was that ecosystem processes are large-scale manifestations of biochemistry and physiology, and changes in biochemistry and physiology should precede or coincide with reduced growth rates and mortality (changes that lead to modifications of ecosystem structure). In corroboration with this hypothesis, forest ecologists reported that during exposures to air pollutants of from several months to a few years, the species composition of tree and shrub communities changed very little. Over the same periods, the leaching of nutrients from soils clearly increased (O'Neill et al. 1977; Van Voris et al. 1980) and there were significant declines in leaf litter decomposition rates (Jackson and Watson 1977) and in the primary productivity of conifers, as manifested by stunted needles and premature needle loss (Williams 1980; Mann et al. 1980).

More recently, Schindler (1987) has argued that the sequences of functional and structural responses to stress that are observed in forests are not typical for most ecosystems but result from the very large sizes, low reproductive rates, and low dispersal capabilities of trees. Many terrestrial plant communities are able to change species composition much more quickly than forests, over periods of weeks to months rather than over decades, and the phytoplankton communities common to aquatic ecosystems often undergo significant structural changes within a time frame of days. Schindler (1987) contends that, in general, ecosystem functions should be highly resistant to chemical stresses. This is because the processes are performed by a myriad of competing species. The demise of one species, therefore, should allow enhanced growth by other less-sensitive species, which will then take on the function abandoned by the stress-affected species. Thus the rate at which a process proceeds may be maintained despite major changes in ecosystem structure. In support of his arguments, Schindler (1987) has described experiments at the Experimental Lakes Area (ELA) of Canada in which sulfuric acid was added to mesocosms and whole lakes. The reductions in pH caused by the additions led to major changes in the species composition of both the plant and animal communities, but had little or no effect on primary productivity, decomposition, or the relative importance of nutrient pools.

Schindler's concepts are logical; yet there is some question as to how accurate they are as generalities. There are many examples of ecosystem process changes under stress, even for aquatic systems. The author of this chapter recently examined about 20 aquatic ecology and toxicology journals in search of articles on the effects of chemical stresses on aquatic ecosystem functions. Of the roughly 100 articles found, a majority reported significant decreases in ecosystem processes after the accidental or deliberate introduction of a chemical to an aquatic ecosystem. Some reported increases in process rates. Of course the generality of ecosystem process response to stress must be weighed against the reluctance of scientists, and journals, to report the results of experiments in which no effects are observed. In addition, methodological problems were apparent in many of the papers examined, and these could explain some of the results. Still it is difficult to accept that widespread findings could be totally artifactual.

In the next section of this chapter, some terms are defined that will be used throughout the remainder of the chapter. Thereafter, the methodological issues alluded to above will be discussed and, finally, new theories regarding why process rates may change under some chemical stresses and not under others will be presented. Although the examples are all from aquatic ecosystems, there is no reason why the derived principles should not apply to terrestrial ecosystems as well.

6.3 Terminology

Most of the definitions included here are original for this work, as the literature is highly erratic in its naming of the phenomena related to stress and ecosystem functioning. The term *ecosystem process* (or sometimes *ecosystem function*) is used to refer to any transfer of material or energy from one location or chemical state to another, and the material or energy being transferred is called the *commodity*. The latter term is chosen because it applies equally well to mass and energy. The definition of ecosystem process encompasses such diverse transformations as photosynthesis, herbivory, respiration, nutrient uptake into cells, and the oxidation reactions of decomposition. Although organisms mediate most ecosystem processes, this definition also allows for abiotic processes (e.g., adsorption onto particles and chemical precipitations). Some eco-system scientists apply the term ecosystem process indiscriminately to both single reactions and large complexes of reactions, such as nutrient cycling. To avoid this practice, the webs of interdependent processes (i.e., nutrient cycles and energy chains) that sequentially transform par-ticular commodities are referred to as *functional networks*.

Sometimes the organisms or abiotic complexes that collectively per-form a process will be referred to as the *functional units* and the organ-elles, biochemical pathways, and abiotic surfaces required for the process will be referred to as the *functional machinery*. These generic terms are invented to avoid encumbering the discussion of concepts by repeatedly listing species and morphological parts.

6.4 Methodological Issues

Methodological issues have complicated the interpretation of the results of ecotoxicological studies and may have contributed to the appearance of more variability in the responses of ecosystem processes to stress than actually exists. Among the more important of these issues have been: poorly defined baselines for processes, experiments that are too short in duration to show "steady-state" effects, and lack of information on the dose–response relationships between chemicals and processes.

Most of the data available on the responses of aquatic processes to chemical stresses are from field studies of polluted water bodies, in which rates that differ significantly from baseline are interpreted as responses (for a review of these studies, see Sheehan 1984). These studies have the advantage of involving natural systems and a sampling time-frame of at least a few weeks. Their principal weakness has been the assessment of baseline activity. Many have been initiated during or after pollution events and, therefore, have lacked information on conditions prior to the introduction of contaminant. Partially to compensate for this drawback, researchers may study neighboring water bodies as "control" systems, but the appropriateness of these extrapolations has rarely been evaluated. Sometimes, neither temporal nor spatial controls are available, and re-searchers rely on values in the literature as estimates of background. For any study with questionable information on baselines, the danger exists that background noise may hide a true response or that a measured change in process rate will be interpreted as a response to stress when it actually was unrelated to the stress, and would have appeared in the background if it had been adequately measured.

One clear incidence of the latter mistake occurred in the Experimental Lakes Area of Canada, where Lake 223 has been experimentally acidified since 1976. Schindler (1987) has discussed this incident as a warning to researchers that both temporal and spatial controls may be necessary during field studies. During the first six years of acid treatment, the transparency of Lake 223 increased, and Schindler and Turner (1982) reported the change as a response to acidification, involving perhaps the coagulation and precipitation of dissolved organic acids. Further analysis, however, revealed that transparency had also increased in other lakes in the area during the same period (Schindler 1987). This was associated with unusually low rainfall and thus with low organic carbon inputs from the watershed. Statistical tests suggested that the reported effect of acidity on transparency probably was real, but it was not nearly as substantial as initially perceived (Schindler 1987).

Unlike field studies, ecotoxicological experiments usually have sufficient controls to define the baseline adequately. The primary methodological pitfall of these studies seems to be a time-span that is too short for the experimental treatments. For safety and for economical reasons, ecotoxicological experiments normally are conducted in microcosms in the laboratory or, less often, in small enclosures *in situ*. To minimize the effects of enclosure, their duration is from a few minutes to a few days. Yet, time-course analyses of process responses to a chemical stress have indicated that the long-term "steady-state" responses of many processes are not achieved over the time-frame of the typical ecotoxicology experiment, but over a period of days, weeks, months, or longer (e.g., Thomas et al. 1977; Griffiths et al. 1982; Kettle and DeNoyelles 1986). Thus, researchers often report ecosystem responses to chemical stresses that are transient and that may not represent the responses that would predominate during an actual pollution event.

A variety of temporal patterns for the responses of ecosystem processes to long-term chemical stress have been observed. The most common of these is a depression of process rate immediately after chemical addition, followed by the partial or total recovery of function. An example is illustrated in Figure 6.1. Here lead and copper were added to Swedish lake sediments at doses of a few micromoles of metal per gram of dry weight of sediment (Broberg 1983a) and the electron transport system activity (ETSA) was measured. ETSA is believed to be a measure of the energy that a biological system has available for metabolic activity, and thus may reflect the status of a composite of functions. ETSA began to recover from an initial depression period of 3–4 days after the metals were introduced, although rates as high as those in control sediments were not achieved during the 11 days of exposure to copper and 287 days of exposure to lead. Larger-scale experiments involving cadmium additions to Kansas ponds (Kettle and DeNoyelles 1986) and additions of copper (10–50 μg/l) to mesocosms in Saanich Bay, B.C. (the CEPEX enclosures; Thomas et al. 1977), have shown complete recovery of pelagic primary production rates after 10 and 20 days of depressed activity, respectively. Sohacki et al. (1969) and McKnight (1981) studied freshwater reservoirs

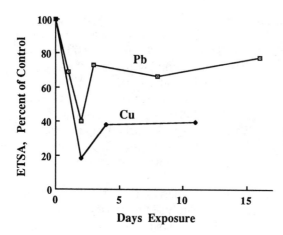

Figure 6.1. Variation of ETSA (usable energy in organisms) with time after additions of lead (22 μmoles g dwt^{-1}) and copper (43 μmoles g dwt^{-1}) to sediments from Lake Strandsjön, Sweden. Modified from Broberg (1983a).

and ponds treated with copper sulfate; they found roughly the same temporal trends for photosynthesis in these freshwater systems as in Saanich Bay.

Organisms with the ability to acclimatize may be involved in the regaining of process rate with time. Initial susceptibility of the process to the chemical probably occurs because organisms require hours or days to produce new enzymes, activate ion pumps, create isolation mechanisms, or invoke other resistance or compensatory tactics (Wood and Wang 1983). As Schindler (1987) pointed out, communities sometimes respond to stress by changing species composition, a process that requires from several hours for bacteria to several years for fish. During the CEPEX experiments with copper, changes in phytoplankton species clearly accompanied the recovery of primary productivity (Thomas and Seibert 1977). Genetic changes within populations subjected to chemical stress have also been documented (see Klerks and Levinton, this book) and may occur over temporal scales similar to those required for species changes.

Not all time-course studies, however, have documented recovery of ecosystem process rates with maintained stress; some have shown increased sensitivity to the chemical with time. For example, Griffins et al. (1982) added crude oil to sediments from off the coast of Alaska and noted that nitrogen fixation was hardly affected during the first hours of exposure. However, over the next few weeks, they measured a steady decline in nitrogen fixation rate; the long-term impact of the oil on nitrogen fixation was complete inhibition (Figure 6.2). Similar temporal trends for oil damage to nitrogen fixation have been reported by Baker and Morita (1983).

Several factors may cause time lags between chemical additions and the full negative responses of processes. Time is required for a chemical to be distributed throughout an environment. In small laboratory microcosms, this process may require several minutes, whereas in large meso-

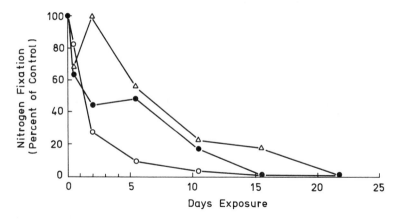

Figure 6.2. Variation of nitrogen fixation rates with time after additions of fresh crude oil (5%) to three marine sediments. Rates are expressed as percent of the activity in untreated sediments. From Griffiths et al. (1982).

cosms and lakes it may take hours or days (longer if the water is thermally stratified). In addition, organisms may be able to avoid chemicals through behavioral mechanisms such as shell closing and burrowing, or through dormancy, and these mechanisms must be superseded before effects will be noted. Some physiological impacts may occur only after bioaccumulation has taken place. Finally, the chemical's impact on a function may not be direct but rather on the production of new functional machinery. Griffiths et al. (1982) did not determine the mechanism by which oil hinders nitrogen fixation, but possible mechanisms are interference with the formation of heterocysts or with the transcription of nitrogenase.

Occasionally, the short-term effect of a chemical is negative, whereas the long-term effect is positive. For example, Griffiths et al. (1982) added crude oil (5%) to marine sediments and found that respiration was inhibited for about 5 days. Thereafter, CO_2 production was stimulated by the oil (Figure 6.3), probably because it was used as a substrate for bacterial growth.

Another factor that has influenced whether researchers have reported minimal or substantial responses to chemical stresses by ecosystem processes has been the dose of the chemical. Traditionally, the results of process studies have been reported in qualitative terms (i.e., a response does or does not occur), while the application of dose–response curves to the analysis of processes under chemical stress is a technique in its infancy. Figures 6.4 and 6.5 provide examples of how data from dose–response studies of ecosystem functions may be used to advance ecotoxicology. The first figure (from Broberg 1982) depicts the relationships between "steady-state" phosphatase activity in Swedish lake sediments and doses of zinc (Zn), lead (Pb), copper (Cu), and cadmium (Cd) (phos-

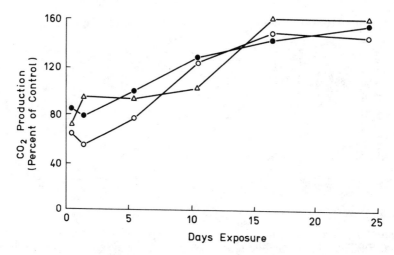

Figure 6.3. Variation in CO_2 production rates with time in three marine sediments after oil addition (5%). Rates are expressed as percent of the activity in untreated sediments. From Griffiths et al. (1982).

phatase activity was measured 75–100 days after dosing; adaptation was assumed). At concentrations of a few micromoles of metal per gram of dry sediment, all four metals degrade phosphatase activity. Cd appears to be the strongest inhibitor, Zn the weakest, and Pb and Cu intermediate in their effects. The high toxicity of Cd may be related to this element's

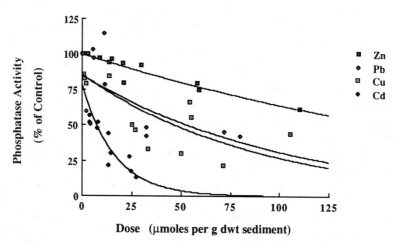

Figure 6.4. Alkaline phosphatase activity in lake sediments exposed to various doses of four different metals. The lines show exponential regressions through the data. Modified from Broberg (1982).

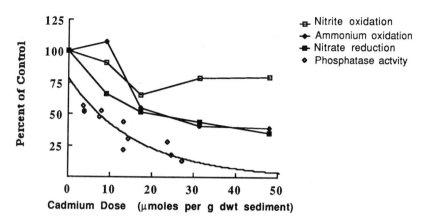

Figure 6.5. Nitrite oxidation, ammonium oxidation, nitrate reduction, and alkaline phosphatase activity in lake sediments exposed to various doses of cadmium. The line for alkaline phosphatase activity shows a regression through the data; other lines are drawn between the data points. Data from Broberg (1982, 1983b, 1984).

interference with the assimilation of iron, which is required for phosphatase activity. Information on the relative toxicities of metals can be used in establishing priorities for regulation and water treatment. The second figure, also from Broberg (1982, 1983b, 1984), compares the responses to cadmium of four different ecosystem processes (nitrite oxidation, ammonium oxidation, nitrate reduction, and phosphatase activity). The processes were measured after sediments had been incubated with Cd for about 20 days. From the curves, the relative sensitivity of each process to Cd stress can be judged: phosphatase activity is the most sensitive, and nitrite oxidation the least.

One drawback of dose–response analysis is that its interpretation requires knowledge of chemical exposure, when in reality the actual concentration reaching the organisms or abiotic units seldom is known. Part of the added chemical may be lost to sediments, the atmosphere, or other sinks, some may be transformed to another chemical form, and some may not have access to physiological sites because of competition with chemical analogs (e.g., sulfate interferes with molybdate assimilation; thus the effective concentration of molybdate may be less than the total concentration when sulfate is present). Alternatively, there may be an internal source of the tested chemical. Detailed analyses of chemical fate and transport, and estimations of internal processes are therefore necessary for good estimates of effective dose. In the absence of this information, the researcher must assume that a dose–response curve for one system may not describe the response of another system.

6.5 Mechanistic Issues

In addition to methodological reasons, there are mechanistic reasons for variability in the responses of ecosystem processes to stresses. These are related to the dual dependency of many ecosystem processes on the availability of substrates (i.e., reactants) and on the resource requirements of organisms. An ecosystem process occurs because one of two conditions are met. First, the products of the reaction may be at a lower energy state than its reactants. In this case, organisms may mediate the reaction to gain energy. The reaction also may occur spontaneously and inorganically, although often at a slower rate than when organisms are involved. Second, the process may provide organisms with a resource necessary for metabolism and growth. To gain a required and scarce resource, organisms will invest energy in an endothermic reaction. Because organisms require many nutrients, some of which may be acquired through more than one mechanism, there are a large number of biologically–mediated endothermic reactions.

An important feature of the resource-gathering activities of organisms is that they can be saturated with substrate. This characteristic stems from the fact that species require resources, and particularly nutrients, at ratios that are relatively fixed (Redfield et al. 1963) (the adjective "relatively" is important here; there is a small amount of within-species variability in intracellular nutrient ratios). The resource whose environmental availability is lowest relative to the physiological demand of a population limits the growth of that species, and thus its requirements for all other resources. Consequently, most of the resources required by organisms are supplied at rates greater than the total biological demand, and the excess is left in solution to accumulate, enter into inorganic reactions, or flush out of the ecosystem. Only for a few nutrients do organisms spend energy on luxury uptake (i.e., the incorporation of nutrient much beyond immediate needs). These nutrients (most notably phosphate) are those ones that often have been limiting during the course of species evolution and are stored as a hedge against the possibility that they will limit growth later in the lifetimes of the organisms or during the next generation. In general, luxury consumption is not sufficient to deplete the environmental pools of a non-limiting resource.

The resource that limits growth of a particular species, and thus places limits on species use of other nutrients, may vary among water bodies, or even within a single water body with time (e.g., season), with depth, or among water masses. In aquatic ecosystems, there is a tendency for many species to be limited by the same resource, usually phosphorus in freshwater and nitrogen in estuarine and coastal ecosystems. (However, during general limitation by one nutrient, some species may be limited by another resource; for example, in mid-summer, diatoms are often limited by silica rather than by phosphorus or nitrogen [Kilham 1971].)

That assimilatory processes follow saturation kinetics as a function of substrate supply is old news to environmental microbiologists and phycologists, who have developed many sets of equations (most importantly, the Michaelis-Menten and Droop equations) to describe the relationships between nutrient uptake rates and substrate concentrations (e.g., see Reynolds 1984). Unfortunately, their equations were designed for chemostats and cannot be applied without modification to natural ecosystems. The Michaelis-Menten equations assume that the concentration of substrate in the medium is an accurate measure of substrate availability. This assumption is often fair enough for closed cultures, but breaks down when applied to ecosystems, where there may be substantial sources and sinks of substrate, and where steady-state conditions rarely apply. In the environment, substrate pools are residual quantities, representing the balance between inputs and outputs at the moment. A dissolved substrate pool may include a large portion of the amount of substrate available to organisms over the course of a day, or a very small portion. There is no way of knowing the relative importance to organisms of pools and inputs without measuring both. The Droop equations use cellular concentrations as estimates of nutrient availability, but this is a rather indirect measure of what should be a forcing function.

To explore the relationships between ecosystem processes and chemical stresses, it is easier to begin with first principles than to modify the Michaelis-Menten or Droop equations to accommodate inputs and outputs. This is what has been done in the next few paragraphs. It should be noted, however, that when one transforms the Michaelis-Menten equations so that substrate availability is given as supply rates (S, with the units mass time^{-1}) rather than as concentrations, the relationships that result are consistent with the conclusions that follow. The primary difference between concentration-based and flux-based equations for assimilatory functions is that while the former predict that at some fixed substrate concentration, process rate will be limited simultaneously by substrate availability and by physiological capacity, the latter do not. Flux-oriented saturation kinetics suggest that at a given time, all ecosystem processes are either capacity limited or substrate limited.

6.5.1 Characterization of the Two Limitation States: Capacity Limitation and Substrate Limitation

Capacity limitation is perhaps the easier sort of limitation to understand. The capacity of an individual for a particular function is the maximum rate that it can achieve in its environment if substrate supply is unlimited. Thus, an ecosystem's total capacity (C) for a biologically mediated process (i.e., the maximum amount of substrate that it can transform per unit time) is a function of both the numbers of organisms participating in the process and the maximum processing rates of these individuals. Of course, considering all individuals is untenable for ecosystem analysis. It is more pragmatic to assume that the functional capacities of individuals are fairly well represented by the means for the species to which they belong (the specific functional capacities). Thus, C may be estimated as $\sum_{i=1}^{n} a_i c_i$, where n is the total number of species present, and a_i and c_i are the abundance and mean functional capacity of species i. For those relatively few ecosystem processes that are not biologically mediated, types of abiotic units (e.g., clay versus detritus) may be substituted for species in the above equation.

For organisms, population densities (the a_i) are related to the availability of limiting resources (which may differ from the substrate in question) and to mortality (e.g., predation and sinking), while specific functional capacities are determined jointly by genetics and environmental history (Figure 6.6). Genetics specify the types of functional machinery that can be produced, while environmental conditions influence gene expression.

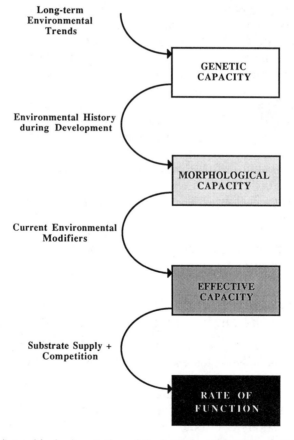

Figure 6.6. A graphical presentation of the interplay of genetics and the environment in determining the rate at which an organism performs a function.

Examples of the effects of environmental history on the morphology and physiology of phytoplankton include the control of chlorophyll *a* content by ambient light intensities (Reynolds 1984), the production of extracellular alkaline phosphatase when dissolved inorganic phosphate has been scarce but not when it is common (Healey 1973), and nitrate reductase production only when nitrate is available and ammonium is not (Morris and Syrett 1963).

A process is capacity limited when its substrate is supplied rapidly enough so that every functional unit (i.e., every organism or, if the reaction is not biologically mediated, every abiotic unit) has enough substrate available to it to saturate its functional machinery (i.e, its biochemical and physiological pathways, etc.): i.e., when $S \geq C$. The existing machinery runs at full tilt ($R = C$; where R, with units mass time^{-1}, is process rate), while additional machinery is not producible because of other limitations on populations sizes and physiology. For example, during nitrogen fixation, access to phosphorus may limit the abundance of heterocystous blue-green algae, and cool temperatures may constrain their fixation rates. The process substrate, nitrogen gas, is so plentiful that fluctuations in its concentration have no effect on fixation rates.

Another example of a capacity limited process is photosynthetic carbon fixation in surface waters at midday. Because the v_{max} (the Michaelis-Menten equivalent of C) for photosynthesis is attained at about one-third of full noon sunlight (Fee 1973), many of the photons that impinge on chloroplasts at this time are not utilized for carbon fixation. Instead, they may drive cyclic phosphorylation and the shunting of ATP to nutrient uptake (Healey 1973) or they may be lost as heat. For photosynthesis, v_{max} is set by algal abundance, usually a function of the availability of limiting nutrients, and by the types and densities of chromoplasts in algae, variables that are related to genetics and previous light exposure regimes. Figure 6.7 depicts in schematic form some of the basic features of capacity limitation.

An important property of the capacity limited process is its low sensitivity to changes in the substrate supply rate. Because the functional machinery is already saturated, the introduction of more substrate cannot augment the rate of this sort of process, but can only increase the flow of substrate to other processes or to environmental sinks (Figure 6.7). Similarly, reductions in substrate supply have no effect on the rates of a capacity limited process, provided that S does not fall below C (in which case the process becomes substrate limited). On the other hand, the capacity limited process is very sensitive to fluctuations in total functional capacity. Because total functional capacity is determined by both the abundances and the specific functional capacities of the functional units, a change in either of these variables will alter the rate of a capacity limited process.

The other kind of process limitation, substrate limitation (Figure 6.8), occurs when the supply of one or more of the reactants for a process is

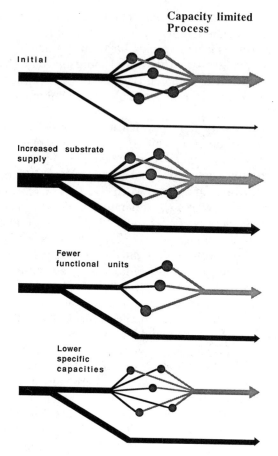

Figure 6.7. Schematic of the flow of substrate (*black arrows*) toward, and products (*grey arrows*) away from, a capacity limited process. Arrow width represents the magnitude of commodity flow; the circles represent functional units; and the degree of circle filling suggests the extent to which process capacity is used. The bottom arrow represents the amount of substrate passing through the environment and not being used by the process. The effects of an increase in substrate supply, a reduction in functional unit abundance, and a reduction in the specific capacities of the functional units are shown.

insufficient to saturate the functional machinery of all the functional units (i.e., when $S < C$). Under these circumstances, some or all of the functional units process less substrate than they are capable of handling and compete with others for the substrate that they acquire. As a result of low supply rates, environmental pools of substrate are exhausted, except perhaps for what cannot be removed from solution because the concentration is below the threshold for uptake by even those species with the highest substrate affinity. Because the environmental pools are small and relatively constant in size, substrate supply rate acts as a ceiling on steady-state process rate. Substrate supply rate is determined by inputs to the

Substrate limited
Process

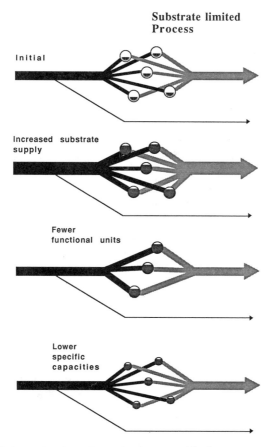

Figure 6.8. Schematic of the flow of substrate (*black arrows*) toward, and products (*grey arrows*) away from, a substrate-limited process. Arrow width represents the magnitude of commodity flow; the circles represent functional units; and the degree of circle filling suggests the extent to which process capacity is used. The bottom arrow represents the amount of substrate passing through the environment and not being used by the process. The effects of an increase in substrate supply, a reduction in functional unit abundance, and a reduction in the specific capacities of the functional units are shown.

ecosystem from the outside and by the rates of internal processes that generate the substance. R may be less than the apparent S if there are chemicals present that interfere with actual substrate availability (e.g., sulfate inhibits molybdate uptake (Howarth and Cole 1985) and arsenate inhibits phosphate uptake (Planas and Healey 1978)) or if the substrate is shared with another ecosystem process (e.g., nitrate is both assimilated by organisms and denitrified).

There are two categories of substrate-limited processes: those in which substrates are growth-limiting resources for the organisms involved and those in which substrates have no direct impact on the relative abundances of organisms. Substrate-limitation of processes providing organisms with growth-limiting resources is easily explained. By competing intensely for these resources, organisms drive environmental pools down to miniscule concentrations. Further population growth then depends on the introduction of substrate via regeneration and external inputs.

The reason for substrate limitation among processes involving resources that are not growth limiting is less obvious. It is related to the fact that many resources are present in aquatic ecosystems in several forms, which are not equally attractive to organisms. Organisms may use to exhaustion those resource forms that require little energy for assimilation (causing assimilation rates to become proportional to substrate supply rates), while they use those forms whose uptake requires more energy just enough to meet the discrepancy between the supply of cheap resource and total requirements. For example, in freshwater systems, where phytoplankton are often phosphorus limited, it is not uncommon for phytoplankton to exhaust first the ammonium, then the nitrate supplies before resorting to substantial use of dissolved organic nitrogen, and finally to nitrogen fixation (Ward and Wetzel 1980). Because through the switching of resource forms total requirements for the resource are met, another resource remains growth limiting.

Substrate limitation also may occur during some abiotic reactions (e.g., during the exchange of certain ions on clays), and during some of the exothermic reactions mediated by bacteria. As examples of the latter situation, Andersen (1977), Cook and Schindler (1983) and Kelly and Rudd (1984) have shown that the sulfate and nitrate reduction rates in lake sediments often are proportional to the supply of sulfate and nitrate in the overlying waters.

The properties of the ecosystem process that is substrate limited but not growth limiting are as follows. The processing rate of any participating species (r_i) depends on its ability to compete with other species for substrate, whereas the abundance of this species is unrelated to the process under consideration and set by other factors (the supply of another nutrient, for example). For the process as a whole, $R = \sum_{i=1}^{n} a_i r_i$, where $r_i \leq c_i$, and is dependent on substrate concentration. This sort of process is sensitive both to small increases in substrate supply, which allow some of the individuals to increase their processing rates (but not to increase their populations), and to small decreases in substrate supply, which slow the processing rates of some of the individuals (but do not decrease their populations). By contrast, small changes in the abundances or the specific capacities of the species do not affect the rates of these processes, provided that C does not fall below S (in which case, the process becomes capacity limited). In short, it makes no difference what the potential for processing is if there is only a small predetermined amount of substrate to process.

The substrate-limited process that brings growth-limiting nutrient into organisms has somewhat different properties. For it, competition influences not only the rates at which species process substrate, but also the abundances of these species ($R = \sum_{i=1}^{n} b_i r_i$, where $r_i \leq c_i$ and $b_i \leq a_i$ and both are dependent on rate of supply; b_i is the substrate-limited abundance of species i and a_i is species i's abundance in the absence of limitation by this specific substrate). Under these circumstances, the competition among the functional units for the resource is intense, and all readily accessible forms of the resource are used as quickly as they are presented. For example, during phosphorus limitation in lakes, phosphate concentration is often less than 0.1 μg/l (Wetzel 1983), and primary productivity can be predicted, at least roughly, from phosphorus supply rates (OECD 1982). If the substrate supply increases, not only can the less competitive organisms take up nutrient at a greater rate (as their original uptake rates were well below capacity), but constraints on their growth and reproduction rates are relieved as well. Therefore, these populations may increase until their new nutrient demands are in line with the greater nutrient supply rate. Decreases in the supply rate of limiting nutrient, on the other hand, will curtail the metabolism of poor competitors for the resource. Consequently, their reproduction will decrease, and mortality may increase until the demands of the biological community no longer exceed the supply of substrate.

It is important to recognize that ecosystem processes are not inherently substrate limited or capacity limited. Any process, at least in theory, can be subject to either limitation state. The fate of any process is entirely dependent on the relationship between the supply rate of usable substrate and the total capacity of the functional units to process substrate. Because both substrate availability and the abundances of different functional unit types may vary seasonally, with water depth and horizontally with water mass, ecosystems may be spatially and temporally patchy with regard to process limitation type.

6.5.2 The Routes of Chemical Impact on Ecosystem Processes

Many anthropogenic chemicals contain elements that are nutrients for organisms and therefore may be degraded and assimilated. When these chemicals enter an environment in amounts beyond what is normally encountered, some ecosystem processes may be stimulated (while others may be inhibited). Other anthropogenic chemicals, especially many of the xenobiotic ones, have no nutritive value. Some of these may participate in abiotic ecosystem processes (e.g., in chemical precipitations or adsorption onto surfaces), but their primary influences on ecosystem processes may be through interactions with substrates or interference with normal physiology.

It is possible then for chemicals to affect an ecosystem process via either of two routes, alteration of the availability of substrate or modification of the functional capacity of the ecosystem for the process. The former assault may occur through direct supplementation of the supply of substrate by the chemical, or by modification of substrate availability indirectly, through interference with substrate production, reaction with the substrate, or blockage of substrate uptake sites. Functional capacity may be affected through the stimulation or depression of growth, with subsequent changes in the abundance of different species, or via the modification of specific functional capacities. Although the impact of environmental chemicals on specific functional capacity is probably primarily on phenotypic expression, there is growing evidence that chronic exposure to chemicals may also induce changes in gene frequencies (see Klerks and Levinton, this book). How an ecosystem process responds to either of the routes of chemical assault depends on the nature of its rate limitation. Table 6.1 provides a summary of the expected responses of capacity limited and substrate limited ecosystem processes to chemical impacts. Details are provided in the next three sections.

6.5.3 Responses of Ecosystem Processes to Changes in the Specific Functional Capacities of Participating Species

A direct assault on the biochemistry and physiology of organisms such that their capacities to perform a function are degraded is what is normally envisioned when a decline in ecosystem process rate is observed under chemical stress. Similarly, when a process is stimulated following chemical inputs, it is tempting to assume that the chemical has somehow augmented specific functional capacities. In fact, such impacts are rarely distinguished from the effects on function caused by changes in biomass or in substrate supply rate.

Table 6.1. Summary of expected responses of capacity limited and substrate limited ecosystem processes

| Process type | Action of the Chemical | | | | | | |
| | Total functional capacity affected | | | | Substrate supply altered | | |
	Decreases capacities of some species for process	Increases capacities of some species for process	Decreases abundances of some species	Increases abundances of some species	Decreases substrate availability	Increases substrate availability	Chemical is a substrate
Capacity-limited	↓	↑	↓	↑	0	0	0
Substrate-limited	0	0	0	0	↓	↑	↑

Responses of process rates to different chemical assaults as influenced by type of process limitation. ↑ = increased rate; ↓ = decreased rate; 0 = no long-term response although transient effects may occur.

Measurements of the responses of the functional capacities of individual species to different doses of a chemical have rarely been made. Such analyses might entail exposing laboratory populations of species to both the test chemical and a surplus of substrate. Process rate would then be measured and the mean rates per individual calculated. However, rates measured in the laboratory may differ from those in the field because of the very different chemical and physical conditions of the two systems.

Even in the absence of real data, one can anticipate the general shapes of the exposure–response curves for specific functional capacities (Figure 6.9). Because anthropogenic chemicals usually do not stimulate ecosystem processes, it is likely that for most processes, the majority of species attain their greatest functional capacities in the absence of the introduced chemical (at zero exposure). With increasing exposure to the chemical and thus with increasing interference with physiology, the specific functional capacities should decline. The exact chemical exposures at which impacts begin to occur and the extent to which additional exposure increments decrease capacity will be related to the mechanism of physiological interference (if there is one) and the population's history of encounter with the chemical. Because of its unique evolutionary and ecological history, each species is expected to have its own relationship between

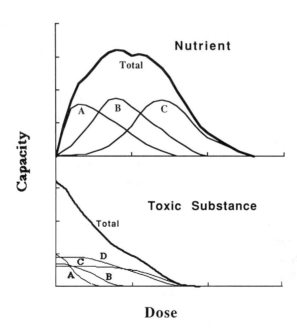

Figure 6.9. The specific functional capacities of the 3 to 4 functional unit types in a fictitious ecosystem (fine lines) and the total functional capacity of the ecosystem (heavy line) at various doses of a toxic substance and a nutrient.

capacity for a particular process and the amount of anthropogenic chemical in the environment. For chemicals with nutritive value, the exposure–response curves of at least some processes may take on a different form. At low concentrations, the chemical may have a positive effect on the biochemistry of the processes; therefore, maximum specific functional capacities will be attained at exposures greater than zero.

Because of the diversity of responses to stress among species, an ecosystem's total capacity for a process may be less sensitive to chemical dose than are the capacities of many of the individual species (Figure 6.9). The relationship between total capacity and intensity of chemical exposure is also more complex than those for specific functional capacities. These characteristics of the dose–response relationship for functional capacity argue against the use of bioassays with one or a few species for examining functional effects.

It is the capacity limited processes in ecosystems that respond readily to chemical stresses that damage the functional machinery, since for them any changes in potential rates are realized changes. Substrate-limited processes resist rate changes associated with injury to capacity through shifts in the relative processing rates of the different species involved. The functional units normally operate at less than full potential, so that when those most sensitive to chemical stress are damaged and their rates diminish, the less sensitive functional units are relieved of competitive constraints, and their r_i increase toward previously unachieved c_i. Chances are that the specific functional capacities of even the less sensitive functional units are lower in the presence of the chemical stress than in its absence, but the enhanced access to substrate counterbalances or overrides this effect (i.e., the r_i are higher, nevertheless). As long as substrate supply rate does not change and $S < C$, all of the substrate not used by the more sensitive functional units may be diverted to the less sensitive ones. For the substrate-limited process that implicates a growth-limiting resource, compensation for the reduced processing of sensitive species may occur through increases in the population densities (the a_i) as well as in the r_i of the less vulnerable species. This is because the relative successes of different species in acquiring growth-limiting resources determines their relative abundance.

It was the relative constancy of this last type of process that attracted Schindler's attention (Schindler 1987) and prompted him to hypothesize that process stability is maintained at the price of changes in community composition and structure. The theory developed here suggests that Schindler's hypothesis is correct only when the process is substrate limited and implicates a growth-limiting factor. Still, it is important to note that the processes to which Schindler's thesis applies are the keystone processes for energy flow and biogeochemical cycling.

The concepts developed above suggest that managers should monitor capacity limited rather than substrate-limited processes if their goal is early detection of ecosystem perturbation through physiological or bio-chemical damage. This approach is not particularly appealing, as capacity limited processes are rarely the key ecosystem processes that interest ecologists most and that the reduction of which would affect the productivity of the ecosystem.

6.5.4 Responses of Ecosystem Processes to Stress Effects on Species Abundance

When the abundance of organisms changes in response to chemical stresses, it is because the availability of a limiting resource has been affected. It may be that another resource has come to limit growth and the species composition has changed to reflect the relative competitive abilities of the resident species for this resource, or the chemical may have affected mortality. Capacity limited processes are sensitive to changes in population sizes just as they are sensitive to changes in specific functional capacities. This mechanism of chemical attack on process rate may be more common than is generally believed. Figure 6.10 illustrates one such situation. Wurtsbaugh and Horne (1982; unpubl. data) measured carbon fixation and chlorophyll *a* concentration in lake water samples exposed to various concentrations of copper. Both variables declined with increased copper dose. Specific carbon fixation (i.e., fixation rate per unit of chlorophyll *a*), on the other hand, was not hindered by copper, but was actually stimulated, probably because competition for a limiting nutrient was less at the smaller algal biomass.

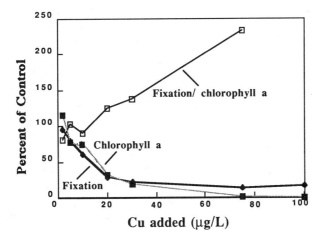

Figure 6.10. Dose–response curves for carbon fixation, chlorophyll a concentration, and specific carbon fixation (fixation per unit chlorophyll a) at various doses of copper in Clear Lake, California. Data from Wurtsbaugh and Horne (1982) and W. Wurtsbaugh, Utah State University (pers. comm.).

For the substrate-limited process, decreases in the abundances of populations cause only transient decreases in rate, if any change occurs at all. The major impact on these processes of the loss of organisms is the same as the impact of diminished specific functional capacities among the more vulnerable species: substrate is freed for use by less sensitive species. The latter species increase their individual processing rates and, when a growth-limiting resource is implicated, their abundances, until they consume all of the substrate that once went to the sensitive populations. Thus there is no long-term impact on total process rate.

6.5.5 Responses of Ecosystem Processes to Stresses that Modify Substrate Supply Rates

Chemical stresses that affect substrate supply rate, but not functional capacity, have no effect on processes that are capacity limited, provided that they continue to be capacity limited at the new supply rate, but they may greatly affect substrate-limited processes. An example of the latter situation has been the stimulation of sulfate reduction in lakes that have received sulfuric acid via acid rain or through experimentation (Rudd et al. 1986). In many lakes, sulfate reduction appears to be constrained by sulfate supply rates (Cook and Schindler 1983). Because sulfuric acid dissociates in water to form sulfate and hydrogen ions, its introduction increases sulfate reduction rates. Sulfate reduction, and its concomitant production of alkalinity, are sometimes so vigorous in lakes exposed to sulfuric acid that lake acidification is greatly delayed or even prevented (Kelly and Rudd 1984).

An example of process rate changes resulting from interference with substrate availability is the addition of sulfate to water from Mirror Lake, New Hampshire (Howarth and Cole 1985). In this case, sulfate interfered with molybdenum uptake by phytoplankton. Molybdenum may limit nitrate reduction and nitrogen fixation rates in some aquatic ecosystems. For example, sulfate additions to Baltic seawaters decreased nitrogen fixation rates, while molybdate additions to the same water enhanced fixation (Howarth and Cole 1985). Interactions among ecosystem functions that are based on substrate production and use will be discussed extensively in the next section of this chapter.

Unfortunately for managerial prediction, processes may switch from one type of rate limitation to another over a chemical gradient. And, given that both C and S may change over the gradient, the shapes of exposure–response curves for processes may sometimes be highly irregular (Figure 6.11 shows three hypothetical curves). If the switching point(s) for limitation are not determined, large errors in prediction can occur. The researcher examining a substrate-limited process may see no change in rate over light-to-moderate exposures. It is tempting to extrapolate to higher exposures and conclude that the chemical has a negligible effect or even

Figure 6.11. A graphic presentation of the interaction of functional capacity and supply in determining process rates at various chemical dose levels. The capacity function is the same for each graph, but the substrate supply rates differ: for **A** and **B**, they are constant, but at a higher and a lower level, respectively; for **C**, supply rate changes with dose level. Rate does not exceed either supply or capacity, but is set by whichever is lower. Substrate supplied in excess of demand is lost as far as the function is concerned.

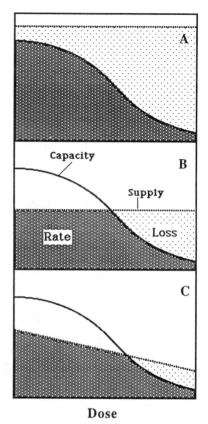

Dose

no effect whatsoever. In fact, total functional capacity probably does change over the observed chemical gradient, but the changes are covert because the process rate reflects substrate supply rather than capacity. At higher exposures than those studied, capacity may fall below substrate supply rate. Further increase in exposure may then initiate severe changes in the process rate.

6.6 Effects of Chemical Stress on Functional Networks

Ecosystem processes are components of functional networks: their substrates are formed by other processes within the ecosystem (or they may be system inputs), and their products fuel further reactions. Thus there is potential for a chemical to affect a process indirectly, via its impact on another process. As pointed out in detail above, capacity limited processes respond directly to the physiological and structural changes induced by chemical stress, but do not respond to changes in substrate supply. By contrast, substrate-limited processes are unresponsive to changes in the abundance or the physiology of their functional units, but they alter their rates to match any change in substrate supply. Therefore, while network disturbances often begin with changes in the rates of capacity limited functions, substrate-limited processes are responsible for passing along the effects of the resulting changes in commodity flow.

The various linkages of capacity limited and substrate-limited processes within a network determines the pathways that disturbances follow. When the rate of a capacity limited process is altered by a change in capacity, there are simultaneous changes in the flow of a commodity to processes or sinks that share substrate with the process, and also to those processes that use the afflicted function's products (Figure 6.12). If the afflicted function shares substrate with a process that is substrate limited (and which maintains substrate limitation throughout the disturbance), the latter function will absorb the entire change in flow. Thus a major

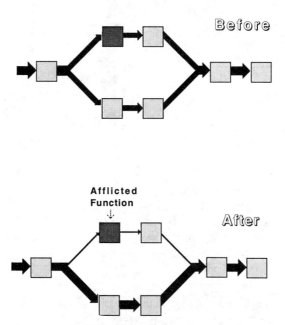

Figure 6.12. A hypothetical example of a change in chain dominance in response to chemical stress. The dark boxes represent capacity limited and the light boxes represent substrate-limited ecosystem processes. The arrows show commodity flow among functions. In this case, the undamaged (lower) chain consists of substrate-limited processes that are able to use all of the flow formerly passed through the afflicted (upper) chain. Processes in this network that lie beyond the reconvergence of the chains are unaffected by the chemical disturbance.

effect of chemical stress on network structure may be the modification of branch dominance within the network. Network branches are formed when two processes use one substrate (e.g., both assimilatory and dissimilatory nitrate reduction require nitrate) or when one process yields two products containing the network commodity (e.g., methane fermentation has two carbon products, carbon dioxide and methane). If network branches later reconverge (as happens when two processes yield the same product, e.g., both ammonium uptake and nitrate uptake by phytoplankton result in protein formation), the impact of the stress on commodity flow may be largely confined to part of the network.

Sinks that share a commodity with capacity limited processes also absorb the disturbance in the commodity flow caused by a change in process rate. However, as Figure 6.13 illustrates, the ultimate effect on network structure of this substrate diversion may be a modification of the relative importance of the different sinks rather than a change in overall loss of commodity from the system.

When the products of the afflicted capacity limited process are used by substrate-limited processes, these processes will change their rates in proportion to the change in substrate flow and thus propagate the change in flow pattern. Sequences of substrate-limited processes yield to disturbance in a domino fashion. If, on the other hand, the afflicted process provides substrate to a capacity limited process, this process will protect

Figure 6.13. A hypothetical example showing that although commodity flow to one sink may increase as a result of chemical stress, the commodity flow to another sink may decline in turn. The dark boxes represent capacity limited and the light boxes represent substrate-limited ecosystem processes. The arrows show commodity flow among functions. In this case, the capacity limited process at the end of the functional chain receives enough substrate to be unaffected by the afflicted function.

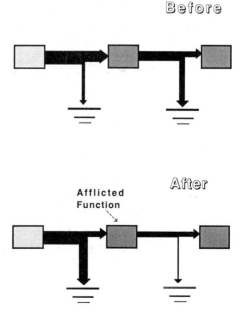

the series of processes beyond it by not responding to a change in substrate supply. The flow of commodity to competing substrate-limited processes or to sinks may be affected instead. Again, there is the possibility of changes in chain dominance or loss patterns.

Chemical stress sometimes alters not just the loss patterns but the inputs of commodity to ecosystems (Schindler et al. 1975; Margalef 1981). Internal control of inputs seems counterintuitive until it is recalled that chemicals move into waterbodies from the atmosphere and sediments along chemical gradients. Organisms modify these gradients by assimilating dissolved chemicals or, alternatively, by adding to the pools through excretion. For example, Schindler and Fee (1973) showed that in lakes fertilized with nitrogen and phosphorus, carbon content steadily increases over a period of several years. The carbon originates in the atmosphere and enters the lake because the partial pressure of carbon dioxide in the epilimnion is depressed by the increased photosynthesis.

6.7 Chemical Stress Effects on Interactions Between Functional Networks

In addition to links among functions, there are links among networks. These occur when processes involve more than one commodity. For example, marl formation (calcium carbonate precipitation) modifies both the calcium and carbon cycles. The best known and perhaps the strongest of network links is primary production, the formation of plant biomass. Because algae require nutrients and energy at rather fixed ratios, the amounts of non-growth-limiting nutrients incorporated into these organisms are proportional to the amount of carbon skeleton produced. Carbon fixation is regulated, in turn, by the availability of a growth-limiting nutrient (usually phosphorus or nitrogen), or, by light when the water body is highly eutrophic or ice-covered.

In addition to direct network linkages, two networks may affect one another indirectly via a third network. For example, methane fermentation and sulfate reduction are processes within the carbon and sulfur cycles, respectively. But they are also competitive reactions within the hydrogen cycle as both require H^+. Thermodynamics favor sulfate reduction over methane fermentation, and the importance of methane fermentation as a link in the carbon cycle of a particular ecosystem is at least partly determined by sulfate concentration. This is why methane fermentation tends to be more significant in lakes than in the sulfate-rich ocean. A major consequence of acidification in lakes appears to be the repression of methane fermentation, as sulfate reducers use hydrogen ions to reduce the newly introduced sulfate (Kelly et al. 1982).

6.8 Indices of Ecosystem Health

Given what is known about functional ecotoxicology, is there a recommendable process-based monitoring scheme for ecosystems under chemical stress? Clearly simultaneous measurement of more than a few processes is impracticable. But which processes will provide the most information about the ecosystem state? One approach is to tailor monitoring to the specific interests of the public. For example, for a lake with a history of blue-green algal blooms, nitrogen dynamics might be monitored, since nitrogen shortages favor blue-green algae (Smith 1983). Similarly, if the bioconcentration of mercury threatens fisheries, measurement of mercury methylation is a high priority.

There is also the question of how to quantify the threat of a chemical to a particular ecosystem process. Is it necessary to compare the shapes of the dose–response curves over wide dose ranges? Some ecotoxicologists (e.g., Babich et al. 1983) have suggested that the dose–response curves be used to estimate EcD_{50} or EcD_{10} values for particular functions in particular ecosystems (the EcD_x index is an estimate of the chemical dose required to reduce the rate of an ecosystem process by $x\%$). Such indices would have the value of being easily incorporated into water quality criteria. One major disadvantage of the specific interest approach to monitoring is that the overall picture is missed, and surprises (i.e., damage to critical functions via indirect routes) are likely. Another disadvantage is that the natural variability of a single monitored process may be so great that the actual extent of the process reduction is uncertain.

An alternative strategy for assessing the impact of stresses on ecosystem functioning is to develop one or more indices of overall "ecosystem health" and monitor these parameters. The appeal of this approach is high, again because indices can serve as endpoints for regulation. Several indices have been suggested: primary productivity, the ratio of photosynthesis to respiration (Woodwell 1962), carbon dioxide spectra (Van Voris et al. 1980), electron transport system activity (ETSA; Ivanovici and Wiebe 1981), the commodity cycling index (i.e., the ratio of commodity cycling to commodity throughput; DeAngelis 1980), and for use with rivers, nutrient spiral length (Rapport et al. 1985). The danger of using a single or even two or three indices to monitor stress lies in the loss of resolution that is incurred. One cannot pinpoint damaged processes or even damaged networks with an index like ETSA. Without information on trouble spots, restoration schemes are likely to be primitive and ineffectual.

Ultimately, the question is confronted of whether functional indicators of ecosystem state are really preferable to structural indicators, especially given that the latter are often more easily measured. I believe that at least for aquatic ecosystems, the answer generally is "no." The reasoning is as follows. In many lakes and coastal regions, the more important and commonly measured ecosystem processes are substrate limited rather than capacity limited. Yet chemical stresses probably affect the physiology and biochemistry (and, thus, the process capacity) of organisms more often than they affect substrate supply rates. Substrate-limited processes do not respond to changes in process capacity, and thus make very poor indicators of the physiological strain on the organisms in an ecosystem. A chemical stress that is severe and toxic may affect capacity enough to cause a substrate-limited process to switch to capacity limitation. A decline in process rate will follow. However, at this degree of stress, changes in community structure probably will be well advanced. Processes that are capacity limited under natural conditions should be very good early warning signals for stress. However, these processes rarely interest ecologists for their own worth. Thus, they are yet to be identified as indicators worth monitoring. A second argument against great reliance on functional indicators is the extent to which ecosystems differ in the types and amounts of substrates available and probably in their capacities to process these substrates. To choose the functional indicators of ecosystem state wisely, a great deal must be known about the functioning of an ecosystem. This sort of information simply is not widely available.

6.9 Conclusions

There has been considerable discussion as of late of the sensitivity of ecosystem processes to anthropogenic stresses (e.g., O'Neill et al.1977; Van Voris et al. 1980; Bormann 1983; Rapport et al. 1985; Odum 1985; Schindler 1987). These discussions have not considered the possible influence of the type of rate limitation (substrate versus capacity limitation) on the responses of the processes. Herein it is argued that process vulnerability to chemical stress depends both on this variable and on the route of chemical assault (i.e., whether the chemical affects physiology directly, or indirectly affects the process through altering population densities or the availability of substrates). Substrate-limited processes occur at rates that are proportional to substrate supply rates and less than the system's total capacity to perform the function. They are very sensitive to chemicals that alter substrate supply through interference with or stimulation of substrate production, reaction with the substrate, or substitution for the substrate in the process. By contrast, they are indifferent to chemicals whose main effects are on the abundances or the physiological abilities of the organisms mediating the reaction. This is because biological species compete for substrate during substrate limitation, and the reduced activity of those sensitive to a chemical is balanced at steady state by the increased activity of less vulnerable species.

Capacity limited processes differ from substrate limited ones in that their rates are not constrained by substrate supply (which is in excess of need) but by the numbers and capabilities of the participating functional units (usually organisms, but sometimes abiotic surfaces). Because every species performs the function at the greatest rate possible for it under ambient conditions, there is no possibility of compensation through increases in the rates of insensitive species if the rates of some of the more vulnerable species are reduced by an anthropogenic chemical. Over gradients in chemical exposure, ecosystem processes may switch from substrate to capacity limitation or vice versa. Thus, an important part of predicting ecosystem response to stress is knowing when to expect these switches.

When the rate of an ecosystem process changes under chemical stress, other processes that depend on it for substrate also are perturbed. Substrate-limited processes pass on such a disturbance to yet other processes, whereas capacity limited processes protect those processes that depend on them for substrate. Thus the pattern of disturbance that a chemical stress induces within a food web or biogeochemical cycle varies with the arrangement of capacity and substrate-limited processes within the network. In addition, the effects of chemical stress on a particular element cycle may be propagated to other cycles. This is possible because chemical processes normally involve two or more elements. Also biological species may respond numerically to changes in substrate supply, and their abundance in turn affects other processes, including some that involve completely different materials.

Changes in ecosystem processes and in functional networks are not always immediately apparent as chemical stress increases, and the rate at which symptoms develop need not be constant. Because of the many linkages among ecosystem processes, indirect effects are an important aspect of an ecosystem's response to stress. Given the complexity of ecosystems, the use of functional parameters as sole indicators of ecosystem state under chemical stress is problematic. Structural indicators of ecosystem condition appear to be more reliable (Ford, this volume), and probably should supplement process-oriented studies.

Acknowledgments. The author thanks M. Braner, R. Howarth, J. Kelly, K. Kimball, K. Limburg, and D. Rudnick for their useful comments on the manuscript.

This publication is ERC-82 of the Ecosystems Research Center, Cornell University, and was supported by the U.S. Environmental Protection Agency Cooperative Agreement Number CR812685–01. Additional funding was provided by Cornell University.

This work and conclusions published herein represent the views of the authors and do not necessarily represent the opinions, policies, or recommendations of the Environmental Protection Agency.

References

Andersen JM (1977) Rates of denitrification of undisturbed sediment for six lakes as a function of nitrate concentration, oxygen, and temperature. Arch Hydrobiol 80:147–159

Babich HR, Bewley RJF, Stotzky G (1983) Application of the "ecological dose" concept to the impact of heavy metals on some microbe-mediated ecological processes in soil. Arch Environ Contam Toxicol 12:421–426

Baker J H, Morita RY (1983) A note on the effects of crude oil on microbial activities in a stream sediment. Environ Poll (Series A) 31:149–157

Barrett GW, Rosenberg R (1981) *Stress Effects on Natural Ecosystems.* Chichester: John Wiley and Sons, 305 pp.

Bormann FH (1983) Factors confounding evaluation of air pollutant stress on forests: pollutant input and ecosystem complexity. In: *Symposium, "Acid deposition, a challenge for Europe."* Karlsruhe, FRD, Sept. 1983, pp. 19–23

Brewer P (1978) Direct observations of the oceanic CO_2 increase. Geophys Res Lett 5:997–1000

Broberg A (1982) Effects of heavy metals on alkaline phosphatase activity in freshwater sediments. In: Bergstrom I, Kettunen J, Stenmark M (eds) *Proceedings from the 10th Nordic symposium on sediment.* Helsinki University of Technology Report 26, Espoo, Finland, pp. 115–126

Broberg A (1983a) Effects of heavy metals on electron transport system activity (ETSA) in freshwater sediments. Environ Biogeochem Ecol Bull (Stockholm) 35:403–418

Broberg A (1983b) Effects of heavy metals on nitrate reduction rates in laboratory experiments with freshwater sediments. In: *Proceedings from the international conference on heavy metals in the environment.* Heidelberg, West Germany, CEP Consultants, Ltd., Edinburg, UK, pp. 784–787

Broberg A (1984) Effects of heavy metals on nitrification in laboratory experiments with freshwater sediments. Environ Biogeochem Ecol Bull (Stockholm) 36:135–142

Broecker WS (1974) *Chemical Oceanography.* New York: Harcourt Brace Jovanovich, 214 pp.

Cook RB, Schindler DW (1983) The biogeochemistry of sulphur in an experimentally acidified lake. Environ Biogeochem Ecol Bull (Stockholm) 35:115–127

Crocker RL (1952) Soil genesis and the pedogenic factors. Quart Rev Biol 27:139–168

DeAngelis DL (1980) Energy flow, nutrient cycling and ecosystem resilience. Ecology 61:764–771

Evans FC (1956) Ecosystem as the basic unit in ecology. Science 123:1227–1228

Fee EJ (1973) A numerical model for determining integral primary production and its application to Lake Michigan. Can J Fish Aquat Sci 30:1447–1468

Forbes SA (1887) The lake as a microcosm. Bull Sci Acad Peoria. Reprinted in Ill Nat Hist Surv Bull (1925) 15:537–550

Griffiths RP, Caldwell BA, Broich WA, Morita RY (1982) The long-term effects of crude oil on microbial processes in subarctic marine sediments. Est Coast Shelf Sci 15:183–198

Healey FP (1973) Inorganic nutrient uptake and deficiency in algae. Crit Rev Microbiol 3:69–113

Howarth RW, Cole JJ (1985) Molybdenum availability, nitrogen limitation, and phytoplankton growth in natural waters. Science 229:653–655

Ivanovici AM, Wiebe WJ (1981) Towards a working 'definition' of 'stress': A review and a critique. In: Barret GW, Rosenberg R (eds) *Stress Effects in Natural Ecosystems*. Chichester: John Wiley and Sons, pp. 13–28

Jackson DR, Watson AP (1977) Disruption of nutrient pools and transport of heavy metals in a forested watershed near a lead smelter. J Environ Qual 6:331–338

Kelly CA, Rudd JWM (1984) Epilimnetic sulfate reduction and its relationship to lake acidification. Biogeochem 1:63–77

Kelly CA, Rudd JWM, Cook RB, Schindler DW (1982) The potential importance of bacterial processes in regulating rate of lake acidification. Limnol Oceanogr 27:868–882

Kettle WD, DeNoyelles F Jr (1986) Effects of cadmium stress on the plankton communities of experimental ponds. J Freshwater Ecol 3:433–444

Kilham P (1971) A hypothesis concerning silica and the freshwater planktonic diatoms. Limnol Oceanogr 16:10–18.

Lean DRS, Liao CF-H, Murphy TP, Painter DS (1978) The importance of nitrogen fixation in lakes. Environ Biogeochem Ecol Bull (Stockholm) 26:41–51

Lovelock JE (1979) *Gaia. A New Look at Life on Earth*. Oxford: Oxford Univ Press, 157 pp.

McKnight D (1981) Chemical and biological processes controlling the response of a freshwater ecosystem to copper stress: A field study of the $CuSO_4$ treatment of Mill Pond Reservoir, Burlington, Massachusetts. Limnol Oceanogr 26:518–531

Mann LK, McLaughlin SB, Shriner DS (1980) Seasonal physiological responses of white pine under chronic air pollution stress. Environ Experim Bot 20:99–105

Margalef R (1981) Stress in ecosystems: a future approach. In: Barret GW, Rosenberg R (eds) *Stress Effects in Natural Ecosystems*. Chichester: John Wiley and Sons, pp. 281–289

Maugh TH III (1983) How many chemicals are there? Science 220:293

Morris I, Syrett PJ (1963) The development of nitrate reductase in *Chlorella* and its repression by ammonia. Arch Mikrobiol 47:32–41

Odum EP (1985) Trends expected in stressed ecosystems. BioScience 35:419–422

O'Neill RV, Ausmus BS, Jackson DR, van Hook RI, van Vorris P, Washburne C, Watson AP (1977) Monitoring terrestrial systems by analysis of nutrient export. Water Air Soil Pollut 8:271–277

OECD (1982) Eutrophication of waters: monitoring, assessment, and control. Paris: Organisation for Economic Co-operation and Development, 154 pp.

Persson GS, Holmgren K, Jansson M, Lundgren A, Nyman B, Solander D, Anell C (1977) Phosphorus and nitrogen and the regulation of lake ecosystems. Experimental approaches in subarctic Sweden. In: *Proceedings from Circumpolar Conference on Northern Ecology*. Sept. 1975. Ottawa National Research Council of Canada, pp. 1–20

Planas D, Healey FP (1978) Effects of arsenate on growth and phosphorus metabolism of phytoplankton. J Phycol 14:337–341

Rapport DJ, Regier HA, Hutchinson TC (1985) Ecosystem behavior under stress. Amer Nat 125:617–640

Redfield AC (1934) On the proportions of organic derivatives in sea water and their relation to the composition of plankton. In: *James Johnstone Memorial Volume*. Liverpool: University Press, pp. 176–192

Redfield AC (1958) The biological control of chemical factors in the environment. Amer Sci 46:205–221

Redfield AC, Ketchum BH, Richards FA (1963) The influence of organisms on the composition of sea water. In: Hill MN (ed) *The Sea,* Vol 2. New York: John Wiley and Sons, pp. 26–77

Reynolds CS (1984) *The ecology of freshwater phytoplankton*. Cambridge: Cambridge University Press, 384 pp.

Rudd JWM, Kelly CA, St. Louis V, Hesslein RH, Furutani A, Holoka MH (1986) Microbial consumption of nitric and sulfuric acids in acidified north temperate lakes. Limnol Oceanogr 31:1267–1280

Schelske CL, Stoermer EF (1971) Eutrophication, silica depletion, and predicted changes in algal quality in Lake Michigan. Science 173:423–424

Schindler DW (1975) Whole-lake eutrophication experiments with phosphorus, nitrogen, and carbon. Int Ver Theor Angew Limnol Verh 19:3221–3231

Schindler DW (1985) The coupling of element cycles by organisms: evidence from whole lake chemical perturbations. In: Stuum W (ed) *Chemical Processes in Lakes*. New York: John Wiley and Sons, pp. 225–250

Schindler DW (1987) Detecting ecosystem responses to anthropogenic stress. Can J Fish Aquat Sci 44, Supplement No. 1:6–25

Schindler DW, Fee EJ (1973) Diurnal variation of dissolved inorganic carbon and its use in estimating primary production and CO_2 invasion of Lake 227. J Fish Res Board Can 30:1501–1510

Schindler DW, Turner MA (1982) Biological, chemical, and physical responses of lakes to experimental acidification. Water Air Soil Pollut 18:259–271

Schopf JW (1982) *Earth's Earliest Biosphere. Its Origins and Evolution*. Princeton: Princeton University Press

Sheehan PJ (1984) Functional changes in the ecosystem. In: Sheehan PJ, Miller DR, Butler GC, Bourdeau Ph (eds) *Effects of Pollutants at the Ecosystem Level*. Chichester: John Wiley and Sons, pp. 101–145

Smith VH (1983) Low nitrogen and phosphorus ratios favor dominance by blue-green algae in lake phytoplankton. Science 221:669–671

Sohacki LP, Ball RC, Hooper FF (1969) Some ecological changes in ponds from sodium arsenite and copper sulfate. Mich Acad 1:149–162

Stanley SM (1986) *Earth and life through time*. New York: W.H. Freeman and Company. 690 pp.

Stuum W, Morgan JJ (1970) *Aquatic Chemistry. An Introduction Emphasizing Chemical Equilibria in Natural Waters*. New York: John Wiley and Sons, 583 pp.

Tansley AG (1935) The use and abuse of vegetational concepts and terms. Ecology 16:284–307

Thomas WH, Seibert DLR (1977) Effects of copper on the dominance and the diversity of algae: Controlled ecosystem pollution experiment. Bull Marine Sci 27:23–33

Thomas WH, Holm-Hansen O, Seibert DLR, Azman F, Hodson R, Takahashi M (1977) Effects of copper on phytoplankton standing crop and productivity: Controlled ecosystem pollution experiment. Bull Marine Sci 27:34–43

Van Voris P, O'Neill RV, Emanuel WR, Shugart HH Jr (1980) Functional complexity and ecosystem stability. Ecology 61:1352–1360

Ward A K, Wetzel RG (1980) Interactions of light and nitrogen source among planktonic blue-green alge. Arch Hydrobiol 90:1–25

Wetzel RG (1983) *Limnology*. New York: Saunders. 767 pp.

Weiss RF (1981) The temporal and spatial distribution of tropospheric nitrous oxide. J Geophys Res 86:7185–7195

Williams WT (1980) Air pollution disease in the California forests. Environ Sci Technol 14:179–182

Wood JM, Wang H-K (1983) Microbial resistance to heavy metals. Environ Sci Tech 17:582A–590A

Woodwell GM (1962) Effects of ionizing radiation on terrestrial ecosystems. Science 138:572–577

Wurtsbaugh WA, Horne AJ (1982) Effects of copper on nitrogen fixation and growth of blue-green algae in natural plankton associations. Can J Fish Aquat Sci 39:1636–1641

Chapter 7

The Effects of Chemicals on the Structure of Terrestrial Ecosystems: Mechanisms and Patterns of Change

David A. Weinstein[1] and Elaine M. Birk[1,2]

Chemicals introduced into natural terrestrial ecosystems can be considered disturbances, similar in mode of action to natural disturbances such as fires, windstorms, and species invasions. The effects of releases of chemicals on the structure of terrestrial ecosystems have not been studied extensively. However, responses of ecosystems to natural disturbances have received substantial scientific scrutiny. Exploration of the parallels between disruptions brought about by chemicals and by natural forces is useful because, in the absence of a large body of experimental evidence or case studies that have followed the effects of chemical releases, we are forced to extrapolate from the numerous studies of the effects of natural disturbances. It is critical, therefore, to evaluate the ways in which ecosystems are likely to respond differently to chemicals than they respond to natural disturbance. We must identify the patterns and mechanisms of change in terrestrial ecosystems exposed to chemical inputs that are unique to this type of anthropogenic disturbance.

The study of natural disturbances in terrestrial ecosystems has concentrated upon destructive events that abruptly change the physical structure of the plant community (Sousa 1984) resulting in the partial or complete removal of plant biomass (Grime 1979) and/or accumulated detritus (Reiners 1983). From these studies, general trends in the response of ecosystem structure have been observed. Table 7.1 lists those trends most recently articulated by Odum (1985) and Woodwell (1983). For example, in a large number of cases the biomass of a terrestrial ecosystem has been

[1] Ecosystems Research Center, Cornell University, Ithaca, New York 14853
[2] Present address: Forestry Commission of New South Wales, PO Box 100, Beecroft, New South Wales, 2119 Australia

Table 7.1. Hypothesized effects of disturbance stress on ecosystem structure

Biomass	Structure is more dramatically altered than are processes; organic matter pools decrease
Species composition	Increased dominance by rapidly reproducing species characteristic of earlier successional stages; sensitive species eliminated
Species richness	Diverse communities become less diverse; diversity increases in low diversity cases
Community size profile	Size decreases
Organism life span	Life span decreases
Trophic web	Shortening of food chains; reduction of number of parallel food chains

Adapted from Odum 1985; Woodwell 1983.

dramatically altered by disturbance. Events such as severe windstorms or fires commonly result in the removal of the large individuals of the ecosystem (e.g., fires removing large conifers in boreal systems). These individuals often belong to species with sensitive life stages or are otherwise more sensitive to disturbance than are smaller individuals. Under these circumstances, the ecosystem becomes dominated by smaller species that reproduce rapidly and which are more characteristic of earlier successional stages. Since these species place a greater amount of growth effort into reproduction than into maintenance and long-term persistence, they do not tend to live as long, causing the average life span of the community to decrease. The life span may also decrease because of direct increases in the rate of mortality of the exposed individuals.

At the community level in these studies, species diversity and food webs change. Under harsh conditions, fewer species can survive the harsh conditions often associated with disturbance. In ecosystems with many species, disturbance results in a decrease in the diversity. Alternatively, in ecosystems dominated by a few species, the increase in mortality rate may increase the availability of light, water, or nutrient resources. Many new species might be capable of taking advantage of these new resources, resulting in an increase in the diversity. Such increases can be observed following the creation of patches in a mature forest canopy. The loss of top predators resulting from their greater sensitivity and lower reproductive rate can cause a decrease in the number of species involved in each food chain and a decrease in the number of chains.

The application of these trends to the responses to disturbances of anthropogenic origin has depended upon the assumption that human-induced disturbance has the same character as natural disturbance. Models of natural disturbance–recovery mechanisms (e.g., White 1979; Bormann and Likens 1979) and patterns of change in ecosystem properties during secondary succession (e.g., Odum 1969; Bormann and Likens 1979; Gorham et al. 1979; Peet 1981) are founded on the assumption that disturbances are discrete, destructive events that occur relatively infrequently. Anthropogenically induced disturbance often has been assumed to be analogous to natural disturbance, i.e., to consist of physical destruction. Woodwell (1983) and Odum (1984) have noted examples where the release of chemical pollutants leads to the elimination of sensitive species, particularly overstory plants, thereby decreasing species diversity, reducing pools of organic matter, and reverting the successional development to earlier seral stages in much the same way as natural disturbance. These generalizations, however, are based largely on observations following exposure to lethal doses of chemicals or radiation. Unfortunately, there are few studies in which effects at lower levels of chemical exposure have been examined systematically at the ecosystem level, particularly cases where the effects from exposure are not accompanied by physical destruction. Further, there is little understanding of the effects of different exposure regimes (i.e., dose, time) or of multiple chemicals on ecosystem properties, at either high or at low dosages.

There are many different forms of human disturbance, and many of them may cause significant changes in ecosystem structure without a significant loss of organisms or biomass. Some forms of human disturbance, such as deforestation and urbanization, do have natural disturbance analogs and, like their natural counterparts, result in direct alteration of ecosystem structure. Exposure to lethal concentrations of phytotoxic chemicals, such as in the air of areas downwind of smelter discharges (Gordon and Gorham 1963; Miller and McBride 1975; Freedman and Hutchinson 1980; Weinstein and Bunce 1982) will cause severe ecosystem deterioration. Ecosystems exposed to sub-lethal concentrations are less likely to show substantial physical deterioration. Alternatively, the direct effect of the release of anthropogenically produced hazardous chemicals may be an increase in stress on exposed organisms, resulting in a gradual degradation of terrestrial ecosystems through cumulative changes, rather than cause an immediate loss of living organisms (see for example, Woodwell 1967, 1970; Sheehan 1984a,b,c; Connell and Miller 1984; Levin and Kimball 1984). If this gradual degradation becomes severe, biomass loss will occur. Some studies indicate that other structural properties, such as species composition and patterns of succession, are altered (see Brandt and Rhodes 1973; Miller 1973; Treshow and Stewart 1973; McClenahan 1978; Westman 1979; Smith 1981). These more subtle effects may be of greater concern because they are much more difficult to detect and may precipitate widespread ecosystem changes that may be irrevocable by the time they are identified. For example, the appearance of injury may be delayed with respect to the onset of the stress itself, as may be the case in the forest die-back and decline in eastern North America and Europe (OTA 1984).

There is, in fact, a continuum of ecosystem response to chemical disturbance, from effects that are undetectable to those resulting in total ecosystem destruction (Smith 1980). Studies of natural disturbance usually involve the physical loss of biomass and, most often, destruction of organisms. Chemical disturbance, by contrast, may involve injury to organisms that does not necessarily result in loss of biomass or organisms, though under extreme conditions it might have these effects. As a result, chemical disturbance cannot be treated as an analog to natural disturbance in many cases. We refer to these sub-lethal disturbances as stress, in contrast to destruction. The concern of this chapter is to trace the differences in structural changes in ecosystems resulting from the sublethal exposures of chemicals versus the lethal consequences of natural destruction. We explore the variety of structural effects that have been seen following chemical release into ecosystems and will identify the similarities and differences between these and physically induced changes in ecosystem structure. We include in our definition of structure the biomass of the biological community, species composition, species richness, community size profile, and the web of trophic interactions. We concern ourselves with the patterns by which these characteristics change over time, both naturally and in response to acute, single, or multiple exposures of toxic substances, and to chronic exposure to both lethal and nonlethal concentrations. Chemical substances here include both beneficial and potentially phytotoxic substances released through human activities. Since the dominant autotrophs determine, in large part, the ecosystem structure, much of the discussion concerns changes to these organisms. Table 7.2 summarizes the similarities and differences between the effects of physical and chemical disturbance on structure that will be discussed in detail in this chapter.

7.1 Mechanisms of Chemical Exposure

One major difference between natural and chemical disturbance is that a chemical must travel from the point of release to the ecosystem and must come in contact or be deposited in the ecosystem at some location where the organisms are vulnerable to perturbation. In order to produce damage, some chemicals, such as heavy metals, must be retained within the system. Through dispersal in the atmosphere and long-range transport, anthropogenically produced air pollutants may be deposited in ecosystems even in quite remote areas (Smith 1981). DDT and PCBs, for example, now are distributed globally, and there is widespread deposition of photochemical oxidants and sulfur and nitrogen oxides.

Substantial amounts of gaseous and particulate substances are released indirectly as by-products of industry (e.g., oxides of nitrogen and sulfur,

Table 7.2. Comparison of the effects of natural physical disturbance with discrete acute, repeated acute, and chronic chemical disturbance

	Natural physical	Discrete acute chemical	Repeated acute chemical	Chronic chemical
Organism effects	Death	Selective death; severe damage	Selective death; severe damage	Metabolic injury; loss of function; behavioral change
Biomass	Decreases	Decreases	Decreases	Decreases or increases
Organic pools	Decreases	Decreases	Increases	Decreases or increases
Recovery	Depends on disturbance of soil media	Depends on rate of disappearance of toxicant, presence of nonsensitive species	Depends on number, intensity, and duration of incidents; repression of invading early successionals; shortness of life cycle of invaders	Depends on completion of life cycle despite presence of chemical
Dominance	Rapid reproducers increase	Release of nonsensitive organisms	Potential for introduction of alien species; increases in species that complete life cycle between events	Shift in competitive dynamics; increase in tolerants
Richness	Declines in diverse communities; increase in poor communities	Depends on chemical released	?	?
Size	Decreases	Decreases	Decreases greatly	Decreases or increases
Life span	Decreases	Decreases	Decreases greatly	Decreases or increases
Food web length	Decreases	Decreases	Decreases	Decreases or increases
Food chain numbers	Decreases	Decreases	Decreases	Decreases
Nutrient supplies	Depends on temperature, organic matter	Depends on role of sensitive organisms	Decreases	Depends on role of sensitive organisms

fluorides, and hydrocarbons) or directly through human domestic and agricultural activities (e.g., pesticides and herbicides). Others, such as photochemical oxidants (e.g., ozone) are synthesized in the air in the process of chemical mixing. Lastly, volatile chemicals can escape into the air from containment (e.g., PCBs from waste sites; Buckley 1982, 1983).

Once they have been transported to an ecosystem, dissolved, particulate, or gaseous substances are either scavenged by the vegetation or reach the ground in precipitation or by settling out (Hill 1971; Guderian and Keuppers 1980). Other chemicals may be deposited directly as anthropogenic waste products in or on the ground. A third class of chemicals, from oil spills, roadside dumps, toxic waste sites, slag heaps, sewage sludge, and fertilizer applications, may flow through surface waters or groundwater to ecosystems.

Retention within the system depends on characteristics of the individual chemicals and the prevailing environmental conditions. Nitrogen and sulfur, for example, naturally cycle through the biosphere, and can be returned to the atmosphere or leached from the soil. Heavy metal particulates, by contrast, are not readily leached, and persist in litter and soil. Synthetic organics, on the other hand, may be degraded by microorganisms, or, like DDT and PCBs, resist attack and accumulate in the biota.

7.2 Effects of Disturbance on Organisms

Alteration of ecosystem structure following disturbance is usually initiated by effects on individual organisms. The severity of natural disturbance typically leads to direct losses of biomass of an ecosystem because the physical destruction (or destructive disturbance) directly kills affected organisms (Figure 7.1). In clear-cutting and grazing, for example, whole organisms are killed and removed. Fires remove and destroy both live organisms and detritus, whereas organisms killed by windstorms accumulate as detritus.

In contrast, chemical substances induce physiological stresses resulting in metabolic injury and, in some cases, loss of some function necessary for growth and/or maintenance (Figure 7.1). In cases of severe damage, this injury may result in organism death and thus will resemble physical disturbance. For example, anthropogenically produced herbicides may cause losses of tissues by enhancing the rate of senescence (Grime 1979). This loss of tissue is similar to the losses that cause death through impaired cell metabolism and physiology following natural environmental disturbances such as frost and drought (Grime 1979). However, most often the injury following chemical exposure is less severe. Reduced growth rates, damaged leaves, and impaired reproduction are more likely responses to low-to-moderate doses of toxic chemicals (Smith

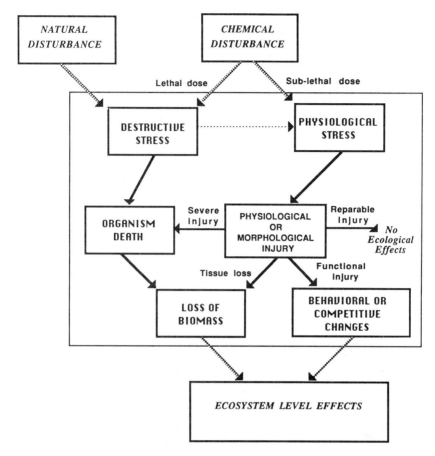

Figure 7.1. Pathways of physical and chemical disturbance. Pathways and components within the box represent the effects of severe natural disturbance and various types of chemical disturbance on organisms, and the pathways by which these effects translate to ecosystem level effects.

1981). These responses are likely to lead to shifts in the behavior patterns of organisms and alterations in the competitive relationships among species. These shifts represent indirect effects that may result in major changes in ecosystem structure.

7.2.1 The Pathway of Chemical Injury

Exposures of high dosage and brief duration are referred to as acute stress. Acute exposures to sensitive individuals cause severe morphological damage, significantly altering metabolism, causing cell collapse and tissue loss. If the tissue loss is too great or if some vital physiological process is permanently impaired, the organism will die. Mortality is induced rapidly if organisms are exposed to highly toxic doses of chemical substances. This is referred to as acute lethal toxicity and results from a rapid breakdown in metabolism. Such high doses are usually associated with localized exposure following accidental spills, or intentional sprayings of herbicides and pesticides. Chronic stress, in contrast, is typified by a continuing series of exposures or lengthy continuous exposures at low concentrations. Effects are also referred to as chronic or acute, although in practice the distinction between these is made based upon whether the injury has manifested as visible damage (acute) or non-visible (chronic). Both chronic and acute injuries are initiated at the subcellular level and propagate through cells, tissues, and organs. Earlier workers associated acute injuries with the appearance of dead tissues and perhaps mortality following rapid absorption of toxic substances, whereas chronic injury was regarded as the delayed appearance of visible injury after exposure to sub-lethal concentrations (Knabe 1976). Today, however, it is more common for all delayed injury resulting in impaired functioning of some biological process within the organism to be described as chronic (Skelly 1980; Skarby and Sellden 1984).

To cause injury, chemical exposure of an organism must occur at a vulnerable location and during a vulnerable period. Particulate sulfuric acid compounds are much more likely to cause damage if leaves and rapidly growing plant parts are exposed than if the compounds fall largely on the soil surface. Toxic, short-lived compounds deposited in an ecosystem on the soil surface in early spring before plants have broken the soil surface may have little effect. In contrast, heavy metals may affect active root growth for long periods because these metals are retained for such a long time within the ecosystem. Depending upon the relationships between chemical deposition and retention and organism vulnerability, the effects of chemicals can deviate greatly from the effects of natural disturbance.

A chemical introduced into the environment will be taken up by biota according to characteristics of dosage, environment, and susceptibility. The dosage exposed organisms receive is a function of the amount of chemical to which the organism is exposed, given climatic, edaphic and biotic factors, the duration of each exposure, and the frequency of exposure episodes (Guderian and Keuppers 1980; Bormann 1983). For example, air pollutant deposition patterns, which in part determine the intensity and duration of exposure, are affected by local variations in topography, climate, and vegetation. High sulfate deposition areas are characterized by orographic rain, fog, cloud contact, and snow (Bormann 1983). The occurrence and frequency of inversions, which prolong exposure to airborne chemicals, are determined by topography (Emberlin 1980). Sensitive trees exposed to high levels of gaseous pollutants may nevertheless receive low doses if environmental conditions cause them to keep their stomates closed during the period of exposure.

Even after the chemical has entered the ecosystem, it may be rapidly transformed by microbial processes, quickly reducing the quantity to which organisms are exposed. The physical structure of organisms can affect the chemical dosage they receive. For example, the height structure of vegetation influences the amount of aerosol deposition on leaf surfaces because the quantity of particles impacted on needles, leaves, and bark is influenced by leaf and canopy surface properties and architectural characteristics (Smith 1973; Lovett et al. 1982).

Once in the biota, there are three potential mechanisms of action of the chemical. Some chemicals are non-toxic, or are quickly converted to non-toxic forms, and consequently pass through the organisms without causing damage. Other chemicals may be present at levels too low to cause damage that cannot be quickly repaired. A chemical causes a stress to the organism if it results in damage to physiological or morphological processes within the organism. The damage may occur directly or as a result of alteration of cell metabolism.

Beginning at a threshold concentration at which injury becomes detectable, increases in dose rate cause increased injury. It is often found that the response to low concentrations over long time periods is not the same as an identical dose delivered at high concentrations for short time periods. Nevertheless, levels of tolerance to toxic chemicals are typically established by laboratory dose–response bioassays in which chemical concentrations are usually varied over constant exposure times. In general, higher concentrations are more likely to be lethal than lower concentrations, although mortality rates at low concentrations may increase with increased exposure time.

The response to low doses may be undetectable or may even stimulate growth. For example, nitrogen and sulfur are essential and are often growth limiting, leading to increased growth rates when deposited in moderate amounts, though they may create ecosystem problems under high levels of deposition. SO_2, NO_2, NH_3, and CO_2 can be converted to nutritive substances and stimulate growth. Non-metabolic substances may be converted to non-toxic substances or accumulate in tissues (Smith 1974; Feder 1973; Amundson and Weinstein 1980). The ability to tolerate toxic chemicals through biochemical processes often can be found in organisms present in high pollutant areas. These organisms may have evolved the mechanisms because of continual exposure of naturally produced pollutants. However, many heavy metals, although essential, can be tolerated only in trace amounts and if present in higher concentrations will cause metabolic injury. Other chemicals, such as ozone, fluoride, and PCBs, are toxic to metabolism in low concentrations.

As an example of the pathway from chemical release to organism injury, consider the case of sulfur dioxide emissions from power plants. Sulfur dioxide can either be absorbed by plants directly as gaseous SO_2 or may become oxidized in the air to H_2SO_4 and absorbed by plants as sulfate (SO_4^{2-}). If absorbed as sulfate, the chemical is either incorporated into amino acids or remains in the ionic form, resulting in no damage and, possibly, enhanced growth. However, sulfur dioxide readily dissolves in cellular fluids and forms SO_3^{2-} through dissociations of sulfurous acid and seriously damages cellular metabolism. Functional injuries, frequently expressed in terms of growth rate reductions, have been reported for many species, both agronomic and native, exposed to air pollutants (Constantis 1971; Feder 1973; Freer-Smith 1984; Harward and Treshow 1975; Kress and Skelly 1982; Skarby and Sellden 1984; Ernst et al. 1983).

7.2.2 Factors Affecting Organism Damage

Few studies are available demonstrating the extent of injury occurring following chemical exposure in the field. Kress and Skelly (1982) measured growth reduction following ozone exposure in ponderosa pine, loblolly pine, and sycamore saplings before there was any visible evidence of injury. Treshow and Stewart (1973) fumigated native species in the field and showed that in grassland, oak, aspen, and conifer communities, over half the perennial forbs and woody species were visibly injured at concentrations that could be anticipated in many ecosystems near anthropogenic sources. Physiological injury thresholds were not assessed, but postulated to occur at even lower concentrations. Morphological injuries, such as lesions and necrotic spots, are evident at somewhat higher concentrations than those that induce functional injuries. Tissues and organs (e.g., leaves, needles) become senescent more rapidly than normal. If feedback effects occur between tissue damage and impairment of resource gathering capabilities, the rate of senescence could be enhanced.

The various organisms in a terrestrial system are differentially susceptible to chemical exposure. Plants, for example, are readily affected by gaseous substances via their gas exchange mechanisms. Microorganisms and soil fauna are more directly susceptible to heavy metal toxicity because particulates accumulate in the litter and soil (Jackson and Watson 1977; Tyler 1984). Organismal tolerance to chemicals is a function of exposure prehistory (Legge 1980), genetic resistance (Bradshaw 1976; Ernst et al. 1983), and the developmental stage of the organisms (Guderian and Keuppers 1980).

Even if the damage does not result in permanent functional loss, it may cause the diversion of a significant amount of resources to the site of injury and subsequently impair the ability of the organism to perform other normal functions. One physiological stress may predispose an organism to increased susceptibility to other stresses. For example, SO_2 is known to decrease frost resistance (Tamm and Cowling 1977; Freer-

Smith 1984) and ozone is thought to predispose ponderosa pine trees to insect attack and disease (Miller 1973; Miller and McBride 1975; Dahlsten and Rowney 1980). In the current acid rain debate, it has been suggested that drought may predispose trees to pollution damage. The synergistic effects that occur when plants are simultaneously exposed to multiple chemical stresses, such as SO_2 with ozone (Knabe 1976) or SO_2 with NO_2 (Freer-Smith 1984), are examples of this phenomenon.

When chemical disturbance causes wide-spread mortality, it will result in effects similar to those under natural disturbance. However, far more commonly the chemical will will cause a chronic stress that does not result in permanent tissue loss; rather, it reduces the ability of the organism to behave normally. Changes in behavior, such as in growth rate, predation ability, or other resource gathering activities, are likely to result in a shift in the competitive balance among species. The inability to compete effectively can be a much more severe problem with which to cope than repairing internal injuries to cells. These shifts in competitive balance invariably lead to large-scale modifications of ecosystem structure. Impaired physiology can result in increased susceptibility to other stresses, such as disease, causing increasing mortality throughout the exposed populations. Alternatively, physiological shifts can alter reproductive rates. These changes significantly alter the long term population dynamics of species, and in turn, dramatically shift the structure of species abundances throughout the ecosystem. Such effects can even have landscape or regional influences if the stock for migration and seed source becomes diminished.

7.3 Consequences of Organism Injury to Alterations in Ecosystem Structure

Although physiological stresses are initiated at the level of individual organisms, manifestation at the whole ecosystem level results from interactions among organisms, and between organisms and their physiochemical environment (Levin and Kimball 1984). Some of the classic studies of succession in terrestrial ecology have identified paradigms for the response and reorganization of ecosystem structure following severe disruption of the existing organisms (Margalef 1968; Odum 1969; Whittaker 1975; Bormann and Likens 1979; Peet 1981). Severe natural disturbances cause large losses in the structural biomass of the terrestrial ecosystem. Following disturbance, there is a net accumulation of living and total biomass (Figure 7.2). This is accomplished through a rapid rise in net primary productivity. Interestingly, the high rate of productivity is associated with an initial rise in species diversity, but the diversity quickly declines without necessarily affecting the productivity rate. This rapid

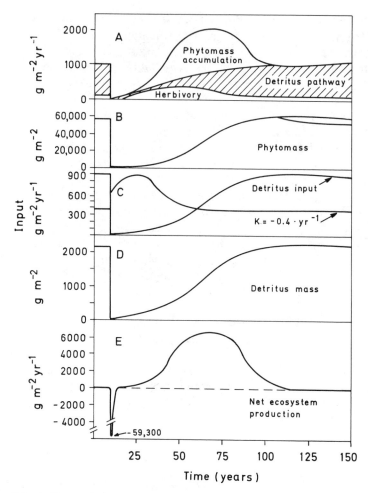

Figure 7.2. **A** Change in annual net primary production and the apportionment of
the energy represented by that mass following a single, non-unique disturbance.
The scaling is for a forest, but the allocation to herbivory is made unusually high
and the period of positive accumulation is made unusually short for illustrative
purposes. **B** The accumulation of phytomass based on the integration of the
phytomass accumulation area in **A**. A shallow, hump-shaped maximum is suggested by the shaded area representing a possible shunting of accumulated biomass to detritus. **C** Detritus input calculated from the integral of the detritus area
in A over the disturbance recovery time course. Also shown is a suggested change
in the decay coefficient (K) with changing conditions following the disturbance.
D Change in detritus mass calculated from input and decay variables in **C**. **E** Net
ecosystem production calculated from the change in phytomass and detritus mass
above. The sudden consumption or export of all phytomass and detritus at the
time of disturbance is considered to be net negative production. Animal biomass
is considered too trivial to consider. (From Reiners 1983.)

accumulation phase is followed by a stabilization, oscillation, or slow decline in biomass with time (Peet 1981). Whichever path the biomass pool follows, there is a change in net ecosystem production from positive during the accumulation phase to near zero during later successional stages (Reiners 1983).

In ecosystems in which nutrient mineralization is limited by temperature or by oxygen availability, physical disturbance is typically accompanied by a pulse of nutrients released from the litter and soil (Bormann and Likens 1979; Vitousek and Reiners 1975). This may occur because the disturbance increases the amount of sunlight and heat energy directly reaching the soil surface or because the disturbance disrupts the integrity of the soil, allowing deeper penetration of oxygen into the medium. Forest systems, especially those in northern latitudes, tend to have large amounts of nitrogen, potassium, and calcium in the litter layer. These forests can lose large quantities of these nutrients following tree cutting (Bormann and Likens 1979). However, forests with low nutrient stock in the litter layer or those in regions of higher ambient temperatures do not necessarily show these losses. Further, when regrowth is rapid following disturbance, the nutrients mobilized by increases in mineralization are quickly absorbed by the new vegetation, resulting in a redistribution of mass and nutrients within the ecosystem, but little net loss.

Species composition changes following destructive disturbances may in some cases follow well-established successional sequences (Noble and Slatyer 1980). Bare ground and open spaces facilitate the invasion of new species, while the elimination of some species releases others from competition. Yet in order to persist in a system disturbed frequently by destructive processes, a species must be adapted to complete its life cycles within the disturbance regime (Noble and Slatyer 1980; Reiners 1983). Frequent disturbance or disruptions covering large regions may cause declines in the sources of new seeds. Some species, such as pin cherry, can exist for long periods as buried seeds in the soil (Marks 1974). Thus, even if reproductive individuals are absent from the region, seedlings can still invade disturbed area. However, many species will decline if the adult seed source is reduced. On the other hand, many species of conifers depend upon fire or other disturbance to prepare a seed bed or to retard the growth of potentially competing species.

7.3.1 Consequences of Acute Lethal Toxicity to Alterations in Ecosystem Structure

The patterns of response to acute lethal chemical toxicity, such as the exposure of a plant community to a toxic dose of herbicide, are similar to those of a single severe episode of physical destruction. Both biomass and species diversity decline following the onset of mortality, and recover to pre-perturbation conditions of biomass structure and species composition as the toxicant concentration declines. Unlike the most physical destruction, where the rate of recovery is a function of the degree of disruption of the soil media, the rate of recovery is a function of the rate of disappearance or decrease in activity of the toxicant and the ability of opportunistic species, unaffected by the exposure, to rebuild biomass. As with physical disturbance, if plant material is lost, it initiates a cascade of decline throughout other trophic levels (Figure 7.3), with the largest and most rapid decline in biomass occurring among the herbivores directly feeding on the plant tissue, and carnivores and top carnivores gradually showing effects as their food supplies are affected. With chemicals that persist and continue to exert stress upon most producer organisms, this recovery will be slow or nonexistant. Examples of these patterns have been reported following oil spills (Hutchinson and Freedman 1978; Deneke et al. 1975) and following old-field herbicide treatment (Wakefield and Barrett 1977). Few studies are sufficiently long term to assess recovery, but Belsky (1982) reported that after 9 years, the effects of a spill in an alpine meadow were undetectable.

Acute lethal toxicity could reduce net productivity of autotrophs and biomass quite rapidly, resulting in a rapid accumulation of detritus. Where a sequential "peeling off" of tree, shrub, and herb strata occurs, for

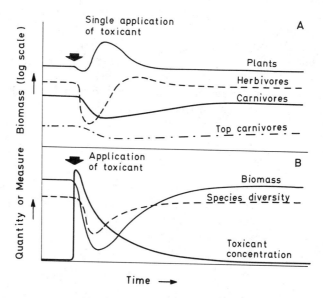

Figure 7.3. Some generalized characteristics of the effects of a toxicant on an ecosystem. Hypothetical changes in biomass of plants, herbivores, carnivores, and top carnivores are shown following a single application of a toxicant (top A). Changes expected in total biomass, diversity, and toxicant concentration with time are shown in bottom. Data from Eschenroeder et al. 1980 and Hynes 1960. (From Connell and Miller 1984.)

instance in areas downwind of smelters discharging SO_2, fluorides, and heavy metals (Gordon and Gorham 1963; Freedman and Hutchinson 1980), the rate of biomass loss apparently declines once the dominant trees are killed. In the event that a community of tolerant herbs and grasses becomes established, the biomass could stabilize once again.

While physical disruption does not tend to be species selective, species often have differential tolerances and sensitivities to chemicals. For example, many herbicides are selective against dicotyledonous species. Sensitivity to oil is a function of growth form (Walker et al. 1978), with woody species with larger underground root systems and latent buds being more resistant than herbs, mosses, and lichens. The consequence of this selectivity can be that species composition will change without any significant change in biomass following exposure.

7.3.2 Repeated Severe Exposures

Chemical exposures are often recurring events, rather than single episodes. Repeated chemical exposures can produce a reduction in the ability of the ecosystem biomass to recover without the physical disruption of the soil structure. Baker (1973) examined the successive inhibition of recovery with increasing duration of oil stress (successive applications) in the salt marsh cordgrass *Spartina angelica* (Figure 7.4). She found that while the amount of biomass reduction, as measured by the number of root tillers produced during the first year after initial exposure, was similar regardless of the number of exposures, the biomass after the second and third years decreased with increasing numbers of exposures. From her data she concluded that approximately 12 successive exposures to oil over 14 months would exceed the threshold of repair of the ecosystem. Hence, with repeated exposure, recovery to the original biomass was likely to be greatly retarded. Recovery to the original state of species composition may be further precluded when less sensitive species are released from competition and become dominant on the site in numbers not found on undisturbed sites. A similar suppression of recovery of biomass and pre-treatment species composition was observed following five years of application of herbicides on sagebrush communities, apparent even 15 years after spraying ended (Department of Defense 1967).

Post-spraying observations in the semi-deciduous forests of Vietnam paralleled those for sagebrush (Tschirley 1969). Repeated sprayings defoliated and subsequently killed trees, those most exposed in the top canopy being initially most susceptible, followed by saplings and seedlings. Living biomass rapidly declined, detritus accumulated, and the trees were apparently unable to recover significantly between successive sprayings. With increased light penetration, grasses and sedges quickly occupied the sites, a pattern typical of early succession. Where bamboo invaded sites, forest regeneration was significantly retarded.

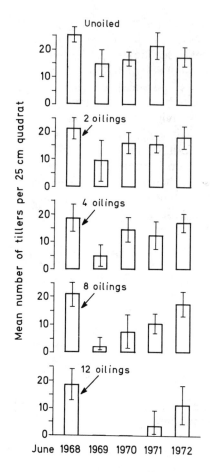

Figure 7.4. Effects of successive oil sprayings on *Spartina angelica* at Crofty, with 95% confidence limits. Number of tillers per quadrat shown in unoiled plots (top bars) and in plots treated with 2,4,8 and 16 oil applications during 1968. Average tiller growth in each treatment recorded in 1968, 1969, 1970, 1971, and 1972. (From Baker 1973.)

The frequency of acute disturbance may be as important a factor as repetition in determining the changes in ecosystem structure. Several patterns of change in ecosystem structure are suggested by an analysis by Reiners (1983) of the effects of changing the frequency of destructive disturbances (Figure 7.5). At low frequencies of disturbance, net primary productivity is increased by slight increases in frequency (Figure 7.5 A). However, when frequencies are already high, increases result in decreases in primary productivity. Regardless of initial frequency, increasing disturbance frequency results in a substantial decline in average plant biomass and detrital mass (Figure 7.5B). The rate of net primary productivity and, consequently, the rate of biomass accumulation, could be maintained at a very high level if the disturbance interval coincides with the length of time it takes slow-growing, late-successional species to replace the early-successional, fast-growing early ones. In essence, the disturbance causes continued dominance by fast-growing early successional

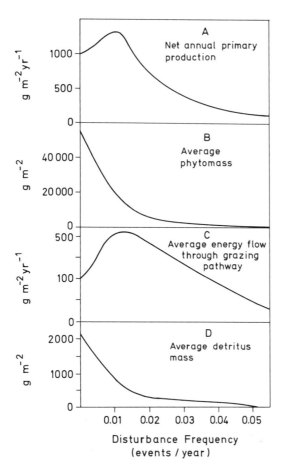

Figure 7.5. A Changes in integrated net primary production over an 180-yr period as a function of disturbance frequency. **B** Changes in average phytomass over the 180-yr period as a function of disturbance frequency. **C** Changes in energy flow through grazing or herbivory pathways over an 180-yr period as a function of disturbance frequency. **D** Changes in average detritus mass over 180-yr period as a function of disturbance frequency. (From Reiners 1983.)

species. Even with high productivity, however, biomass in the ecosystem is likely to be lower than if unperturbed. Under conditions of greater frequency of disturbance, vegetation may not have sufficient development time to reach maximum rates of productivity. In addition, in many cases frequent disturbance causes site conditions to deteriorate, resulting in a decline in the potential productivity achievable on the site following each disturbance. Reiners (1983) suggested that such a progressive decline would be reflected in an acceleration in the rate of decline in plant biomass and detritus. This model of the effects of disturbance frequency, while developed for natural disturbance, would be expected to apply to cases of repeated acute chemical exposure.

Exposure of the same ecosystem at two different times may produce differences in responses. For example, the effectiveness of pesticides in destroying pests and thus altering species composition has decreased over time because of their damage to predatory insects. Insect pests frequently evolve pesticide-resistant strains quite rapidly. However, predators exposed to the same pesticides and similarly susceptible to the toxic effects of those pesticides often do not adjust as rapidly. Pesticides used in cotton production in the United States enhanced production during the early years of application. However, resistant strains of the pests evolved in the early 1960s. By that time the predator populations, themselves sensitive to the pesticide, had been eliminated. The organophosphate insecticide, Azodrin, proved to be more lethal to cotton bollworm predators than to the bollworm for which it was targeted. The result of the use of this pesticide was a proliferation of pests and a drop in cotton productivity to one-third of the previous high value (Adkisson et al. 1982). The introduction of integrated pest management, a program desired to encourage the abundance of natural predators, has since enabled high production levels to once again be attained.

7.3.3 Chronic Exposures

A chronic stress exerts a lower level of impairment on organisms for a more lengthy period of time than those discussed above. It may be a repeated succession of episodes or a steady, continuous force. There are many natural stresses that clearly fall into the category of chronic stress. However, since these stresses tend to be a consistently ubiquitous part of the environment for decades or centuries, natural communities have developed a tolerance for them. However, a chronic xenobiotic chemical exposure is likely to be a stress not previously experienced in a terrestrial ecosystem. Many organisms would be sensitive to the chemical, as the processes of evolution would not have sufficient time to eliminate those sensitive individuals. Microorganism populations capable of detoxifying the chemicals, if they exist at all, are not likely to have had sufficient time to build significant population sizes.

A key difference between acute and chronic stresses, both physical and chemical, occurs in the dynamics of recovery. Between repeated destructive events, organisms may experience no residual restraint on growth and consequently may reestablish or recover from the stress. The same may be true between repetitions of acute chemical disturbances that disappear quickly from the environment. Those species able to complete their life cycles between events are those most likely to persist or invade a stressed area. Under a chronic stress, however, persistence is a function of being able to complete one's life cycle despite the stress. The rate of life-cycle development may be retarded by continued chemical exposure, causing a lengthening of the time required for life-cycle completion. Tolerant genotypes are rapidly selected in systems dominated by short-lived species, e.g., annual grasslands. Genotypic selection has certainly been shown to occur rapidly in species exposed to heavy metals accumulations (Bradshaw 1976; Ernst 1983), but it is not clear how productivity recovery is affected under these conditions.

Several potential responses of ecosystem biomass to chronic dosages of chemical substances have been suggested (Trudgill 1977; Smith 1981). Established stands exposed to low doses may well function effectively as "sinks" (Smith 1981) and show no measurable change in structure or function. Alternatively, productivity and biomass accumulation could be stimulated in ecosystems exposed to low concentrations of some chemicals, particularly growth-limiting substances like nitrogen (Abrahmsen 1980; Laurence and Weinstein 1981). Low ambient levels of fluoride, a theoretically toxic material, have also been reported to increase growth (Bunce 1985). In some cases, however, depending on the particular chemicals, the rate of deposition, and site conditions, indirect effects associated with shifts in resource allocation within the ecosystem, shifts in competitive ability among species, or other abiotic or biotic stresses could lead to a deterioration in biomass and biomass accumulation. However, where substances such as heavy metals accumulate in the system, it is likely that structural changes could be delayed and perhaps arise indirectly from effects in mineral nutrition, mediated by microorganisms and soil fauna.

The response to intermediate dosage levels is difficult to predict because of the possibility of multiple organism-level effects occurring at sub-lethal concentrations. In some cases, subtle shifts in species interactions through altered growth rates and competitive relationships occur, resulting in no significant alteration in net primary production and biomass accumulation (Laurence and Weinstein 1981). Injuries and increases in mortality in one or more species, particularly the dominants, could be induced, facilitating a decline in net primary production and biomass. In other cases, organism health, growth rates, and pest and pathogen defenses are diminished, making organisms susceptible to other stresses. Chronic oxidant pollution in the San Bernadino mountains is a good example of this scenario (see Miller 1973; Miller and McBride 1975). Ozone rendered the pine forests susceptible to insect invasion, eventually causing widespread tree mortality and biomass declines.

According to Woodwell (1983) and Odum (1984), chronic chemical exposure simplifies ecosystems by decreasing species richness and diversity. This pattern of decline is in principle similar to that caused by gamma radiation (Woodwell 1970) and has been observed in a number of sites in proximity to point-source discharges involving sizeable exposures. Yet results from other studies suggest that with less extreme exposure, species richness may either decrease or increase. For example, McClenahan (1978) found a decrease in species richness with increasing pollution load in hardwood stands, while Westman (1979) reported a significant increase in annuals and total species number in place of *Salvia* in a coastal sage scrub exposed to SO_2. An increase in diversity was observed in ponderosa pine stands in the San Bernadino mountains exposed to ozone, where scrubby species may become established once gaps in the canopy open (Miller and McBride 1975).

Effects on species richness have been observed where the release of chemicals has increased the nutrient capital in an ecosystem. The amount of influence of these releases depends in part on the initial levels of site fertility. The effects of nitrogen and sulfur enrichment via acidic deposition can result in a decline in richness in fertile sites, as observed in fertilization tests in the Park Grass experiments (Tillman 1982). Sludge applications in relatively low fertility stands of hardwoods and conifers, on the other hand, have been shown to increase species richness in the understory (David and Struchtmeyer 1982; Birk 1983). Such increases may not occur unless a long-term deterioration in the canopy facilitates an invasion of weedy species. Further, since terrestrial ecosystems show natural patterns of increases and decreases in species richness during stand development, the direction of change in richness resulting from nutrient input will depend on the stage of the stand development prior to exposure.

The better known examples of air pollutant effects on species composition and succession (summarized by Smith 1980, 1981; Sheehan et al. 1984) indicate that polluted systems exhibit replacement of more advanced communities by species of early seral stages (Whittaker and Woodwell 1974). This is clearly evident in areas around point-source discharges, particularly in response to exposures that injure or eliminate dominant species. For example, a 25-year-old coastal sage community reportedly resembled a 7-year-old stand after years of exposure to sublethal concentrations of SO_2 (Westman and Preston 1980). In aspen communities highly sensitive to low doses of ozone, exposure opens gaps in the canopy sufficiently large in size to favor the reestablishment of aspen instead of the normal transition to white pine (Harward and Treshow 1975). The white pine seedlings will be unable to become established, even though they are more resistant to ozone than the aspen, because they are outcompeted in the larger gaps.

The mechanisms that induce compositional changes vary in response to the intensity of exposure. Acute, lethal exposures killing both early and late successional species, result in communities characteristic of early seral stages. These are composed of tolerant, weedy species that are much faster invaders. Relatively infrequent, destructive events remove species, preparing a site for reinvasion and recovery along a natural successional trajectory. Changes induced by chronic chemical exposure, on the other hand, reverse the direction of compositional development by selectively eliminating the more sensitive species. Though chronic stress results in the same type of replacement community, the success of the weedy species comes not from the speed of seed distribution and establishment, but by their ability to maintain a faster growth rate than the original species. For example, in coastal sage scrub ecosystems exposed to air pollutants, *Salvia* is gradually replaced, though not eliminated, through competitive interactions. Functional injuries reduce plant size and shape, leaf area, growth rate, and reproduction (Westman and Preston 1980) facilitating invasion by more opportunistic species. Compositional changes occur quite rapidly (within a few years after the initiation of exposure) if the dominant species are very susceptible to injury (Treshow and Stewart 1973; Harward and Treshow 1975). With sufficient time and exposure, these retrogressive trends may become irreversible even after removal of the stress. If the disturbance leads to the breakdown of the soil or results in soil contamination sufficient to render it a poor rooting medium, the resulting ecosystem would remain permanently altered.

Our ability to predict changes in species composition based on species tolerance levels is complicated by the fact that species that show growth rate reductions in the laboratory may not do so in the field. Release from competition can more than make up for pollutant effects (Shugart et al. 1980). Kercher et al. (1980) predicted that, in western coniferous ecosystems exposed to SO_2, white pines with a decreased growth rate from pollution injury would replace ponderosa pines because the exposed ponderosa pines succumb to bark beetle attack following a weakening by the SO_2. In another example, along an SO_2 gradient at Wawa, Ontario, the understory species exhibit increases in abundance with SO_2 exposure because of the removal of their overstory competitors, even though these understory species were also damaged by the pollutant (Scale 1980).

7.3.4 Effects of Chemicals on Food Webs and Nutrient Supplies

Toxins have been observed to produce a breakdown in the tight coupling of food web interactions in ecosystems. Disruptions can occur when a few species have a much higher tolerance for a pollutant being released into the ecosystem than do their competitors or predators. As a result of the loss of these later groups, some of the remaining species are able to grow unchecked by predation. Alternatively, breakdowns in the food webs can occur when toxins alter behavioral patterns of organisms or increase the susceptibility to disease. Even if laboratory studies indicate that a given level of chemical exposure will not affect a population, detrimental effects may result from similar exposures in the field because of a shift in the competitive balance among species. Several cases have been reported in which a prey species not significantly affected in bioassays nonetheless declined in abundance because the ability to escape predators was affected (Hatfield and Anderson, 1972). In addition, the impairment of escape mechanisms can result in predators being exposed to higher than average concentrations of toxics because of their preference for prey weakened by toxicants.

Toxicant effects on critical species will disrupt population controls for the entire community. For example, exposure of mixed conifer–deciduous forests of western Canada to hydrogen fluoride gas resulted in an infestation of a number of insects which defoliated the trees. This infestation may have occurred because of damage to the predators feeding on these insects (Weinstein and Bunce 1981). The net result is that these trees, without a direct injury observable in response to the pollutant, have been severely damaged by the indirect effects of the pollutant upon the tree-pest-predator community. Further, trees exposed to sublethal levels of the hydrogen fluoride gas during the 1960s may have been weakened sufficiently to allow an insect population explosion to begin that did not become fully manifest until a decade later.

There are many populations within an ecosystem that perform a vital function in the cycles of physical and biological processes essential for providing all organisms within the ecosystem with sufficient supplies of energy and nutrients. As a consequence, the ability of the ecosystem to continue to maintain energy and nutrient pools is dependent on the effects on these populations. The recovery of these ecosystem properties from toxic exposure is closely linked to the recovery of these critical species.

At this time few of these critical species have been identified. As a consequence, our understanding of the ability of most ecosystems to accommodate toxic stress or to quickly recover from it is limited. It is clear, however, that ecosystems are less vulnerable than some constituent species; that is, they continue to provide resources in the presence of exposures that impair some non-critical member species (O'Neill et al. 1980). It is also clear that once the breakpoint at which these resources are no longer provided is past, a catastrophic loss of many groups of species is unavoidable.

Often the ability of a system to retain scarce quantities of nutrients is dependent on the coordinated activities of producers, consumers, and decomposers. When nutrients are mineralized from organic matter, producers must be in an active growth stage and able to take up the available nutrients or risk mobile ions being lost from the system. Cadmium has been demonstrated to disrupt the coordination of organisms and result in loss of calcium and nitrogen from microcosms (O'Neill et al. 1977). Inhibition of plant growth with herbicides resulted in large losses from the watershed system of the Hubbard Brook Experimental Forest (Bormann and Likens 1979). Chemical additions can also increase nutrient loss by physically disrupting the soil matrix or by indirectly causing elevated soil temperatures because of inhibition of the development of the plant canopy. In some cases, such as in Copper Hill, Tennessee, and Sudbury, Ontario, toxic emissions of sulfur dioxide are high enough to disrupt all biotically mediated processes within the system and prevent regeneration of any biota whatsoever. Persistent chemicals prevent the ecosystem from trapping newly mobilized nutrients in rapidly growing new vegetation.

7.4 Conclusions

Severe cases of physically and chemically induced stresses in ecosystem have similar consequences with regard to retarding the development of ecosystem biomass and composition. Chronic levels of chemical exposure create a very different perturbation than physical disturbance. As a consequence, the mechanisms that operate following these perturbations differ, as do the changes in structural properties that result from those mechanisms. The general models of ecosystem response to chemical stress proposed by Woodwell (1983) and Odum (1984) are appropriate to the case of exposure to lethal doses of chemicals, but do not adequately account for changes in ecosystem structure induced by sub-lethal doses. It appears that chronic exposure to low doses of air pollutants, for example, could alter the species mix of a forest, by inducing subtle changes in competitive interactions, with or without a concomitant decline in productivity or biomass. The question still remains, however, as to whether systems exposed to low-level chemical stresses will eventually show the same structural changes observed in ecosystems exposed to toxic concentrations. To examine the ecosystem-level effects of chemical doses, we must construct ecosystem models that incorporate both competitive interactions and species sensitivities to single and multiple chemicals.

Given the paucity of ecosystem data, conclusions drawn from the responses discussed above are perhaps rather speculative. Several reasons for caution in extrapolating from the above examples are listed below:

Rarely are ecosystems subjected to a single chemical input of anthropogenic origin. Systems typically are exposed to a suite of substances simultaneously.

Many of the effects of chemicals may take a long time to become manifested at the ecosystem level. As a consequence, exposures that took place decades ago may have effects that have not yet been recognized or even searched for.

There are species differences in sensitivity to different chemicals. Consequently, exposure of an ecosystem to two different chemicals may produce two very different responses.

Competitive interactions, coincident natural environmental stresses such as fire and drought, and inherent variability and change in an ecosystem over time are further confounding factors that make understanding the effects of chemical stress a difficult task.

Nevertheless, it seems that some broad generalizations regarding structural change are possible. There is substantial evidence to indicate that direct terrestrial ecosystem responses, such as biomass accumulation, vary according along a dose gradient from low to high (Smith 1981). However, indirect effects, such as food web alteration or nutrient supply alteration, can occur at any point on this gradient, with manifestations in ecosystem structure that vary depending on the role played by the damaged species. It is clear that the rates and degree of change in terrestrial ecosystem structure will vary with the intensity of exposure to chemicals, but it is not clear how rapidly different communities exposed to the same dose will change. Serious deficiencies in our understanding of the effects of chemical stress in terrestrial ecosystems concern the degree to which developing ecosystems are deflected from their normal trajectories over the long-term, and their capacity to recover once a chemical stress is removed.

The chemical substances that may have the most substantial effect on terrestrial ecosystems over the long term are those that are dispersed over wide regions at concentrations that induce sub-lethal, chronic, physiological stress. These substances are the hardest to detect. The changes in structure that such chronic exposures can induce may not be manifested for long periods. As a consequence, the danger of the release of these chemicals may go unrealized until the ecosystem alteration is fully expressed. The worry is that at such a point the change may be virtually irreversible.

Acknowledgments. This publication is Number ERC-83 of the Ecosystems Research Center, Cornell University, and was supported by the U.S. Environmental Protection Agency Cooperative Agreement Number CR812685–01. Additional funding was provided by Cornell University.

The work and conclusions published herein represent the views of the authors, and do not necessarily represent the opinions, policies, or recommendations of the Environmental Protection Agency.

References

Abrahamsen G (1980) Acid precipitation, plant nutrients, and forest growth. In: Drabløs D, Tollan A (eds) *Ecological Effects of Acid Precipitation,* Mysen, Norway, pp. 58–63

Adkisson PL, Niles GA, Walker JK, Bird LS, Scott HB (1982) Controlling cotton's insect pests: a new system. Science 216: 19–22

Amundson RG, Weinstein LH (1980) Effects of airborne F on forest ecosystems. In: *Proceedings of Symposium on Effects of Air Pollutants on Mediterranean and Temperate Forest Ecosystems.* U.S.D.A. Forest Service Pacific SW Forest Experiment Station General Technical Report PSW-43, Berkeley, CA, pp. 63–78

Baker JM (1973) Recovery of salt marsh vegetation from successive oil spillages. Environ Pollut 4: 223–230

Belsky J (1982) Diesel oil spill in a sub-alpine meadow: Nine years of recovery. Can J of Bot 60: 906–910

Birk EM (1983) Nitrogen availability, nitrogen cycling, and nitrogen use efficiency in Loblolly pine stands at the Savannah River Plant, South Carolina. Dissertation, University of North Carolina, Chapel Hill, NC, USA

Bormann FH (1983) Factors confounding evaluation of air pollution stress on forests: pollution input and ecosystem complexity. Paper prepared for the symposium *"Acid deposition, a challenge for Europe."* Commission of the European Communities, Karlsruhe, FRG, September 19–23 , 1983

Bormann FH, Likens GE (1979) *Pattern and process in a forested ecosystem.* New York: Springer-Verlag, 253 pp.

Bradshaw AD (1976) Pollution and evolution. In: Mansfield TA (ed) *Effects of Air Pollutants on Plants.* Cambridge: Cambridge University Press, pp. 135–159

Brandt CJ, Rhodes RW (1972) Effects of limestone dust accumulation on composition of a forest community. Environ Pollut 3: 217–225

Buckley EH (1982) Accumulation of airborne polychlorinated biphenyls in foliage. Science 216: 520--522

Buckley EH (1983) Decline of background PCB concentrations in vegetation in New York state. Northeast Environ Sci 2: 181–187

Bunce HWF (1985) Apparent stimulation of tree growth by low ambient levels of fluoride in the atmosphere. J Air Pollut Control Assoc 35: 46–48

Connell DW, Miller GJ (1984) *Chemistry and Ecotoxicology of Pollutants.* New York: Wiley, 444 pp.

Constantis AC (1971) Effects of ambient sulfur dioxide and ozone on eastern white pine in a rural environment. Phytopathology 6: 717–720

Cowling EB, Dochinger LS (1980) Effects of acidic precipitation on health and the productivity of forests. In: *Proceedings of Symposium on Effects of Air Pollutants on Mediterranean and Temperate Forest Ecosystems.* U.S.D.A. Forest Service Pacific SW Forest Experiment Station General Technical Report PSW-43, Berkeley, CA, pp. 165–173

Dahlsten DL, Rowney DL (1980) Influence of air pollution on population dynamics of forest insects and on tree mortality. In: *Proceedings of Symposium on Effects of Air Pollutants on Mediterranean and Temperate Forest Ecosystems.* U.S.D.A. Forest Service Pacific SW Forest Experiment Station General Technical Report PSW-43, Berkeley, CA pp. 125–130

David MB, Struchtemeyer RA (1982) Vegetation response to sewage effluent disposal on a hardwood forest. Can J For Research 12: 1013–1017

Deneke FJ, McCown BH, Coyne PI, Rickard W, Brown J (1975) Biological aspects of terrestrial oil spills. *USA CRREL Oil Research in Alaska, 1970–1974.* Cold Regions Research and Engineering Laboratory, Hanover, NH, Research Report 346, December 1975

Department of Defense (1967) Assessment of the effects of extensive or repeated use of herbicides. Final Report 15 August –1 December 1967. Sponsored by the Advanced Research Projects Agency ARPA Order No. 1086

Emberlin JC (1980) Smoke sulfur dioxide concentrations in relation to topography in a rural area of central southern England. Atmosph Environ 14: 1381–1390

Ernst WHO, Verkleij JAC, Vooijs R (1983) Bioindication of a surplus of heavy metals in terrestrial ecosystems. Environ Monitor and Assess 3: 297–305

Eschenroeder A, Irvine E, Lloyd A, Tashima C, Khanh Tran (1980) Computer simulation models for assessment of toxic substances. In: Haque R (ed) *Dynamics, Exposure, and Hazard Assessment of Toxic Chemicals*. Ann Arbor, MI: Ann Arbor Sciences, pp. 323–386

Feder WA (1973) Cumulative effects of chronic exposure of plants to low levels of air pollutants. In: Naegle JA (ed) *Air Pollution Damage to Vegetation*. Adv Chem Series No. 122, American Chemical Society, Washington DC, U.S.A., pp. 21–30

Freedman B, Hutchinson TC (1980) Long-term effects of smelter pollution at Sudbury, Ontario, on forest community composition. Can J Bot 58: 2123–2140

Freer-Smith PH (1984) The responses of six broadleaved trees during long-term exposure to SO_2 and NO_2. New Phytol 97: 49–61

Gordon AG, Gorham E (1963) Ecological effects of air pollution from an iron-sintering plant at Wawa, Ontario. Can J Bot 41: 1063–1078

Gorham E, Vitousek PM, Reiners WA (1979) The regulation of chemical budgets over the course of terrestrial ecosystem succession. Ann Rev Ecol Sys 10: 53–88

Grime JP (1979) *Plant Strategies and Vegetation Processes*. Chichester, UK: Wiley, 222 pp.

Guderian R, Keuppers K (1980) Response of plant communities to air pollution. In: *Proceedings of Symposium on Effects of Air Pollutants on Mediterranean and Temperate Forest Ecosystems*. U.S.D.A. Forest Service Pacific SW Forest Experiment Station General Technical Report PSW-43, Berkeley, CA, pp. 187–199

Harward M, Treshow M (1975) Impact of ozone on the growth and reproduction of understory plants in the aspen zone of western U.S.A. Environ Conserv 2: 17–23

Hatfield CT, Anderson JM (1972) Effects of two insecticides on the vulnerability of Atlantic salmon parr to brook trout predation. J Fish Res Board Can 29: 27–29

Hill AC (1971) Vegetation: a sink for atmospheric pollutants. J Air Pollut Contr Assoc 21:341–346

Hutchinson TC, Freedman W (1978) Effects of experimental spills on subarctic boreal forest vegetation near Norman Wells, N.W.T., Canada. Can J Bot 56: 2424–2433

Hynes HBN (1960) *The biology of polluted waters*. Liverpool, UK: Liverpool University Press, 202 pp.

Jackson DR, Watson AP (1977) Disruption of nutrient pools and transport of heavy metals in a forested watershed near a lead smelter. J Environ Qual 6: 331–338

Kercher JR, Axelrod MC, Bingham GE (1980) Forecasting effects of SO2 pollution on growth and succession in a western conifer forest. In: *Proceedings of Symposium on Effects of Air Pollutants on Mediterranean and Temperate Forest Ecosystems*. U.S.D.A. Forest Service Pacific SW Forest Experiment Station General Technical Report PSW-43, Berkeley, CA, pp. 200–202

Knabe W (1976) Effects of SO$_2$ on terrestrial vegetation. Ambio 5: 213–218

Kress LW, Skelly JM (1982) Response of several eastern forest tree species to chronic doses of ozone and nitrogen dioxide. Plant Disease 66:1149–1152

Laurence JA, Weinstein LH (1981) Effects of air pollutants on plant productivity. Ann Rev Phytopathol 19: 257–71

Legge AH (1980) Primary productivity, sulfur dioxide, and the forest ecosystem: an overview of a case study. In: *Proceedings of Symposium on Effects of Air Pollutants on Mediterranean and Temperate Forest Ecosystems.* U.S.D.A. Forest Service Pacific SW Forest Experiment Station General Technical Report PSW-43, Berkeley, CA, pp. 51–62

Levin SA, Kimball K (1984) New perspectives in ecotoxicology. Environ Manage 8: 375–442

Lovett GM, Olson RK, Reiners WA (1982) Cloud droplet deposition in sub-alpine balsam fir forests: hydrological and chemical inputs. Science 218: 1303–1304

Margalef R (1968) *Perspectives in ecological theory.* Chicago: University of Chicago Press, 111 pp.

Marks PL (1974) The role of pin cherry (*Prunus pennsylvanica* L.) in the maintenance of stability in northern hardwood ecosystems. Ecol Monogr 44: 73–88

McClenahan JR (1978) Community changes in a deciduous forest exposed to air pollution. Can J For Research 8: 432–438

Miller PL (1973) Oxidant-induced community change in a mixed conifer forest. In: Naegle JA (ed) *Air Pollution Damage to Vegetation.* Adv Chem Series No 122, American Chemical Society, Washington DC, U.S.A., pp. 101–117

Miller PR, McBride JR (1975) Effects of air pollutants on plants. In: Mudd JB, Kozlowski TT (eds) *Responses of Plants to Air Pollution.* New York: Academic Press, pp. 195–235

Noble IR, Slatyer RO (1980) The use of vital attributes to predict successional changes in plant communities subject to recurrent disturbances. Vegetatio 43:5–21

Odum EP (1969) The strategy of ecosystem development. Science 164: 262–270

Odum EP (1985) Trends expected in stressed ecosystems. Bioscience 35(7):419–422

O'Neill RV, Ross-Todd BM, O'Neill FG (1980) Synthesis of terrestrial microcosm studies. In: Harris WF (ed) *Microcosms as Potential Screening Tools for Transport of Toxic Substances.* ORNL TM-7028. Oak Ridge National Laboratory, Oak Ridge, TN, pp. 239–257

O'Neill RV, Ausmus BS, Jackson DR, Van Hook RI, Van Voris P, Washburne C, Watson AP (1977) Monitoring terrestrial ecosystems by analysis of nutrient export. Water Air Soil Pollut 8: 271–277

OTA, Office of Technology Assessment (1984) Acid rain and transported air pollutants: implications for public policy. Washington D.C., U.S. Congress, OTA-O-204, June 1984

Peet RK (1981) Changes in biomass and production during secondary forest succession. In: West DC, Shugart HH, Botkin DB (eds) *Forest Succession: Concepts and Application.* New York: Springer-Verlag, pp. 324–338

Reiners WA (1983) Disturbance and basic properties of ecosystem energetics. In: Mooney HA, Godron M (eds) *Disturbance and Ecosystems.* New York: Springer-Verlag. pp. 83–98

Scale PR (1980) Changes in plant communities with distance from an SO$_2$ source.

In: *Proceedings of Symposium on Effects of Air Pollutants on Mediterranean and Temperate Forest Ecosystems*. U.S.D.A. Forest Service Pacific SW Forest Experiment Station General Technical Report PSW-43, Berkeley, CA, p. 248–258

Schofield CL (1976) Acid precipitation: effects on fish. Ambio 5: 228–30

Sheehan PJ (1984a) Effects on individuals and populations. In: Sheehan PJ, Miller DR, Butler GC, Bordeau P (eds) *Effects of Pollutants at the Ecosystem Level, SCOPE No. 22*. New York: Wiley, pp. 23–50

Sheehan PJ (1984b) Effects on ecosystem and community structure and dynamics. In: Sheehan PJ, Miller DR, Butler GC, Bordeau P (eds) *Effects of Pollutants at the Ecosystem Level, SCOPE No. 22*. New York: Wiley, pp. 50–100

Sheehan PJ (1984c) Functional changes in the ecosystem. In: Sheehan PJ, Miller DR, Butler GC, Bordeau P (eds) *Effects of Pollutants at the Ecosystem Level, SCOPE No. 22*. New York: Wiley, pp. 101–145

Shugart HH, McLaughlin SB, West DC (1980) Forest models: their development and potential application for air pollution effects research. In: *Proceedings of Symposium on Effects of Air Pollutants on Mediterranean and Temperate Forest Ecosystems*. U.S.D.A. Forest Service Pacific SW Forest Experiment Station General Technical Report PSW-43, Berkeley, CA, pp. 203–214

Shugart HH (1984) *A Theory of Forest Dynamics*. New York: Springer-Verlag, 278 pp.

Skarby L, Sellden G (1984) The effects of ozone on crops and forests. Ambio 13: 68–72

Skelly JM (1980) Photochemical oxidant impact on mediterranean and temperate forest ecosystems: real and potential effects. In: *Proceedings of Symposium on Effects of Air Pollutants on Mediterranean and Temperate Forest Ecosystems*. U.S.D.A. Forest Service Pacific SW Forest Experiment Station General Technical Report PSW-43, Berkeley, CA, pp. 38–50

Smith WH (1973) Metal contamination of urban woody plants. Environ Sci Technol 7: 631–636

Smith WH (1974) Air pollution-effects on the structure and function of the temperate forest ecosystem. Environ Pollut 6: 111–129

Smith WH (1980) Air pollution - a 20th century allogenic influence on forest ecosystems. In: *Proceedings of Symposium on Effects of Air Pollutants on Mediterranean and Temperate Forest Ecosystems*. U.S.D.A. Forest Service Pacific SW Forest Experiment Station General Technical Report PSW-43, Berkeley, CA, pp. 79–87

Smith WH (1981) *Air pollution and forests. Interactions between air contaminants and forest ecosystems*. New York: Springer-Verlag, 379 pp.

Sousa WP (1984) The role of disturbance in natural ecosystems. Ann Rev Ecol Syst 15: 353–392

Tamm CO, Cowling EB (1977) Acidic precipitation and forest vegetation. Water Soil Air Pollut 7: 503–511

Tilman D (1982) *Resource competition and community structure*. Monographs in Population Biology 17. Princeton: Princeton University Press, 296 pp.

Treshow M, Stewart D (1973) Ozone sensitivity of plants in natural communities. Biol Cons 5: 209–214

Trudgill ST (1977) *Soil and Vegetation Systems*. Oxford: Clarendon Press, 180 pp.

Tschirley FH (1969) Defoliation in Vietnam. Science 167: 779–786

Tyler G (1984) The impact of heavy metal pollution on forests: A case study of Gusum, Sweden. Ambio 13:18–24

Vitousek PM, Reiners WA (1975) Ecosystem succession and nutrient retention: a hypothesis. Bioscience 25: 376–381

Wakefield NG, Barrett GW (1979) Effects of positive and negative perturbations on an old field ecosystem. Amer Mid Nat 101:159–169

Walker DA, Weber PJ, Everett KR, Brown J (1978) Effects of crude and diesel oil spills on plant communities at Prudhoe Bay, Alaska, and the derivation of oil spill sensitivity maps. Arctic 3: 242–259

Weinstein LH, Bunce HWF (1981) Impact of emissions from an alumina reduction smelter on the forests at Kitimat, B.C.: A synoptic review. In: *Proceeding of the 74th Annual Meeting of the Air Pollution Control Association,* Philadelphia, PA, June 21–26, 1981, pp. 2–16

Westman WE (1979) Oxidant effects on Californian coastal sage scrub. Science 205: 1001–1003

Westman WE, Preston KP (1980) Sulfur dioxide and oxidant effects on Californian coastal sage scrub. In: *Proceedings of Symposium on Effects of Air Pollutants on Mediterranean and Temperate Forest Ecosystems.* U.S.D.A. Forest Service Pacific SW Forest Experiment Station General Technical Report PSW-43, Berkeley, CA, p. 256

White PS (1979) Pattern, process, and natural disturbance in vegetation. Botan Rev 45: 229–299

Whittaker RH (1975) *Communities and Ecosystems*. New York: Macmillan, 387 pp.

Whittaker RH, Woodwell GM (1974) Retrogression and coenocline distance. In: Whittaker RH (ed) *Ordination of Plant Communities*. The Hague, The Netherlands: Dr. W. Junk, pp. 51–70

Woodwell GM (1967) Radiation and the patterns of nature. Science 156: 461–470

Woodwell GM (1970) Effects of pollution on the structure and physiology of ecosystems. Science 168: 429–433

Woodwell GM (1983) The blue planet: of whole and parts of man. In: Mooney HA, Godron M (eds) *Disturbance and Ecosystems*. New York: Springer-Verlag, pp. 1–10

Part III
Methods and Models

Prediction, as with any form of generalization or extrapolation, must build on a body of experiences. In the case of ecosystems and chemical stress, understanding of empirical responses and underlying mechanisms may be incorporated into models providing the basis for wider extrapolation. Thus, models are an essential facet of ecotoxicology.

Yet models vary and predictive capability is not always an explicit goal. Levin (Chapter 8) reveals a milieu of purposes and approaches for models, but also offers some fundamental technical challenges that often cut across form. As examples, the issue of spurious correlation without causation; decisions as to level of detail and dimension; and the recognition of ecosystems as dynamic, non-equilibrium systems with scales of heterogeneity and variance all complicate the modeling of biological–physical interactions.

Some models can be derived from basic physical principles, such as those emphasizing fate and transport of elements. But biological involvement in these processes complicate the problem, creating needs to fit parameters to a specific situation, to alter the scales considered by the model, or extrapolate from one type of biological model to the greater variety of species that exist in the ecosystem (O'Connor et al., Thomann, and Farrington, Chapters 9, 10, and 11, respectively). Because of the biological diversity, biological effects modeling remains most difficult, and biomonitoring approaches may be instrumental to incorporate into environmental decision-making models (Herricks et al., Chapter 13).

Other types of models may be mathematical in form, but vastly different in detail and structure. For example, in the global arena of biogeochemical cycles, the detail of physical principles emphasizing diffusion and advection are less relevant to provide rough estimates for time constants for large-scale perturbation scenarios. Here, the results rest more on empirical determination of sizes of compartments and assumptions as to flow between major environmental reservoirs (Lerman et al., Chapter 12).

Finally, there are also physical models, such as microcosms or meso-cosms (Gillett, Chapter 14; Gearing, Chapter 15). These models can serve as a bridge between single-species tests and *in situ* field manipulations of entire ecosystems. They provide the proper framework for assessing a complex set of ecological interactions and thus aid in gaining mechanistic understanding of certain processes to be included in mathematical de-scriptions of system behavior. Physical models have limitations of scale; the degree to which results can be extrapolated, and to what ecosystems, needs explicit consideration. This principle concern with scale, however, is ubiquitous throughout modeling in general; and the development of a more predictive ecotoxicology depends heavily on fundamental under-standing of how biological processes and biological–physical interactions change with scales of observation and disturbance.

Chapter 8
Models in Ecotoxicology: Methodological Aspects

Simon A. Levin[1]

The science of ecotoxicology is based to a large extent on extrapolation—extrapolation from one system and one stress to another; extrapolation from laboratory tests and microcosm studies to field situations; extrapolation across scales. Such extrapolation must be based on some underlying model or models; thus, models are an essential and ineluctable component of ecological risk assessment.

Models come in a variety of forms, and serve a variety of purposes. They may be the means for explicit prediction, may serve as screening tools, may provide understanding of the mechanisms underlying observed patterns, or may be part of an integrated adaptive management scheme. Because models are called on to serve so many purposes, it would be foolhardy to expect a simple model to meet all objectives. Different criteria must be applied depending on the uses to which one intends to put the model. Some models are deliberately oversimplified in order to isolate parts of a complex system and to provide a means for investigating the implications of hypothetical relationships. Often, to understand why a particular ecological relationship is what it is, it is necessary to embed that relationship within a broader framework of possibilities, and then to ask why one form of a relationship is observed more often than another. In some situations, such as models used for prediction of fish stock dynamics, parameter estimation is a fundamental consideration, and the ease and reliability of estimating parameters must influence the form of the model. Other population models, however, including some fisheries models, are intended primarily for pedagogical purposes, and are de-

[1] Section of Ecology and Systematics, Ecosystems Research Center, and Center for Environmental Research, Cornell University, Ithaca, New York 14853

signed without regard for the parameter estimation problem. It is when such models have been misapplied, taken outside of the contexts in which they were developed, that abuses have occurred. The result has been misunderstanding about the power of models (Limburg et al. 1986). The most important conclusion is that the multiplicity of purposes that models serve must be recognized and the assumptions underlying particular models, and their capabilities, must be understood. One must select models that are appropriate to the purposes at hand rather than trying to fit a square peg into a round hole.

Ecosystem models are perhaps the best choices for illustrating a familiar problem for modelers – that of appropriate detail—but it is a problem that pertains to any class of models. Many of the models developed for ecosystems are highly detailed, and include hundreds of components. Such models serve as tableaux, upon which can be laid out a panoply of interactions and flows; but such models are useless as tools for prediction. Apart from the fact that the level of detail in such models must necessarily be arbitrary—any grouping could be subdivided further and further—they contain too many parameters to be estimated robustly and to provide reliable prediction. The problem of appropriate detail is a fundamental one in modeling, and one that will be addressed later. The large ecosystem models do serve a purpose in displaying relationships; but more simplistic relationships such as the Vollenweider loading curves (Vollenweider 1968) have been far more useful and influential as predictive devices (Peters 1986). A challenge to modelers, however, is to replace these regressions with mechanistic models that do not err on the side of mindless detail, but provide sufficient biological and physical detail to explain why the observed relationships hold and predict when they will not (Lehman 1986).

In general, in ecotoxicology, it is essential to recognize that no single model can meet all objectives, and that one must interface models that contain different levels of detail. This is evident in the models of fate and transport of chemicals discussed by O'Connor et al. and Thomann in the following two chapters of this section. Generic models of fate and transport incorporate basic mechanisms, such as diffusion, advection, reaction, and depuration. These components must always be present, and generic models provide the basic structure for incorporating them. However, application of these models to specific cases involves fine-tuning them to particular environments and fixing the parameters either from basic considerations or by calibration. Many general questions are best answered by consideration of the more abstract and generic models; others require detailed consideration of local geometry and topography, and can be investigated only by computer simulations. Both kinds of models, and hybrid versions, are essential in ecotoxicology. Much has been learned from the full spectrum of investigations, ranging from abstract treatment of partial differential equations to very detailed site-specific, three-dimensional simulations.

8.1 Physical and Biological Scales

The earliest applied models in mathematical ecology, those introduced by Volterra (1926) in considering the fisheries of the Adriatic, shared a number of features:

1. They considered only the interactions among populations, ignoring details of the physical environment.
2. They were deterministic and autonomous; that is, they expressed relationships that did not depend on time.
3. They treated time as continuous.

Each of these features subsequently has been relaxed by various investigators, but it is the first restriction to which attention is directed here. The dichotomy between models that emphasize biological versus physical factors has divided population ecologists from ecosystem ecologists, and unnecessarily so (Kingsland 1985). As is usually the case with scientific dichotomies, the truth lies somewhere between the extremes, and is inseparable from the question of scales. In the oceanographic literature, it is often taken as a starting point that temporal scales are biologically determined while spatial scales are physically determined (Denman and Powell 1984). But the situation is more complicated than that. Oceanic species are patchily distributed on almost every scale of investigation. On broad scales, consideration of Fourier spectra indicates coherence between the distribution of physical markers such as temperature and the distribution of biological species. This suggests that on the broadest scales the spatial distribution of the biota indeed is determined by physical factors. On finer spatial scales, however, on the order of up to 50 km, such coherence often breaks down, especially at higher trophic levels (Weber et al. 1986) indicating that biological processes and interactions must be important in determining fine scale patchiness. Such correlations, often viewed as being synonymous with prediction (Peters 1986), cannot be the stopping point. Lehman (1986) elegantly has reminded us of the need to move from correlation to causation, for otherwise we cannot know the limits of predictability. However, correlations certainly are suggestive, and can serve an important role in helping to structure a system of models, especially with regard to choosing appropriate scales.

In the development of predictive models, two factors have been paramount in preventing the integration of physics and biology. One is the desire for simplicity and reliability: a model that has only two or three parameters has little to hide, whereas a complex computer code is virtually impenetrable. Furthermore, as already discussed, detailed models are harder to parameterize and may be less reliable for prediction than simpler models. Secondly, highly detailed models can be investigated only via computer simulations, and exhaustive simulations can be expensive and cumbersome to carry out. Recent advances in computing, however, and in hierarchical approaches to modeling provide us the opportunity to overcome these obstacles, and to develop the requisite combined models. Component biological models, such as those describing the movements and searching behavior of organisms on small scales, can be integrated to provide the behavior of aggregates at higher scales; these can be coupled with models of physical processes. Parallel processing facilitates incorporation of environmental heterogeneity and systematic exploration of parameter regimes. It is to be expected that the next few years will see great advances in the ability to develop integrated physical and biological models, and improvements in methods for examining the interrelationships among phenomena on different scales.

8.2 Aggregation, Simplification, and the Problem of Dimensionality

There is no correct level of description for an ecosystem. Just as Mandelbrot (1983) has educated us to the endless spectrum of detail that can be observed at finer and finer levels of description, so can we find an endless spectrum of detail in the taxonomic or functional organization of an ecosystem (O'Neill et al. 1986; Levin 1987). The analysis of the structure of food webs, one of the hottest topics in ecological theory, is frustrated by the arbitrariness with which descriptions of food webs must be amassed. Any investigator will, according to personal taste, lump certain groups of species while breaking others into age classes. The decision is based inevitably on the particular investigator's perspective, and on what is of special interest in a particular study. There can be no unique way to describe an ecosystem. Each biological grouping is an attempt at simplification, focusing on the average properties of an ensemble rather than on the variance within the ensemble. Partitioning an ensemble into subunits accounts for some of the heterogeneity within the ensemble, but at the cost of a reduced capability to make deterministic statements about the subunits. This paradox is not unique to biology; it confronts investigators studying any system, and underlies such basic concepts as the measurement of temperature in physical systems.

Given that there is no correct level of detail, it is important to learn how variability changes across scales, and what the consequences are of lumping components previously treated as independent.

For the dynamical systems common in ecological modeling, in which the dynamics of system components are assumed to be governed by autonomous equations of the form

$$dX_i/dt = f_i(X_1, \ldots, X_n), i = 1, \ldots, n \qquad (8.1)$$

the formal aggregation problem is that of finding a reduced set of variables.

$$Y_j = g_j(X_1, \ldots, X_n), j = 1, \ldots, m, m < n \qquad (8.2)$$

that satisfy their own autonomous dynamics

$$dY_j/dt = F_j(Y_1, \ldots, Y_m), j = 1, \ldots, m. \qquad (8.3)$$

In general, such reduction is not possible; the dynamics of any reduced set of variables Y_j is likely to depend not only on its own values, but also on other (hidden) variables. However, in some cases such reduction can be made (Iwasa et al. 1987). If one defines the Jacobian matrix $B = (B_{jk})$, where $B_{jk} = \partial g_j/\partial X_k$, and the second matrix $A = (A_{jk})$, where $A_{jk} = \partial(\Sigma_l B_{jl} f_l)/\partial X_k$, then (Iwasa et al. 1987) the formal condition that (2) leads to an autonomous (perfect aggregation) scheme is that

$$AB^+B = A \qquad \text{for all } X, \qquad (8.4)$$

where B^+ indicates the generalized inverse of the matrix B (Penrose 1955).

In general, it is unreasonable to expect such perfect aggregation to be possible, and one turns to approximate schemes that minimize the error associated with dimensional reduction. This is closely related to the investigations of Schaffer and his collaborators (Schaffer et al. 1986), who apply the techniques of Takens (1981) to determine the fractal dimensionality of data sets. To the best of my knowledge, no attempt has been made to relate these two lines of inquiry, which are complementary to one another.

The need for simplification is widely acknowledged, but techniques for achieving it are scarce. Levin (1987) argues that "the choice of a useful model must be governed by the dynamic behavior of candidate models, by their parametric sensitivity, and by the tradeoffs between uncertainty in knowledge of model structure and uncertainty in acquiring data. . . . The general approach is to begin with some particular system description, to study the behavior of the dynamics associated with that description, and to compare the behavior with that of reduced (e.g., aggregated) descriptions, with reference to some set of indicator variables." For ecological models, a few such studies have been carried out (e.g., Ludwig and Walters 1985; Gardner et al. 1982); but we are just beginning to learn how to proceed. With the need in ecotoxicology to focus on ecosystem processes, and on the ecosystem context for population processes, in assessing the responses of ecosystems to stress, we need to develop simplified descriptions of the critical processes and components. We need to know which species are crucial to maintaining system structure and function, and which are exchangeable parts of functional groupings. Paine's perceptive Tansley Lecture to the British Ecological Society (Paine 1980), elaborating on notions of tight linkage and the modular organization of ecosystems, introduces a valuable way to think about this problem. Ultimately, such questions cannot be resolved without performance of experimental manipulations of natural systems; but computer and analytical experiments with mathematical models can suggest and guide such experiments.

8.3 Equilibrium and Variability

A final point must be made about the interrelationship between variability and scale, and about the need to escape from the straitjacket imposed by the assumption that ecological systems are homogeneous and equilibrial. The idea that a system, once disturbed, will tend asymptotically to an equilibrium state has its roots both in classical vegetation theory (Clements 1916) and in the mathematical theory of dynamical systems. Indeed, it is congruent with our intuition that the effects of initial conditions should become less and less important as time passes. Mathematically, this is equivalent to the observation that the asymptotic dynamics of the system (Equation 8.1) should have lower fractal dimension than does the full space of initial states, and this principle underlies the methods of Takens (1981) and Schaffer et al. (1986) referred to earlier. Yet the classical mathematical ecological literature has carried this observation to the extreme by focusing virtually entirely on the lowest dimensional attractors—equilibrium points and limit cycles—and by ignoring more complicated autonomous behavior, non-autonomous influences, stochasticity, and delays.

Implicit in the classical mathematical approach, based on systems of the form Equation 8.1, also is the notion of homogeneity, since spatial distributions are ignored entirely. A substantial recent literature exists on distributed systems (see Levin 1976), with a considerable focus on regional coexistence through local variability. As emphasized earlier, in oceanographic systems, patchiness can be detected on virtually every scale of examination. Similar observations hold for terrestrial, freshwater aquatic, and intertidal systems, in many of which localized disturbances continually disrupt competitive interactions and underlie the maintenance of ever-changing, spatio-temporal mosaics (Watt 1947; Levin and Paine 1974; Whittaker and Levin 1977). Systems may seem highly variable on small scales and less so on larger scales, although the interplay between temporal and spatial variability complicates this expectation.

The principal conclusion from this discussion is that variability is a property of scale, and thus that observed variability is affected by the level of description the observer imposes on the system. We must proceed beyond the homogeneous descriptions implicit in Equation 8.1 (or discrete-time analogs), by considering their spatially distributed relatives. These may still take the form of systems of ordinary differential equations, formally indistinguishable from Equation 8.1 if the geometry is fixed and patchy; or may be partial differential equations of the form discussed in the next two chapters; or may take some other form if the patch itself is taken as the unit of description. Whatever description is chosen, methods must be developed for measuring and examining spatial patterns on multiple scales, both to understand how our conclusions depend on the choice of scale and to relate to techniques ranging from remote sensing to detailed small-scale investigations that provide information on widely differing scales (Cushman 1986; Levin 1988; Milne 1988). Furthermore, only by understanding how processes change across scales will we develop the requisite ability to scale up from laboratory and microcosm studies, and provide the basis for a predictive ecotoxicology.

Acknowledgments. This publication is ERC-163 of the Ecosystems Research Center, Cornell University, and was supported by the U.S. Environmental Protection Agency Cooperative Agreement Number CR812685–01. Additional funding was provided by Cornell University.

The work and conclusions published herein represent the views of the author, and do not necessarily represent the opinions, policies, or recommendations of the Environmental Protection Agency.

References

Clements FE (1916) *Plant succession: an analysis of the development of vegetation.* Carnegie Inst Wash Publ 242. 512 pp.

Cushman JH (1986) On measurement, scale, and scaling. Water Resources Research 22:129–134

Denman KL, Powell TM (1984) Effects of physical processes on planktonic ecosystems in the coastal ocean. Oceanogr Mar Biol Ann Rev 22:125–168

Gardner RH, Cale WG, O'Neill RV (1982) Robust analysis of aggregation error. Ecology 63:1771–1779

Iwasa Y, Andreasen V, Levin SA (1987) Aggregation in model ecosystems. I. Perfect aggregation. Ecological Modelling 37:287–302

Kingsland SE (1985) *Modeling Nature.* Chicago: The University of Chicago Press, 267 pp.

Lehman JT (1986) The goal of understanding in limnology. Limnol Oceanogr 31:1160–1166

Levin SA (1976) Population dynamic models in heterogeneous environments. Ann Rev Ecol System 7:287–311

Levin SA (1987) Scale and predictability in ecological modeling. In: Vincent TL, Cohen Y, Grantham WJ, Kirkwood GP, Skowronski JM (eds) *Modeling and Management of Resources Under Uncertainty.* Lecture Notes in Biomathematics 72. Berlin: Springer-Verlag, Berlin. pp. 2–8

Levin SA (1988) Challenges in the development of a theory of community and ecosystem structure and function. In: Roughgarden J, May RM, Levin SA (eds) *Perspectives in Ecological Theory*. Princeton, NJ: Princeton University Press. In press.

Levin SA, Paine RT (1974) Disturbance, patch formation, and community structure. Proc Natl Acad Sci USA 71:2744–2747

Limburg KE, Levin SA, Harwell CC (1986) Ecology and estuarine impact assessment: lessons learned from the Hudson River (U.S.A.) and other estuarine experiences. J Environ Manage 22:255–280

Ludwig D, Walters CJ (1985) Are age-structured models appropriate for catch-effort data? Can J Fish Aquat Sci 42:1066–1072

Mandelbrot BB (1983) *The Fractal Geometry of Nature*. San Francisco: WH Freeman & Co, 468 pp.

Milne BT (1988) Measuring the fractal geometry of landscapes. Appl Math Computation (in press)

O'Neill RV, DeAngelis DL, Waide JB, Allen TFH (1986) *A Hierarchical Concept of Ecosystems*. Monographs in Population Biology 23, Princeton, NJ: Princeton University Press, 253 pp.

Paine RT (1980) Food webs: linkage, interaction strength and community infrastructure. The Third Tansley Lecture. J Animal Ecol 49:667–685

Penrose R (1955) A generalized inverse for matrices. Proc Cambridge Philos Soc 51:496–513

Peters RH (1986) The role of prediction in limnology. Limnol Oceanogr 31:1143–1159

Schaffer WM, Ellner S, Kot M (1986) Effects of noise on some dynamical models in ecology. J Math Biol 24:479–523

Takens F (1981) Detecting strange attractors in turbulence. In: Rand DA, Young LS (eds) *Dynamical Systems and Turbulence*. Warwick 1980 Lecture Notes in Mathematics 898. Berlin: Springer-Verlag, pp. 366–381

Vollenweider RA (1968) *Scientific Fundamentals of the Eutrophication of Lakes and Flowing Waters, with Particular Reference to Nitrogen and Phosphorus as Factors in Eutrophication*. Rep. Organisation for Economic Cooperation and Development, DAS/CSI/68.27, Paris. 274 pp.

Volterra V (1926) Variazioni e fluttuazioni del numero d'individui in specie animale conviventi. Mem R Accad Nazionale del Lincei (Ser. 6) 2:31–113

Watt AS (1947) Pattern and process in the plant community. J Ecology 35:1–22

Weber LH, El-Sayed SZ, Hampton I (1986) The variance spectra of phytoplankton, krill, and water temperature in the Antarctic Ocean south of Africa. Deep-Sea Research 33:1327–1343

Whittaker RH, Levin SA (1977) The role of mosaic phenomena in natural communities. Theor Pop Biol 12:117–139

Chapter 9
Mathematical Models—Fate, Transport, and Food Chain

Donald J. O'Connor,[1] John P. Connolly,[1] and Edward J. Garland[1]

Organic chemicals, heavy metals, and radionuclides are present in all phases of the environment—air, water, and land. When these substances are released in the water phase of the environment, they are transported by the fluid motion, are transferred to the atmosphere and bed, are subject to various physicochemical and biochemical reactions, and are assimilated by all levels of the aquatic food chain. They are also transmitted by direct ingestion through the food chain to higher organisms and ultimately to humans. It is the purpose of this chapter to present a mathematical analysis of the transport and fate of trace constituents in estuarine systems.

The basic principle on which this analysis rests is the conservation of mass, coupled with the hydrodynamic equations of momentum and state and with the ecological principle of the structure of the food chain. Given the spatial and temporal distribution of the mass input of a toxicant to a water system, the overall framework of the model consists of four distinct units, as shown in Figure 9.1: the input(s) of toxicant, the hydrodynamic component, the fate and transport element, and the food chain entity. The model is applied to the analysis of Kepone in the James River Estuary (shown in Figure 9.2), defining the spatial and temporal distribution of this substance from the initiation of its production through the period following its cessation. Projections are made to estimate the time required for the physical system to achieve non-toxic levels and for the biological system to depurate to acceptable concentrations. The various components of the model are calibrated and validated with the extensive data set

[1] Manhattan College, Environmental Engineering and Science Program, Bronx, New York 10471

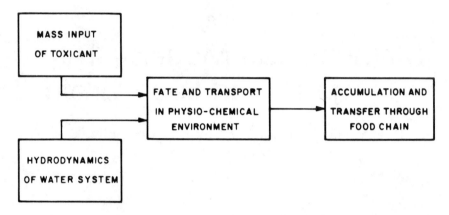

Figure 9.1. Overall structure of analysis.

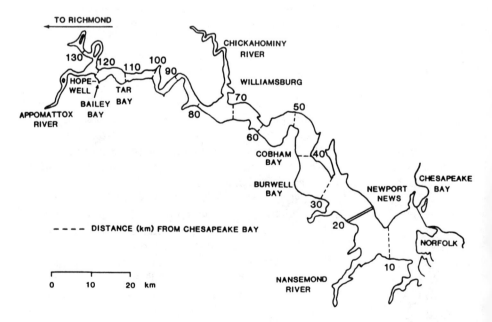

Figure 9.2. James River study area.

collected over the period from the late sixties to the early eighties (Battelle 1978; Bender et al. 1977; Connolly 1980; Garnas et al. 1978; Gregory, unpublished data; Lunsford et al. 1980; Lunsford et al. 1982; Nichols 1972; Huggett 1979).

9.1 Components of Model

The components of the model are shown schematically in Figures 9.3 and 9.4. Figure 9.3 presents the hydrodynamic features of the estuarine flow and the transfer routes between the water and the bed. The top panel of the figure illustrates the net intertidal water motion characteristic of stratified estuaries. Denser saline water moves upstream in the lower layer (u_2)

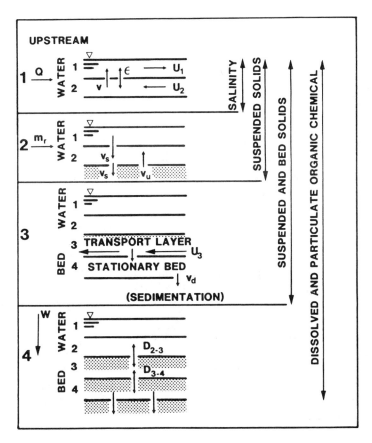

Figure 9.3. Schematic of fluid, solids, and toxicant transport in the water column and bed.

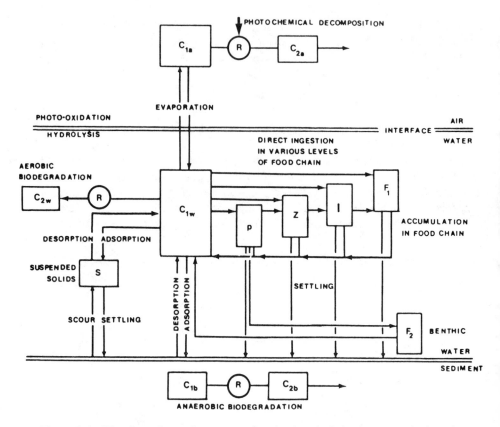

Figure 9.4. Kinetic and transfer routes of toxic chemicals in the water, bed, and food chain.

while freshwater moves downstream in the upper layer (u_1). Vertical transport includes an upward advective transport of water (v) necessary to maintain continuity and mixing between the layers (ε). Panels 2 and 3 indicate the additional transport processes affecting the movement of particulate material including suspended and bed solids and chemical adsorbed to the solids. These processes include settling (v_s), resuspension (v_u), bed movement (u_3), and sedimentation (v_d). Chemicals in the estuarine system are also subject to transport through interstitial water diffusion as illustrated in panel 4.

The physicochemical and food chain interactions and decay mechanisms are portrayed in Figure 9.4. They include both transfer and reaction processes. Transfer is the movement of the chemical between the air, water, and solid phases of the system. It includes the physicochemical processes of volatilization and adsorption, and the biological processes of bioconcentration–bioaccumulation. In bioconcentration–bioaccumulation, all members of the food chain accumulate the chemical directly from the water. In addition, there is a transfer of chemical from prey to predator through assimilation of ingested chemical. The rates of transfer are controlled by the metabolic and growth rates of each species and characteristics of the chemical that affect its movement across the gill and gut membranes. Reaction is the transformation or degradation of the chemical. It includes biodegradation and the chemical reaction processes of photolysis, hydrolysis, and oxidation. Each element is analyzed separately in a sequential fashion, proceeding from the hydrodynamic analysis to the physicochemical model and finally to the food chain. Each is decoupled from its contiguous element and calibrated independently. The justification for the decoupling of the hydrodynamic analysis is evident since the subsequent steps do not influence the transport component. The decoupling of the physicochemical and food chain models is based on the fact that the mass of the chemical is contained primarily in the physical environment, the mass in the food chain being a very small fraction of the total.

9.2 Transport, Salinity, and Solids Analyses

The type of analysis which is described in this chapter is appropriate for those cases in which the system has been previously subjected to an input of a toxic substance, and essentially consists of four steps, shown diagrammatically in Figure 9.3. First, equations of continuity, momentum, and state are employed in a steady-state, tidally averaged mode to generate the horizontal and vertical velocities of the estuarine circulation. A two-layer, steady-state model is then used to determine the vertical dispersion coefficient between the two water layers using ocean salt as a tracer. Secondly, settling velocities of solids to the bed and resuspension of solids from the bed are estimated to fit observed solids concentrations in the water column. Using observed organic chemical concentrations in the surficial sediments as a boundary condition, the dissolved and particulate organic chemical fractions in the water column are calculated.

In the third step, the solids transport and the stationary bed sediment are addressed. Net horizontal velocities of the transport layer are estimated, and the flux of sediment between the two sediment layers is determined. With selected values of solids concentrations for both the transport and stationary beds, the net sedimentation rate of the bed is calculated and compared to observed rates in the estuary.

The fourth step consists of calculating the dissolved and particulate concentrations of the organic chemical in the water and bed. The molecular diffusivity of the dissolved component in the interstitial water and partition coefficients are selected for the water and sediment phases.

Evaporative transfer and reaction coefficients supplemented by laboratory measurements, if relevant, are assigned in accordance with the properties of the chemical.

Note that all coefficients indicated in Figure 9.3 are used for the final
four-layer analysis. In the sketch of each step, only the additional coefficients selected in that step are shown. A detailed discussion of each of the
four steps follows.

The circulation within the saline zone of estuaries is due to the interaction of a number of factors: tidal action, density differentials, freshwater
flow, winds, and the characteristics of the channel. The most fundamental
approach to the analysis of this type of estuarine circulation is based on
the simultaneous solutions of the fundamental equations of momentum,
continuity, and state (Blumberg 1977; Festa and Hansen 1976; Leendertse
et al. 1973). It involves a numerical solution of the basic equations, which
is generally quite complex. On a tidally averaged basis, the circulation is
characterized by a horizontal seaward velocity in the upper layer and a
landward velocity in the lower layer, with a compensating vertical velocity pattern to maintain hydraulic continuity. The final equation of salinity
of a two-dimensional estuary of constant width and depth is

$$0 = \frac{\partial}{\partial y}\left(\varepsilon_y \frac{\partial s}{\partial y}\right) - \frac{\partial}{\partial x}(us) - \frac{\partial}{\partial y}(vs) \tag{9.1}$$

in which

u = tidally averaged longitudinal velocity (l/t)
s = tidally averaged salinity concentration (M/l^3)
v = tidally averaged vertical velocity (l/t)
ε_y = vertical eddy diffusion coefficient (l_2/t)

A simplified analytical solution has been developed which defines the
vertical distribution of horizontal velocity (O'Connor and Lung 1981). It
is based on the condition that the salinity distributions in both the longitudinal and vertical planes are known or may be assigned. With the plane of
no net motion defined by the depth at which the net horizontal velocity is
zero, integrated surface layer and bottom layer flows are calculated,
which subsequently yield the vertical flows. Vertical dispersion coefficients of mass are then estimated from vertical eddy viscosities and Richardson numbers developed in the hydrodynamic analysis. These are
input to a salinity model of the estuary and adjusted until calculated
salinity profiles for both surface and bottom waters agree with observed
values. An example of the results of this analysis is shown in Figure 9.5
for an upstream freshwater flow of 1000 cfs (28 m³/s) (cfs = cubic feet per
second).

Because the concentration of solid material is an important factor as an
accumulation site for Kepone, the temporal and spatial distribution of
suspended solids within the estuarine system is a necessary element in the
overall analysis. The solids distribution, in turn, is related to estuarine

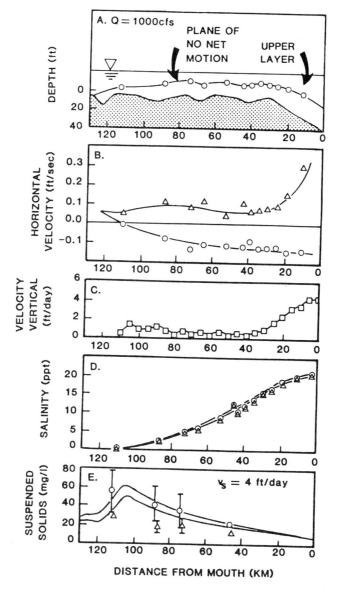

Figure 9.5. Transport, salinity, and solids analysis in the water column.

circulation. A steady-state mass transport equation for suspended sediment, analogous to that for salinity, Equation 9.1, may be written for a tidally averaged, two-dimensional estuary of constant width and depth:

$$0 = \frac{\partial}{\partial y}\left(\varepsilon_y \frac{\partial m}{\partial y}\right) - \frac{\partial(um)}{\partial x} - \frac{\partial}{\partial y}\left[(v - v_s)m\right] \qquad (9.2)$$

in which

m = tidally averaged suspended sediment concentration (M/l^3)
v_s = averaged settling velocities of suspended sediment (l/t)

Organic particles and fine-grained material, such as clays, remain primarily in suspension and follow the residual tidally averaged flow. A fraction of these particles and the silts are subject to deposition at slack water and to erosion at maximum current. The coarser-grained sands are located primarily in the bed and if eroded travel along the bed, being less susceptible to alternate erosion and deposition. Solids eroded from the bed move upstream if retained in the lower layer and downstream if they are dispersed to the upper layer.

In coastal plain estuaries, the landward residual flow near the bed, in conjunction with the solids load brought in by the freshwater flow, causes a net accumulation of solids within the estuary. This accumulation is described by a sedimentation velocity whose order is tenths or units of centimeters per year. There is in general very little sediment from the rivers reaching the continental shelf and most of the solid material is accumulated within the estuary proper.

The spatial distribution of this material, as observed in many partially mixed estuaries, such as the James, is usually characterized, in the vicinity of the salinity intrusion limit, by a peak concentration higher than that of either the river source or the oceanic waters. This phenomenon, referred to as the "turbidity maximum," decreases with decreasing river flow and may not occur under relatively high flow conditions. Other factors, which affect the distribution of suspended and bed solids, include the settling velocity (particle size) of the sediment, the amount of sediment introduced at both the ocean and river sources, the strength of the estuarine circulation, flocculation and deflocculation regulated by salinity variation, and local resuspension by tidal currents or waves.

The longitudinal boundary conditions specified for the evaluation of Equation 9.2 are the concentrations of suspended sediment associated with the freshwater inflow at the upstream limit and with the oceanic waters in the lower layer at the downstream location. In addition, a vertical boundary condition is specified at the water–sediment interface. The resulting flux is the net effect of settling of particles from the water to the bed and entrainment of particles from the sediment layer. A net flux from the water to the sediment layer implies that the mean settling is greater than the resuspension of particles and conversely a net flux from the sediment layer to the water stipulates that uplift is greater than settling.

The suspended sediment distribution is dependent on four transport coefficients: the horizontal and vertical velocities, vertical dispersion coefficients, and the settling velocity. The first three coefficients are determined in the salinity analysis, previously described. The fourth coefficient, the settling velocity, is a function of the size, shape, and density of the suspended particles. Furthermore, the density of cohesive sediments depends on the water content and nature of the particle. Recent work on marine particles from various sources provides a basis to relate particle density to inorganic composition and to particle size (McCave 1975), which yields the distribution of suspended solids in the water, as shown in Figure 9.5E for the 1000 cfs (28 m³/s) flow.

The transport components of the bed consist of a longitudinal advection and a vertical dispersion or mixing. The advective component of transport in the bed is induced by the shearing stress of the overlying flowing water, with which it is in contact. The horizontal velocity of the bed is a maximum at the water interface and diminishes in depth to a point at which the bed may be regarded as stationary. Vertical uniformity over the transport layer is assumed, and the concentration of this layer is assigned as constant. The concentration of the underlying stationary bed is assigned as an order of magnitude greater.

The rate of change of the bed elevation is referred to as a sedimentation velocity. It may be envisioned as the velocity at which the water–bed interface approaches the stationary bed if erosion is predominant or at which the interface moves away from that datum if settling is the significant term.

Assigning the solids concentrations in the transport and stationary bed layers, a depth of the transport layer and a net velocity of the transport layer equal to one one-hundredth of the bottom water layer tidally averaged horizontal velocity, entrainment rates, and sedimentation rates are computed. Entrainment rates range from approximately 40 g/m²-day upstream, to a peak value just downstream of the salinity tail of 175 g/m²-day. These values are consistent with the range of 50–100 g/m²-day reported by Partheniades (1965) who used estuarine clays, similar to those in the James River, in a set of controlled laboratory experiments. Sedimentation rates alternate between scour and deposition zones in the main channel, whereas the side bays of the estuary are deposition zones with sedimentation rates of from 1–2 cm/yr.

9.3 Organic Chemicals in the Water Column

The distribution of an organic chemical in an estuary, under steady-state tidally averaged conditions may be written as follows:

$$0 = \frac{\partial}{\partial y}\left(\varepsilon_y \frac{\partial C_T}{\partial y}\right) - u\frac{\partial C_T}{\partial x} v\frac{\partial C_T}{\partial y} - v_s\frac{\partial p}{\partial y} \tag{9.3}$$

in which

C_T = total concentration of chemical (M/l³)
p = concentration of chemical-particulate component (M/l³)

and the remaining terms have been previously defined following Equations 9.1 and 9.2.

The first two terms represent the vertical dispersion and horizontal advection; the third and fourth terms, the vertical advection. The fourth term, which includes the settling velocity, represents the settling of the particulate component adsorbed to suspended solids. The vertical boundary conditions are assigned at the air–water and water–bed interfaces, at which concentration and flux conditions are specified. The flux may be negative or positive, representing net deposition or scour at the bed, respectively, or exchange at the water surface. An additional source due to precipitation may be effective under certain conditions. A zero-flux condition indicates equilibrium.

To be noted is the fact that, although the suspended and bed solids may be in equilibrium, the same condition does not necessarily apply to the organic chemical, due to the adsorption–desorption interaction. If the coefficients defining this process are much greater than the other transfer and kinetic coefficients, the assumption of instantaneous equilibrium is appropriate.

The equilibrium relationship between the dissolved and particulate components is expressed in terms of a partition coefficient:

$$\P = \frac{r}{c} \tag{9.4}$$

in which

$\P$ = partition coefficient (l^3/M)
r = solid phase concentration (M/M)
c = dissolved concentration (M/l^3)

Equation 9.4, which represents the linear portion of a Langmuir isotherm, is a valid representation of the equilibrium conditions when the concentration levels are well below the saturation value of the adsorbing solids, a condition which invariably holds in natural water systems.

The solid phase concentration, r, is related to the particulate concentration, p, by the suspended solids:

$$p = rm \tag{9.5}$$

Since the total concentration of the toxicant is the sum of the dissolved and the particulate, it may be expressed, after substituting Equation 9.5, as

$$C_T = c + rm \tag{9.6}$$

Substituting Equation 9.4, the total may be expressed in terms of each of these components and the dissolved and particulate fractions are therefore

$$f_d = \frac{c}{C_T} = \frac{1}{1 + \P m} \tag{9.7a}$$

$$f_p = \frac{p}{C_T} = \frac{\P m}{1 + \P m} \tag{9.7b}$$

Substitution of the total concentration for the particulate fraction (Equation 9.7b), assuming instantaneous equilibrium, yields the final form:

$$0 = \frac{\partial}{\partial y}\left(\varepsilon_y \frac{\partial C_T}{\partial y}\right) - u\frac{\partial C_T}{\partial x} - v\frac{\partial C_T}{\partial y} - v_s \frac{\partial}{\partial y}(f_p C_T) \tag{9.8}$$

Expressing the dissolved fraction in terms of the total, the air–water boundary flux is given as

$$j_d = \varepsilon_y \frac{\partial}{\partial y}(f_d C_T)\Big|_{y=0} = \frac{K_L}{1 + \P m} C_T \tag{9.9}$$

in which K_L = the evaporative transfer coefficient. The flux of the particulate is evidently zero at the air–water interface.

The boundary condition at the water–bed interface for the particulate component is:

$$j_p = \varepsilon_y \frac{\partial}{\partial y}(f_p C_T)\Big|_{y=H} - v_s f_p C_T \tag{9.10}$$

The exchange coefficient between the two layers for the particulate fraction may be expressed in terms of an entrainment velocity v_u, since the particulate concentration in the bed is much greater than that in the water column:

$$j_p = v_u f_p C_{Tb} - v_s f_p C_{Tw} \tag{9.11}$$

in which $v_u = \varepsilon_y/\Delta_y$, the entrainment velocity. Similarly, the boundary condition for the dissolved component is:

$$j_d = -D\frac{\partial c}{\partial y}\Big|_{y=H} = \frac{D}{\Delta y}(f_d C_{Tb} - f_d C_{Tw}) \tag{9.12}$$

in which D = the diffusivity of the dissolved chemical in the interstitial water, and where the subscripts w and b = the water and bed, respectively.

9.4 Application to Kepone in the James River

The procedure as described is applied to the calculation of the organic chemical, Kepone, in the James River. In addition to the transport, settling, and kinetic coefficients previously used in the two-layer model for

the water column, the bed system is added with the appropriate coefficients. These include the resuspension rate, the advective coefficients for
the transport sediment layer, and the sedimentation rate in the stationary
bed. A partition coefficient of 500 μg/kg per μg/L, derived from laboratory and field measurements, is assigned. The vertical dispersion coefficient for the dissolved component is assigned as 5×10^{-6} cm^2/sec between
the active and stationary layers, whereas the diffusion from the interstitial
waters of the active bed to the lower water layer, expressed as a transfer
velocity, is set equal to 0.01 cm/s (Gregory unpubl.).

A time trace of the Kepone concentration in the bed is calculated,
beginning in January 1966, prior to the manufacture of Kepone in the
Hopewell, Virginia area. Thus, at the start of the analysis, both water
column and bed are free of contamination and the initial conditions are
zero for all segments. The modeled estuary is then subjected to a discharge of Kepone assumed to be one percent of the mass produced, until
1975, when production of Kepone was discontinued. The discharge after
1975 was assumed to decrease exponentially from the 1975 input, as
shown in Figure 9.6A.

The actual monthly average hydrograph for this period is shown as a
dashed line in Figure 9.6B. In order to make the calculation tractable, the
hydrograph was approximated by a series of constant flows. Six flow

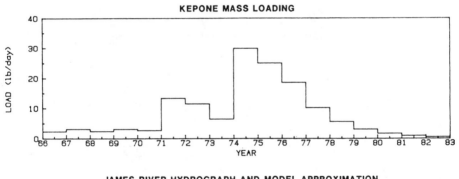

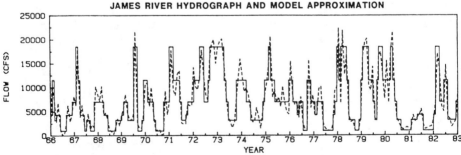

Figure 9.6. Kepone mass input and James River hydrograph.

conditions ranging from 1,000 to 18,500 cfs (28 to 520 m3/s) were selected on the basis of available data on salinity intrusion (Nichols 1972). For each year these conditions were arranged sequentially for specific durations to represent the actual hydrology, as indicated by the solid line in Figure 9.6B.

Calculated longitudinal distributions of Kepone and the observed data in the transport layer of the bed are shown in the upper panels of Figure 9.7 for 1980 and 1982. The solid lines represent the distributions calculated at the end of June, which better reproduce the observed data (Lunsford et al. 1980; Lunsford et al. 1982), collected from March to October, than do the end of year distributions, shown as dashed lines. The calculated concentrations in the stationary bed, which vary little between the middle and end of the year, are presented in the lower panels of Figure 9.7 for the same years, and reproduce the data reasonably well.

The calculated Kepone distributions in the water column for 1978 to 1981 are shown in Figure 9.8. The solid lines represent the concentration in the upper water layer at the end of each year. The dashed lines for 1978 and 1980 represent the concentration in the lower water column, indicating no significant vertical gradient. The dashed lines for 1979 and 1981 correspond to the end-of-June calculated values. The calculated concen-

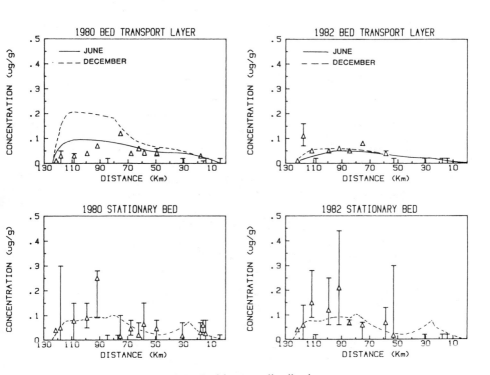

Figure 9.7. Bed kepone distributions.

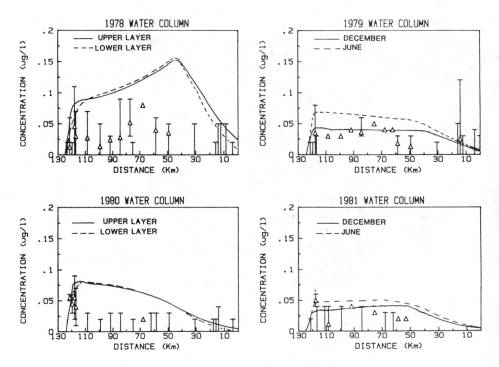

Figure 9.8. Water column kepone distributions.

trations in the water column are very sensitive and respond quickly to changes in the constant flow conditions used in the model. Therefore, the calculated distribution is representative of the conditions at the flow used to approximate the hydrograph at the end of June or the end of December, while the data (Lunsford et al. 1980; Lunsford et al. 1982) are representative of the conditions at the time of the year that the samples were obtained.

The loading estimates used in the model are constant for one year periods. This approximation may be appropriate for the calculation of the Kepone distribution in the bed (Figure 9.7), which responds slowly, but it is too crude to adequately represent the time variable loading which rapidly affects the water column.

9.5 Food Chain

The basic equation which defines the concentration of a toxic substance in the food chain is developed by the principle of continuity similar to that used in the development of the above equations. The element about which the mass balance is taken is an organism in the food chain, rather than a discrete volume of the water body. The factors which increase the concentration of the toxicant in the organism are direct uptake from the water and assimilation by predation of contaminated prey; those factors which decrease the concentration are desorption and depuration from body tissue. The toxicant is assumed to be uniformly distributed over the entire mass of organism and consequently the growth of the various species must be taken into account.

The accumulation of toxic chemicals by an organism in a food chain may be described by the following equation (Thomann and Connolly 1984):

$$\frac{dv_i}{dt} = K_{ui}c + \sum_{j=1}^{n} \alpha_{ij}C_{ij}v_j - K_i'v_i \qquad (9.13)$$

in which the state variables are:

v = concentration of chemical in the organism (M/M)
c = dissolved chemical concentration (M/l^3)

the coefficients are:

K_u = rate of uptake of chemical from water ($l^3 - t^{-1} - M^{-1}$)
 α = assimilation efficiency of chemical in ingested prey
 C = consumption rate of a particular prey (M $-$ M^{-1} $-$ t)
K' = $k_i + (dw_i/dt)w_i(t^{-1})$
 k = excretion rate (t^{-1})
 w = weight of the organism (m)

and the subscripts i and j refer to the organism being modeled and prey species of that organism, respectively. The first term of Equation 9.13 represents the direct uptake of chemical by the organism from the water. The second term represents the flux of chemical into the organism through feeding. The third term is the loss of chemical due to desorption and excretion from body tissue plus the change in concentration due to growth of the individual. This equation has been applied to single species by Norstrom et al. (1976) and Weininger (1978) and to an entire food chain by Thomann (1981), Thomann and Connolly (1984), and Connolly and Tonelli (1984). The values of the coefficients depend on the bioenergetics of the species and the physical and chemical characteristics of the chemical and are parameterized as specified by Connolly and Tonelli (1984).

The dissolved chemical concentration, c, to which the organism is exposed is dependent on time and the location of the organism. Nonmigratory fish species and the lower levels of the food chain are assumed to remain within some arbitrary region that may be represented by a single, time-variable chemical concentration. Several regions may be considered to account for spatial gradients of chemical concentration. Anadromous fish species are assumed to migrate through the various regions thereby being sequentially exposed to the dissolved chemical and prey species in each region.

Equation 9.13 is applied to each species or level of the food chain, which is separated into discrete age classes in order to represent the predator–prey relationships that characterize the food chain. Constant assimilation efficiencies and growth rate are specified for each age class. All other bioenergetic parameters vary continually in relation to body weight.

The lower levels of the food chain such as plankton and invertebrates generally attain equilibrium with the chemical rapidly and exhibit little variation in concentration with age. Therefore, it is appropriate to assume a steady-state concentration for these levels in equilibrium with the chemical and independent of size. This is accomplished by setting the left side of Equation 9.13 to zero. The concentration of chemical is then given by the following equation:

$$v_i = \frac{K_{ui}c + \sum_{j=1}^{n} \alpha_{ij}C_{ij}v_j}{K'_i} \tag{9.14}$$

Average values for the species food ingestion, growth, and respiration rates are used. Note that for the phytoplankton–detritus level, Equation 9.14 simplifies further since there is no uptake through feeding.

9.6 Application to James River Striped Bass Food Chain

Accumulation of Kepone in the biological component is modeled for the food chain leading to striped bass. Four trophic levels are used to describe this food chain (Figure 9.9). The phytoplankton–detritus level is the base of the food chain. The invertebrate level is represented by *Neomysis* and *Nereis,* reflecting the importance of both pelagic and benthic species to the higher levels. Atlantic croaker (*Micropogan undalatus*) and white perch (*Morone americana*) are the fish species representing the level immediately below the striped bass (*Morone saxatilis*).

The phytoplankton–detritus level, *Neomysis,* and *Nereis* are assumed to be in dynamic equilibrium with Kepone in the water column and in their food (Equation 9.14). The white perch, the Atlantic croaker, and the striped bass are separated into year classes to which Equation 9.13 is applied. Growth rate and predator–prey relationships are assumed to be constant within any age class.

The parameters that describe the interaction of Kepone and each species include growth rate, respiration rate, the assimilation efficiencies of food and Kepone in food, and the bioconcentration factor for Kepone.

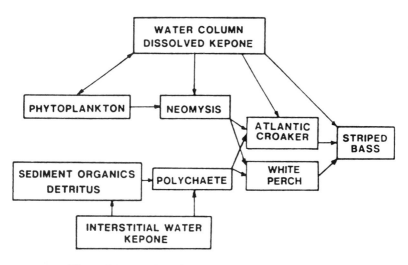

Figure 9.9. Food chain structure used in the model.

For *Neomysis* and *Nereis* an average value is used for each parameter consistent with the assumption of steady-state Kepone concentrations for these species. Parameter values are specified for each age class of Atlantic croaker, white perch, and striped bass.

James River striped bass and Atlantic croaker migrate from the Chesapeake Bay and the Atlantic coast to the limit of salinity intrusion in the estuary. Consequently, they are mainly found in the lower 60 km (37.26 miles) of the estuary. Observed water and bed Kepone concentrations do not vary significantly within this section (Bender et al. 1977; Connolly and Tonelli 1984). Thus, it is appropriate to consider this section as a single region. Striped bass and Atlantic croaker are then assumed to migrate between this region and a region of zero Kepone concentration representative of the Chesapeake Bay and the Atlantic coast. Kepone concentrations in the water and bed were assigned based on observed data (Lunsford et al. 1980; Lunsford et al. 1982) and the results of the physicochemical Kepone model as shown in Figure 9.10.

The comparison between observed data (Lunsford et al. 1980; Lunsford et al. 1982) and calculated Kepone concentrations in Atlantic croaker, white perch, and striped bass is shown in Figure 9.11. The data and calculated values are averages over all age classes. In the calibration procedure the Kepone assimilation efficiency and excretion rate were adjusted within their range of observed values to provide the best comparison of observed and computed Kepone concentrations. The model reproduces the observed within-year and year-to-year concentration profiles for all three species. The oscillation in concentration computed for Atlantic croaker and striped bass reflects the migration of these species between the James River and the uncontaminated Chesapeake Bay and

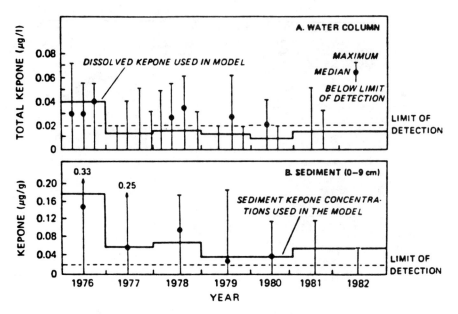

Figure 9.10. Kepone concentrations observed in the lower James River Estuary (0–60 km, 0–37.26 miles) and the values used in the model for (*A*) the water column and (*B*) the surface sediment.

Atlantic Ocean. This oscillation indicates that the fish respond rapidly to changes in exposure level. For Atlantic croaker there is very little carry-over of concentration from year-to-year. The average concentration in striped bass reflects previous years more strongly because bass have lower excretion rates than the croaker.

The calibrated model was used to project the relationship between the maximum concentration in each species and the exposure concentration. A constant ratio between water column and sediment Kepone concentration of 0.23 µg/l per µg/g was assumed, based on observed concentrations in the saline portion (lower 60 km or 37.26 miles) of the estuary. Prediction uncertainty was assumed to be equal to calibration uncertainty. Calibration uncertainty was assessed from the difference between calculated and observed average concentrations in each species (Connolly and Tonelli 1984).

Figure 9.12 shows the range of calculated maximum striped bass, white perch, and Atlantic croaker Kepone concentrations in relation to sediment and water column exposure concentrations. The FDA action limit of 0.3 µg/g, the level used to determine suitability of the fish for human consumption, is indicated on the plots. Maximum concentrations will be below this level at water column dissolved concentrations in the range of 3–5 ng/l for striped bass, 6–9 ng/l for Atlantic croaker, and 4–8 ng/l for white perch. The associated sediment concentration ranges are 13–22 ng/g

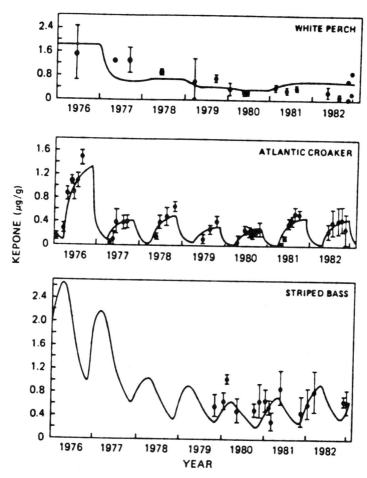

Figure 9.11. Comparison of observed and calculated kepone concentrations in the Atlantic croaker, white perch, and striped bass.

for striped bass, 26–39 ng/g for Atlantic croaker, and 17–35 ng/g for white perch. Because these ranges reflect the uncertainty of the calibration, which is the lower limit of prediction uncertainty, the actual ranges may be greater.

Based on the 1980–1982 data and calibration, exposure concentrations may be assumed to be in the order of 10 ng/l in the water column and 40 ng/g in the sediment. To achieve the action limit the projection then indicates exposure concentrations must be reduced 50–70% for striped bass, 10–40% for Atlantic croaker, and 20–60% for white perch. These reductions are significant in view of the apparent slow decline in exposure concentration indicated in Figure 9.10. Atlantic croaker is the only one of these species likely to reach the action limit in the near future.

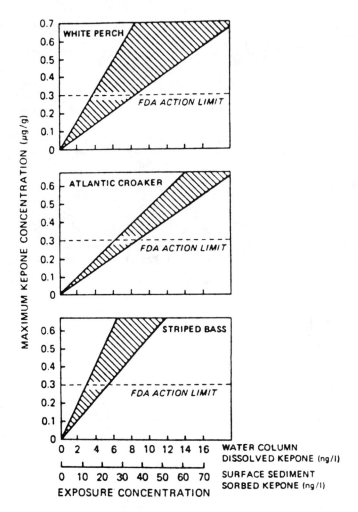

Figure 9.12. Projected relationship between kepone exposure concentration and the maximum kepone concentration in white perch, Atlantic croaker, and striped bass.

A preliminary assessment of the temporal response of the fish species in the river was performed by assuming future reductions in exposure concentration based on the trends indicated by the data and the physico-chemical model. Figure 9.13 shows the assumed water column dissolved Kepone profile and the resulting maximum concentrations in white perch, Atlantic croaker, and striped bass. As before, the range shown in the plots reflects the calibration uncertainty. The FDA action limit is indicated by the dashed lines in the figures. Atlantic croaker were projected to be below the action limit by 1987. White perch should be below the limit by

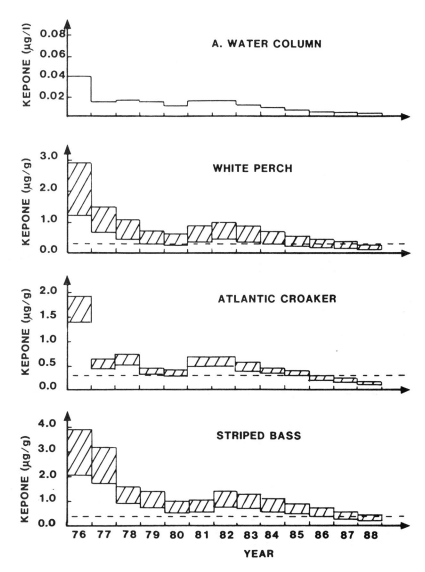

Figure 9.13. Projected water column dissolved kepone concentration and the maximum kepone concentration in white perch, Atlantic croaker, and striped bass.

1989. Striped bass will be at or near the limit by 1989. Again, these projections are preliminary, but these suggest a time scale of 3–6 years to reach levels at which the fishery may be opened.

9.7 Conclusion

An analysis framework which consists of hydrodynamic, transport, fate, and ecosystem components has been presented for computing the response of the water column, sediment, and food chain of an estuary to the input of a toxic chemical. Application of these models to Kepone in the James River Estuary illustrates both their utility in correlating field and laboratory data and in providing a means of projecting environmental concentrations. The latter step requires engineering judgments. Significant among these were an estimate of the discharge of Kepone to the estuary during and after its manufacture and the approximations correlating the hydrologic and ecologic conditions. In spite of the uncertainty inherent in such assumptions the overall framework is proposed as a rational method for projections, which could be used in assessment of ameliorating alternates.

Acknowledgments. This work was supported by a National Science Foundation Grant (CEE-8207319) and a cooperative agreement (R807827) with the USEPA-Environmental Research Laboratory, Gulf Breeze, Florida. Support for Donald J. O'Connor and Edward J. Garland is from the former and for John P. Connolly from the latter.

The assistance of Robert Guena, research assistant, supported in part by a grant from the Thomas J. Lipton Foundation, is acknowledged.

Charles Lunsford and Gail Walton of the Virginia State Water Control Board provided the Kepone data for the water column, sediment, and fish. Their cooperation and support are appreciated.

The careful typing of this work, by Margaret Cafarella, is gratefully acknowledged.

References

Battelle Pacific Northwest Laboratories (1978) The feasibility of mitigating Kepone contamination in the James River basin. Appendix A to the *EPA Kepone Mitigation Project Report,* EPA 44015–78–004, Washington DC

Bender ME, Huggett RJ, Hargis WJ Jr (1977) Kepone residues in Chesapeake Bay biota. In: *Proceedings of the Kepone Seminar II,* September 19–21, 1977. U.S. Environmental Protection Agency, Region II, pp. 14–65

Blumberg AF (1977) Numerical model of estuarine circulation. J Hydraul Div, ASCE 103:295–310

Connolly JP (1980) The effect of sediment suspension on adsorption and fate of Kepone. Austin, TX: University of Texas. Dissertation

Connolly JP, Tonelli R (1985) Modeling Kepone in the striped bass food chain of the James River Estuary. Est Coast Shelf Sci 20:347–366

Festa JF, Hansen DV (1976) A two-dimensional numerical model of estuarine circulation: The effects of altering depth and river discharge. Est Coast Marine Sci 4: 275–289

Garnas RL, Bourquin AW, Pritchard PH (1978) The fate of 14C-Kepone in estuarine microcosms. In: *Kepone in the Marine Environment,* Appendix C to the *EPA Kepone Mitigation Project Report,* EPA 440/5–78–004, Washington, DC

Huggett RJ (1979) The role of sediments in the storage, movement and biological uptake of Kepone in estuarine environments. In: *Kepone in the Marine Environment,* Appendix C to the *EPA Kepone Mitigation Project Report,* EPA 440/5–78–004C, Washington DC

Leendertse JJ, Alexander RC, Liu SK (1973) A three-dimensional model of estuaries and coastal seas. *Principles of Computation Report,* vol. 1. R-1417-OWRR, Santa Monica, CA: Rand Corp.

Lunsford CA, Todd BG, Soldano CE (1982) Summary of Kepone study results - 1979–1981. Virginia State Water Control Board Basic Data Bulletin No. 49

Lunsford CA, Walton CL, Shell JW (1980) Summary of Kepone study results - 1976–1978. Virginia State Water Control Board Basic Data Bulletin No. 46

McCave IN (1975) Vertical flux of particles in the ocean. Deep-Sea Research Great Britain 22:491–502

Nichols MM (1972) Sediments of the James River estuary, Virginia. The Geological Society of America Inc. Memoir 133

Norstrom RJ, McKinnon AE, DeFreitas ASW (1976) A bioenergetics-based model for pollutant accumulation by fish. Simulation of PCB and methylmercury residue levels in Ottawa River yellow perch (*Perca flavescens*). J Fish Res Bd Can 33:248–267

O'Connor DJ, Lung WS (1981) Suspended solids analysis of estuarine systems. J Env Eng Div ASCE 107: No. EE1: 101–120

Partheniades E (1965) Erosion and deposition of cohesive soils. J Hydraul Div ASCE 91: No. HY1: 105–140

Thomann RV (1981) Equilibrium model of fate of microcontaminants in diverse aquatic food chains. Can J Fish Aquat Sci 38:280–296

Thomann RV, Connolly JP (1984) An age-dependent model of PCB in the Lake Michigan lake trout food chain. Env Sci Tech 18:65–71

Weininger D (1978) Accumulation of PCBs by lake trout in Lake Michigan. Madison, WI: University of Wisconsin. Dissertation

Chapter 10
Deterministic and Statistical Models of Chemical Fate in Aquatic Systems[1]

Robert V. Thomann[2]

This chapter has several purposes, among them: to summarize the basic models of the steady-state transport and fate of chemicals in aquatic systems including uptake and distribution in the aquatic food chain; to illustrate the deterministic time variable behavior of chemical fate models with several applications to the Great Lakes; to develop some statistical models of chemical variability in aquatic organisms, specifically, fish.

The ability to analyze and predict the transport of potentially toxic chemicals is one of the central requirements of risk assessment and subsequent risk management. Steady-state models can be of specific value in the early stages of chemical screening for generic problem contexts and to elucidate basic principles of chemical fate and uptake into the food chain. Time variable models are particularly useful for predicting recovery times of aquatic systems following some abatement program of chemical control. These steady-state and time variable models essentially estimate the average or deterministically varying chemical exposure concentration to aquatic organisms. Risk assessment also requires some evaluation of the stochastic behavior of chemicals both in the water and in fish. This chapter is therefore divided into four parts: 1) the basic theory and associated equations; 2) steady-state simplifications; 3) deterministic time variable models; 4) analytical and numerical models of statistical behavior of chemicals in fish.

---

[1] This article also appears in "Applied Mathematical Ecology," edited by S.A. Levin et al., Springer-Verlag, Berlin, Heidelberg, New York, 1989.
[2] Manhattan College, Environmental Engineering and Science Program, Bronx, New York 10471

10.1 Theory

10.1.1 Physical-Chemical Fate and Transport Model

The principal components of the physical-chemical fate and transport model framework are reviewed by Thomann and Mueller (1987), Delos et al. (1984), Thomann and Di Toro (1983), and Di Toro et al. (1981) among others.

The development can begin by considering a simple one-dimensional river as shown in Figure 10.1. The chemical in the water column is transported by the flow Q. Losses of chemical may occur as a result of microbial degradation, volatilization, or other pathways. The sediment however in all of the models discussed in this paper is not considered to be moving. There is a transfer of chemical from the sediment to the water column and vice versa via settling and resuspension of particulate chemical forms and sediment diffusion of dissolved chemical.

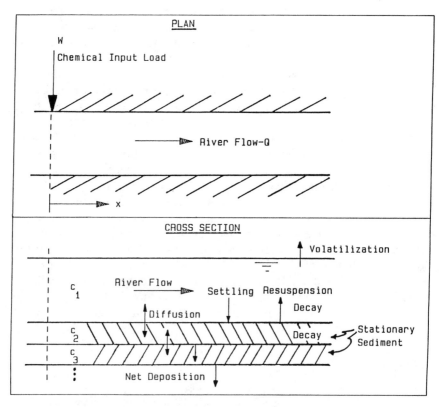

Figure 10.1. Notation for physico-chemical fate model in streams.

The one-dimensional mass balance equation for any form of the chemical (dissolved or particulate) is for the water column

$$\frac{\partial c_1}{\partial t} = -\frac{1}{A}\frac{\partial}{\partial x}(Qc_1) + \frac{1}{A}\left[\frac{\partial}{\partial x}EA\frac{\partial c_1}{\partial x}\right] + sources - sinks \quad (10.1)$$

and for the surface sediment

$$\frac{\partial c_2}{\partial t} = sources - sinks \quad (10.2)$$

where c_1 and c_2 are the chemical concentrations in the water column and sediment (m_T/l^3; m_T = mass of toxicant, l^3 = bulk volume of solids plus water); Q is the river flow (l^3/t); A is the cross-sectional area (l^2); E is the longitudinal dispersion coefficient (l^2/t); x is distance downstream and t is time.

The chemical in the models discussed herein is assumed to be composed of two forms: 1. the dissolved form, c_d' (m_T/l_w^3; l_w^3 = volume of water); and 2. the particulate form, c_p (m_T/l^3), i.e., the toxicant sorbed onto particulate matter in the water column or sediment. The total chemical concentration is then

$$c_T = c_p + \phi c_d' \quad (10.3)$$

where ϕ is the porosity [l_w^3/l^3].

Equation 10.3 is

$$c_T = c_p + c_d \quad (10.4)$$

where

$$c_d = \phi c_d' \quad (10.5)$$

for c_d (m_T/l^3) as the porosity corrected dissolved concentration.

With the general framework described, the detailed equations for the various forms of the chemical can be presented.

Dissolved Chemical

An explicit finite differencing of Equation 10.1 together with sources and sinks of the dissolved chemical in a temporally constant control volume (V_1) of the water column is given by

$$V_1 = \frac{dc_{d1}}{dt} = [(Qc_{d1}^+) - Q_1 c_{d1} + E'((c_{d1})^+ - c_{d1})$$

$$+ E'((c_{d1})^- - c_{d1})] \quad \text{(Transport)}$$

$$+ k_{d1} V_1 c_{p1} - k_{u1} m_1 V_1 c_{d1} \quad \text{(Sorption-desorption)}$$

$$+ K_{f12} A(c_{d2}' - c_{d1}') \quad \text{(Sediment diffusive exchange)}$$

$$- K_{d1} V_1 c_{d1} \quad \text{(Decay \& losses)}$$

$$- k_{l1} A(c_g/H_e - c'_{d1}) \quad \text{(Volatilization)}$$

$$+ W_{d1} \quad \text{(Input)} \tag{10.6}$$

The group of terms in brackets represents the transport and dispersion of the dissolved toxicant. Superscript + indicates the upstream direction and superscript − indicates the downstream direction. The net transport flows, Q, are written in an equivalent backward difference approximation to the underlying partial differential equation (Equation 10.1). The dispersion or mixing between segments of length Δx is given by the bulk dispersion coefficient which in turn is related to the dispersion coefficient by

$$E' = \frac{EA}{\Delta x} \tag{10.7}$$

The second group (sorption-desorption) of Equation 10.6 is the balance between the desorption of the chemical in the particulate phase ($k_{d1} V_1 c_{p1}$) which increases the dissolved form (the desorption rate is k_{d1} [1/t]), and the adsorption from the dissolved phase onto the particulates given by $k_{u1} m_1 V_1 c_{d1}$. The sorption rate is $k_{u1}(l^3/m_s - d)$ and the solids concentration is m_1 (m_s/l^3). (m_s = mass of solids.) Note that this latter term depends on the mass of solids available for sorption from the dissolved phase.

The third group of Equation 10.6 represents the diffusive exchange between the sediment dissolved chemical concentration c'_{d2} in the interstitial water and the dissolved chemical concentration in the water column, c'_{d1}. The sediment–water diffusive transfer coefficient, $K_{f12}(l/t)$ can be considered as an overall interfacial transfer coefficient relating to the diffusion of the toxicant across the sediment–water interface.

Decay and loss mechanisms such as biodegradation, photolysis etc. of the dissolved form are included in the fourth group of the equation. Therefore, $K_{d1}(1/t)$ represents the sum of individual rates, some of which in turn may represent rather complex mechanisms. Note that for this model all the loss rates are assumed to be first order.

Volatilization of the dissolved toxicant is given by the fifth group of Equation 10.6 where c_g represents the gas phase of the chemical (m_T/l_g^3; l_g^3 = volume of gas) which may or may not be zero, and H_e is the Henry's constant for the chemical ($m_T/l_g^3 \div m_T/l_w^3$).

The last line represents all external sources or inputs of dissolved chemical, $W_{d1}(m_T/t)$ from point direct discharge sources as well as nonpoint and tributary inputs.

An equation similar to Equation 10.6 can be written for the dissolved chemical in the sediment layer underneath the typical water column segment 1. This layer is designated with the subscript 2. Thus,

$$V_2 \frac{dc_{d2}}{dt} = k_{d2} V_2 c_{p2} - k_{u2} m_2 V_2 c_{d2}$$

$$+ K_{f12} A (c'_{d1} - c'_{d2})$$

$$- K_{d2} V_2 c_{d2}$$

$$- v_{d2} A c_{d2}$$

$$+ K_{f23} A (c'_{d3} - c'_{d2}) \tag{10.8}$$

The first three lines of the right side of Equation 10.8 have already been discussed relative to the water column. The fourth line of Equation 10.8 expresses the "burial" or transfer down into the sediment of the dissolved toxicant due to net sedimentation or build-up of the sediment layer at a net sedimentation rate of $v_{d2}(l/t)$. The last line of Equation 10.8 is the diffusive exchange of dissolved toxicant between the first and second sediment layers under the water column. Similar equations can be written for each successive sediment layer. Note that there are no dissolved transport terms for the sediment thereby indicating that the sediment is assumed to be stationary in the horizontal direction. Also, mechanical mixing of sediment layers (due, for example, to bioturbation) is not included, but is readily added with an additional mixing term.

Particulate Chemical

The mass balance equation for the chemical sorbed onto the particulates in the water column segment 1 is given by

$$V_1 = \frac{dc_{p1}}{dt} = [(Q c_{p1})^+ - Q_1 c_{p1} + E'((c_{p1})^+ - c_{p1})$$

$$+ E'((c_{p1})^- - c_{p1})] \quad \text{(Transport)}$$

$$- k_{d1} V_1 c_{p1} + k_{u1} m_1 V_1 c_{d1} \quad \text{(Desorption-sorption)}$$

$$- v_s A c_{p1} \quad \text{(Particulate settling)}$$

$$+ v_u A c_{p2} \quad \text{(Particulate resuspension)}$$

$$- K_{p1} V_1 c_{p1} \quad \text{(Decay)}$$

$$+ W_{p1} \quad \text{(Input)} \tag{10.9}$$

The first group of this equation is the transport of the particulate chemical due to net advection (Q) and dispersion (E'). The particulate chemical is assumed to be transported in the same manner as the dissolved form. The second group is the desorption-sorption mechanism discussed above and as can be noted for the particulate form, sorption is a source and desorption is a sink of toxicant. The third and fourth groups are respectively the particulate settling of the chemical from the water column and the resuspension of particulate chemical from the sediment into the water column. The settling velocity, $v_S(l/t)$ and the resuspension velocity $v_u(l/t)$ are functions of particle type (sand, silt, organics) and the hydrodynamics of the water–sediment interface. The fifth group represents any decay mechanisms (e.g. bacterial degradation) of the chemical on/in the particulates at a rate $K_{p1}[l/t]$ and the last line is the external mass input of particulate toxicant, $W_{p1}[m_T/t]$.

The particulate chemical in the sediment is given by an equation similar to Equation 10.9 except that, as noted, the sediment is assumed to be stationary in the horizontal direction. That is, bed load transport or sediment movement horizontally throughout the water body is not considered.

The particulate chemical equation for the sediment segment underlying the water column segment 1 is then given by

$$V_2 \frac{dc_{p2}}{dt} = -k_{d2}V_2c_{p2} + k_{u2}m_2V_2c_{d2}$$

$$+ v_sAc_{p1} - v_uAc_{p2}$$

$$- K_{p2}V_2c_{p2}$$

$$- v_dAc_{p2} \qquad (10.10)$$

The first three lines of this equation parallel the equivalent mechanisms in the water column (sorption-desorption, settling-resuspension, and decay; at rate K_{p2}). The fourth line represents the net downward flux of sediment particulate toxicant due to the net sedimentation velocity v_d. Again, mixing of the sediment due to factors such as bioturbation or deep sediment mixing is not included, but can be added as an additional mixing term.

10.1.2 Local Equilibrium Equations

Equations 10.6 and 10.8 for the dissolved component and Equations 10.9 and 10.10 for the particulate component in the water column segments and sediment segments, respectively, represent a set of interactive, differential equations, one for each control volume of the finite difference grid. Note that the coupling of the dissolved and particulate components is through the reaction kinetics of sorption and desorption. For some chemicals, these reaction kinetics tend to be "fast" (i.e. completion times on the order of hours) compared to the kinetics inherent in other mechanisms of the problem. These latter mechanisms include bacterial decay, net loss rates to the sediment, and sedimentation rates that have reaction times on the order of days to years.

The "fast" kinetics of sorption-desorption indicate that for time scales of days to years, there will be a virtually continuous equilibration of the dissolved and particulate forms depending on the local solids concentration. This partitioning between the two components permits the specification of the fraction of dissolved and particulate chemical to the total. The dissolved and particulate chemical are therefore assumed to be always in a "local equilibrium" with each other. Assuming that the kinetics are reversible and that the sorption/desorption kinetics are linear, then a partition coefficient π [$m_T/m_S \div m_T/l_w^3$] can be defined as follows:

$$\pi = r/c_d' \tag{10.11}$$

or since $c_d' = c_d/\phi$

$$\pi' = \pi/\phi = r/c_d \tag{10.12}$$

for π' as [$m_T/m_s \div m_l/l^3$] and r as the chemical concentration on a solids basis [m_T/m_s].

The particulate toxicant concentration relative to the bulk volume is given by (for m as the solids concentration, m_s/l^3)

$$c_p = rm \tag{10.13}$$

The fraction of the total that is dissolved, f_d, is given by

$$f_d = (1 + \pi'm)^{-1} \tag{10.14}$$

and the particulate chemical as a fraction of total chemical (f_p) is given by

$$f_p = \frac{\pi'm}{1 + \pi'm} \tag{10.15}$$

The local equilibrium assumption therefore permits specification at all times and places of the fraction of the total toxicant in the dissolved and particulate form. It should be stressed again here that this local equilibrium assumption assumes complete reversibility between the solid and liquid phases. There is evidence (e.g., Di Toro et al. 1982a; Di Toro 1985) that this is not the case for certain chemicals.

Also in these relationships it is assumed that the partition coefficient does not depend on the concentration of the sorbing solids. There is considerable evidence, however, as given by O'Connor and Connolly (1980) and Di Toro (1985) who indicate that the partition coefficient does apparently depend on the concentration of solids. The development continues here on the assumption of a constant partition coefficient.

With this assumption, attention can then be focused solely on the mass balance equation for the total chemical. The total chemical in the water column or sediment is given by Equation 10.3. Adding the water column equations for dissolved chemical (Equation 10.6) and particulate chemical (Equation 10.9) and using Equations 10.14 and 10.15 gives

$$V_1 \frac{dc_{T1}}{dt} = [Qc_{T1}^+ - Q_1 c_{T1} + E'(c_{T1}^+ - c_{T1}) + E'(c_{T1}^- - c_{T1})$$

$$+ k_f A(f_{d2} c_{T2}/\phi_2 - f_{d1} c_{T1}) - (K_1) V_1 c_{T1}$$

$$+ k_{l1} A[(c_g/H_e) - f_{d1} c_{T1}] - v_s A f_{p1} c_{T1} + v_u A f_{p2} c_{T2}$$

$$+ W_{T1} \tag{10.16}$$

where $K_1 = K_{d1} + K_{p1}$

Note that the kinetics of sorption-desorption do not appear in this equation because it represents a mass balance of the total. The net loss rates and exchanges that are dependent on the form of the toxicant do however, remain.

A total chemical equation for the sediment segment (subscript 2) can be obtained in a similar manner. Thus adding Equations 10.8 and 10.10 gives:

$$V_2 \frac{dc_{T2}}{dt} = -K_f A(f_{d2} c_{T2}/\phi_2 - f_{d1} c_{T1}) - (K_2) V_2 c_{T2}$$

$$+ v_s A f_{p1} c_{T1} - v_u A f_{p2} c_{T2} - v_d A f p_2 c_{T2}$$

$$+ K_f A(f_{d3} c_{T3}/\phi_3 - f_{d2} c_{T2}/\phi_2) \tag{10.17}$$

where $K_2 = K_{d2} + K_{p2}$

Equations 10.16 and 10.17 are the fundamental equations used in the succeeding analyses. These equations are coupled parametically to the suspended solids and sediment solids concentrations (see Equations 10.14 and 10.15). These concentrations can be specified externally as an input or the mechanisms of solids settling, resuspension, and deposition can be explicitly modeled. In addition, an independent tracer can be used to calibrate these parameters; see Thomann and Di Toro (1983) for the use of plutonium-239,240 as a tracer.

10.1.3 Food Chain Model

The transfer of a chemical in the aquatic food chain occurs through two principal routes:

1. direct uptake from the water.
2. accumulation due to consumption of contaminated prey.

The uptake of a chemical directly from water through transfer across the gills as in fish or through surface sorption and subsequent cellular incorporation as in phytoplankton is an important route for transfer of chemicals. This uptake is often measured by laboratory experiments during which test organisms are placed in aquaria with known (and fixed) water concentrations of the chemical. The accumulation of the chemical over time is then measured and the resulting equilibrium concentration in the organism divided by the water concentration is termed the bioconcentration factor (BCF). A simple representation of this mechanism is given by a mass balance equation around a given organism. Thus,

$$\frac{dv'}{dt} = k_u wc - Kv' \tag{10.18}$$

where v' is the whole body burden of the chemical (m_T); k_u is the uptake sorption and/or transfer rate ($l^3/t \cdot m(w)$); $m(w)$ = mass of organism, wet weight; w is the weight of the organism, $m(w)$; c is the dissolved water concentration (m_T/l_w^3); K is the desorption and excretion rate ($1/t$) and t is time. This equation indicates that the mass input ($\mu g/d$) of toxicant given by $k_u wc$ is offset by the depuration mass loss rate ($\mu g/d$) given by Kv'. The whole body burden v' is given by

$$v' = vw \tag{10.19}$$

where v is the concentration of the chemical ($m_T/m(w)$). Substitution of Equation 10.19 into Equation 10.18 gives, after simplification

$$\frac{dv}{dt} = k_u c - K'v \tag{10.20}$$

where $K' = K + G$

for $G(1/t)$ as the net growth rate of the weight of the organism. At equilibrium or steady state,

$$v = \frac{k_u c}{K'} \tag{10.21}$$

and the BCF is given by

$$N_w = \frac{v}{c} = \frac{k_u}{K + G} \tag{10.22}$$

The ratio N_w, the bioconcentration factor, is in units $m_T/m(w) \div m_T/l^3$, e.g., $\mu g/kg \div \mu g/l$ (= l/kg).

For organic chemicals, the BCF is conveniently defined on a lipid-normalized basis, i.e., $m_T/m(lip) \div m_T/l^3$, e.g., $\mu g/kg(lipid) \div \mu g/l$. The lipid normalization assumes that the lipid compartment of the organism is the principal receptor of the hydrophobic organic chemical.

The octanol-water partition coefficient (K_{ow}) of a chemical is a useful ordering parameter to express the tendency of organic chemicals to partition into the lipid pool.

At equilibrium then for organic chemical BCF, to first approximation,

$$N_w = K_{ow} \tag{10.23}$$

for the laboratory case of no organism growth and N_w now defined as the lipid normalized BCF. Thomann (1987, 1988) suggests the following expression for the field BCF, as a function of K_{ow}

$$N_w = K_{ow} \left[1 + \frac{10^{-6} K_{ow}}{E(K_{ow})} \right]^{-1} \qquad (10.24)$$

where $E(K_{ow})$ is an efficiency of chemical transfer across the gills as a function of K_{ow} and for fish can be approximately expressed as (Thomann 1987, 1988)

$$\begin{array}{lll}
\log E = -1.5 + 0.4 \log K_{ow} & \text{for } \log K_{ow} = 2\text{--}3 \\
E = 0.5 & \text{for } \log K_{ow} = 3\text{--}6 & (10.25) \\
\log E = 1.2 - 0.25 \log K_{ow} & \text{for } \log K_{ow} = 6\text{--}10
\end{array}$$

Age-Dependent Model

The general age-dependent model utilizes a mass balance of chemical around a defined compartment of the aquatic ecosystem. In the most general case, a compartment is defined as a specified age class of specified organism or in steady-state simplification, a compartment is considered as an "average" age class or range of ages for a given organism. Figure 10.2 schematically shows the compartments. As indicated in Figure 10.2a, each age class of a given trophic level is considered as a compartment and a mass balance equation can be written around each such age class. The zero trophic level is considered to be the phytoplankton–detritus component representing one of the principal sorption mechanisms for incorporating toxicants into the food chain.

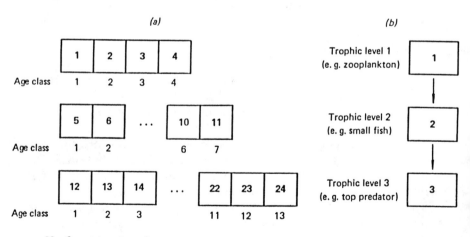

Figure 10.2. Schematic of compartment definition for (a) age dependent model, and (b) simplified steady-state model.

Consider then the phytoplankton, detrital organic material, and other organisms, all of size approximately $< 100~\mu m$ as the base of the food chain. An equation for this compartment is given by a simple reversible sorption-desorption linear equation as:

$$\frac{dv_0}{dt} = k_{u0}c - K_0 v_0 \tag{10.26}$$

where all terms have been defined, the subscript zero refers to the base of the food chain, and t is real time.

For a compartment above the phytoplankton/detritus level, the mass input of the toxicant due to ingestion of contaminated food must be included. This mass input will depend on: toxicant concentration in the food; rate of consumption of food; and the degree to which the ingested toxicant in the food is actually assimilated into the tissues of the organisms.

The general mass balance equation for the whole body burden for a given compartment, i, is then similar to Equation 10.18 for water uptake but with the additional mass input due to feeding. Therefore,

$$\frac{dv_i'}{dt} = \frac{d(vw)_i}{dt} = \frac{w_i dv_i}{dt} + \frac{v_i dw_i}{dt} = k_{ui}w_i c - K_i v_i'$$

$$+ \sum_j p_{ij}\alpha_{ij}C_i v_j w_i \qquad i = 1 \ldots m \tag{10.27}$$

where α_{ij} is the chemical assimilation efficiency (m_T absorbed/m_T ingested); C_i is the weight-specific consumption of organism i, $m(w)$ predator/$m(w)$ prey-d; p_{ij} is the food preference of i on j; and t is real time (days). Consider now a simple case of a sequential food chain where predation is only on the next lowest trophic level.

An equation for the individual organism weight is

$$\frac{dw_i}{dt} = (a_{i,i-1}C_i - r_i)w_i, \qquad i = 1 \ldots m \tag{10.28}$$

where $a_{i,i-1}$ is the biomass assimilation efficiency, $m(w)$ predator/$m(w)$ prey, and r_i is the respiratory weight loss ($1/t$) due to routine metabolism, swimming, and other activities. The weight change is therefore

$$G_i = \frac{dw_i}{dt} \bigg/ w_i = (a_{i,i-1}C_i - r_i) \tag{10.29}$$

Equation 10.27, for a food chain in contrast to a food web, can then be written as

$$\frac{dv_i}{dt} = k_{ui}c + \alpha_{i,i-1}C_i v_{i-1} - K_i' v_i, \qquad i = 1 \ldots m \tag{10.30}$$

where i is the predator and $i - 1$ is the prey.

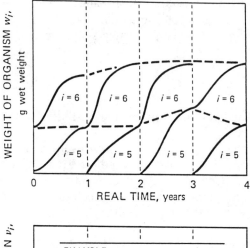

Figure 10.3. Illustration of meaning of $w_i(t)$ and $v_i(t)$ showing as an example a 0–1-year-old and 1–2-year-old alewife.

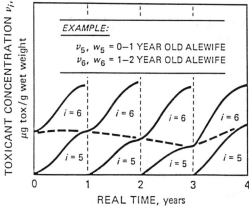

The interpretation of w_i and v_i in Equations 10.28 and 10.30 is further explained in Figure 10.3. The variation of the weight (and chemical organism concentration) of a given compartment (i.e., a given age class of a given organism) is shown with real time. If as an example, w_5 is a 0–1-year-old alewife then it is seen that the weight of this age class may vary from year to year. Similarly, the distribution of the chemical may change from year to year for a given compartment depending on, for example, the variation in the water column toxicant concentration. The specification of the boundaries of a given compartment depends on the life cycle of the organism.

It should be recognized therefore that several of the biological and chemical parameters of the weight change equation 10.29 and the food chain equation 10.30 are functions of organism weight within the time interval defined by the compartment. See Thomann and Connolly (1984) for an application of the age-dependent model to PCB accumulation in the lake trout of Lake Michigan.

10.2 Steady-State Simplification

10.2.1 River Physico-Chemical Fate

In spite of the complexity of Equations 10.16 and 10.17, it can be shown (see e.g., Thomann and Mueller 1987) that under steady-state and constant spatial parameters that for a single source to first approximation, the maximum concentration c_o is at the outfall. Thus,

$$c_o = (Q_u c_u + Q_e c_e)/Q \tag{10.31}$$

where Q_u and c_u are the upstream flow and concentration, respectively; Q_e and c_e are the effluent flow and concentration, respectively; and Q is the total river flow.

Since the maximum generally occurs at the outfall, there are generally two situations to be considered in estimating the downstream fate of a discharged chemical:

1. There is a critical water use point downstream of a single discharge and the concentration at the point of use (e.g., a water supply withdrawal) needs to be estimated.
2. There are several inputs of the same chemical along the length of the river and the total concentration must be estimated.

The downstream fate of the chemical or mixture depends on:

1. The properties of the river such as the depth, velocity, and dilution downstream due to groundwater infiltration or tributary inflow.
2. The chemical properties, such as volatilization, biodegradation, or partitioning onto the solids.

Thomann and Mueller (1987) discuss these factors in some detail. A simplified summary from Thomann and Salas (1986) is given here. An important point for the computation of the downstream fate of a chemical or chemical mixture is that the calculation is very similar, indeed for preliminary analyses, the computation is identical to classical stream water quality calculations. The basic equation under steady-state conditions is given by

$$c_T = c_{T0} \, exp \left[-\left(\frac{v_T}{HU} + q \right) x \right]$$
$$= c_{T0} \, exp[-K_T + q')t^*] \tag{10.32}$$

where c_T is the concentration as a function of distance downstream; K_T is the overall net loss rate $(1/t)$; v_T is the net loss of chemical expressed as velocity, l/t; t^* is the time of travel $(= x/U)$; q is the slope of the natural logarithm of river flow with distance (the flow dilution rate); H is river depth; U is river velocity; and q' is the river flow slope on a travel time basis.

The calculation of the downstream fate of a toxic substance or toxicity of an effluent or effluents depends then on the estimation of the dilution of the river and the loss rate of the chemical. The discharge of a conservative substance with no dilution would result in a constant concentration in the downstream direction, i.e., from Equation 10.32 with K_T and $q' = 0$, $c = c_o$.

The determination of whether there is a downstream infiltration of groundwater or overland drainage can be made by examining the downstream distribution of a known conservative substance such as chloride or total dissolved solids or other tracer.

If dilution exists then a conservative substance may exhibit apparent non-conservative behavior. Then for $K_T = 0$, but $q' \neq 0$, Equation 10.32 is

$$c = c_o \, exp(-qt^*) \tag{10.33}$$

which indicates an exponential decline in the conservative substance due to distributed downstream dilution.

Finally, if the toxic substance is non-conservative, i.e., the chemical undergoes biodegradation or volatilization or other losses and dilution is also occurring, then the net loss rate of the chemical, K_T must be estimated.

Table 10.1 provides some guidelines for a preliminary assessment of downstream fate. Toxicity in this table is whole-effluent toxicity of the chemical mixture to standard organisms and is expressed on a toxic-unit basis. A toxic unit is defined (USEPA, 1985)

$$Tu = 100/LC_{50} \text{ or NOEL} \tag{10.34}$$

where LC_{50} is the lethal concentration to 50% of the organisms and NOEL is the no observed effect level. As indicated, both the heavy metals and the toxicity measure are assumed to be a conservative variable for first approximations. Thus, only dilution needs to be considered for these variables. For the organic chemicals a somewhat arbitrary division has been made based on the water solubility of the chemical.

Table 10.1. Guideline for estimating downstream loss rate of chemicals and toxicity

Group	Guideline[a]
Heavy metals	Conservative ($K_T = 0$) and additive
Toxicity	Conservative ($K_T = 0$) and additive
Organic chemicals, water solubility > 1 $\mu g/l$	Conservative ($K_T = 0$) and additive
Organic chemicals, water solubility > 1 $\mu g/l$	Estimate loss rate (Eq. 10.34)

[a] In all cases, dilution in the downstream direction must be included. Source: Thomann and Salas 1986.

The rationale is that at solubilities less than about 1 μg/l, the chemical will partition onto the solids because of a relatively high partition coefficient (about 10^4–10^6 l/kg). Also, the general tendency will be for such chemicals to biodegrade and volatilize to a lesser degree than the more soluble chemicals. For first approximations, then, such low solubility chemicals may be assumed conservative.

For organic chemicals with solubilities greater than about 1 μg/l, the loss rate must be estimated. For first approximations, the net loss velocity v_T ($=K_T H$) can be estimated from

$$v_T = K_T H = (K_d H + k_l)f_d + v_n f_p \qquad (10.35)$$

where $K_d[T^{-1}]$ is the decay rate of the chemical due to processes such as biodegradation or photolysis; k_l is the loss due to volatilization; v_n is the net loss velocity of the solids in the river, l/t; and f_d and f_p are the dissolved and particulate fractions of the total (see Equations 10.14 and 10.15).

For rivers, the net loss of solids often, although not always, can be assumed equal to zero. Thus, $v_n = 0$ and the chemical loss rate depends only on the degradation rate, volatilization rate, and fraction dissolved.

Table 10.2 provides some guidelines for estimating the fraction of the chemical that is in the dissolved form. These guidelines employ a more complex interaction of chemical partitioning and solids concentration as given in Di Toro (1985) and discussed in Thomann and Mueller (1987). The f_d can also be approximated with Equation 10.14. Figure 10.4 shows

Table 10.2. Approximate fraction of total chemical in dissolved form and on particulates

Organic chemicals				
Chemical solubility (μg/l)	(f_d) fraction of chemical in dissolved form[a]		Ratio of particulate conc. to total conc. (r/c_T [μg/g ÷ μg/l])[b]	
	Range	Approx. mean	Range	Approx. mean
>100	0.5–1.0	0.7	0–50	10
10–100	0.3–0.9	0.5	0.1–70	50
<10	0.3–0.8	0.4	0.1–70	60
Heavy metals				
	0.6–1.0	0.8	0.4–16	2.5

[a] Approximated from solids dependent partition coefficient relationships of Di Toro (1985), solids range 10 → 1000 mg/l
[b] $r/c_T = (1 - f_d)/(.01 → 1.0)$
Source: Thomann and Salas 1986.

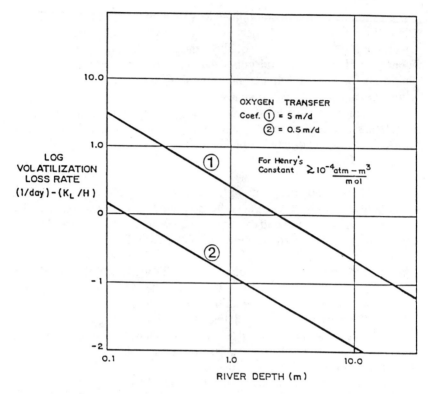

Figure 10.4. Range of volatilization loss rate as a function of river depth for different oxygen transfer rates.

the range of the volatilization loss rate as a function of the river depth and reaeration characteristics. This figure is for substances with Henry's constant $> 10^{-4}$ atm-m^3/mol for which the volatilization rate is estimated from the oxygen transfer rate K_L. Thus,

$$k_l \approx \left(\frac{32}{M}\right)^{1/4} K_L \tag{10.36}$$

where $K_L = \left(D_L \dfrac{U}{H}\right)^{1/2}$

for $D = $ *oxygen diffusivity* $(.000181\ m^2/d)$

The loss rate, K_d, is generally site-specific and chemical specific and no general simplification is available.

Table 10.2 also gives the approximate ratio of the chemical adsorbed to the suspended particulates to the total water concentration. Therefore,

$$\Pi = r/c_T \tag{10.37}$$

for Π in $\mu g/g(d) \div \mu g/l$, r in $\mu g/g(d)$ and c_T in $\mu g/l$, $g(d) = $ grams dry weight.

It can also be shown that for $\P_2 = \P_1$ and $K_{d2} = 0$, the sediment particulate concentration is equal to the water column particulate concentration, i.e.,

$$r_2 = r_1 \qquad (10.38)$$

where r_2 is the sediment particulate concentration and r_1 is the water column particulate concentration both in $\mu g/g(d)$.

In summary, it is seen that the calculation of the fate of the chemical or mixture is similar to the procedure for conventional water quality variables. For many chemicals including the toxicity measure, the assumption that the chemical is conservative is appropriate for first approximations. If such an assumption cannot be made, then an estimate of the downstream loss rate must be made using the preliminary guidelines discussed here.

10.3 Deterministic Time Variable Models

10.3.1 Application of Time Variable Model to Benzo(a)pyrene and Cadmium in the Great Lakes

In this section, the fully time-variable model (Equations 10.16 and 10.17) is applied to two chemicals: 1. benzo(a)pyrene, a polycyclic aromatic hydrocarbon (PAH), and 2. cadmium, a representative metal. The model uses the Great Lakes segmentation of Thomann and Di Toro (1983) shown in Figure 10.5 and details are in Thomann and Di Toro (1984).

Benz(a)pyrene

The distribution of this chemical, one of the PAH compounds resulting from incomplete combustion of organic materials has been widely studied (e.g., Neff 1979) because of its potential carcinogenicity. The fate of benzo(a)pyrene (BaP) in the Great Lakes has been evaluated in a series of papers by (Eadie 1983; Eadie et al. 1982, 1983). In that work, data are presented for the range of concentration of BaP in the water column and surficial sediments as well as preliminary data on the BaP concentration in the pore water of the sediment. It is those data (together with estimates of loading) that can be used as an application of the physico-chemical model.

BaP is sparingly soluble in water 0.172 $\mu g/l$ (Neff 1979) and as such would be expected to have an affinity for solids. The partitioning onto particulates and in addition, the extent of volatilization of the BaP must be

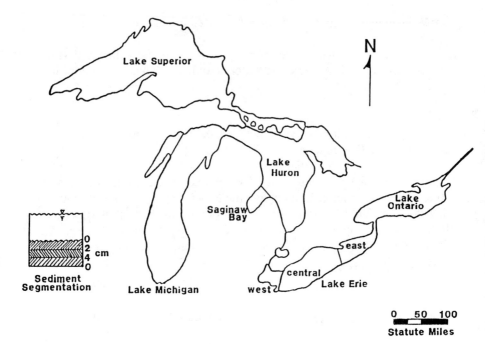

Figure 10.5. Great Lakes and Saginaw Bay segmentation used in model.

estimated. Although BaP is known to undergo photolysis (Neff 1979) this pathway is not considered in this application.

The estimated atmospheric loading on an area basis is about 95 g BaP/km²-yr across all of the lakes but as noted by Eadie (1983), all BaP load estimates are based on quite limited data and therefore may vary as additional information becomes available. From the data given in Eadie (1983) for Lake Michigan, and other data, the model for BaP was run for ¶ from 10,000 to 100,000 l/kg, thereby providing a range of one order of magnitude in the solids partition coefficient.

The time variable calculation using a steady loading for 20 years is shown in Table 10.3 and in Figures 10.6 and 10.7. As shown in Table 10.3, the lake-to-lake variation in BaP concentrations either in the water column or the sediment differs by less than about a factor of two. The highest concentration of BaP in the surface sediments (¶ = 100,000) is in Saginaw Bay. It can also be noted in Table 10.3 and Figure 10.6 that at ¶ = 10,000 the calculated surface sediment concentration for Lakes Michigan and Erie is about 45 ng/g(d) or about one order of magnitude lower than the observed data. Figure 10.7 shows the calculated time history under the two partition coefficients. For ¶ = 100,000 l/kg, the water column and sediment are at about 80 and 60% of steady state, respectively, while for ¶ = 100,000 l/kg, the water column and sediment are about 20% of steady state.

Table 10.3. Comparison of calculated and observed BaP for Great Lakes under different solids partition assumptions

	Calculated range of BaP across all lakes[a]			
	$\P = 10{,}000$ l/kg	$\P = 100{,}000$ l/kg	Observed mean of BaP	Reference
Total water conc. (ng/l)	5–6	1–2	12 ± 8[b]	Eadie et al. 1983
Surficial sediment conc. (ng/g(d))	38–60	46–133	Michigan 480 ± 246(7)[c] Erie 255 ± 152(3) Superior 28(1) Huron 294(1) Ontario 306(1)	Eadie 1983
Sediment pore water conc. (ng/l)	3–5	0.5–1.3	850 ± 1260	Eadie et al. 1983
Particulate conc. in water column (ng/g(d))	46–64	46–165	Michigan 200–400	Eadie 1983

[a] After 20 years of loading
[b] Mean ± SD
[c] ( ) = no. of samples

Figure 10.6 and Table 10.3 indicate that a more favorable (but not totally desirable) comparison to observed data is obtained at the higher BaP partition coefficient of 100,000 l/kg. The results also indicate the need to determine the partition coefficient for BaP, as a representative PAH for Great Lakes solids concentrations. On the basis of this application of the physico-chemical model to BaP in the Great Lakes, it is concluded that:

1. The estimate of the BaP partition coefficient, obtained from published empirical relationships, is probably low by about an order of magnitude for the Great Lakes system.
2. With an increased BaP partition coefficient and assuming loss due to volatilization, the physico-chemical toxic substances model of the Great Lakes approximate observed BaP water column and sediment data only to order of magnitude.
3. The model confirms that on a lake-wide scale, the principal external source of BaP is the atmosphere.
4. For larger lakes such as Lake Michigan, the 50% response time of the lake to external loads is about 6–10 years for the water column–sediment system while for Lake Erie the response time is about 2 years.
5. Lake-to-lake variations in BaP water column and sediment concentrations are less than a factor of two.

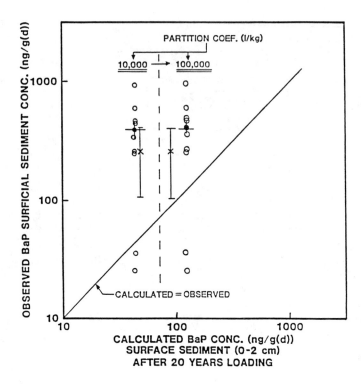

Figure 10.6. Comparison between calculated surface sediment BaP concentration after 20 years and observed concentration.

Cadmium

Time variable model calculations for cadmium were made using the low and high load estimates of Table 10.4 and two assumptions on the cadmium partition coefficient: a. variable partition with solids concentration as given in HydroQual (1982) by

$$\P = (3.52 \cdot 10^6)m^{-0.92} \tag{10.39}$$

and b. a constant partition coefficient of $2 \cdot 10^5$ l/kg. Cadmium was assumed to be conservative. For all calculations, zero initial conditions were assumed and the loads were input as constant over time. It became apparent from initial runs that the time to steady state especially for the

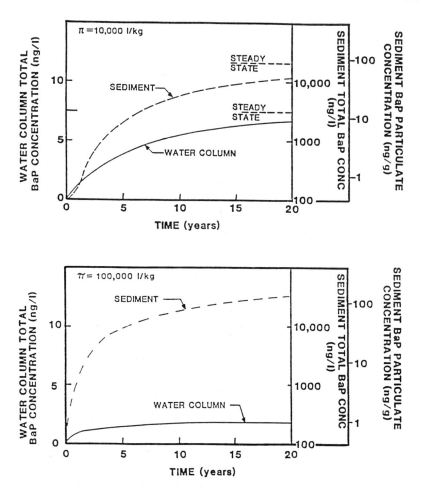

Figure 10.7. Time variable benzo(a)pyrene (BaP) response in Lake Michigan under two partition coefficients with 20 year constant loading.

upper lakes is long so that the computation was carried out for a 100-year period. The computation therefore represents the response of the Great Lakes system to a constant loading and such loading can be viewed as a loading in addition to the background loading and cadmium concentrations shown in Table 10.5. If increased loading of cadmium is assumed to have begun in approximately the 1920s, then the output from the model calculation at $t = 50$ years would be representative of the 1970s. This is the period for which some reliable data are available.

Figure 10.8 shows the comparison of calculated and observed surface sediment cadmium data for $t = 50$ years. As shown, the calculation is reasonable to order magnitude. Figure 10.9 (high-load estimate) shows the full 100-year calculation for Lake Michigan and central Lake Erie as

Table 10.4. Summary of contemporary external cadmium loads (not including upstream loads)

Lake/Region	Atmospheric[a] mt/yr	Tributary[c] mt/yr	Mun. + Ind.[d] (mt/yr)	Total mt/yr	Total g/km²-yr
Superior	41–108	13–126	0–0.3	54–234	651–2823
Michigan	12–120	9–92	0.4–4	21–216	363–3739
Huron	23–57	14–136	0.1–1	37–194	644–3378
Saginaw Bay[b]	3	2	0	5	1184
Erie West	1–18	2–20	1–8	4–46	1321–15202
Central	3–94	2–16	0.2–2	5–112	318–7126
East	1–38	1–12	0.1–0.5	2–51	320–8155
Ontario	10–44	7–66	0.6–6	18–116	924–5953

[a] From Allen and Halley 1980.
[b] From Dolan and Bierman 1982.
[c] Assuming tributary flow at total cadmium conc. of 0.2–1.0 μg/l.
[d] At 0.5–5.0 μg/l for direct municipal point sources.

illustrations. The sensitivity of the calculation to the assumptions in the partition coefficient is shown. As indicated, the effect of the solids-dependent partition coefficient is to greatly increase the time to steady state as a result of the diffusive flux of cadmium from the sediment due to the lower partition coefficient. For Lake Michigan, the surface sediment concentration decreases with the variable partition coefficient but for Lake Erie the

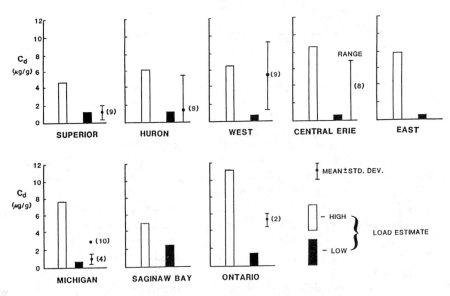

Figure 10.8. Calculated surface sediment concentration, μg/g(d), at $t = 50$ years with partition coefficient as a function of solids concentration.

Table 10.5. Estimate of approximate background cadmium concentration for Great Lakes

Lake/Region	Range in total background water col. conc. (ng/l)[a]	Background conc. used in load estimate (ng/l)	Background total load (mt/yr)[b]	External background load (mt/yr)[c]	External background load g/km²-yr	Ratio of contemporary load to background load
Superior	2.8– 5.6	4.0	6.1	6.1	37	18–76
Michigan	2.8– 5.6	4.0	3.1	3.1	54	7–69
Huron	2.8– 5.6	4.0	5.2	4.8	83	8–41
Saginaw Bay	6.5–13.0	10.0	0.6	0.4	105	11
Erie West	12.5–25.0	20.0	10.3	8.6	283	5–54
Central	5.0–10.0	8.0	22.0	10.0	635	0.5–11
East	5.0–10.0	8.0	12.0	4.7	754	0.4–11
Ontario	2.8– 5.6	4.0	4.0	2.6	132	7–45

[a] At assumed background sediment conc. of 05–1.0 μg Cd/g(d); $r_2 = r_1$ and $\P = 200{,}000$ l/kg.

[b] Includes any upstream or exchange loads.

[c] Not including any upstream or exchange loads.

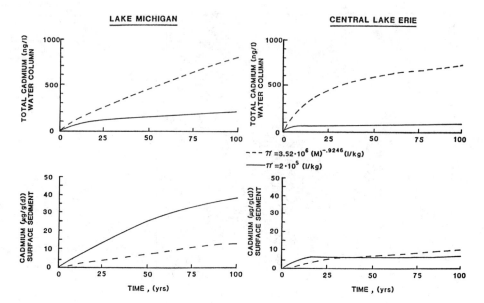

Figure 10.9. Comparison of calculated cadmium concentration under two assumptions on partition coefficient. High load estimate.

surface sediment concentration increases. The continual increase in concentration for central Lake Erie reflects the non-equilibrium condition of the upstream lakes.

If a constant partition coefficient is assumed (as in Muhlbaier and Tisu 1981), then it is seen that for the water column a steady state is reached after about 25 years for Lake Michigan and after about 10 years for Lake Erie. A calculation then for Lake Michigan that attempts to calibrate to a mean concentration of 27 ng/l (Muhlbaier and Tisue 1981) then is simply a matter of estimating the approximate average load and may not reflect a non-steady-state condition as concluded by Mulhbaier and Tisu (1981). However under a solids-dependent partition coefficient for cadmium, the Great Lakes are not in equilibrium with the external load and for all practical purposes never reach a steady-state condition. Clearly then, under this model construct, it is important to determine the solids dependence of cadmium for the range of solids encountered in the Great lakes water column and sediments (i.e., 0.5–240,000 mg/l). If however it is assumed that a solids-dependent partition coefficient is applicable to the Great Lakes, then the system is not in equilibrium with respect to the external load.

It is concluded from this application of the time variable physico-chemical model to cadmium in the Great Lakes that

1. The degree of any dependence of the cadmium partition coefficient with solids has a marked effect on time to steady state and interstitial cadmium concentration.

2. Under a solids-dependent cadmium partition assumption, the Great Lakes, especially the upper Lakes, do not reach a steady-state condition after 100 years of constant loading.
3. Under a constant partition coefficient for cadmium, the Great Lakes do reach an equilibrium condition varying from about 25 years for Lake Michigan to 10 years for Lake Erie.
4. The concentration of cadmium in the Lakes would be expected to increase by about 60% over the next 50 years if the average cadmium loading for the preceding 50 years continues.

10.4 Statistical Variation in Fish

The statistical behavior of a chemical in a fish population is of interest for at least three reasons:

1. U.S. Food and Drug Administration action limits for fish have resulted in the closing of commercial fisheries because of excessive concentrations of chemicals in predators such as the striped bass in the Hudson estuary and surrounding waters. The ability to predict not only the mean value of a chemical in a fish but also the variance of the concentration is of importance in control strategies needed to reopen a fishery.
2. The relationships between variable exposure concentrations in the water column and resulting variability in fish and other organisms is also related to the subsequent acute and chronic toxicity effects on the organism, and, as such, a framework for predicting organism chemical variance is of value in elucidating toxic effects on aquatic animals.
3. The integration of the physico-chemical modeling framework (which includes external inputs of the chemical to the water column) and the biological modeling framework in a time variable sense to predict statistical properties in the water and fish is of particular value in a generalized risk assessment determination.

The variability of chemicals in water over time is large. Figure 10.10 from the Mississippi River is an example. Recognizing this variability in the water chemical concentration of the food chain model of Section 10.1.3, analytical models of expected response in the fish can be developed.

10.4.1 Analytical Model of Statistical Variation of an Organic Chemical in Fish–Water Uptake Only

Consider the case where the water concentration to which the fish is exposed is represented by a first-order autoregressive process. Thus, the autocorrelation function for the water concentration is assumed as

$$\rho_{ck} = exp(-ak); \; k = 0, 1, 2, \ldots m \qquad (10.39)$$

where ρ_{ck} is the normalized autocovariance; k is the lag number; m is the maximum number of lags; and a is the exponential decay rate of the autocorrelation.

Now

$$\rho_{ck} = \rho_{cl}^{k} \qquad (10.40)$$

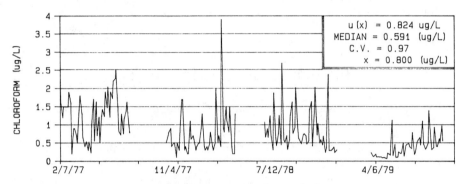

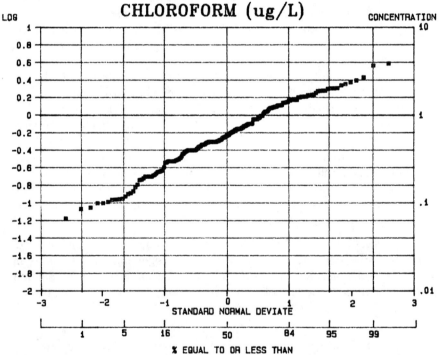

Figure 10.10. Organic chemical data from Mississippi River, Jefferson Parish station. (*Top*) Chloroform time series; (*bottom*) cumulative frequency distribution. (Compiled from data of USEPA 1986.)

and then from Equations 10.39 and 10.40,

$$\rho_{c1} = exp(-a) \tag{10.41}$$

For the synthetic generation of a time series x_t with a first-order autoregressive input, Bras and Iturbe (1985) give

$$(x_t - \mu) = \rho_1(x_{t-1} - \mu) + \sigma_x(1 - \rho_1^2)^{1/2}Z(t) \tag{10.42}$$

for mean μ, variance σ_x^2 and $Z(t)$ a standard normal deviate (mean $= 0$; variance $= 1$).

Thus if $a \gg 0$, i.e., the autocorrelation function drops rapidly to zero at about lag one, $\rho_1 \to 0$ and Equation 10.42 gives

$$x_t - \mu = \sigma_x Z(t)$$

or a normally distributed uncorrelated random variable, the "white noise" case.

For the autocorrelation given by Equation 10.39, Bendat and Piersol (1971) show that the spectrum for frequency f is

$$G_c^N(f) = \frac{4a}{a^2 + 4\pi^2 f^2} \tag{10.43}$$

where $G_c^N(f)$ is a normalized spectrum. Therefore,

$$G_c(f) = \sigma^2 G_c^N(f) \tag{10.44}$$

The variance of the fish concentration can then be shown to be (Thomann 1987)

$$\sigma_v^2 = \sigma_c^2 k_u^2 4a \int_0^\infty (K'^2 + 4\pi^2 f^2)^{-1/2}(a^2 + 4\pi^2 f^2)^{-1/2} \, df \tag{10.45}$$

Completing the integration, the ratio of the coefficients of variation between the fish and the water is

$$r = \sqrt{\frac{K'}{a + K'}} \tag{10.46}$$

where $r = \dfrac{v_v}{v_c}$ and $K' = K + G$,

for v_v and v_c as the coefficients of variation of the chemical in the fish and in the water, respectively. This remarkably simple result indicates the significance of the depuration rate plus growth rate, K', as the principal controlling factors in generating relative variability in the fish concentration.

In Equation 10.46, r depends primarily on K' because at low excretion rate, pulses in water concentration tend to be retained by the fish over a longer period of time than at high excretion rates. Similarly, as a fish increases in weight, concentration variability will tend to shift into the low frequency end of the variance spectrum.

Figure 10.11 shows the behavior of r from Equation 10.46 and indicates that correlated water concentrations increases the variability of the fish concentration relative to water. This is a consequence of correlated inputs introducing more "low frequency" variations in water concentration which are not dampened by the fish. Thus additional variance propagates through the fish and is reflected in the increase in the coefficient of variation.

Laboratory data for evaluating this analytical development come from two papers of Oliver and Niimi (1983, 1985). In this work, rainbow trout were exposed to chlorinated and brominated organic chemicals for up to about 100 days. In Oliver and Niimi (1983), the data reported include the individual fish concentrations and associated water concentrations and the accompanying statistics of mean and standard deviation. For Oliver and Niimi (1985) only the BCF are reported. To use the data in this research, the fish concentrations were calculated from the individual BCF values and the mean water concentration. These data are therefore only an approximation to the actual fish concentration.

As noted frequently by the authors, several of the chemicals did not reach equilibrium during the test. The preceding statistical development

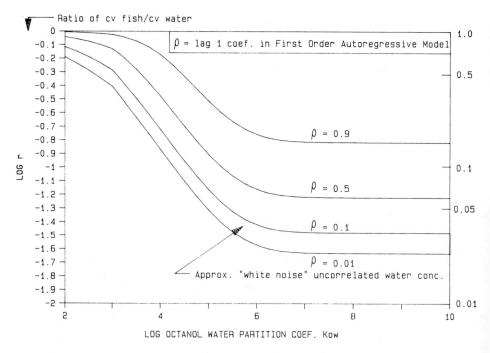

Figure 10.11. The effect of the first order autocorrelation parameter of the water concentration on the relationship between r and K_{ow}. Growth rate variable, wt = 1000 g.

assumes a dynamic equilibrium has been reached. The computation of the variance would be severely biased by the increase in concentration to a dynamic equilibrium. With a few exceptions, those chemicals excluded by the authors have also been excluded herein.

Figure 10.12 (top) shows the results of compiling these data and calculating the r ratio: v_ν/v_c. As a function of log K_{ow}, it is seen that there is a general downward trend to the data from initial values of >1 at low log K_{ow}. The four circled points at the high log K_{ow} value of >6 are exceptions. The chemicals are pentabromotoluene, pentabromoethylbenzene, hexabromobenzene, and octachloronapthalene. These points were also noted by Oliver and Niimi (1985) in their evaluation of the BCF data. No ready explanation is available for why these chemicals exhibit less relative variability than expected, although there are no data in this log K_{ow} range where r values are low. Indeed, the data indicate a trend towards decreasing r values to a minimum at about log K_{ow} of about 5.5 and then an increase to higher r values with increasing log K_{ow}. This is not consistent with the previously developed theory as shown in Figure 10.12 (bottom).

This figure shows the calculation of the theoretical r ratio under three different assumptions. For these calculations the growth rate of the rainbow trout was calculated from the reported data at an average level of about 0.01/d. Also, the excretion rate is given in Equations 10.22 and 10.23 and the uptake rate as a function of the efficiency of transfer E. Thus,

$$K = \frac{10^3 w^{-0.2} E/p}{K_{ow}} \tag{10.47}$$

where p is the fraction lipid and E is given by Equation 10.25.

The first calculation assumed the water concentration was statistically "white noise," i.e., uncorrelated in time. As noted, this calculation underestimates r by a significant amount.

The second calculation using Equation 10.46 assumed a correlated first order process ($a = 0.9$) for the water concentration. Now, the lower bound of the r ratios is captured although at log $K_{ow} > 6$ the model deviates from the data.

The third calculation assumes some metabolism at a high rate of 1.0/d. This is essentially an additive factor to the excretion rate and as seen, the calculation now approximates the data somewhat better. As noted however, this calculation is simply an hypothesized mechanism for increasing the loss rate of the chemical. In the theoretical development, the only parameter that influences the shape of the r function is the excretion (plus metabolism) rate. To capture the four circled points with the theory would require an increase in the effective excretion plus metabolism rates to levels of >1.0/day.

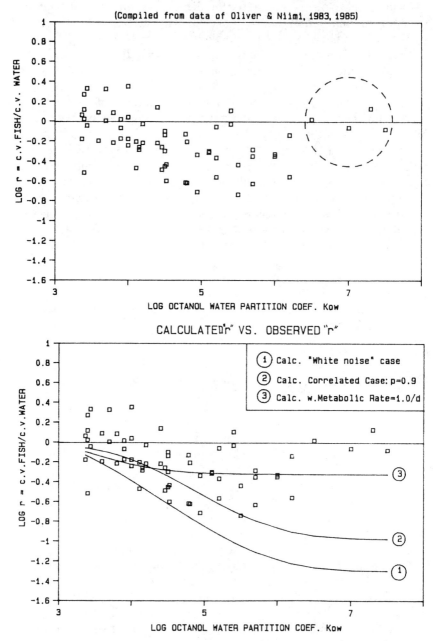

Figure 10.12. (*Top*) Compiled laboratory *r* values from experiments of Oliver and Niimi (1983, 1984); (*bottom*) comparison of calculated and observed *r* values under assumptions on water correlation structure and metabolic rate.

In summary, the laboratory data of Oliver and Niimi show a general downward trend of the r ratio with log K_{ow}, but with several notable exceptions at high log K_{ow}. The theory approximately duplicates the observed data only with a high level of first order correlated water input to about log K_{ow} of about 5.5–6.0. At higher log K_{ow}, increased metabolism and/or excretion would be necessary to reproduce the observed data.

10.5 Conclusions

The physio-chemical and food chain model structures discussed herein provide a basis for understanding and predicting the fate and transport of chemicals in surface water systems. While the fully time variable equations appear formidable, steady-state simplifications permit rapid assessment of chemical fate, both in the water column and in the food chain. Such initial screening models have the same structure as "classical" water quality models; notably a linear response to external inputs of chemicals.

The interaction of the sediment and water column plays a significant role and again under some reasonable assumptions, rapid estimates can be made of chemical concentrations in the sediment.

In a similar setting, the age-dependent food chain models can be simplified at steady state and rapid screening assessments of chemical bioconcentration can be made. For organic chemicals, the octanol water partition coefficient is a useful ordering parameter. Long-term time variable deterministic calculations for the Great Lakes indicate the significance of the sediment as a reservoir of the chemical. Response times to changes in external load are long (e.g., years) and are significantly increased when sediment partition coefficients are assumed to be different from water column partition coefficients.

Stochastic variability in chemical concentration is high. Analytical models of the resulting variability of chemical concentration in fish indicates the importance of the excretion rate of the chemical. However, additional research is necessary to more fully describe the observed variance in laboratory experiments, especially for chemicals with log octanol water partition coefficients greater than about 6.5.

Acknowledgments. This research was supported in part by a grant from the National Science Foundation, Washington, DC and by a Cooperative Agreement with the USEPA, Large Lakes Research Station, Grosse Ile, Michigan. Grateful thanks are given to my colleagues at Manhattan College for their insights and discussions especially Dominic M. Di Toro, John P. Connolly and Donald J. O'Connor. Thanks are also offered to Eileen Lutomsky for her patient typing of the manuscript.

References

Allen HE, Halley MA (1980) Assessment of airborne inorganic contaminants in the Great Lakes. In: *Appendix B of 1980 Annual Report Appendix-Background Reports*. Grt Lks Sci Adv Bd, IJC, Windsor, Ont, 160 pp.

Bendat JS, Piersol AG (1971) *Random Data: Analysis and Measurement Procedures*. New York: Wiley-Interscience, 407 pp.

Bras RL, Rodriguez-Iturbe I (1985) *Random Functions and Hydrology*. Reading, MA: Addison-Wesley, 559 pp.

Delos CG, Richardson WL, DePinto JV, Ambrose RB, Rodgers PW, Rygwelski K, St. John JP, Shaughnessy WJ, Faha TA, Christie WN (1984) *Technical Guidance Manual for Performing Waste Load Allocations, Book II. Streams and Rivers*. Chapter 3, Toxic Substances. Off of Water Reg and Stds, Monitoring and Data Support Div, Water Qual Anal Br, USEPA, Washington DC, 203 pp. ∫ Appendix, EPA-440/4–84–022

Di Toro DM, O'Connor DJ, Thomann RV, St. John JP (1981) *Analysis of Fate of Chemicals in Receiving Waters, Phase 1*. Washington DC: Chemical Manufact Assoc. Prepared by HydroQual, Inc, Mahwah, NJ, 8 Chapters + 4 Appendixes

Di Toro DM, Horzempa LM, Casey MM, Richardson W (1982a) Reversible and resistant components of PCB adsorption-desorption: Adsorbent concentration effects. J Great Lakes Res 8(2):336–349

Di Toro DM, O'Connor DJ, Thomann RV, St. John JP (1982b) Simplified model of the fate of partitioning chemicals in lakes and streams. In: Dickson KS, Maki AW, Cairns J Jr (eds) *Modeling the Fate of Chemicals in the Aquatic Environment*. Ann Arbor, MI: Ann Arbor Science, pp. 165–190

Di Toro DM (1985) A particle interaction model of reversible organic chemical sorption. Chemosphere 14(10):1503–1538

Dolan DM, Bierman VJ Jr (1982) Mass balance modeling of heavy metals in Saginaw Bay, Lake Huron. J Grt Lks Res 8(4):676–694

Eadie BJ, Faust W, Garnder WN, Nalepa T (1982) Polycyclic aromatic hydrocarbons in sediments and associated benthos in Lake Erie. Chemosphere 11(2):185–191

Eadie BJ, Rice CP, Frez WA (1983) The role of the benthic boundary in the cycling of PCBs in the Great Lakes. In: Mackay D, Paterson S, Eisenreich SJ, Simmons MS (eds), *Physical Behavior of PCBs in the Great Lakes*. Ann Arbor, MI: Ann Arbor Science, pp. 213–228

Eadie BJ, Faust WR, Landrum PF, Morehead NR, Gardner WS, Nalepa T (1983) Bioconcentrations of PAH by some benthic organisms of the Great Lakes. 7th Polycyclic aromatic hydrocarbons (PAH) Vol, Battelle Mem Inst, Richland, Wash, 14 pp., manuscript

HydroQual (1984) Water sediment partition coefficient for priority metals, attachment I in book II, Streams and Rivers, Chapt 3, Toxic Substances, USEPA, Washington DC. Delos, CG et al (eds), pp. I-1 to I-72. EPA-440/4–84–022

Muhlbaier J, Tisu GT (1981) Cadmium in the southern basin of Lake Michigan. Water Air Soil Poll 15:45–59

Neff JM (1979) *Polycyclic Aromatic Hydrocarbons in the Aquatic Environment*. London: Applied Sci Pub Ltd, 262 pp.

O'Connor DJ, Connolly JP (1980) The effect of concentration of adsorbing solids on the partition coefficient. Water Research 14:1517–1523

Oliver BG, Niimi AJ (1983) Environ Sci Technol 17:287–291

Oliver BG, Niimi AJ (1985) Environ Sci Technol 19:842–849

Thomann RV, Di Toro DM (1983) *Physico chemical model of toxic substances in the Great Lakes.* J Great Lakes Res 9(4):474–496

Thomann RV, Di Toro DM (1984) *Physico Chemical Model of Toxic Substances in the Great Lakes.* Project Report to USEPA, Large Lakes Res Sta, Grosse Ile, MI, 163 pp.

Thomann RV, Connolly JP (1984) Age-dependent food chain model of PCB in Lake Michigan lake trout. Env Sci Tech, 18:65–71

Thomann RV, Salas HJ (1986) *Manual on Toxic Substances in Surface Waters.* Section 3, Preliminary Problem Evaluation. Pan American Health Organization, CEPIS, Lima, Peru, 39 pp.

Thomann RV (1987) Statistical model of environmental contaminants using variance spectrum analysis. Final Report to National Science Foundation, NSF/ENG-87053, NTIS No. PB88-253130/A09, 161 pp. + Append

Thomann RV, Mueller JA (1987) *Principles of Surface Water Quality Modeling and Control.* New York: Harper and Row, 644 pp.

Thomann RV (1988) Bioaccumulation model of organic chemical distribution in aquatic food chains. Draft manuscript, submitted for publication.

USEPA (1985) *Technical Support Document for Water Quality-based Toxics Control.* Off of Water Enforce and Permits, Off of Water Regs and Stds, Washington DC, 74 pp. + Append

USEPA (1986) Private Correspondence, Municipal Environmental Research Lab, Cincinnati, OH

Chapter 11

Bioaccumulation of Hydrophobic Organic Pollutant Compounds

John W. Farrington[1]

This chapter provides a brief overview of the biogeochemistry of organic chemicals of environmental concern in aquatic ecosystems and the key aspects important to uptake, retention, release, and metabolism by aquatic organisms. The literature is not exhaustively reviewed, but key review articles or books are cited within the overview context, and a few illustrative examples are discussed in more detail. An integrative mathematical modeling approach to this topic is discussed by O'Connor et al. in Chapter 9 of this book.

Organic chemicals of environmental concern are those with known or potentially deleterious effects on natural resource populations and on humans. Rather than use more cumbersome terminology, these compounds are referred to as pollutant organic chemicals. This is done with the explicit recognition that many of the chemicals have not been proven to have deleterious effects except in a limited number of circumstances, and many organic chemicals have never been tested for deleterious effects but are suspect because of a chemical structure similar to compounds proven to have deleterious effects.

Many processes and conditions of modern societies introduce pollutant organic compounds into aquatic ecosystems. Some of these processes are associated with production, use, and disposal of synthetic organic chemicals. Many of these chemicals are anthropogenic in that prior to synthesis by humans, these chemicals were not present; these are often termed

[1] Woods Hole Oceanographic Institution, Chemistry Department and Coastal Research Center, Woods Hole, Massachusetts 02543. Now at Environmental Sciences Program, University of Massachusetts-Boston, Harbor Campus, Boston, Massachusetts 02125–3393

xenobiotic compounds. Examples of some of the more commonly cited xenobiotics are chlorinated pesticides such as DDT or chlordane, and compounds that have been problems in groundwater contamination such as tetrachloroethylene, a common dry-cleaning agent and industrial solvent. Others, such as phenols and xylenes, are synthesized and released by anthropogenic processes, but are also present in low concentrations as part of non-anthropogenic processes in the environment. A third category of organic pollutants includes those natural organic compounds mobilized by anthropogenic activities and released into the environment at rates far in excess of natural release rates or in locations not normally receiving inputs other than trace amounts. Examples of this category are oil spills, chronic oil releases, and effluents from coal liquifaction, and shale retorting operations. There is a fourth category of compounds, those synthesized as inadvertent byproducts of human activities. Examples are reaction products of chlorination of drinking water and chlorination of effluents (de Leer et al. 1985, 1986; Kringstad et al. 1985).

The era of multinational companies and rapid multinational trade, coupled with varying socioeconomic and political arrangements, make it very difficult to assess production and release of organic pollutants to the environment. The total number of organic compounds synthesized easily exceeds 100,000, with an estimated 60,000 in common use and approximately 1,000 being added every year as of 1978 (CEQ 1984).

The vast majority of these compounds do not pose serious environmental threats. However, Butler (1978) estimated that approximately 1,000 substances are produced in amounts that potentially could cause environmental concerns if released indiscriminantly to the environment. A summary of knowledge about environmental dangers and human health concerns of chemicals, mostly organic, is given in Table 11.1.

Characteristics of chemicals of environmental concern are persistence, mobility, and short-term or long-term adverse biological activity (Miller 1984). Examples of partial lists of such chemicals are taken from the often-quoted U.S. EPA priority pollutants list (Keith and Telliard 1979) and presented in Table 11.2. A few examples of chemical structures of relevance to the remainder of this chapter are given in Figures 11.1 and 11.2. These compounds enter aquatic ecosystems by various means: accidental spills, effluent releases, natural releases (e.g. oil seeps), release to atmosphere and subsequent direct deposition to bodies of surface water, and runoff of dispersed releases (e.g. pesticides sprayed on crops, or road washings of oil from automobile crankcase drippings). The mode of entry and the physical-chemical form at time of release are key considerations when assessing eventual uptake by aquatic organisms, as discussed in later sections of this chapter.

Given the large and growing number of chemicals involved and the several modes of entry, the development of a sufficient body of knowledge is necessary to provide, at the very least, rudimentary predictive capabilities concerning: 1) *Routes of entry* of compounds to the environment and routes of movement through ecosystems; 2) *Reactions* of these compounds to the environment, either chemical, photochemical, or biological transformations/degradations; 3) *Rates* of movement and reactions; 4) *Reservoirs* of short-term (days-months) and long-term (years to decades) accumulations (Farrington and Westall 1986; Sheehan et al. 1984). These are the 4 R's of the biogeochemistry of organic pollutants in aquatic ecosystems: Routes, Reactions, Rates, Reservoirs, and they are incorporated into quantitative and qualitative descriptions of biogeochemical cycles. A schematic of the biogeochemical cycle of one group of compounds, polynuclear aromatic hydrocarbons, in oceanic ecosystems is given in Figure 11.3 to illustrate the main features of biogeochemical cycles of pollutant organic compounds in aquatic ecosystems.

Aquatic organisms take up organic pollutant compounds via two general processes: direct uptake from water across membrane surfaces and via ingestion. Aquatic mammals and birds are exposed to a third pathway

Table 11.1 Hazard not assessable for most of select 65,725 chemicals

	Size of category	Complete assessment possible (%)	Partial assessment possible (%)	Minimal toxicity data available (%)	Below minimal toxicity available (%)	No toxicity data available (%)
Pesticides and inert ingredients of formulations	3,350	10	24	2	26	38
Cosmetic ingredients	3,410	2	14	10	18	56
Drugs and inert vehicles used in formulations	1,815	18	18	3	36	25
Food additives	8,627	5	14	1	34	46
Chemicals in commerce: ~ 1 million lb per yr	12,860	0	11	11	0	78
Chemicals in commerce: ~ 1 million lb per yr	13,911	0	12	12	0	76
Chemicals in commerce: production unknown or inaccessible	21,752	0	10	8	0	82

Adapted from Chemical and Engineering News, 2/13/1984, American Chemical Society, Washington, D.C.

Table 11.2 Organic compounds in the U.S. Environmental Protection Agency's list of 129 priority pollutants

---

Volatiles-33

Acrolein	Methylene chloride
Acrylonitrile	Tetrachloroethylene
Benzene	Toluene
Bis(chloromethyl)ether	Trichloroethylene
Bis(2-chloroethyl)ether	Trichlorofluoromethane
Bis(2-chloroethoxy)methane	Vinyl chloride
Bromoform	1,1-Dichloroethane
Carbon tetrachloride	1,1-Dichloroethylene
Chlorobenzene	1,1,1-Trichloroethane
Chlorodibromomethane	1,1,2-Trichloroethane
Chloroethane	1,1,2,2-Tetrachloroethane
Chloroform	1,2-Dichloroethane
Dichlorobromomethane	1,2-Dichloropropane
Dichlorofluoromethane	1,2-Trans-dichloroethylene
Ethyl benzene	1,3-Dichloropropene
Methyl bromide	2-Chloroethyl vinyl ether
Methyl chloride	

Pesticides and PCSs-28

Aldrin	Hexachlorocyclopentadiene
α-BHC	Hexachlorobutadiene
β-BHC	Hexachlorobenzene
Δ-BHC	PCB 1221
γ-BHC Lindane	PCB 1232
Chlordane	PCB 1242
Dieldrin	PCB 1248
Endosulfan sulfate	PCB 1254
Endosulfan, β	PCB 1260
Endosulfan, α	PCB 1016
Endrin	Toxaphene
Endrin aldehyde	4,4'-DDD
Heptachlor, epoxide	4,4'-DDE
Heptachlor	4,4'-DDT

Other Neutrals-10

Bis-2-chloroisopropyl ether	1,4-Dichlorobenzene
Hexachloroethane	2-Chloronaphthalene
1,2-Dichlorobenzene	4-Bromophenyl phenyl ether
1,2,4-Trichlorobenzene	4-Chlorophenyl phenyl ether
1,3-Dichlorobenzene	Isophorone

Phenols-11

p-Chloro-m-cresol	2,4-Dimethylphenol
Pentachlorophenol	2,4-Dinitrophenol

Table 11.2 (continued)

Phenol	2,4,6-Trichlorophenol
2-Chlorophenol	4-Nitrophenol
2-Nitrophenol	4,6-Dinitro-o-cresol
2,4-Dichlorophenol	

N-Containers

1,2-Diphenyl hydrazine	2,4-Dinitrotoluene
Nitrobenzene	2,6-Dinitrotoluene

Hetero-Carcinogens-6

N-nitroso-di-n-propyl amine	2,3,7,8-Tetrachlorodibenzo(p)dioxin
N-nitroso-diphenyl amine	Benzidine
N-nitroso-dimentyl amine	3,3'-Dichlorobenzidine

Phthalates-6

Bis(2-ethyl hexyl)phthalate	Diethyl phthalate
Butyl benzyl phthalate	Dimethyl phthalate
Di-n-butyl phthalate	Di-n-octyl phthalate

Hydrocarbons-16

Acenaphthylene	Indeno(1,2,3-cd)pyrene
Acenaphthene	Naphthalene
Anthracene	Phenanthrene
Benz(e)acephenanthrylene	Pyrene
Benzo(k)fluoranthene	Benzo(ghi)perylene (1,12 benzo-
Benzo(a)pyrene	perylene)
Chrysene	Benzo(ghi)anthracene (1,2 benzan-
Fluoranthene	thracene)
Fluorene	1,2,5,6-Dibenzanthracene

Source: Reprinted with permission from Environmental Science and Technology, Vol. 13, L. H. Keith and W. A. Telliard. ES&T Special Report on Priority Pollutants I. A Perspective View. Copyright 1979, American Chemical Society.

from breathing air, although this mode of uptake would, *a priori,* appear to be less significant than ingestion for most organic pollutants of long-term environmental concern. Mammals and birds will not be considered further in this chapter. The reader is referred to Nisbet and Reynolds (1983) for further discussions of organic pollutants and birds and to Addison et al. (1984), Knap and Jickells (1983), and Tanabe et al. (1983) for further discussion and references pertinent to marine mammals.

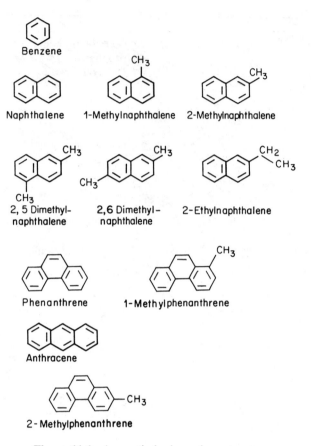

Figure 11.1. Aromatic hydrocarbon structures.

11.1 Physical-Chemical Considerations and Bioavailability

Solubility is an obvious key characteristic of chemicals in aquatic systems. Yet solubilities in water for many organic chemicals of environmental concern have only recently become available because these chemicals were classed as insoluble according to classical definitions in chemistry. In reality these chemicals have very low solubilities. The challenge was to determine their solubility. This required new or modified techniques to determine solubilities, and the challenge has been met by innovative techniques. Details are beyond the scope of this chapter. A few examples are provided in May et al. (1978), May (1980), Whitehouse (1984), and Mackay and Shiu (1977).

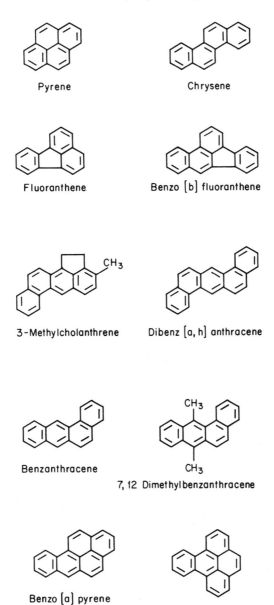

Pyrene

Chrysene

Fluoranthene

Benzo [b] fluoranthene

3-Methylcholanthrene

Dibenz [a, h] anthracene

Benzanthracene

7, 12 Dimethylbenzanthracene

Benzo [a] pyrene

Benzo [e] pyrene

Triphenylene

Figure 11.1 (*Continued*)

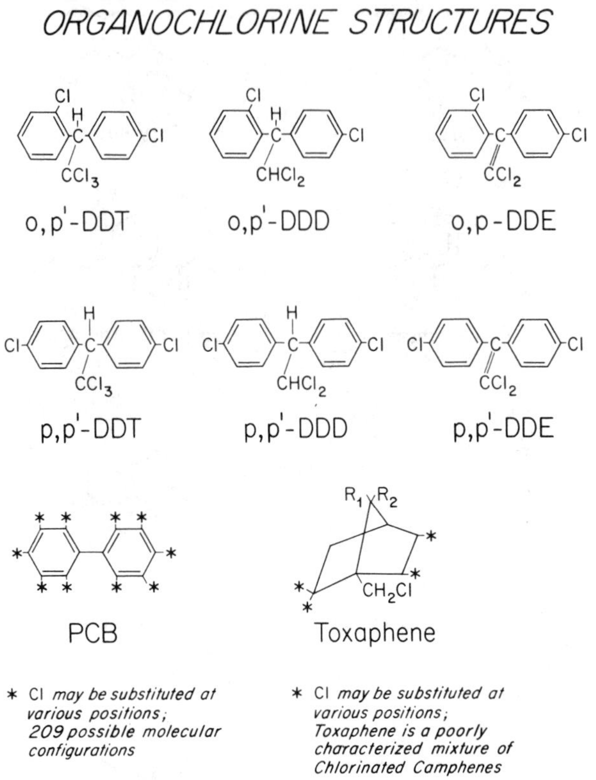

ORGANOCHLORINE STRUCTURES

o,p'-DDT   o,p'-DDD   o,p-DDE

p,p'-DDT   p,p'-DDD   p,p'-DDE

PCB   Toxaphene

* Cl *may be substituted at various positions;* 209 *possible molecular configurations*

* Cl *may be substituted at various positions;* Toxaphene *is a poorly characterized mixture of Chlorinated Camphenes*

Figure 11.2. Organochlorine xenobiotic chemical structures.

Methods of calculating solubilities in aqueous systems from molecular structure considerations have evolved (e.g. Banerjee 1985; Lyman et al. 1982). Thus it is rapidly becoming possible to assess solubility for new synthetic chemicals and natural compounds. Often these solubilities are for relatively pure water only. The influence of physical parameters such as temperature have been studied, and extrapolations to a range of temperature conditions can be extended by analogy to many compounds, although more work is needed to verify these extrapolations. A more serious problem is the lack of data concerning the influence on solubility of ionic strength and composition—particularly saline conditions in seawater. An example of the influence of seawater and temperature on solubility of an important class of compounds, polynuclear aromatic hydrocarbons (PAHs), comes from the recent studies of Whitehouse (1984). Three important points are illustrated by his results: the influence of temperature on solubility is no more than a factor of 2 to 5 in the range of temperatures encountered in most aquatic ecosystems (4 to 28°C); the influence of salinity is almost a factor of 2, in the range of 0 to 36‰; and

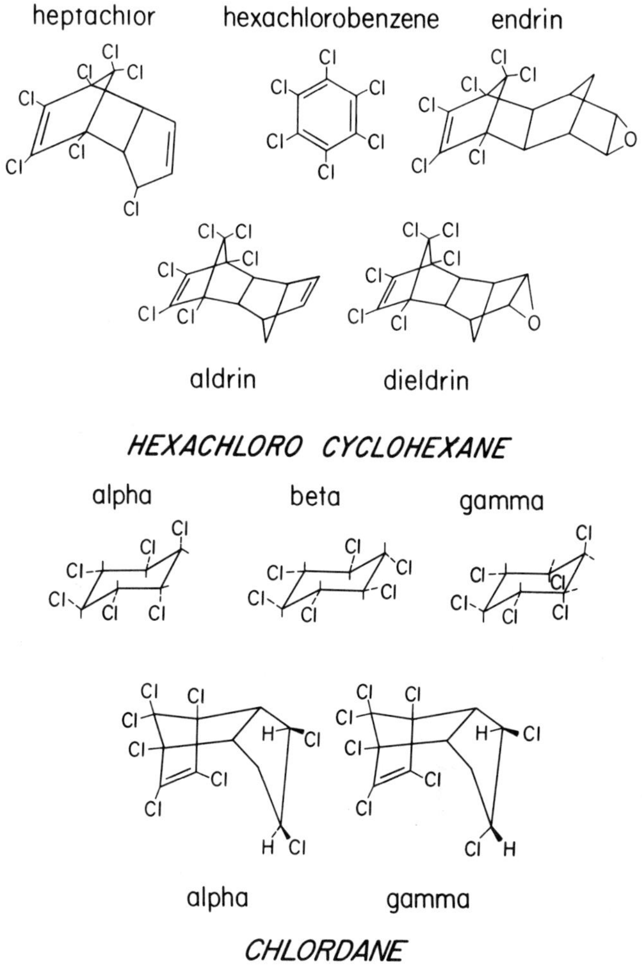

Figure 11.2 (*Continued*)

the benzanthracene solubility as a function of salinity and temperature departs from the other PAHs studied. This latter point is important because there is no completely satisfactory explanation for this observation given the limited number of PAHs studied, the importance of solubility data for understanding biogeochemical cycles of these and similar compounds, and the need to understand biogeochemical cycles of these compounds in estuarine waters near urban and industrial areas. It is clear that more work is needed on solubilities and factors influencing solubilities in aquatic ecosystems. The wide range of actual solubilities encountered within the general class of less soluble organic compounds of environmental concern is illustrated by a compilation of data by Banerjee (1985) (Table 11.3).

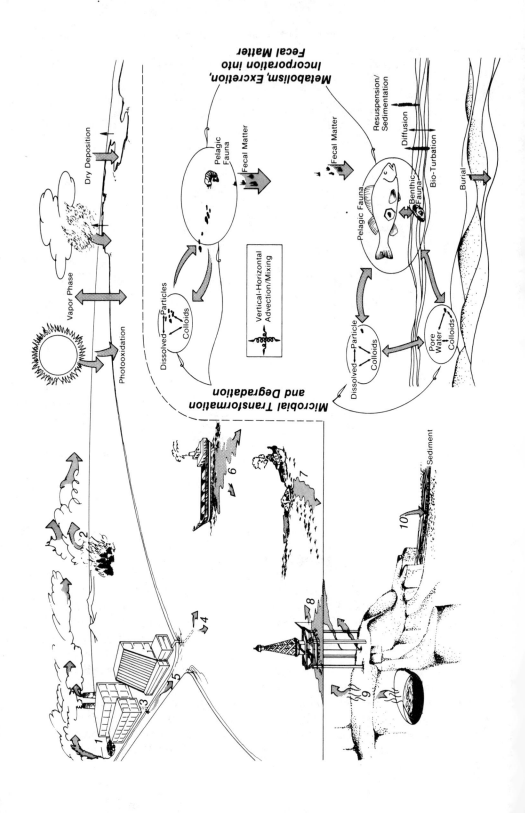

Vapor Phase

Dry Deposition

Photooxidation

Metabolism, Excretion,
Incorporation into
Fecal Matter

Pelagic Fauna

Fecal Matter

Fecal Matter

Dissolved → Particles
Colloids

Vertical-Horizontal
Advection/Mixing

Microbial Transformation and Degradation

Pelagic Fauna

Dissolved → Particle
Colloids

Benthic Fauna

Resuspension/
Sedimentation

Diffusion

Bio-Turbation

Pore
Water → Colloids

Burial

Sediment

Table 11.3. Water solubilities of selected organic compounds at 25°C

Compound	Log S
Acenaphthene	−3.86
Acrylonitrile	0.15
Aniline	−0.15
Benzene	−1.68
Benz(a)pyrene	−6.26
Biphenyl	−3.88
1,3-Butadiene	−1.86
Carbon tetrachloride	−2.31
Chlorobenzene	−2.35
Dibenz(ah)anthracene	−6.28
m-Dichlorobenzidine	−3.01
3,3'-Dichlorobenzidine	−3.84
2,4'-Dichlorobiphenyl	−5.32
1,2-Dichloroethane	−1.09
7,12-Dimethylbenzanthracene	−5.71
Ethylbenzene	−2.80
Hexachlorobenzene	−5.48
2,2',4,4',5,5'-Hexachlorobiphenyl	−7.66
3-Methylcholanthrene	−6.54
Naphthalene	−3.04
Pentachlorobenzene	−4.85
Phenanthrene	−4.42
Styrene	−2.57
1,2,3,5-Tetrachlorobenzene	−4.33
2,2',5,5'-Tetrachloroethane	−1.75
Toluene	−2.22
1,3,5-Trichlorobenzene	−4.09
1,1-Trichloroethane	−2.00

Adapted from Banerjee 1985, who compiled the data from several sources.
S = molar solubility.

Figure 11.3. Biogeochemical cycle of PAH in the ocean. Some sources of PAH input to ocean: *1*, incomplete combustion of fossil fuels; *2*, forest and grass fires; *3*, industrial effluents; *4*, sewage effluents; *5*, river borne material from interior areas; *6*, oil tanker operation–routine and accidents; *7*, waste barges; *8*, offshore oil production–routine and accidents; *9*, natural oil seeps; *10*, diagenisis of organic matter in sediments.

Solubility is a key factor in determining bioavailability, because aquatic organisms that obtain their oxygen for respiration needs by passage of water across gills take up several xenobiotic compounds. Hamelink et al. (1971) were among the first to formulate the concept of "exchange equilibria," an idea they used to understand the observed distributions between concentrations of certain organic pollutant compounds in aquatic organisms and the water of their habitat. Compounds with low solubilities in water are hydrophobic and if presented with an opportunity will partition into another substance, or as discussed later, undergo sorbtion.

Aquatic organisms constitute a pool of proteins, carbohydrates, lipids, and a few other classes of biochemicals. A first approximation of the equilibrium distribution between an organism and the water of its habitat for a hydrophobic chemical such as a PAH or a polychlorinated biphenyl (PCB) can be obtained by designating the organism as a pool of lipophilic, or mainly hydrophobic, material. The equilibrium partitioning becomes mainly a lipid–water partitioning. Neeley et al. (1974) were among the first to introduce a parameter to estimate the equilibrium partitioning, the octanol–water partition coefficient K_{ow}. Octanol is a convenient surrogate for lipids, and the partitioning of a chemical between octanol and water in laboratory measurements provides a useful estimate of hydrophobic behavior in equilibrium partitioning. Recent compilations of data show that

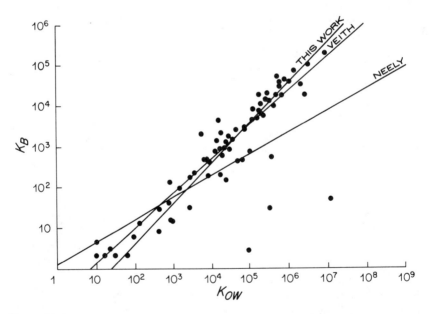

Figure 11.4. K_B versus K_{ow} on log scales. This work refers to Mackay; Neeley and Vieth are references to other sets of data in Mackay 1982. Taken with permission from Mackay 1982, *Environmental Science and Technology*, Vol. 6. Copyright 1982, American Chemical Society. See text for definitions of K_{ow} and K_B.

K_{ow} provides a first approximation of biological concentration factor (K_B), as shown in Figure 11.4 (Mackay 1982) and discussed elsewhere (Lyman et al. 1982). (K_B = the concentration of a given chemical, at equilibrium, in an organism on a wet weight or lipid weight basis divided by the concentration of the chemical in water of the organism's habitat.) There are departures from the lines of correlation even for log–log plots. There are several reasons why this can occur and they will be discussed later. An important reason to state now is that equilibrium is an important assumption if the estimation of K_B *based on* K_{ow} is to be accurate.

K_{ow} can be related to aqueous solubility by an equation of the form

$$\log K_{ow} = a - b \, (\log S) \tag{11.1}$$

where S is solubility and a and b are empirical constants calculated by regression of an initial set of measured K_{ow} and S. A detailed treatment of relationships between K_{ow} and aqueous solubility has been published (Miller et al. 1985).

Sorption to surfaces such as suspended particulate matter and sediments influences the concentration of an organic chemical in the water-soluble phase of an aquatic ecosystem and can influence the uptake of a chemical by an organism. Hydrophobic compounds tend to sorb with surfaces. This results because almost all particles in contemporary aquatic ecosystems have an organic coating (Thurman 1986) which can be thought of as a liquid film or gel-like coating. It is possible to provide a first approximation of this tendency using K_{ow}. This is expressed in equation form as follows (Karickhoff et al. 1979; Means et al. 1979; Schwarzenbach and Westall 1981; Chiou et al. 1983):

$$K_p = f_{oc} K_{oc} \tag{11.2}$$

where K_p is the partition coefficient between particulate matter and water, f_{oc} is the fraction of the particulate matter that is organic carbon, and K_{oc} is the organic carbon normalized partition coefficient.

Several field observations and experimental studies suggest the following relationships for non-ionic hydrophobic pollutants:

$$\log K_{oc} = a \, (\log K_{ow}) + b \tag{11.3}$$

where K_{oc} is the carbon normalized partition coefficient, and a and b are empirical constants calculated from regressions of actual data. The values of a and b seem to be related to the type of compounds and possibly the type of organic coating on the sorbant (Means et al. 1979; Schwarzenbach and Westall 1981; Chiou et al. 1983).

Recent research has established the importance of dissolved organic matter (DOM) in the behavior of non-ionic hydrophobic compounds. DOM is an operational definition often defined by a filtration process (Thurman 1986). Colloids are part of dissolved organic matter and are often a major portion of DOM in locations with high concentrations of DOM (Thurman 1986). The term "colloid" is itself operational, but generally refers to any particle with some linear dimension between 10^{-7} cm and 10^{-4} cm (Himenez 1977). Colloidal organic matter has a substantial capacity to sorb hydrophobic pollutants (Means and Wijayaratne 1982; Wijayaratne and Means 1984; Brownawell 1986). The K_{oc} for colloids is within the general range of K_{oc} for large particles. Brownawell and Farrington (1985; 1986) and Baker et al. (1986) have shown that observed distributions of PCBs, a class of non-ionic hydrophobic compounds, in waters rich in organic matter (e.g., pore waters of sediments) are best described by a three-phase model: soluble, colloidal, and particulate matter. A substantial portion of the pollutant can be associated with the colloid phase, as illustrated by the prediction nomograph from Brownawell (1986) (Figure 11.5).

Macromolecules such as humic-like materials account for a substantial portion of DOM (Thurman 1986; Rashid 1985). Experiments with the uptake by organisms of non-ionic, hydrophobic compounds, such as petroleum compounds and PAHs, have demonstrated reduced uptake when DOM, humic-like material is present compared to aqueous systems with low DOM concentrations (e.g., Boehm and Quinn 1976; McCarthy et al.

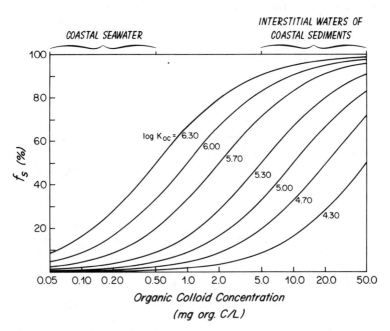

Figure 11.5. Example of idealized nomograph of log K_{oc}, colloid concentration in water and sorbtion to colloid. f_s (%) is percent of compound with given K_{oc} that is sorbed by colloid. From Brownawell 1986 with permission.

Table 11.4. Bioconcentration factors (±SD) observed after 30-hr exposure of *D. magna* to 3-methylcholanthrene in the presence of different concentrations of DHM

DHM concentration (mg C/l)	f, Fraction of MC bound to DHM	Observed BCF	Fractional decrease in BCF due to presence of DHM
0	0	13.209 (±653)	—
0.15	0.05	12.171 (±194)	0.07
1.5	0.32	8.121 (±161)	0.38
15.0	0.82	2.311 (±127)	0.82

The fraction of MC bound to DHM was calculated from eq. 1, based on data on the K_p (3.3 × 10^5; McCarthy et al. 1985). DHM refers to dissolved humic material.

1985). Table 11.4, taken from the McCarthy et al. (1985) study, illustrates this point. This phenomenon is expected, based on principles discussed in the preceding paragraph if the uptake process by marine organisms is primarily a true solution-to-membrane process. The DOM colloids compete with the across membrane uptake for non-ionic hydrophobic chemicals.

An important factor in physical-chemical considerations is sequestering, or entrapment, of an organic pollutant within a particle. Farrington et al. (1983a,b) noted that comparison of PAH compositions for bivalve molluscs, sediments, and polychaetes in sediments from the same general locations revealed substantial differences between the sediments and the biota. The causative factor was hypothesized to be different sources and physical chemical forms of PAH. Petroleum PAH entering aquatic ecosystems in the water-accommodated form in effluents or via oil spills would be partitioned among soluble, colloidal, and particulate phases as discussed above. However, pyrogenic PAH, entering the environment incorporated into combustion process particulate matter from atmospheric release in automobiles, home heating, or power plants, (e.g., Figure 11.3) are more strongly bound by entrapment or "baking" into the particles. Chemical extraction of sediment will remove these pyrogenic PAH, whereas passage of particles through the water or through an animal's digestive track will not remove as much of the pyrogenic PAH in comparison to the petroleum PAH entering the habitat in water accommodated form. This hypothesis is consistent with recent observations by Varanasi et al. (1985) concerning bioavailability of PAH in sediments and the observations of Rheadman et al. (1984) concerning physical-chemical speciation of PAH in an estuarine ecosystem. Thus, for organisms that are filter feeders or detritivores, it is important to take into account that chemical analysis methods for sediments may measure tightly bound or entrapped pollutant compounds which are not bioavailable to filter feeder and detritovore organisms. Estimates of contamination of organisms may be much too high if no account is taken of what portion of the compound present in sediments or in particles in the water column is bioavailable.

11.2 Biological Uptake, Retention, Metabolism, and Release

The log *BCF* (or log *B*) bioconcentration factor, as estimated from log K_{ow} and discussed in the previous section is a useful first-order approach for estimating biological uptake and bioconcentration of organic chemicals of environmental concern. However, several points on the log K_{ow} vs. log *BCF* plot depart from a general correlation (Figure 11.4), and numerous field data and experimental studies demonstrate the need to extend the equilibrium theory model to assess a variety of factors in a kinetic approach to biological uptake, retention, and release. A specific example of field data illustrates this point.

Representative data for individual chlorobiphenyls that have been measured in water and for several species in the outer harbor area of the Acushnet River Estuary-New Bedford Harbor, U.S. EPA Superfund site in Buzzards Bay, Massachusetts, U.S.A., are presented in Table 11.5. Weaver (1984) and Farrington et al. (1985, 1986) described the general aspects of severe pollution of this site by polychlorinated biphenyls. The chlorine substitution positions on the biphenyl molecules considered here

Table 11.5. Octanol–water partition coefficients and bioaccumulation concentration factors on a wet weight basis of selected chlorobiphenyls

IUPAC No.	Log K_{ow}	Mussel	Clam	Lobster A		Lobster B		Flounder
				Muscle	Viscera	Muscle	Viscera	
28	5.69	3.74	3.66	4.14	5.37	3.87	4.95	3.35
52	6.09	4.42	4.06	3.95	5.44	3.47	4.86	3.67
49	6.22	4.47	4.08	3.88	5.35	3.13	4.50	3.95
44	5.81	4.48	4.11	3.80	5.09	3.11	4.44	3.72
70	6.23	4.84	4.17	4.20	5.63	3.69	5.07	4.66
95	6.55	4.78	4.07	4.75	6.28	4.44	5.81	4.56
101	7.07	5.10	4.41	4.70	6.28	4.10	5.72	5.01
87	6.37	5.12	4.54	4.57	6.13	4.13	5.56	5.29
60	5.84	4.74	4.16	4.72	6.15	4.28	5.71	4.49
153	7.75	5.48	4.83	5.75	7.46	5.37	7.08	5.90
138	7.44	5.45	4.93	5.72	7.38	5.26	7.01	5.85
128	6.96	5.42	4.81	5.78	7.39	5.42	7.12	5.33

Log BCF from data of Farrington et al. 1986.

Table 11.6. Individual chlorobiphenyls

IUPAC No.	Chlorine substitution
8	2,4'
28	2,4,4'
29	2,4,5
44	2,2',3,5'
49	2,2',4,5'
52	2,2',5,5'
60	2,3,4,4'
70	2,3',4',5
86	2,2',3,4,5
87	2,2',3,4,5'
95	2,2',3,5',6
101	2,2',4,5,5'
105	2,3,3',4,4'
110	2,3,3',4',6
118	2,3',4,4',5
128	2,2',3,3',4,4'
129	2,2',3,3',4,5
137	2,2',3,3',6,6'
138	2,2',3,4,4',5
143	2,2',3,4,5,6'
153	2,2',4,4',5,5'
156	2,3,3',4,4',5
180	2,2',3,4,4',5,5'

are given in Table 11.6 with their corresponding IUPAC (International Union of Pure and Applied Chemistry) congener number. Species for which data are presented are: common blue mussel (*Mytilus edulis*), lobster (*Homarus americanus*) muscle and viscera; and flounder (*Pseudopleuronectic americanus*) tissue. Bioconcentration factors have been calculated using tissue data and water column data from different dates and assuming a reasonably uniform water column concentration over time. This can be challenged as a reasonable assumption, but water column PCB data were sparse and exceedingly difficult to obtain because of the state of the art at the time of these studies.

Representative plots of log BCF vs. log K_{ow} are given in Figures 11.6, 11.7, and 11.8. There are significant departures from a simple correlation for several chlorobiphenyls. Log K_{ow}/log BCF relationships apparently vary among species and among tissues within a given species. Varying lipid type and concentration are thought to be influential (Chiou 1985), but no specific data are available at this time.

Farrington et al. (1986) presented high resolution analytical data for PCBs in the tissues of fish and crustacea from this area that illustrated substantial differences in the pattern of PCBs in tissues compared to the

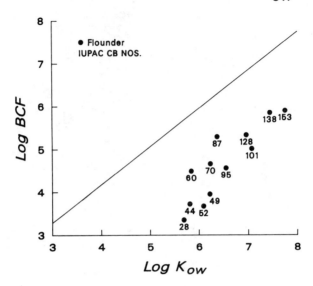

Figure 11.6. Log *BCF* versus log K_{ow} for PCBs in *Pseudopleuronectes americanus* from outer New Bedford Harbor (Farrington et al. 1986a). Numbers refer to IUPAC chlorobiphenyl congener numbers.

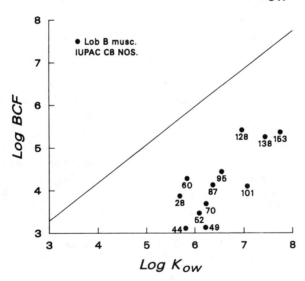

Figure 11.7. Log *BCF* versus log K_{ow} for PCBs in *Homarus americanus* (lobster) tail muscle. Outer New Bedford Harbor sample (Farrington et al. 1986a). Numbers refer to IUPAC chlorobiphenyl congener numbers.

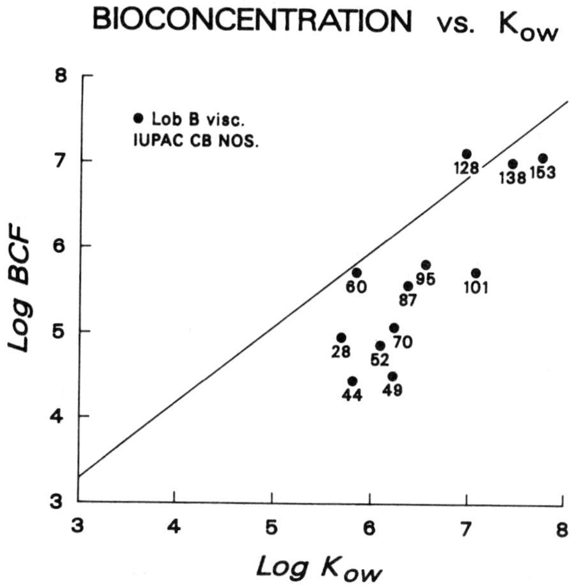

BIOCONCENTRATION vs. K_{ow}

Figure 11.8. Same as for Figure 11.7 except that data are from lobster viscera.

original mixture of PCBs in industrial mixtures discharged to the area (Figures 11.9–11.11). Selective metabolism or biotransformation by the species was postulated as a major factor controlling the observed differences. Microsomal enzymes known as mixed function oxidases or cytochrome P450 enzyme systems are known to be capable of metabolizing several chlorobiphenyls and induction of such enzymatic activity has been attributed to selected chlorobiphenyls and mixtures of chlorobiphenyls (NRC 1979; Stegeman 1981). Selected chlorobiphenyls appear to be more resistant to metabolism or microbial degradation than other chlorobiphenyls (NRC 1979; Schulte and Acker 1974).

The preceding is but one example of several studies illustrating the need to take the next step beyond log K_{ow} versus log BCF equilibrium considerations, especially in field situations where more detailed understanding or predictive capabilities are needed to protect human health and a valuable natural resource population. O'Connor and Pizza (1987) among others have stated the need to understand factors such as differential uptake and release rate, food input, bioavailability, metabolism, and excretion within the context of a kinetic modeling approach.

A simple mathematical expression for consideration of uptake and elimination kinetics for a given chemical is described by Farrington and Westall (1986) among several others:

$$C_1 \underset{k_2}{\overset{k_1}{\rightleftharpoons}} C_2 \tag{11.4}$$

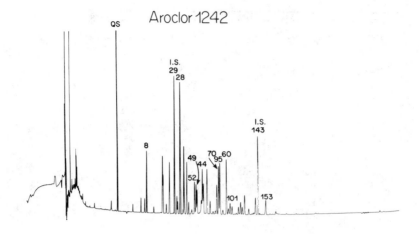

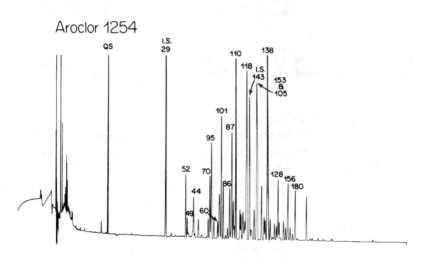

Figure 11.9. Glass capillary gas chromatograms, electron capture detector. I.S. indicates internal standard. Peak numbers refer to IUPAC chlorobiphenyl congener number eluting at that retention time. Aroclor 1242 and Aroclor 1254 are industrial mixtures of PCBs. QS refers to quantitation standard for gas chromatography.

C_1 concentration in compartment 1 (e.g., food, water, sediment, etc.)
C_2 concentration in compartment 2 (body, tissue, etc.)
  First order rate equations:

$$\frac{dC_1}{dt} = -k_1C_1 + k_2C_2 \tag{11.5}$$

$$\frac{dC_2}{dt} = +k_1C_1 - k_2C_2 \tag{11.6}$$

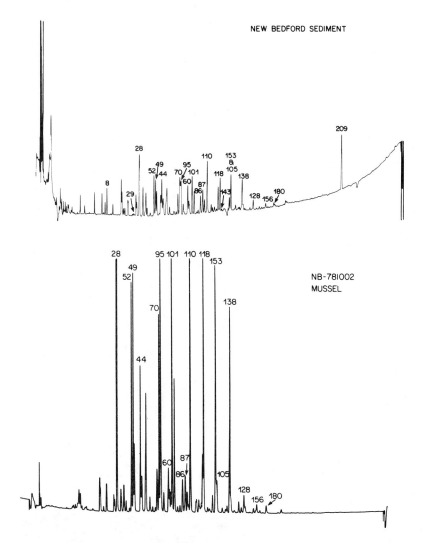

Figure 11.10. PCBs extracted from New Bedford Harbor surface sediment and *Mytilus edulis* (blue mussel) sampled in the outer harbor of New Bedford. NB-781002 is code number for year, month, day of sampling. Also, see Figure 11.9 legend.

Time course of concentrations:

$$C_1(t) = [C_1(o) - C_1(\infty)]exp[-(k_1 + k_2)t] + C_1(\infty) \qquad (11.7)$$

$$C_2(t) = [C_1(o) - C_2(\infty)]exp[-(k_1 + k_2)t] + C_2(\infty) \qquad (11.8)$$

$C(o)$: initial concentration

$C(\infty)$: equilibrium concentration

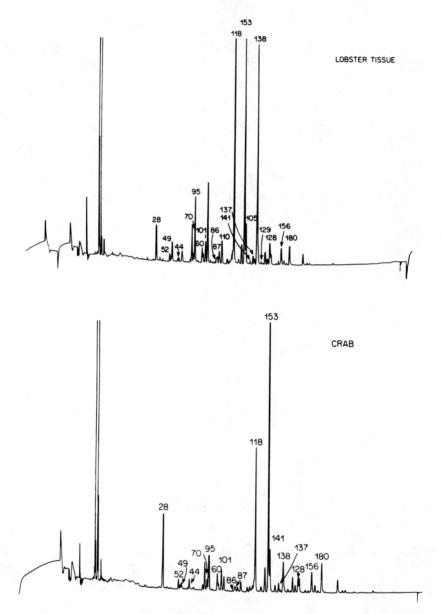

Figure 11.11. PCBs extracted from *Homarus americanus* (lobster) tail muscle and from whole *Neopanope taxons* (crab) from outer harbor of New Bedford.

At Equilibrium

$$\frac{C_1}{C_2} = \frac{k_2}{k_1}$$

(11.9)

11.3 Bivalve Molluscs

Bivalve molluscs have been studied by many researchers in terms of chemical pollution of tissues (e.g., reviews by Phillips 1980; NRC 1980; Farrington et al. 1983). Many characteristics of several bivalve mollusc species make them good candidates as sentinel organisms or biomonitors of coastal pollution by several types of chemicals (Phillips 1980; Farrington et al. 1983b). Several prototype regional and national monitoring programs utilizing the bivalve as sentinel organism concept have been completed (e.g., Farrington et al. 1983b; NRC 1980; Topping 1983). Operational national programs are now underway (Matta et al. 1986; IFREMER 1985) and a coordinated international effort is in the planning stages (Kullenberg 1986). Thus, it is important to understand factors controlling the uptake, retention, and release of organic chemicals in bivalve molluscs. Application of equations 11.4–11.9 for three specific examples utilizing bivalves as sentinel organisms or biomonitors illustrates consideration of several factors that need to be taken into account when applying these equations.

In the first example blue mussels, *Mytilus edulis,* were transplanted from a clean site off Nantucket Island to the following locations: sites in Buzzards Bay, Massachusetts; a site in the New Bedford Harbor-Acushnet River Estuary (referred to in the example above); a site off Penikese Island; and a site near Cleveland Ledge. The latter two sites are much less polluted. The uptake curves are illustrated in Figure 11.12. Application of equations to the data yield parameters tabulated in Table 11.7. Data from a second set of transplants are also provided in Table 11.7, as well as data for PCB release after several months of exposure to high concentrations in the habitat and transplanting to relatively clean seawater.

The recent work of Pruell et al. (1986) provides a second example. They conducted a very interesting experiment exposing blue mussels, *M. edulis,* to PCBs and PAHs associated with particulate matter obtained from suspended sediments for a forty-day period of exposure. This was followed by exposure to clean seawater. Biological half-lives for the release of chemicals after transfer to clean seawater were calculated using equations identical to 11.4–11.9 and are presented in Table 11.8, taken from Pruell et al. (1986).

The third example is from oil spill studies. Release of petroleum hydrocarbons from mussels or other bivalves contaminated by accidental oil spills has been studied by many researchers (e.g., NRC 1985; Farrington et al. 1982; Marchand and Cabane 1980). Figures 11.13 and 11.14 contain a selection of data comparing release of No. 2 fuel oil hydrocarbons from a natural population of blue mussels, *M. edulis,* in the Cape Cod Canal, Massachusetts. Two small oil spills occurred three years apart within one week of the same date, thereby providing a "duplicate experiment" under

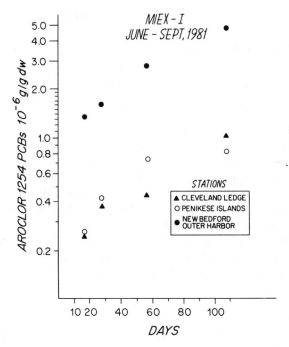

Figure 11.12. Total Aroclor 1254 PCB uptake by *Mytilus edulis* (blue mussel) after transplanting to stations in Buzzards Bay (see text for further details). MIEX refers to Mussel Implant Experiment.

natural conditions. Biological half-lives calculated using equations 11.4–11.9 and data exemplified by Figures 11.13 and 11.14 are given in Table 11.9.

Close examination of the data for uptake or release of chemical contaminants by bivalve molluscs demonstrates that the relatively simple equations (11.4–11.9) do not provide a good fit with actual data after the first 14 to 21 days. Exposure concentrations in the habitat and duration of exposure have been shown to be important considerations. A combination of higher exposures and longer times of exposure apparently results in much slower overall release kinetics (Boehm and Quinn 1977; NRC 1980). Collectively the results of several studies support the concept of multiple compartments within the bivalves, as set forth by Stegeman and Teal (1973). This multiple compartment model is not difficult to visualize given the many types of tissues present in bivalves. Although it is beyond the scope of this chapter, equations can be written for each compartment and then connected together. Stated simply, if we consider a three-compartment "bivalve" of gills, circulatory fluid, and energy storage lipid, then we can propose the following: initial uptake of an organic pollutant across the gills is rapid, followed by slightly less rapid transfer to circula-

Table 11.7. Uptake and elimination rates for PCBs in *Mytilus edulis* transplanted in Buzzards Bay

Experiment I—Uptake June–September 1981 (Day 17 to Day 108)	$-k_a{}^a$	r^2	$t_{1/2}{}^a$
New Bedford	.01	.98	69
Cleveland Ledge	.01	.96	69
Penikese Island	.01	.75[b]	69
Experiment II—Uptake September–October 1982 (Day 6 to Day 27)			
New Bedford	.03	.99	23
Cleveland Ledge	.02	.99	25
Penikese Island	.02	.93	35
Experiment III—Elimination February–June 1982 (120 days)	$-k_e{}^a$	r^2	$t_{1/2}{}^a$
New Bedford transplanted to ESL-WHOI (Vineyard Sound)	-0.1	.66[b]	69

[a] Calculated from equations 11.4–11.7 under the assumption that the system is far from equilibrium and the back reaction is negligible:

Absorption (assume $k_1C_1 \gg k_2C_2$)

$$\frac{dC_2}{dt} = k_1C_1$$

$$C_2(t) = C_2(o)\ exp\ (k_1t)$$

$$ln\ C_2(t) = ln\ (C_2(o)) + k_1t$$

Elimination (assume $k_2C_2 \gg k_1C_1$)

$$\frac{dC_2}{dt} = k_2C_2$$

$$C_2(t) = C_2(o)\ exp\ (-k_2t)$$

$$ln\ C_2(t) = ln\ (C_2(o)) - k_2t$$

[b] Poor fit of data to equations.
Adapted from Farrington et al., unpubl.

tory fluid, followed by much slower transfer to energy storage lipid. The longer this process takes place, the larger the reservoir of the pollutant in the storage lipids, until the maximum storage capacity or equilibrium is reached. If the organism is transferred to clean water, the process is reversed. Pollutant organic compounds present in the gill tissue will be exchanged quite rapidly, but those present in the storage lipid may be

Table 11.8. Biological half-lives of some individual PAHs and PCBs in *Mytilus edulis*

Compound	$T_{1/2}$ (days)
Fluoranthene	29.8
Benz(a)anthracene	17.8
Benzo(e)pyrene	14.4
Benzo(a)pyrene	15.4
2,3,4'-Trichlorobiphenyl	16.3
2,2',4,5,5'-Pentachlorobiphenyl	27.9
2,2',4,4',5,5'-Hexachlorobiphenyl	45.6
2,2',3,3',4,4'-Hexachlorobiphenyl	36.5

Adapted from Pruehl et al 1986.

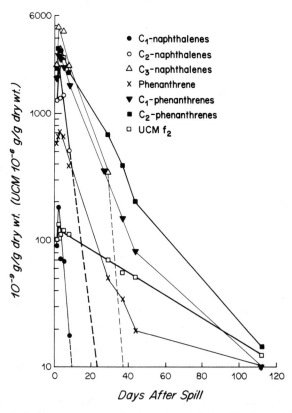

Concentrations of Aromatic Compounds After Spill
(Quantitative GCMS Analysis)

Figure 11.13. Release of aromatic hydrocarbons from *Mytilus edulis* contaminated by a No. 2 fuel oil spill. C_1 refers to methyl groups; C_2 refers to dimethyl or ethyl groups; totals of all isomers in each case (from Farrington et al. 1986b).

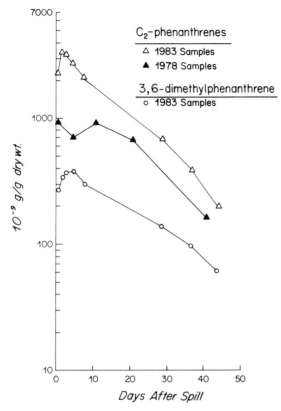

Figure 11.14. Legend same as for Figure 11.13. Data compared from two different oil spills (from Farrington et al. 1986b).

released very slowly. This scenario explains why bivalve molluscs taken from chronically polluted urban harbors and transferred to clean waters do not rapidly release their organic pollutant chemicals such as PAH (Boehm and Quinn 1977).

As a general rule, physiological state of the organism, such as spawning, post-spawning, or pre-spawning conditions, temperature in the habitat, food availability, as well as exposure concentration and duration affect uptake, retention, and release of organic pollutants (Phillips 1980; NRC 1980; Langston 1978).

Metabolism of PAHs, PCBs, and similar organic pollutants is thought to be much less in bivalves than in fish, crustacea, or polychaetes (Stegeman 1981; NRC 1985). However, it is important to note that there are indications of enzymatic capabilities to metabolize PAH and perhaps similar compounds in bivalve mollusc tissues (Livingstone et al. 1985; Stegeman 1985). The importance of these considerations is indicated by the recent data of Farrington et al. (1986b) for the release of aromatic hydrocarbons by *M. edulis* following contamination under oil spill conditions.

Table 11.9. Comparison of biological half-lives of no. 2 fuel oil compounds: 1978 and 1983 oil spills

Compounds	$T_{1/2}$ in Days	
	1978[a] Day 1–day 21	1983 Day 3–day 29
n-C_{16}	.99	8.7
Pristane	7.7	6.3
Phytane	6.9	5.8
n-C_{23}	4.6	5.8
Σ C_2-Naphthalenes	5.8	1.5
Σ C_3-Naphthalenes	7.7	5.8
Phenanthrene	17	5.8
Σ C_1-Phenanthrenes	9.9	6.9
Σ C_2-Phenanthrenes	69[b]	9.9
UCM f_1	14	17
UCM f_2	14	35

[a] Recalculated for 1978 data. These are slightly different from those reported in Farrington et al. (1982) because of an error in a calculator program discovered since that publication.
[b] Correlation fit of data to equation is poor.
Adapted from Farrington et al. 1986b.

The C_2 and C_3 alkylated phenanthrene isomers (e.g., dimethylphenanthrenes and trimethylphenanthrenes) show relative concentration decreases over time that may indicate isomer-specific metabolism.

11.4 Fish, Crustacea, and Polychaetes

Fish, crustacea, and polychaetes, like bivalves, are capable of reducing concentrations of several pollutant organic chemicals by exchange with water. In addition, fish, crustacea, and polychaetes have active enzyme systems capable of metabolizing PAHs, PCBs, and similar compounds. The metabolites can be conjugated with biochemicals and excreted (e.g., NRC 1979, 1985; Stegeman 1981; Varanasi et al. 1985).

Thus, concentrations in tissues can be a function of a variety of conditions controlling enzymatic activity. These conditions and the resulting enzymatic activity for metabolizing PAHs can be species specific. For example, Reichert et al. (1985) exposed two species of deposit-feeding amphipods to sediment-associated, radioactively labeled [³H]benzo[a]pyrene. Both species accumulated the compound, but one species converted a higher proportion to metabolites. It is important to add that metabolism does not necessarily mean reduction in the biological potency of the compound to harm the organism. Some of the metabolites may themselves be carcinogens or mutagens (NRC 1979, 1985; Stegeman 1981). Varanasi (1987) extensively reviews metabolism of PAHs in the environment.

11.5 Dietary Source of Organic Pollutants

The hypothesis of Hamelink et al. (1971), since expanded by others (e.g., NRC 1979), that water–organism interactions were the major factors controlling pollutant organic chemical concentrations (e.g. PAHs, PCBs) in whole aquatic organisms has prevailed for almost fifteen years (Levin et al. 1984; NRC 1979, 1985). Rosenberg (1975) presented data and reasoning to the effect that when concentrations of organic pollutant chemicals in the water were much reduced, then suspended matter and associated chemicals become sources of pollutant for the organism. Other researchers have shown that organisms living in or on polluted sediments maintain higher concentrations than organisms living in pelagic portions of the ecosystems (e.g., Roesijadi et al. 1978). In these experiments, the release of organic pollutant by desorption from sediments was thought to cause elevated concentrations of the pollutant chemical, and water–organism interactions were described as important to controlling the concentration in the organism (Roesijadi et al. 1978). Similar reasoning prevailed in an experiment where sediment was a source of Mirex for uptake by the hogchoker *Trinectes maculatus*, a small flounder (Kobylinski and Livingston 1975).

An elegant experiment by Rubinstein et al. (1984) examined bioaccumulation of PCBs by a demersal fish exposed to PCB-contaminated sediment and worms fed contaminated sediment in a matrix, multi-phase experimental design (Figure 11.15). This work conclusively demonstrated the importance of predation of the fish *Leiostomus xanthurus* on the polychaete *Nereis virens*, resulting in a significant dietary source of PCBs for the fish. This has important implications for the modeling of biogeochemical behavior of hydrophobic organic pollutants and uptake by aquatic organisms. If the primary source of input to an aquatic ecosystem is slow release from sediments contaminated by previous inputs, now reduced or eliminated, or because sediments have been dredged and deposited elsewhere, then it is plausible to put forward the hypothesis that uptake by benthic organisms such as polychaetes, small bivalves, and crustacea followed by predation by bottom feeding fish may be a significant source of uptake of pollutant organic chemicals such as PCBs and PAHs. McElroy et al. (1987) have identified this as a major important area for current research in aquatic biogeochemical cycles. Lack of better knowledge of this phenomena hampers further success in mathematical modeling of biogeochemical cycles in many aquatic ecosystems.

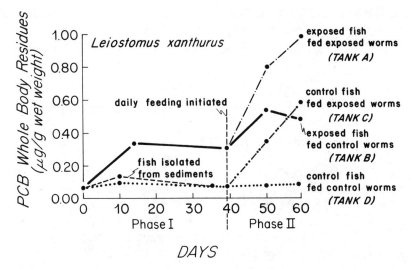

Figure 11.15. Average PCBs whole body residues in spot (*Leiostomus xanthurus*) during Phase I (exposure to sediments) and Phase II (exposure to sediments plus fed worms). Adapted with permission from Rubinstein et al. 1984.

11.6 Conclusion

The large number of organic chemicals of environmental concern and the wide variety of aquatic organisms and environmental exposure conditions require an understanding of factors controlling concentrations in aquatic organisms to the extent that predictive equations tied to molecular structures and characteristics of organisms can be derived. Substantial progress in all aspects (solubility, sorbtion, uptake, metabolism, retention, and release of organic chemical contaminants) has been made during the 1970's and 1980's. It is now possible to incorporate this knowledge into dynamic models of entire ecosystems to predict chemical concentrations in particular target species (e.g., Cohen 1986). Chapter 9 by O'Connor et al. in this volume is a fine example of this type of endeavor.

Care, however, must be exercised to avoid overconfidence in the predictive capabilities of science in this regard. Thus far these concepts have been tested on relatively few chemicals and organisms for relatively short periods of time (days to months). High precision predictive capabilities over long time periods of years are limited by the complex phenomena involved with input and transport of organic chemicals in aquatic ecosystems and the uptake, retention and release in aquatic organisms, as illustrated by examples in this chapter. Nevertheless, an optimistic view is in order as research is narrowing the uncertainties. It is encouraging that there is now sufficient knowledge to prevent wholesale ubiquitous releases of DDT and PCB-type compounds to the environment, and to support concern about continuing releases of compounds such as PAHs (NRC 1983); this knowledge is being used by industry and by agencies responsible for protecting the environment.

Acknowledgments. Much appreciation is extended to Dr. Simon Levin for his patience in awaiting this manuscript. Apologies to co-authors in the volume for the delay caused by factors only partly beyond my control. Discussions with Drs. Judy Capuzzo, Ann McElroy, Bruce Brownawell, John Stegeman, Judy Grassle, John Teal, John Westall, and Rene Schwarzenbach; and Alan Davis, C. Hovey Clifford, and Bruce Tripp over the past few years were very helpful in formulating thoughts expressed in this chapter. Ms. Peggy Chandler deserves thanks for typing this chapter during trying times.

Financial support to the Coastal Research Center, Woods Hole Oceanographic Institution by the Andrew W. Mellon Foundation and Richard King Mellon Foundation made possible the preparation of this chapter. This is Contribution No. 6437 of Woods Hole Oceanographic Institution.

References

Addison RF, Brodie PF, Zinck ME, Sergeant DE (1984) DDT has declined more than PCBs in Eastern Canadian seals during the 1970s. Environ Sci Technol 18:935–937

Baker JE, Capel PD, Eisenreich SJ (1986) Influence of colloids on sediment-water partition coefficients of polychlorobiphenyl congeners in natural waters. Environ Sci Technol 20:1136–1143

Banerjee S (1985) Calculation of water solubility of organic compounds with UNIFAC-derived parameters. Environ Sci Technol 19:369–370

Boehm PD, Quinn JG (1976) The effect of dissolved organic matter in seawater on the uptake of mixed individual hydrocarbons and No. 2 fuel oil by a marine filter feeding bivalve (*Mercenaria mercenaria*). Est Coast Mar Sci 4:93–105

Boehm PD, Quinn JG (1977) The persistence of chronically accumulated hydrocarbons in the hard shell clam, *Mercenaria*. Mar Biol 44:227

Brownawell BJ (1986) The role of colloidal organic matter in the marine geochemistry of PCBs. Ph.D. Dissertation, Woods Hole Oceanographic Institution/Massachusetts Institute of Technology Joint Program in Oceanography, Woods Hole, MA, 318 pp.

Brownawell BJ, Farrington JW (1985) Partitioning of PCBs in marine sediments. In: Sigleo AC, Hattori A (eds) *Marine and Estuarine Geochemistry*. Chelsea, MI: Lewis Pub Inc, 97 pp.

Brownawell BJ, Farrington JW (1986) Biogeochemistry of PCBs in interstitial waters of a coastal marine sediment. Geochim Cosmochim Acta 50:157

Butler GC (ed) (1978) *Principles of Ecotoxicology, SCOPE 13*. New York: John Wiley and Sons, 350 pp.

CEQ (1984) *Environmental Quality 1983*. 14th Annual Report of The Council on

Environmental Quality, Executive Office of the President of the United States, Washington, DC

Chiou CT (1985) Partition coefficients of organic compounds in lipid-water systems and correlations with fish bioconcentration factors. Environ Sci Technol 19:57–62

Chiou CT, Porter PE, Schmedding DW (1983) Partition equilibria of nonionic organic compounds between soil organic matter and water. Environ Sci Technol 17:227–231

Cohen Y (ed) (1986) *Pollutants in a Multimedia Environment*. New York: Plenum Press, 338 pp.

de Leer EWB, Baggerman T, van Schaik P, Zuydeweg CWS, de Galan L (1986) Chlorination of ω-cyanoalkanoic acids in aqueous medium. Environ Sci Technol 20:1218–1223

de Leer EWB, Sinninghe Damste JS, Erkelens C, de Galan L (1985) Identification of intermediates leading to chloroform and C-4 diacids in the chlorination of humic acid. Environ Sci Technol 19:512–522

Farrington JW, Davis AC, Frew NM, Rabin KS (1982) No. 2 fuel oil compounds in *Mytilus edulis*. Retention and release after an oil spill. Mar Biol 66:15–26

Farrington JW, Goldberg ED, Risebrough RW, Martin JH, Bowen VT (1983b) U.S. "Mussel Watch" 1976–1978: An overview of the trace metal, DDE, PCB, hydrocarbon, and artificial radionuclide data. Environ Sci Technol 17:490–496

Farrington JW, Teal JM, Tripp BW, Livramento JB, McElroy A (1983a) Biogeochemistry of petroleum components at the sediment-water interface. Final report prepared for DOE, BLM, March 1983, U.S. Dept of Energy, Office of Energy Research, Ecological Research Division, Washington, DC

Farrington JW, Tripp BW, Davis AC, Sulanowski J (1985) One view of the role of scientific information in the solution of enviro-economic problems. Proceedings of the International Symposium on Utilization of Coastal Ecosystems, Planning, Pollution, Productivity, 22–27 November, 1982, Rio Grande, RS, Brazil. Vol. I (Chao NL and Kirby-Smith W, eds). Duke University Marine Laboratory, Beaufort, NC, pp. 73–102

Farrington JW, Davis AC, Brownawell BJ, Tripp BW, Clifford CH, Livramento JB (1986a) Some aspects of the biogeochemistry of polychlorinated biphenyls in the Acushnet River Estuary. In: Sohn M (ed) *Organic Marine Geochemistry*. ACS Symposium Series No. 305, Washington, DC: American Chemical Society, pp. 174–197

Farrington JW, Jia X, Clifford CH, Tripp BW, Livramento JB, Davis AC, Frew NM, Johnson CG (1986b) No. 2 fuel oil compound retention and release by *Mytilus edulis:* 1983. Cape Cod Canal oil spill. Woods Hole Oceanog Inst Tech Rep WHOI-86-8 (CRC-86–1), Woods Hole, MA

Farrington JW, Westall J (1986) Organic chemical pollutants in the oceans and groundwater: A review of fundamental chemical properties and biogeochemistry. In: Kullenberg G (ed) *The Role of the Oceans as a Waste Disposal Option*. Proceedings NATO Advanced Research Workshop. Boston: D. Reidel Publ. Co., pp. 361–425

Hamelink JL, Waybrant RC, Ball RC (1971) A proposal: exchange equilibria control the degree chlorinated hydrocarbons are biologically magnified in lentic environments. Trans Am Fish Soc 100:207

Himenez PC (1977) *Principles of Colloid and Surface Chemistry*. New York: Marcel Dekker, Inc., 516 pp.

IFREMER (1985) Reseau national d'observation de la qualité du milieu marin. Institut Français de Recherche pour l'Exploration de la Mer. Département Milieu et Ressources, B.P. 1049, 44037 Nantes, Cedex, France

Karickhoff S, Brown DS, Scott T (1979) Sorption of hydrophobic pollutants on natural sediments. Water Res 13:241

Keith LK, Telliard WA (1979) Environmental science and technology special report: priority pollutants. I. A perspective view. Environ Sci Technol 13:416–423

Kobylinski GJ, Livingston RJ (1975) Movement of mirex from sediment and uptake by the hogchoker, *Trinectes maculatus*. Bull Environ Contam Toxicol 14:692–698

Knap AH, Jickells TD (1983) Trace metals and organochlorines in the goose-beaked whale. Mar Poll Bull 14:271–274

Kringstad KP, de Souza F, Stromberg LM (1985) Studies on the chlorination of chlorolignins and humic acid. Environ Sci Technol 19:427–431

Kullenberg G (ed) (1986) The IOC programme on marine pollution. Mar Poll Bull 17:341–352

Langston WJ (1978) Persistence of polychlorinated biphenyls in marine bivalves. Mar Biol 46:35–40

Levin SA, Kimball KD, McDowell WH, Kimball SF (1984) New perspectives in ecotoxicology. Environ Manage 8:375–442

Livingstone DR, Moore MN, Lowe DM, Casci C, Farrar S (1985) Responses of the cytochrome P-450 monoxoygenase system to diesel oil in the common mussel *Mytilus edulis* L. and the periwinkle, *Littorina littorea* L. Aquat Toxicol 7:79–91

Lyman WJ, Reehl WF, Resenblatt DH (1982) *Handbook of Chemical Property Estimation Methods: Environmental Behavior of Organic Compounds*. New York: McGraw-Hill Book Co., 960 pp.

Mackay D (1982) Correlation of bioconcentration factors. Environ Sci Technol 16:274–278

Mackay D, Shiu WY (1977) Aqueous solubility of polynuclear aromatic hydrocarbons. J Chem Eng Data 22(4):399–402

Marchand M, Cabane F (1980) Hydrocarbures dans les moules et les huitres. Rev Int Oceanogr Med 59:3–30

Matta MB, Mearns AJ, Buchanan MF (1986) The National Status and Trends Program for Marine Environmental Quality. Trends in DDT and PCB's in US West Coast Fish and Invertebrates. Ocean Assessments Division, Office of Oceanography and Marine Assessment, National Ocean Service, NOAA, U.S. Department of Commerce, Seattle, WA

May WE, Wasik SP, Freeman DH (1978) Determination of the solubility behavior of some polycyclic aromatic hydrocarbons in water. Anal Chem 50:997

May WE (1980) The solubility behavior of polycyclic aromatic hydrocarbons in aqueous systems. In: Petrakis L and Weiss FT (eds) *Petroleum in the Marine Environment*. Washington: American Chemical Society, pp. 143–192

McCarthy JF, Jimenez BD, Barbee T (1985) Effect of dissolved humic material on accumulation of polycyclic aromatic hydrocarbons: structure-activity relationships. Aquat Toxicol 7:15

McElroy AE, Farrington JW, Teal JM (1987) Bioavailability of polynuclear aromatic hydrocarbons in the aquatic environment. In: *Metabolism of Polynuclear Aromatic Hydrocarbons (PAHs) in the Environment*. Boca Raton: CRC Press, in press

Means JC, Hassett JJ, Woods SG, Banwart WL (1979) Sorption properties of energy-related pollutants and sediments. In: Jones PW, Leber P (eds) *Polynuclear Aromatic Hydrocarbons*. Ann Arbor: Ann Arbor Sci Pub Inc, 327 pp.

Means JR, Wijayaratne R (1982) Role of natural colloids in the transport of hydrophobic pollutants. Science 215:968

Miller DR (1984) Chemicals in the environment. In: Sheehan PJ, Miller DR, Butler GC, Bourdeau P (eds) *Effects of Pollutants at the Ecosystem Level, SCOPE 22*. New York: John Wiley and Sons, pp. 7–14

Miller MM, Wasik SP, Huang GL, Mackay D (1985) Relationship between octanol–water partition coefficient and aqueous solubility. Environ Sci Technol 19:522–529

Neeley WB, Branson DR, Blau GE (1974) Partition coefficient to measure bioconcentration potential of organic chemicals in fish. Environ Sci Technol 13:1113

Nisbet ICT, Reynolds L (1983) Organochlorine residues in common terns and associated estuarine organisms, Massachusetts, USA,1971–1981. Mar Environ Res 11:33–66

NRC (1979) Polychlorinated Biphenyls. U.S. National Research Council, Publications Office. Washington DC: National Academy of Sciences

NRC (1980) The International Mussel Watch: Report of a Workshop. National Research Council, Publications Office. Washington DC: U.S. National Academy of Sciences

NRC (1983) Polycyclic Aromatic Hydrocarbons: Evaluation of Sources and Effects. U.S. National Research Council, Publications Office. Washington DC: U.S. National Academy of Sciences

NRC (1985) Oil in the Sea. National Research Council, Publications Office. Washington DC: U.S. National Academy of Sciences

O'Connor JM, Pizza JC (1987) "Pharmacokinetic model for the accumulation of PCBs in marine fish." In: Capuzzo JM, Kester DR (eds) *Oceanic Process in Marine Pollution*, Vol 1. (Biological Processes and Wastes in the Ocean). Malabar, FL: RE Krieger Publishing Co, pp. 119–129

Phillips JDH (1980) *Quantitative Biological Indicators. Their Use to Monitor Trace Metal and Organochlorine Pollution*. London: Applied Science Publishers

Pruell RJ, Lake JL, Davis WR, Quinn JG (1986) Update and depuration of organic contaminants by blue mussels, *Mytilus edulis*, exposed to environmentally contaminated sediment. Mar Biol 91:497

Rashid MA (1985) *Geochemistry of Marine Humic Compounds*. New York: Springer-Verlag, 300 pp.

Reichert WL, Le Ederhart B-T, Varanasi U (1985) Exposure of two species of deposit-feeding amphipods to sediment-associated (^{3}H) benzo(a)pyrene: uptake, metabolism and covalent binding to tissue macromolecules. Aquat Toxicol 6:45–56

Rheadman JW, Mantoura RFC, Rhead MM (1984) The physico-chemical speciation of polycyclic aromatic hydrocarbons (PAH) in aquatic systems. Fresenius Z Anal Chem 319:126–131

Roesijadi G, Anderson JW, Blaylock JW (1978) Uptake of hydrocarbons from marine sediments contaminated with Prudhoe Bay crude oil: influence of feeding type of test species and availability of polycyclic aromatic hydrocarbons. J Fish Res Bd Can 35:608

Rosenberg DM (1975) Fate of dieldrin in sediment, water, vegetation, and invertebrates of a slough in central Alberta, Canada. Quaest Entomol 11:69–96

Rubenstein N, Gilliam WT, Gregory NR (1984) Dietary accumulation of PCBs from a contaminated sediment source by a demersal fish (*Leiostomus xanthurus*). Aquat Toxicol 5:331–342

Schulte E, Acker L (1974) Identifizietung und Metabolisierbarkeit von polychlorierten Biphenylen. Naturwiss 61:79–81

Schwarzenbach RP, Westall J (1981) Transport of nonpolar organic compounds from surface water to groundwater. Laboratory sorption studies. Environ Sci Technol 15:1360

Sheehan PJ, Miller DR, Butler GC, Bourdeau P (eds) (1984) *Effects of Pollutants at the Ecosystem Level, SCOPE 22.* New York: John Wiley and Sons

Stegeman JJ, Teal JM (1973) Accumulation, release, and retention of petroleum hydrocarbons by the oyster *Crassostrea virginica.* Mar Biol 22:37–44

Stegeman JJ (1981) Polynuclear aromatic hydrocarbons and their metabolites. In: Gelboin HV, Ts'O POP (eds) *Polycyclic Hydrocarbons and Cancer,* vol. 3. New York: Academic Press, pp. 1–60

Stegeman JJ (1985) Benzo(a)pyrene oxidation and microsomal enzyme activity in the mussel (*Mytilus edulis*) and other bivalve mollusc species from the western North Atlantic. Mar Biol 89:21–30

Tanabe S, Mori T, Tatsukawa R, Miyazaki N (1983) Global pollution of marine mammals by PCBs, DDT, and HCHs (BHCS). Chemosphere 12:1269–1275

Thurman EM (1986) *Organic Geochemistry of Natural Waters.* Boston: Martinus Nijhoff/Dr. W. Junk Publishers, 497 pp.

Topping G (1983) Guidelines for the use of biological material in first order pollution assessment and trend monitoring. Department of Agriculture and Fisheries for Scotland. Scottish Fisheries Research Report Number 28, 1983, ISSN 0308 8022. The Director, Marine Laboratory, DAFF, Aberdeen Scotland

Varanasi U, Reichert WL, Stein JE, Brown DW, Sanborn HR (1985) Bioavailability and biotransformation of aromatic hydrocarbons in benthic organisms exposed to sediment from an urban estuary. Environ Sci Technol 19:836

Varanasi U (ed) (1987) *Metabolism of Polynuclear Aromatic Hydrocarbons (PAHs) in the Environment.* Boca Raton: CRC Press, in press

Weaver G (1984) PCB contamination in and around New Bedford, Mass. Environ Sci Technol 18:22A–27A

Whitehouse BC (1984) The effects of temperature and salinity on the aqueous solubility of polynuclear aromatic hydrocarbons. Mar Chem 14:319

Chapter 12
Environmental Chemical Stress Effects Associated with Carbon and Phosphorus Biogeochemical Cycles

Abraham Lerman[1], Fred T. Mackenzie[2],
and Robert J. Geiger[1]

Primary biological productivity on land and in waters forms organic materials made of six main elements—C, H, O, N, S, and P—and about a dozen minor elements that are important to the maintenance of organic structures and physiological functions of living organisms. The main elements are present in different proportions in aquatic and land plants. In the surface environment of the Earth, the elements carbon, sulfur, nitrogen and phosphorus mostly are found as separate chemical forms. Primary biological productivity represents a coupling mechanism that joins the biogeochemical cycles of the individual elements one to another. The following ratios denote atomic proportions of C, N, S, and P in terrestrial and aquatic plants representative of this coupling:

marine plankton	$C:N:S:P = 106:16:1.7:1$	(Redfield et al. 1963)
marine benthos	$C:N:P = 550:30:1$	(Atkinson and Smith 1983)
land plants	$C:N:S:P = 882:9:0.6:1$	(Deevey 1973)
	$= 510:4:0.8:1$	(Delwiche and Likens 1977)

Among the four elemental components, carbon, nitrogen, and sulfur represent natural chemical redox couples, because each element occurs both in an oxidized and a reduced state. The process of formation of organic matter by photosynthesis constitutes chemical reduction of C, N, and S from their inorganic forms. The oxidized inorganic forms of the four elements are such commonly occurring species as carbon dioxide (CO_2),

[1] Department of Geological Sciences, Northwestern University, Evanston, Illinois 60208
[2] Hawaii Institute of Geophysics, University of Hawaii at Manoa, Honolulu, Hawaii 96822

dissolved carbonate (HCO_3^- and CO_3^{2-}), nitrate (NO_3^-) and nitrogen oxides, sulfate (SO_4^{-2}), and phosphate. The reduced inorganic forms of these biologically important elements—methane CH_4, hydrogen sulfide H_2S, ammonia NH_3, and phosphine PH_3—are thermodynamically unstable in the presence of oxygen, and eventually they become oxidized.

The net photosynthetic reactions that produce organic matter with the stoichiometric proportions of C, N, S, and P as given above can be written for the mean composition of marine plankton and land plants in a form that balances oxidized inorganic reactants with organic matter and free oxygen as the reaction products. The two reactions are:

In water: $106\ CO_2 + 16\ HNO_3 + 2\ H_2SO_4 + H_3PO_4 + 120\ H_2O$

$$\underset{\text{photosynthesis}}{\overset{\text{decay}}{\rightleftharpoons}} C_{106}H_{263}O_{110}N_{16}S_2P + 141\ O_2$$

On land: $882\ CO_2 + 9\ HNO_3 + H_2SO_4 + H_3PO_4 + 890\ H_2O$

$$\underset{\text{photosynthesis}}{\overset{\text{decay}}{\rightleftharpoons}} C_{882}H_{1794}O_{886}N_9SP + 901.5\ O_2$$

In the aquatic photosynthesis reaction, for convenience, the proportion of sulfur was raised from 1.7 to 2, and for land photosynthesis the proportion of sulfur was rounded up to 1. The proportions of hydrogen and oxygen in organic matter are nearly $H:O = 2:1$; this is the reason for the abbreviated chemical notation of organic matter usually written as CH_2O.

Many natural chemical substances circulate through the environment and are important to the chemistry and biology of the Earth. The circulation of a particular substance—as defined by its reservoirs, processes affecting it, and fluxes—is termed its *biogeochemical cycle*. Biogeochemical cycles vary in time and spatial scale. The long-term circulation of *earth* materials, the *exogenic cycle,* represents one extreme in which materials are transported through the atmosphere to the land, and through the soils to streams that carry materials to the oceans. In the oceans, stream-borne solids and some originally dissolved substances that are now part of solids, sink and become sea floor sediments. Other substances are returned to the atmosphere.

Thus, it is important to understand the natural circulation of a substance to assess the impact of society's release of that substance on the natural system, and eventually, perhaps, on humanity. In particular, it is important to determine the pre-human state of a system and its response to perturbations. An increased rate of addition of a potential pollutant to the environment could result in negative or positive feedbacks in the natural systems that relieve or enhance the magnitude of the perturbation. Unfortunately, we are still somewhat naive about how natural biogeochemical cycles function, and particularly, about their responses to perturbations.

This chapter concentrates on the biogeochemical cycles of carbon and phosphorus, and on their coupling in the biosphere, because of the major role that carbon plays in global ecosystem functioning, and because of the critical role of phosphorus as a nutrient and, sometimes, as a limiting nutrient element in aquatic ecosystems. Both carbon and phosphorus in the earth's surface environment are represented by large sedimentary reservoirs that exchange their contents with the global aquatic and atmospheric reservoirs. Thus, the elements carbon and phosphorus, being the most and the least abundant of the main components of aquatic organic matter, are involved both in biological and inorganic cycles.

The background or baseline structures of the C and P biogeochemical cycles will be discussed early in this chapter, to be followed by an analysis of several environmental stresses that are affecting, or may affect, ecosystems on a global scale. Responses to perturbations of the biogeochemical C and P cycles will be analyzed in terms of simple mathematical models on a time scale of years to centuries. Because such relatively short time scales are of concern to humans, the environmental system that might noticeably change on a time scale of 10^0 to 10^2 years must be a system inhabited by humans and/or immediately affecting their habitats. Such a system obviously includes land, soils, atmosphere, river water, terrestrial and aquatic biota, and sections of the ocean in proximity to the land.

The main conclusions of the model analyses that apply to a number of environmental stress scenarios are that the global environmental system as a whole changes slowly because of the large size of the reservoirs. The potential geochemical consequences of large changes taking place in one part of the global environment are to a variable extent smoothed out when they pass to other environmental reservoirs. On a time scale of human generations, biogeochemical perturbations in the carbon and phosphorus cycles—barring the most drastic effects of a catastrophic global change—do not uniformly affect all parts of the global environmental system.

12.1 Carbon Cycle

12.1.1 Background

A diagram of the present-day global cycle of carbon is shown in Figure 12.1. This representation is an updated and revised version of that of Garrels et al. (1975). The four major reservoirs of carbon in the surface environment are the atmosphere, land, ocean, and sediments. The sediments are by far the biggest reservoir of oxidized and reduced carbon, in $CaCO_3$ and in buried organic matter, followed by the land or continental surface, the ocean, and the atmosphere. In Figure 12.1 the fluxes of carbon and their nature are indicated by arrows; fluxes include both natural background and human-produced fluxes. (The latter fall in the category of environmental stresses that will be discussed in a subsequent

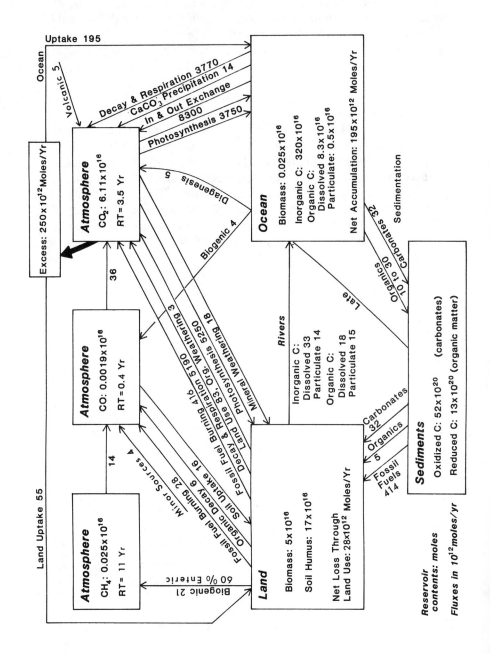

section.) Each of the major C reservoirs is subdivided into smaller compartments. The three main C-containing species in the atmosphere are CO_2, CO, and CH_4. The occurrences of carbon in such reservoirs as living biomass, dead organic matter in soils, and in solution are indicated for the land, ocean and sediments.

The main features of the carbon cycle, and the magnitudes of the reservoirs and fluxes are summarized below.

Note in Figure 12.1 that the mass of carbon contained in the living biomass is much greater on land than in the ocean. Because oxygen is produced by photosynthesis of the living biomass, its global production is primarily supported by photosynthesis on land, but consumption of oxygen depends on the living and dead (humus) organic matter on land and on other sinks. The mass of carbon in ocean water and in the atmosphere (dissolved inorganic carbonate and organic carbon, and CO_2) is comparable to the mass of carbon contained in land plants. However, the amount of carbon in dead organic matter on land (humus) is, on a global scale, much greater, by about a factor of three, than in living plants. A comparison of the global numbers, as given in Figure 12.1, does not reflect the actual picture of the very uneven distribution of living and dead organic matter on land: most of the living biomass plants occur in tropical forests, where the amount of humus in the forest soils is very small, owing to its rapid oxidation by bacteria; the amounts of humus preserved in soils are particularly large in temperate climates where the vegetation density is smaller than in the humid tropics.

The carbon cycle shown in Figure 12.1 is one of several possible ways of showing the exogenic pathways of the element. A recent review by Sundquist (1985) compiles estimates of reservoir masses and fluxes of carbon.

12.1.2 Environmental Stresses

Within the global carbon cycle, some of the fluxes are natural background fluxes that are driven by physical, chemical, and biological processes always operating in the environment. Some other fluxes are the products of human agricultural and industrial activity. The more important of the human-produced fluxes include the fluxes of CO_2 and CO because of the burning of fossil fuels and deforestation; the flux of CH_4 that is in part related to agricultural animals and field cultivation; greater transport of dissolved and particulate carbon from land to the oceans because of locally intensive agriculture and deforestation; and greater rates of lime-

Figure 12.1. Schematic global biogeochemical cycle of carbon, based on data compiled by Mackenzie et al. (1983) and Sundquist (1985). Reservoir contents are in moles C, fluxes in units of 10^{12} moles C y^{-1}, and RT is carbon residence time in years. 1 mole = 12 g C.

stone weathering in deforested and acid-rain exposed areas. Reverse fluxes, from the atmosphere down to other reservoirs, are affected by eutrophication of natural waters and a potentially greater rate of storage of carbon in sediments.

An important natural stress on the carbon cycle is addition of CO_2 to the atmosphere from volcanoes. Although even the most powerful volcanic eruptions recorded in human history were capable of adding only a negligibly small amount of CO_2 to the atmospheric reservoir, periods of geologically prolonged volcanic activity might have been responsible for considerable increases in the CO_2 fluxes from the land and ocean reservoirs to the atmosphere (compare with the fluxes shown in Figure 12.1). The volcanic CO_2 fluxes and the subsequent redistribution of the added CO_2 downcycle lend themselves to a qualitative comparison with CO_2 emissions of pre-industrial and modern times.

In 1984, fossil fuel combustion resulted in the release of 4.4×10^{14} moles C (1 mole C = 12 g) to the atmosphere (Rotty 1987), or about 5% of the total CO_2 released by natural processes of decay and respiration. The magnitude of CO_2 release from land use practices is still somewhat controversial; estimates range from less than 10^{14} moles yr^{-1} to a rate equivalent to fossil fuel CO_2 production (Woodwell et al. 1978; Houghton et al. 1987; Siegenthaler and Oeschger 1987). Because of the small size of the atmospheric CO_2 reservoir, these releases can have a substantial effect on the CO_2 concentration of the atmosphere. The actual measured annual rate of increase of atmospheric CO_2 today is about 1.4 ppmv. Figure 12.2 shows the pattern of increase of atmospheric CO_2 obtained from air sampled at Mauna Loa, Hawaii, for the period 1958–1983, and also the centuries-long change as obtained from ice core data (Sundquist 1985; Siegenthaler and Oeschger 1987).

Projections based on large-scale modeling efforts show that by the middle of the 21st century, atmospheric CO_2 concentration may be two times its late 1800's value of about 290 ppmv. Simulations of some other physical models suggest that a doubling of the CO_2 concentration can result in a 1.5–4.5°C increase in earth's surface temperature, with increases as large as 8°C near the poles. Some researchers (e.g., Idso 1982) predict much less of a temperature effect. If such a global CO_2 warming occurs, the consequences can only be crudely estimated from current climate models and paleoclimatic data. It is likely that large regional climatic changes will occur as a result of the warming. Precipitation patterns and the distributions of deserts, fertile, and marginal lands may be affected. The nature and distribution of soil biota, including bacteria that catalyze reactions such as those involving nitrogen, also may change as a result of the predicted warming. Floating polar sea ice may disappear, or be present only during winter, depending on the degree of warming. In response to thermal expansion of the oceans, a sea level rise of one to four meters could occur late in the next century, flooding many lowlands

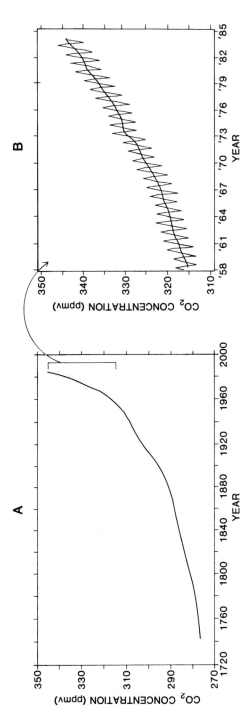

Figure 12.2. Atmospheric CO_2 content through time. **A** Record from ice cores since 1760, from Siegenthaler and Oeschger (1987). **B** Record from the Mauna Loa Observatory atmosphere (Hawaii), showing seasonal variations since 1958, from compilation of Sundquist (1985).

throughout the world (Hoffman et al. 1983). There may be an increase in the length of the growing season in certain regions and countries (for example, parts of the Soviet Union), and an increase in C_3 plant biomass production and agricultural productivity.

Conventional wisdom has it that about one-half of all the fossil fuel CO_2 released has remained airborne; the rest has been taken up by other reservoirs, primarily the oceans. As shown in Figure 12.1, however, if land use practices (primarily deforestation and cultivation) have led to release of CO_2 to the atmosphere, then it is likely that some anthropogenic carbon may be taken up in the land reservoir, perhaps in unperturbed forest areas in trees, litter, and soil humus, as discussed by Houghton et al. (1987) and Siegenthaler and Oeschger (1987). The figure shows a net loss of carbon from the land reservoir. The magnitude of this loss depends primarily on the relative magnitudes of the carbon fluxes owing to land use and storage, and streams (organic C). In particular, if the land use flux is high, then the compensatory increase in the fluxes (over their values in ancient times) related to the sinks of CO_2 dissolution in ocean water, storage in the atmosphere, CO_2 reaction with shallow-water carbonate minerals, organic carbon burial in sediments, and storage on land must be enhanced over that required simply to accommodate fossil fuel CO_2. The modern carbon cycle is difficult to balance. The problems are to define the magnitudes of today's fluxes related to potential major and minor sinks of anthropogenic CO_2. Evaluation of the course of future anthropogenic fluxes and sink strengths is based on assumptions and empirical- or process-level models.

The record of CO_2 increase in the atmosphere for the last two centuries is summarized in Table 12.1. The 130 years from about 1760 to 1890 were

Table 12.1. Atmospheric CO_2 concentrations ast different times and linearized rates of increase for time periods from Figure 12.2A,B

Year	CO_2 (ppmv)	$\Delta M/\Delta t$ (ppmv yr^{-1})	$(\Delta M/M)\Delta t^*$) (yr^{-1})
1760	277		
1890	291	0.11	3.8×10^{-4}
1958	314	0.34	11.2×10^{-4}
1974	329	0.94	29.2×10^{-4}
1983	344	1.4	41.6×10^{-4}
1958 (January)	315		
1974 (January)	329	0.88	27.3×10^{-4}
1984 (January)	343	1.4	41.7×10^{-4}

* M is mean of the CO_2 values at the beginning and end of a time period.

characterized by a relatively slow rate of increase. Subsequently, the rate of increase has been accelerating. The total increase achieved since the late 18th century is about 24%.

The terrestrial biota may be a significant source of CO_2 release to the atmosphere (Woodwell et al. 1978; Houghton et al. 1987). Figure 12.3 compares the annual carbon flux to the atmosphere because of fossil fuel burning with that resulting from land use activities of deforestation and cultivation. Land use fluxes are based on model calculations whereas fossil fuel fluxes are obtained from estimates of the world's consumption of fossil fuels and their carbon content. It can be seen from this figure that annual carbon release from the terrestrial biosphere may have been greater than that from fossil fuels until about 1960 (for further discussion see Siegenthaler and Oeschger 1987; Bolin 1986). Care must be taken in interpretation of this figure in that the land use flux may not be a net flux to the atmosphere, as it is based on model calculations. These calculations may underestimate the role of temperate forests, particularly in the Soviet Union, in sequestering organic carbon, and the possibility of increased C_3 plant production because of increased atmospheric CO_2 concentrations. Nevertheless, clarification of the role of the terrestrial biosphere as a net source or sink of CO_2 is important to future considerations of atmospheric CO_2 change. Large scale deforestation and cultivation activities occurring today, particularly in tropical areas, may be substantially increasing CO_2 release to the atmosphere.

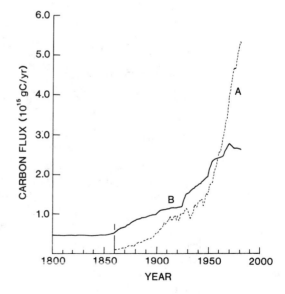

Figure 12.3. Carbon fluxes to the atmosphere from fossil fuel burning (curve A), and from deforestation and land use activities (curve B), based on Houghton et al. (1985).

12.2 Phosphorus Cycle

12.2.1 Background

The global cycle of phosphorus (Figure 12.4) consists of the land, ocean, and sediment reservoirs, with the atmosphere as the fourth reservoir containing a very small mass and having a short residence time of P in particulate materials.

The mass of P in the land biomass is some 20 times greater than in the oceanic biomass—a relationship similar to the one discussed for carbon in the preceding section.

Because there are no gaseous phosphorus compounds (with the exception of the oxidizable phosphine PH_3) in significant quantities, the atmosphere plays a role comparable to that of rivers in the global phosphorus cycle: it affords rapid transit of P-containing particles from land, but in much smaller quantities than the rivers.

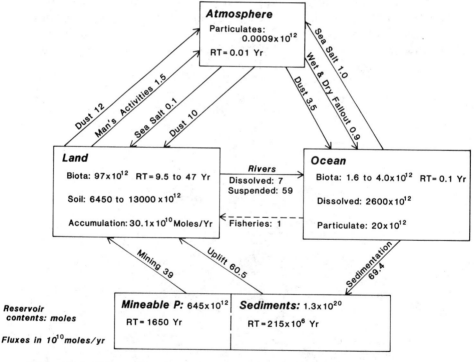

Figure 12.4. Schematic global biogeochemical cycle of phosphorus, based on data compiled by Mackenzie et al. (1983) with later additions. Reservoir contents are in moles P, fluxes in units of 10^{10} moles P yr^{-1}, and RT is phosphorus residence time in years (cf. Figure 12.1). 1 mole P = 31g P.

The soil reservoir on land contains P in organic matter and in inorganic forms, as phosphate ions absorbed on clays, as the calcium-phosphate mineral apatite, and as Fe- and Al-phosphate bearing minerals.

Concentrations of phosphate minerals in sediments make them mineable deposits. A phosphorus-rich deposit may be technologically and economically worth mining at some time, but not worth mining at another time. Thus technological and economic factors primarily determine the magnitude of the P flux in fertilizers from the sediments to the soils (Lerman et al. 1975).

The mass of phosphorus contained in weathered materials on land (that is, in soils) is very large in comparison to the phosphorus in the standing crops of the land and oceanic biota. The soils and sediments represent a very big reservoir of phosphorus, whose role as a growth-limiting nutrient in ecosystems reflects not its global shortage, but its rate of supply to ecosystems that may fall behind biological demand.

12.2.2 Environmental Stresses

The major impact of humans on the phosphorus cycle results from the use of phosphorus-bearing fertilizers. As is true also for nitrogen on a global scale, streams today may be transporting more dissolved P than they did in ancient times (Meybeck 1982). This phosphorus is introduced to streams via sewage and by leaching of P from fertilizers applied to the land's surface. The increased P burden of streams, lakes, and coastal marine waters generally results in increased rates of eutrophication of these systems.

The C and P cycles are obviously interwoven. One interesting tie involving society's effect on these cycles is eutrophication. For example, an estimate of phosphorus added to aquatic systems from municipal wastes and leaching of phosphate fertilizers is 0.09×10^{12} moles P yr^{-1} (1 mole P = 31 g) (Mackenzie 1981; Mackenzie et al. 1983). If this P is used in plant photosynthesis with an average atomic ratio of C : P of 250 : 1, 23×10^{12} moles of carbon would be utilized, or about 10% of the yearly atmospheric CO_2 increase from fossil fuels (Figure 12.1). The excess organic carbon accumulating in sediments of aquatic systems because of fertilization by phosphorus represents an additional sink of fossil fuel CO_2. If this carbon were "permanently" stored in sediments, the global sink of organic carbon would be increased by this amount (see also Peterson and Melillo 1985). The range of values associated with this flux depends on the average C : P ratio used, and the amount of phosphorus released to aquatic systems via fertilizers and municipal wastes, plus the percent of fixed C that is permanently stored in biomass and sediments.

Additional environmental stresses, but of smaller magnitudes, in the P cycle are the landward bound flux owing to transfer of fishing catches and the outward bound fluxes of P-containing solids to the atmosphere owing to erosion, deforestation, and fuel burning. The P-flux from ocean to land owing to fisheries is a relatively small addition from the fish bones and scales from the catch remaining on land.

12.3 Simple Cycle Models

12.3.1 Mass Balance in a Cycle

A model of a cycle consisting of three reservoirs is shown in Figure 12.5. Each reservoir receives fluxes from the other two and delivers fluxes to them. If this cycle is thought of as representing a biogeochemical cycle of an element, then the chemical or biological form of the element may differ from one reservoir to another. Any transformations or chemical reactions, such as oxidation or reduction, dissolution or precipitation, are not explicitly represented in the diagram of the cycle. The material flows between the reservoirs deal with the masses transported, and the rates of transport are measured in units of mass per unit of time. The nature of the chemical processes that are responsible for the transformation of the components from one state to another, and the physical and biological processes that are responsible for the flows determine in detail the rates of

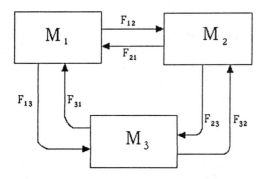

$$\frac{dM_1}{dt} = F_{21} + F_{31} - (F_{13} + F_{12})$$
$$= k_{21}M_2 + k_{31}M_3 - M_1(k_{12} + k_{13})$$

$$\frac{dM_2}{dt} = k_{12}M_1 + k_{32}M_3 - M_2(k_{21} + k_{23})$$

$$\frac{dM_3}{dt} = k_{13}M_1 + k_{23}M_2 - M_3(k_{31} + k_{32})$$

Figure 12.5. Schematic diagram of a three-reservoir cycle and flows. Reservoir contents (mass M_i) and fluxes (F_{ij}) are interrelated as given in Equations 12.4 and 12.5.

transport. Knowledge of the fundamental processes and of the driving forces behind the flows, however, may not be sufficiently sound to enable us to relate them quantitatively to the material fluxes. Thus, the fluxes of materials (denoted F_{ij} in Figure 12.5) are commonly measured and related to the conditions in the system according to some chosen model. In the present case, the two simplest flux models are of zeroeth- and first-order fluxes. The zeroeth-order flux is a constant,

$$F_{ij} = \text{constant} \tag{12.1}$$

and a first-order flux is directly related to the reservoir mass,

$$F_{ij} = k_{ij} M_i \tag{12.2}$$

where M_i is the mass of reservoir i, and k_{ij} is a rate parameter for the flux going from reservoir i to j. Although, in general, k_{ij} may vary with time and reservoir size, and may be a function of environmental conditions within the system, in Equation 12.2 it is treated as a constant.

In a steady-state system, reservoir masses do not change with time, corresponding to an equality of input and output fluxes for every reservoir:

$$\underset{\substack{\text{input from} \\ \text{all } j \text{ to } i}}{\sum_j k_{ji} M_j} \;=\; \underset{\substack{\text{output from} \\ i \text{ to all } j}}{M_i \sum_j k_{ij}} \qquad (i \neq j) \tag{12.3}$$

If one of the fluxes changes, the steady state of the system becomes perturbed, resulting in changes in all the reservoir masses. In this case, the mass balance equation for each reservoir becomes (Figure 12.5):

$$\frac{dM_i}{dt} = \sum_j k_{ji} M_j - M_i \sum_j k_{ij}, \tag{12.4}$$

where the term dM_i/dt indicates the rate of change in the mass of reservoir i.

For a system containing several reservoirs, there are, in general, an equal number of equations of the type of (12.4). The cyclic system in Figure 12.5 is a closed system, because no material enters or leaves the system. In an open system, there may be fluxes from the outside entering the reservoirs, and fluxes taking material out of the system. Clearly, the definition of a system as either open or closed commonly is a matter of convenience or a choice based on the problem at hand. Portions of the cycles of carbon and phosphorus, as shown in Figures 12.1 and 12.4, respectively, may be considered closed systems on geologically short time scales, as long as addition or subtraction of materials through global tectonic processes is not significant. On geologically long time scales, however, the system would be affected by flows of materials from the Earth's crust and by the rates of uplift and weathering of crustal rocks.

For an open system, by analogy with the mass-balance equation (Equation 12.4), a mass balance may be written as the algebraic sum of external and internal fluxes:

$$\frac{dM_i}{dt} = (F_{in} - F_{out})_{ext} + (F_{in} - F_{out})_{int} \tag{12.5}$$

External fluxes (subscript *ext*) bring materials from the outside or remove them from the system, whereas the internal fluxes are responsible for material redistribution within the system.

12.3.2 Residence Time

The residence time of a substance in a reservoir is defined as the ratio of its mass (M_i) to the sum of either input or output fluxes at steady state. The mean residence time (τ_i) with respect to output is

$$\tau_i = \frac{M_i}{\Sigma_j F_{ij}} = \frac{1}{\Sigma_j k_{ij}}, \tag{12.6}$$

where F_{ij} are the fluxes out of reservoir i.

A perturbation of a cycle caused by a change in an input flux will result in a change in reservoir mass. For a fixed change in input, up or down, the reservoir mass will come to within 5 percent of the new steady-state value after three residence times have elapsed. Thus most of the change (95%) caused by a perturbation in input would be complete in time (T_{ss})—time to achieve steady state—equal to

$$T_{ss} = 3\tau \tag{12.7}$$

(e.g., Lerman 1979) where residence time τ is as defined in Equation 12.6. Equation 12.7 reflects the fact that environmental reservoirs of short residence times respond rapidly to external perturbations, whereas in reservoirs of long residence time perturbations require more time to work their way through.

From an environmental point of view, a short residence time of an element in an environmental reservoir could be both detrimental and advantageous: detrimental, because a relatively small change in either input or output would produce a rapid change in reservoir contents; advantageous, because it would take only a relatively short time for conditions to return to their previous state, once a perturbation has been removed. With reservoirs having long residence times, small changes in input and/or output may become significant only after a long period of time, as the cumulative effect of such changes grows. It would also take a long time, however, for the conditions to return back to normal after the input or output fluxes had been restored to their previous states. As a rule of thumb, discussed above, it takes 3 to 4 residence times for a steady perturbation to work its way through an environmental reservoir and for the reservoir to come close to a new steady state. This statement holds for changes in either an increase or decrease of reservoir mass.

12.3.3 Measure of Resistance to Change

From the preceding discussion of residence time and the times required for changes owing to external perturbations in input, a measure of resistance to change may be defined for biogeochemical reservoirs as time needed to achieve a certain fractional increase in mass. An arbitrary change of 10 percent in reservoir mass can be achieved relatively fast if there is a perturbation in input that is large in comparison to the existing reservoir mass. Conversely, a perturbation in input that is small relative to reservoir mass would require a longer time for a 10 percent change in reservoir mass to be achieved. A system where it takes longer to achieve a change in a reservoir mass, other conditions being equal, is a system that has a greater environmental resistance to perturbations in input.To obtain a quantitative measure of stability, one can write mass-balance Equation 12.4 for one reservoir as:

$$\frac{dM}{dt} = F_{in} - \frac{M}{\tau} \qquad (12.8)$$

where F_{in} denotes all the input fluxes into reservoir. The time Δt needed to produce a change $\Delta M/M$ in the reservoir mass is

$$\Delta t = \frac{\Delta M/M}{F_{in}/M - 1/\tau} \qquad (12.9)$$

Thus, if input flux F_{in}, mass M, and residence time with respect to output are known, Δt can be computed for any value of a change in mass ΔM that is small relative to M.

A linearized form of Δt in Equation 12.8 may be also written as a rate of relative change in mass:

$$\frac{\Delta M/M}{\Delta t} = \frac{1}{\tau_{in}} - \frac{1}{\tau} \qquad (12.10)$$

$$= \frac{R - 1}{\tau} \qquad (12.11)$$

where $\tau_{in} = M/F_{in}$ is instantaneous residence time with respect to input, and $R = F_{in}\tau/M$ is the ratio of a steady-state mass ($F_{in}\tau$ as given by the model of Equation 12.8) to the instantaneous mass M.

Any change in M (that is, $\Delta M/M$) is constrained by the value of R: a change cannot exceed a steady-state mass, when $R = 1$.

Alternatively, a measure of stability, defined as above, is from Equation 12.11

$$\Delta t = \frac{\tau \Delta M/M}{R - 1} \qquad (12.12)$$

A more accurate and general relationship for the reservoir stability measure Δt may be derived as follows:

Starting from an instantaneous mass value M_o, input flux F_{in} increases the reservoir mass to M that is some fraction (α) of the difference between the steady-state and initial masses:

$$M = M_o + \alpha(M_{ss} - M_o) \qquad (12.13)$$

Substitution for M from Equation 12.13 in the solution of Equation 12.8 gives:

$$\Delta t = -\tau \ln (1 - \alpha) \qquad (12.14)$$

The latter indicates that for any value of fractional change in the reservoir mass (α), the longer the residence time, the more stable is the system with respect to a new environmental stress F_{in}.

The values of $\Delta t/\tau$ from the two Equations 12.12 and 12.14 are plotted in Figure 12.6. For small values of the fractions $\Delta M/M$ and α, the following approximate equality holds in Equations 12.12 and 12.14:

$$\alpha \approx \frac{\Delta M/M}{R - 1} \qquad (12.15)$$

From Equation 12.14, a change of $\alpha = 0.1$ is a change in the reservoir mass by 10% of the difference between its present value and a model derived steady-state value. Thus for $\alpha = 0.1$, the time required for the change is

$$\Delta t = 0.1\tau$$

and it is directly related to the residence time of the substance in the

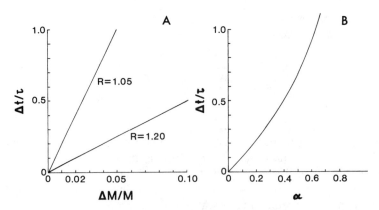

Figure 12.6. Time (Δt) needed to attain fractional change in reservoir mass shown as $\Delta M/M$ or α. Parameters explained under Equations 12.12–12.14.

reservoir. The time is independent of the reservoir mass or the steady-state mass value. If the system is far from a steady state, then a 10% change corresponds to a relatively large increment ΔM; in a system close to a steady state, $\alpha = 0.1$ would correspond to a relatively small ΔM value.

The simple model (Equation 12.10), with a constant input F_{in} and residence time with respect to removal τ indicates that as M becomes bigger, the rate of change $(\Delta M/M)/\Delta t$ becomes smaller. If a measured rate of change behaves differently, either increasing with time or remaining constant, then this may indicate that either the parameters F_{in} and/or τ are functions of time, or the model as a whole does not apply to the system.

Constant or nearly constant rates of increase in atmospheric CO_2 are identifiable in the CO_2 atmospheric concentrations of the last 200 years (Table 12.1). If the mass increase rate, as given in Equations 12.10 and 12.11 was constant for some period of time, this implies that the difference

$$\frac{F_{in}}{M} - \frac{1}{\tau} = \text{constant} \qquad (12.16)$$

Because M is increasing with time, the difference can be constant only if input flux F_{in} and/or residence time τ also increases.

In general, the rate of change of dM/dt can be approximately linear when the time considered is much shorter than the residence time in the reservoir, Equation 12.8 ($t/\tau \ll 1$). This, however, is not the case for atmospheric CO_2: its residence time $\tau \approx 3.5$ years is much shorter than the periods of years to decades of the nearly constant increase rates.

12.4 Analysis of Environmental Stresses in Carbon and Phosphorus Cycles

For our analysis of environmental stresses in the coupled land–ocean C and P global cycles we constructed the cycle diagrams shown in Figures 12.7 and 12.8, respectively, that emphasize those organic and inorganic reservoirs most likely to be affected globally. In the figures, the masses (in moles) and residence times (in years) of C and P are given for each reservoir.

Each cycle consists of a terrestrial and an oceanic part. The terrestrial part includes the reservoirs of inorganic soil, humus, and land plants. The oceanic part is made of the coastal and open-water sections. In the near-shore section, the reservoirs include coastal water, coastal sediments, and coastal biota. In the open-water section, the surface ocean, deep ocean, and oceanic plankton are connected by two-way flows with the corresponding reservoirs of the coastal section. The atmospheric reser-

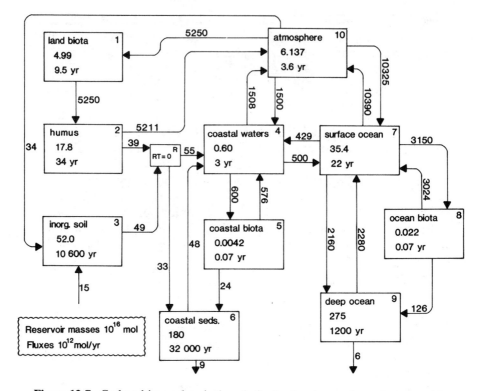

Figure 12.7. Carbon biogeochemical cycle for land and oceanic environment in a steady state. Notation: reservoir masses in units of 10^{16} moles C, fluxes in units of 10^{12} moles C yr^{-1}, and residence times of carbon in years. Data summary in Tables 12.2 and 12.3.

voir is shown only in the carbon cycle, because the atmospheric reservoir of phosphorus is negligible in the present context (cf. Figure 12.4).

The masses of the individual reservoirs of carbon and phosphorus are summarized in Table 12.2. In the biotic reservoirs on land and in the ocean (reservoirs 1, 5, and 8), the residence times of carbon and phosphorus are equal, because the cycles of the two elements are coupled in the living biota. In the inorganic and dead organic matter reservoirs, the residence times of the two elements are different, owing to the differences in their geochemical pathways in the environment. The relatively short residence time of C and P in the land biota may be noted (9.5 years; Bolin and Cook 1983). A residence time of about 40 years for land plants is a longer estimate based on the amount of carbon contained in tree trunks, leaves, and other types of global vegetation.

The flows between the reservoirs are indicated by arrows in Figures 12.7 and 12.8. The magnitude of the flux is shown next to the arrow. In the carbon cycle, land plants (reservoir 1) obtain their carbon from the atmo-

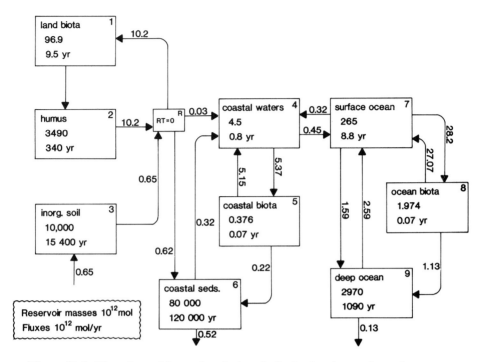

Figure 12.8. Phosphorus biogeochemical cycle for land and oceanic environment in steady state. Notation: reservoir masses in units of 10^{12} moles P, fluxes in units of 10^{12} moles P yr^{-1}, and residence times in years. Data summary in Tables 12.2 and 12.3.

Table 12.2. Reservoir sizes for land and ocean biogeochemical cycles of carbon and phosphorus

Reservoir	Carbon (moles)	Phosphorus (moles)
1. Land biota	4.99×10^{16}	9.69×10^{13}
2. Humus	17.8×10^{16}	3.49×10^{15}
3. Inorganic soil	5.2×10^{17}	1.0×10^{16}
4. Coastal waters	6.0×10^{15}	4.5×10^{12}
5. Coastal biota	4.2×10^{13}	3.76×10^{11}
6. Coastal sediments	1.8×10^{18}	8.0×10^{16}
7. Surface ocean	3.54×10^{17}	2.65×10^{14}
8. Ocean biota	2.205×10^{14}	1.974×10^{12}
9. Deep ocean	2.75×10^{18}	2.97×10^{15}
10. Atmosphere	6.137×10^{16}	

See Figures 12.7 and 12.8.

sphere (reservoir 10). Dead plants become humus (flux from 1 to 2), and both humus and inorganic soil supply carbon to the rivers, box R. The rivers are treated as a flux but not as a reservoir; therefore no residence time value is assigned to box R. Some of the humus is oxidized, returning carbon to the atmosphere. The flux from rivers to coastal sediments (flux from R to 6) is the flux of suspended material; the flux of carbon in solution carries it from rivers to coastal waters.

In the oceanic section of the cycle, the water reservoirs are connected by two-way fluxes that are responsible for exchange of dissolved carbon among them and with the atmosphere. The coastal biota and oceanic biota derive their carbon from the water reservoirs and return part of it to them in the process of respiration. The remaining part of the dead biota sinks to the bottom as particle fluxes transporting carbon from the biotic reservoirs to the sediments.

Table 12.3. Rate constants (k_{ij}) for fluxes in the carbon and phosphorus cycles

	k_{ij}	C	P
	k_{12}	0.1053	0.1053
	k_{2R}	2.19×10^{-4}	2.92×10^{-3}
	k_{210}	0.0293	—
	k_{3R}	9.42×10^{-5}	6.50×10^{-5}
	k_{45}	#	1.1933
	k_{47}	0.0833	0.1
	k_{410}	0.2513	—
	k_{54}	13.7	13.7
	k_{56}	0.571	0.585
	k_{64}	2.67×10^{-5}	4.0×10^{-6}
	k_{74}	1.21×10^{-3}	1.21×10^{-3}
	k_{78}	#	0.1064
	k_{79}	6.1×10^{-3}	6.0×10^{-3}
	k_{710}	0.0294	—
	k_{87}	13.7	13.7
	k_{89}	0.571	0.572
	k_{97}	8.29×10^{-4}	8.72×10^{-4}
	$k_{9\text{-out}}$	2.18×10^{-6}	4.38×10^{-5}
	k_{103}	5.54×10^{-4}	—
	k_{104}	0.0244	—
	k_{107}	0.1682	—
(#) Constant/fluxes:	F_{101}	5.25×10^{15}	
(moles/year)	F_{45}	6.00×10^{14}	
	F_{78}	3.15×10^{15}	
	$F_{\text{out-3}}$	1.50×10^{13}	6.5×10^{11}
	$F_{6\text{-out}}$	9.00×10^{12}	5.2×10^{11}

Units of k: yr^{-1}; refer to Figures 12.7 and 12.8.

The flux from outside to the inorganic soil reservoir (15×10^{12} moles C yr^{-1}) is an estimated rate of addition of carbon from sedimentary limestones ($CaCO_3$) to soils. The flux of CO_2 from the atmosphere to soils is responsible for the chemical weathering of the soil reservoir.

The oceanic section of the phosphorus cycle (Figure 12.8) is in principle similar to the carbon cycle. In the land section, the flux to inorganic soils (0.65×10^{12} moles P yr^{-1}) from outside the system represents addition of phosphorus to soils from the weathering of sediments on the continents. The uptake of phosphorus by land plants (reservoir 1) is at the expense of the humus and inorganic soil reservoirs, from which phosphorus is released to soil and ground waters, to be taken up by the biotic reservoir before its entry into the rivers.

The external fluxes of C and P to the inorganic soil reservoir were taken as constant and independent of the sizes of the soil and source sediment reservoirs. At present, the growth of land plants and of aquatic plankton is not limited by concentrations of CO_2 in the atmosphere and of HCO_3^- in ocean waters. Therefore, the carbon fluxes supporting the biotic reservoirs on land, in coastal waters and in the surface ocean are shown in Figure 12.7 as constant and independent of the reservoir mass. All the remaining fluxes in the two cycles are first-order fluxes, the rate constants of which are listed in Table 12.3.

The cycles drawn in Figures 12.7 and 12.8 are balanced: the sums of the flows into each reservoir are equal to the flows out. The extent of adjustment of the raw data that was involved in "creating" balanced cycles lies well within the uncertainties in estimates of the masses of the individual reservoirs and fluxes (e.g., Sundquist 1985). A balanced cycle, however, is a basic structure of the models that make it possible to weigh individual processes within the system and to analyze the effects of environmental stresses on it.

12.5 Stresses and Perturbations in the Carbon and Phosphorus Cycles

For the C and P cycles we chose to analyze four major perturbations:

1. Increased rate of weathering on land;
2. Cessation of new plant growth on land;
3. Cessation of new growth in the ocean; and
4. Increased weathering and reduced net primary productivity on land.

The four perturbations may seem extreme at first glance. However, such drastic changes for our linear-flux models make it easier in an analysis to focus on the directions and rates of change within the individual element cycles. An extreme perturbation used in model analysis strives to isolate the cycle from some of the processes, and it helps identify their controlling role in the cycle. A choice of gradual changes, rather than abrupt, for any of the four processes listed above might have given our analysis a greater semblance of realism. But, it would have also required considerable greater detail in definition and justification of the rates of change in weathering, and in net primary production on land and in the ocean. By introducing "instantaneous" changes in these processes in our models of the carbon and phosphorus cycles, we produce a picture that helps us identify the controlling roles of the individual processes (weathering, net primary productivity) and some of their combinations. For a more advanced analysis of such perturbations, considerable mathematical nonlinearity may have to be built into the model. For example, extreme changes in the rates of new plant growth on land or in the ocean, and extreme changes in the rates of chemical weathering, may affect the fundamental nature of some of the fluxes in the cycles: the rate constants (k_{ij}) may be different and the fluxes may no longer be controlled by the first-order mechanisms (Equation 12.2) that are used in our model.

An increased rate of weathering on land (an increase by a factor of three was taken in the model) can, in principle, be a result of natural causes as well as of human activity. The natural causes, such as an increase in the mean elevation of land above sea level, have manifested themselves many times throughout the geological history of the earth. The human causes are likely to be, primarily, deforestation, cultivation, and open strip mining.

Two drastic perturbations denoted above as cessation of new plant growth on land and in the oceans are extreme cases of possible changes in the biotic reservoirs that also produce chemical stresses in the reservoirs connected to the biota.

The fourth perturbation combines high rates of weathering on land with a lower rate of biological productivity on land. Conceivably, these perturbations could be the result of stresses induced by acidic precipitation from the atmosphere that adversely affects plant growth, but at the same time causes a faster weathering of soils and rocks and mobilization of potentially toxic trace metals into aquatic systems. On a smaller local scale these conditions might also result from overharvesting, analogously to the conditions that were discussed for perturbation 1.

The mathematical models used for the analysis of each of the four perturbations are based on Equation 12.4, the initial reservoir masses given in Table 12.2, and the rate constants k_{ij} given in Table 12.3. The procedure consists of solving sets of simultaneous ordinary differential equations, and the results give the reservoir masses as a function of time. The individual fluxes can be computed from the values of the reservoir masses and the rate constants using the definition of flux in Equation 12.2.

Discussion of the results obtained from perturbations of the cycle models follows.

12.5.1 Increased Rate of Weathering on Land

Increased rates of weathering in the carbon and phosphorus cycles correspond to greater fluxes involving inorganic soils and humus, the atmosphere and coastal water. An increase in the weathering fluxes by a factor of 3 is represented in the mathematical model by multiplying by 3 the rate constants of the three land reservoirs (1, 2 and 3; the weathering rate constants are k_{3R}, k_{2R} and k_{210}, listed in Table 12.3). The resulting changes in the terrestrial, aquatic, and biotic reservoirs of carbon and phosphorus are plotted in Figure 12.9A and B for a length of time of about 250 years.

In the carbon cycle, the higher weathering rate leads to withdrawal of more CO_2 from the atmosphere, and an initial decrease in the size of the humus reservoir.

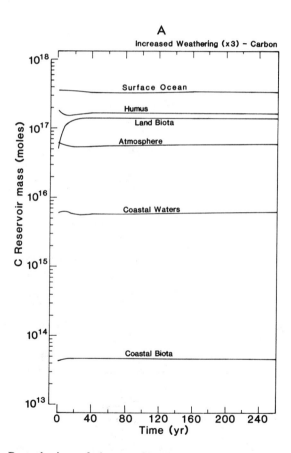

Figure 12.9. Perturbation of the steady-state carbon and phosphorus cycles: model of increased weathering rate on land. **A** Carbon. **B** Phosphorus (see p. 338). Changes computed from the model analyses shown for the reservoirs most affected by the perturbation

338   A. Lerman, F.T. Mackenzie, and R.J. Geiger

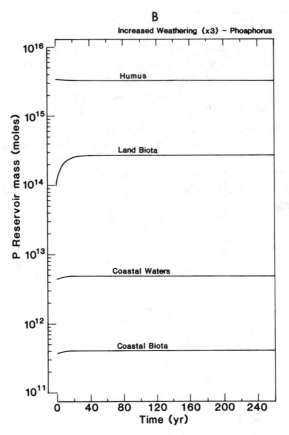

Figure 12.9 (*Continued*)

The surface ocean and coastal waters (reservoirs 4 and 7) show at first a slight decrease (less than 10 percent) in their carbon contents, owing to a fast exchange of CO_2 between surface waters and the atmosphere. Subsequently, the carbon contents of the aquatic reservoirs and of the humus begin to rise toward a new steady state.

The most pronounced changes occur in the land biota. A greater flux of phosphorus released from the soils promotes plant growth at the expense of atmospheric CO_2, resulting in a nearly threefold increase in the standing crop of land plants.

Higher fluxes of dissolved carbon and phosphorus into the coastal waters (reservoir 4) also produce an increase in the standing crop of the coastal biota, but by a much smaller amount than on land.

The different magnitudes of responses of the biotic and inorganic reservoirs under the stress of a higher weathering rate indicate different degrees of reservoir stability to the environmental perturbation. The time required to achieve a 10 percent change in the four phosphorus-cycle reservoirs under the new weathering conditions are (from Equation 12.14 and the residence times given in Figure 12.8):

Land biota	$\Delta t = 0.1 \times 9.5 = 1$ yr
Humus	$\Delta t = 0.1 \times 340 = 34$ yr
Coastal water	$\Delta t = 0.1 \times 0.8 = 0.1$ yr
Coastal biota	$\Delta t = 0.1 \times 0.1 = 0.01$ yr

The shortness of these estimates reflects a feature peculiar to the model: rivers transport materials "instantaneously" from land to ocean, as shown in Figures 12.7 and 12.8.

It follows from the preceding results that the P reservoir of humus is the most stable of the four with respect to a change in the global weathering rate. The standing crop of land plants shows low stability, in the sense of experiencing a rapid increase in mass owing to much higher fluxes of carbon and phosphorus. If the residence time of land plants were taken as 30 yr, instead of 9.5 yr (see Figure 12.4), then the reservoir stability would have been greater, about 3 years. The coastal reservoirs of water and biota respond very rapidly to the new perturbation, as might have been expected on the basis of their very short residence times.

12.5.2 Cessation of New Plant Growth on Land

Analysis of this extreme perturbation enables us to focus on the controlling role of the land plants in the C and P cycles. (An even more extreme case could be represented by total cessation of primary production. Then the rates of decay of the biotic reservoirs would have been faster than those derived for net primary production.) The reservoirs most affected under these conditions are the standing crop of land plants and the dead organic matter in soils. The reservoirs of land biota and humus decay without being replenished. The decay of each reservoir, in its carbon and phosphorus contents, is plotted as a function of time in Figure 12.10A and B. A faster rate of loss of carbon than phosphorus from the humus reservoir reflects the different pathways of the two elements: carbon is oxidized to CO_2 whereas phosphate is retained in soils.

The reservoirs that receive flows from the land biota and humus initially all experience increases in their masses of P and C. The increase results from redistribution of phosphorus and carbon from the land plants and humus among the reservoirs located downcycle. Carbon is transported to the atmosphere, coastal waters, and surface ocean, and phosphorus goes to the oceanic reservoirs. The increase in atmospheric CO_2 (Figure 12.10A) may be noted as an analogy for land deforestation, overharvesting, and other methods of reduction of the vegetation cover.

As a result of higher flows to the ocean, coastal and oceanic biota increase at the expense of phosphorus from the land reservoirs. Subsequently, however, as the land reservoirs become exhausted and the supply of phosphorus to coastal waters diminishes, the coastal reservoirs readjust themselves downward. A factor that could modify these results

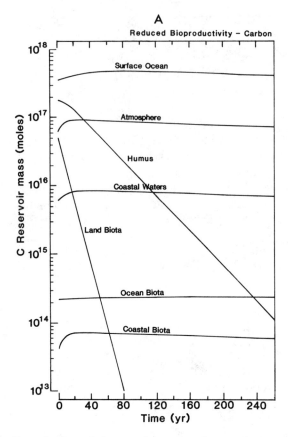

Figure 12.10. Perturbation of the steady-state carbon and phosphorus cycles: model of cessation of new growth on land. **A** Carbon. **B** Phosphorus. Note the decay of the land plants and humus reservoirs

would be greater retention of phosphorus in inorganic soils. As the pool of phosphorus in decaying humus is released to running waters, according to the model in Figure 12.8, there is no greater retention of phosphorus in the land reservoirs. If the phosphorus released from the biota and humus reservoirs, however, were to be retained for a longer period of time in inorganic soils, then the increase rate in the phosphorus content of the coastal reservoirs might be much smaller.

12.5.3 Cessation of New Growth in the Ocean

The biotic reservoirs in the ocean are much smaller than on land. Thus, cessation of new growth of phytoplankton in the coastal and open ocean reservoirs would release only relatively small amounts of phosphorus and

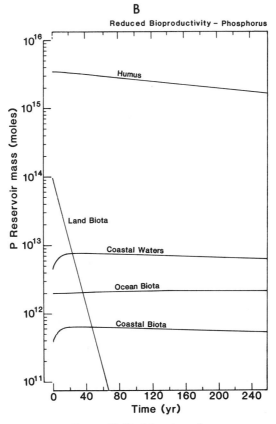

Figure 12.10 (*Continued*)

carbon to surrounding waters. In the schematic cycles of Figures 12.7 and 12.8, the masses of biotic reservoirs 5 and 8 are small in comparison to the water reservoirs 4 and 7. For phosphorus, however, sedimentation of dead organic matter is an important mechanism of its removal from the oceanic reservoirs as indicated by the fluxes F_{56} and F_{89} in Figures 12.7 and 12.8. This mechanism would no longer operate after bioproductivity in water had ceased. Consequently, phosphorus brought in from land will accumulate up to some limit in surface ocean waters. Increase in the mass of phosphorus in the aquatic reservoirs, shown in Figure 12.11B, is a result of weakening of its flow to the biological sinks.

On a global scale, the mass of oceanic primary producers is very small, and its decay would not amount to any significant drain of oxygen from the atmosphere. A different picture may arise in ocean water: the mass of oxygen dissolved in surface ocean waters is the same order of magnitude as the mass of carbon in coastal and open ocean biota, whose decay may locally become a significant drain on dissolved oxygen that would be replenished through exchange with the atmosphere only after some time.

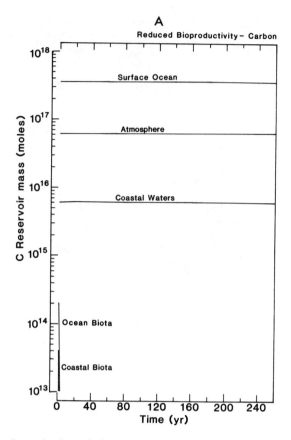

Figure 12.11. Perturbation of the steady-state carbon and phosphorus cycles: model of cessation of new growth in ocean waters. **A** Carbon. **B** Phosphorus. Changes computed from the model shown for the reservoirs most affected by the perturbation.

12.5.4 Increased Weathering and Reduced Net Primary Productivity on Land

In this perturbation, discussed at the beginning of the section, the rate of weathering on land is increased by a factor of 1.5, and the rate of net primary production is decreased by a factor of 4. The results of the model analysis are plotted in Figure 12.12A and B. This perturbation combines in it elements of 1 and 2 and, as such, may be expected to produce some intermediate results between the two.

Associated with a decrease (but not cessation) of biological productivity on land, the land, the biota, and humus reservoirs decrease in size,

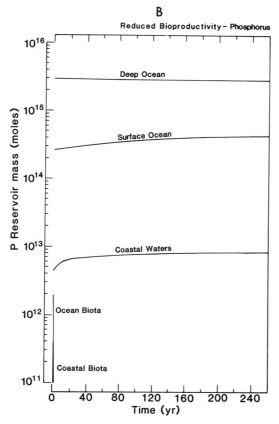

Figure 12.11 (*Continued*)

stabilizing at lower levels. This effect is pronounced in the carbon reservoirs (Figure 12.12A), where the residence time of carbon in the organic matter of soils (humus) is shorter than the residence time of phosphorus. The downcycle reservoirs—atmosphere and the aquatic reservoirs—absorb the excess of carbon and phosphorus released by the reduction in size of the land reservoirs. The biotic reservoirs in the ocean benefit from the higher fluxes of nutrients from land, similarly to the first two perturbations discussed above. As the supply of nutrients from the land diminishes, the aquatic reservoir contents go through maxima and decline toward new lower values. Some elements of this scenario may have taken place during the last 200 years. During this period of time the earth may have experienced higher global denudation rates (Judson 1968; Garrels and Mackenzie 1971). There are also some arguments that the terrestrial biomass may have diminished (e.g., Siegenthaler and Oeschger 1987). Whatever the case, future human influences on the environment could make the direction of this scenario correct, even if not its magnitude.

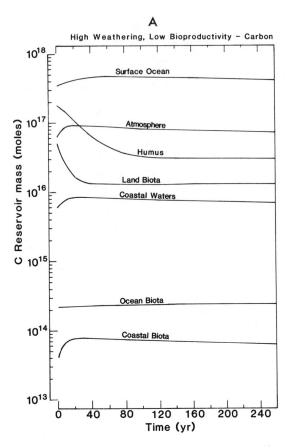

Figure 12.12. Perturbation of the steady-state carbon and phosphorus cycles: model of higher weathering rates and lower net primary production on land. **A** Carbon. **B** Phosphorus.

12.6 Sensitivity of Nutrient Flows to Biotic and Mineral Controls

In the preceding discussion of the carbon and phosphorus cycles, and of the effects of their perturbations, the conclusions were based on sets of representative values characterizing reservoir masses and their residence times, derived from a wide range of estimates of numerous investigators (e.g.; Garrels et al. 1975; Lerman et al. 1975; Sundquist 1985). In reality, the reservoir masses and fluxes within a biogeochemical cycle may be expected to vary in time within some bracketing limits, irrespective of whether the estimates used in the model are, or are not, representative of the true means.

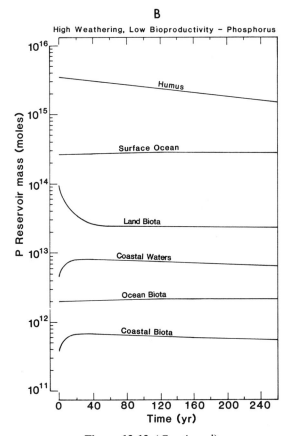

Figure 12.12 (*Continued*)

The flux of phosphorus from land is derived from the mineral and organic fractions of rocks and soils. To estimate the contributions of each to the riverine flux of P, one possible approach would be to make a sensitivity analysis of the cycle model for different values of residence times (τ) or rate constants (k) for the land reservoirs of P. Another approach is to start not with fixed values of reservoir masses (M) and residence times (τ), but with bracketing upper and lower limits of their ranges.

The reservoir masses of phosphorus in plants, humus and inorganic soils, as drawn in Figure 12.13, and their residence times determine the magnitude of the phosphorus flux out of three reservoirs, according to Equation 12.2:

$$\text{Flux out} = (M/\tau)_{\text{plants}} + (M/\tau)_{\text{humus}} + (M/\tau)_{\text{minerals}} \qquad (12.17)$$

A sought value of a *maximum* flux, when the values of M and τ are known to lie between some limits, can be obtained by a numerical optimi-

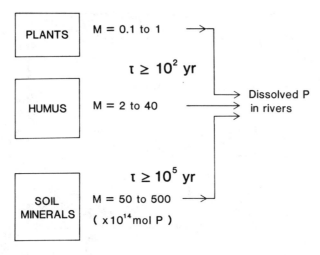

Figure 12.13. Model for riverine flux of phosphorus derived from land plants, humus, and mineral fraction of soils. Equations 12.16 and 12.17. Reservoir masses (M) shown as bracketed by lower- and upper-limit estimates (units of 10^{14} moles P). Lower limits of residence times (τ) discussed in the text.

zation procedure (known as a "first problem in linear programming"). The value of the phosphorus flux in rivers, fed by the release from the plants, humus, and soil minerals, is obtainable by maximizing the sum of the terms on the right-hand side of Equation 12.17:

$$F_{rivers} = \max \left\{ \sum_i M_i k_i \right\} \tag{12.18}$$

The numerical solution is based on a set of M_i values and a set of k_i values or their combinations that are specified only as bracketed by upper and lower limits: for example, $10^{-7} < k_{mineral} < 10^{-4}$ yr^{-1}.

Figure 12.13 shows a diagram of the three land reservoirs and the bracketing lower and upper estimates of their phosphorus contents and residence times. The residence times of the living biomass and humus reservoirs are combined in one value, for two reasons. First, the mass of phosphorus cycled between the standing crop of land plants and the humus reservoir is much greater than the amount released by the mineral components of the soils (Figure 12.8). Thus, if the short residence time (9.5 to 40 years) were characteristic of the rates of release of phosphorus to the streams, then the entire flow from land to the coastal zone would be dominated by the biotic component. Second, some fraction of dead land plants can move through the cycle without becoming humus; occurrences of complete plant remains, parts of tree trunks, and leaves are common in river, lake, and coastal ocean sediments. This undecayed residual of the standing crop is some small fraction of the total, and its release of phosphorus and of other chemical constituents is likely to take place on a time scale comparable to the rates of release from humus. Thus, for a long-term picture, it is of interest to learn how the variations in the rates of release from the mineral soil and from the organic components of the soils (humus plus undecayed plants) control the flow of phosphorus from terrestrial to coastal oceanic ecosystems.

The question "what reasonable combinations of M_i and k_i values for the P reservoirs on land may bracket the present-day flux of P in rivers, about 3×10^{10} moles P yr^{-1}," is answerable by solving Equation 12.18. The combinations of different ranges of the rate constants used for numerical solution and the resulting fluxes are given in Table 12.4.

The results in the table show that values comparable to the river flux are obtainable under two very different sets of conditions. One, the river flux value of between 3.0 and 4.5×10^{10} moles P yr^{-1} can be maintained by a combination of very low release rate from the organic reservoirs and a high release rate from the soil mineral reservoir. Other conditions that produce a phosphorus river flux similar to the present are a lower rate of release from the soil mineral reservoir and a higher rate of release from the organic reservoirs. Extreme conditions—that is, any combination of either very low or very high release rates from the organic and mineral reservoirs—produce flux values that are far removed from the present day value. Surprisingly, the computed fluxes that best match the phosphorus river flux, indicate that the residence times of phosphorus in the organic and the mineral reservoirs are generally longer than the residence times used in this chapter and in the literature on the modern day biogeochemical cycles of phosphorus. One of the possible conclusions that may be drawn from this analysis is that the effective reservoirs of available

Table 12.4. Phosphorus flux in rivers owing to release from the organic (plants and humus) and inorganic (soil minerals) reservoirs, Figure 12.13. Fluxes given in units of 10^{10} moles P yr^{-1} for different combinations of the decay and weathering rates, as given in equations (12.17) and (12.18). Rate parameters (yr^{-1}): k_p for the plant reservoir, k_h for humus, and k_m for soil minerals.

$k_p + k_h$ ＼ k_m	10^{-7}	10^{-6}	10^{-5}
$0 - 10^{-8}$		0.45	**3.9 – 4.5**
10^{-5}		0.85	5.0 – 5.5
10^{-4}	**3.0 – 4.5**	**3.5 – 4.8**	7.7 – 9.4
10^{-3}	30.0 – 45.0	32.0 – 45.0	36.0 – 48.0

phosphorus on land are much smaller than the total masses of P contained in the standing crop of land plants, humus and soil minerals. From the results of Table 12.4, the residence times of P in the organic reservoirs are of an order of 10^4 to 10^5 years, and in soils about 10^6 years; both values are much longer than the residence times computed from the reservoir contents and the flows (Figure 12.8). Insofar as the flows out of the organic and soil mineral reservoirs include both dissolved and particulate materials, the residence times of phosphorus in them, as given in Figure 12.8, are of the order of 10^2 to 10^4 years. For the flux of dissolved phosphorus alone ($F_{R4} = 3 \times 10^{10}$ moles P yr^{-1}), the residence times of the reservoirs become similar to those estimated by the numerical optimization procedure: for plants plus humus, 1×10^5 moles yr^{-1}, and for the inorganic soils, 3×10^5 moles yr^{-1}.

12.7 Conclusion

The global biogeochemical cycles of carbon and phosphorus (and of other biologically important elements) are coupled by the land and aquatic biota. The coupling of the cycles is reflected in the chemical composition of the photosynthetically produced organic matter, the main six elemental components of which are C, H, O, N, S, and P. The choice of the C and P cycles for a mathematical model analysis is based on the roles of the two elements in global ecosystems: carbon is the main structural chemical component of the biota, and phosphorus, albeit less abundant, is an important and usually growth-limiting component because of its low abundance in natural waters.

Starting with the balanced-model cycles of carbon and phosphorus in the terrestrial and coastal oceanic environment, the response of the cycles to model perturbations was studied by means of numerical analysis. The perturbations assumed to affect the cycle were the following: high weathering rates on land; cessation of new biological growth on land; cessation of new biological growth in the ocean; and higher rates of weathering combined with lower rates of productivity on land. These perturbations and their results, as obtained from the models, should be viewed as limiting cases of such environmental stresses as human industrial and agricultural activity, deforestation and cultivation, reduction of vegetation cover on land, increase in acidity of atmospheric precipitation, and massive kills of photosynthetic communities either on land or in the ocean.

It is important to emphasize that our models of environmental stress based on increased weathering rates and/or lower bioproductivity on land indicate an increase in atmospheric CO_2 on a time scale of decades. The magnitude of the increase (Figures 12.10–12.12) obtained from an oversimplified linear response model is only a guideline for the direction of the process as a whole.

In terms of the cycle models, higher weathering rates on land result in greater fluxes of dissolved phosphorus and, consequently, in increases in the land and aquatic biomass. The greater uptake of carbon from the atmosphere, needed for the greater weathering of soil mineral components and for the greater standing crop of land plants, results initially in some decrease in the amount of atmospheric and surface-ocean CO_2.

Cessation of new plant growth on land causes redistribution of phosphorus from the plants and humus among the downcycle reservoirs, resulting in an increase in productivity in coastal and surface ocean water.

Cessation of new plankton productivity in the ocean adds very little of carbon and phosphorus from the biotic pool to the surrounding waters. A more important effect of this perturbation is curtailment of the sedimentation flux of dead organic matter to the floor of the coastal zone. Interruption of the sedimentation fluxes of the dead plankton results in a gradual increase in concentrations of dissolved phosphorus in ocean water. Ultimately, such increases in concentration would be controlled either by solubilities of inorganic phases or by restoration of biological productivity.

The present-day flux of dissolved phosphorus in rivers is a net result of phosphorus release from plants, humus, and soil minerals. The masses of the reservoirs of the living plants, dead organic matter and soil minerals are known only within some ranges bracketed by lower and upper limit estimates. The same uncertainty applies to estimates of the residence times of phosphorus in the individual reservoirs, or to the rate constants of the fluxes. A numerical optimization technique was used for lower and upper limit estimates of the reservoir masses and residence times in order to compute an optimal flux of dissolved phosphorus under different combinations of mass and residence time values. The results show good agreement between residence times of P in the organic and mineral reservoirs, that underlie the present day flux of dissolved phosphorus in rivers, with the global estimates used in construction of the cycle models.

Acknowledgments. Hawaii Institute of Geophysics Contribution No. 1907 ACS-PRF Grant 17371–17C2 (FTM). We thank Mark A. Harwell and Jack Kelly for a thorough review of the initial manuscript of this paper.

References

Atkinson MJ, Smith SV (1983) C:N:P ratios of benthic marine plants. Limnol Oceanogr 28(3):568–575

Bolin B (1986) How much CO_2 will remain in the atmosphere? The carbon cycle and projecting for the future. In: Bolin B, Jager J, Doos BR, Warrick RA (eds) *The Greenhouse Effect, Climatic Change and Ecosystems, Scope 29.* Chichester, UK: Wiley and Sons, 574 pp.

Bolin B, Cook RB (1983) (eds) *The major biogeochemical cycles and their interactions, Scope 21.* Chichester, UK: Wiley and Sons, 554 pp.

Deevey ES Jr (1973) Sulfur, nitrogen, and carbon in the biosphere. In: Woodwell GM, Peacan EV (eds) *Carbon and the Biosphere*. USAEC, Washington, DC pp. 182–190

Delwiche CC, Likens GE (1977) Biological response to fossil fuel combustion products. In: Stumm W (ed) *Global Chemical Cycles and Their Alterations by Man*. Dahlem Konferenzen, Berlin, pp. 73–88

Garrels RM, Mackenzie FT (1971) *Evolution of Sedimentary Rocks*. New York: Norton, 397 pp.

Garrels RM, Mackenzie FT, Hunt C (1975) *Chemical Cycles and the Global Environment*. Los Altos, California: Kaufmann Wm, 206 pp.

Hoffman JS, Keyes D, Titus JG (1983) *Projecting Future Sea-Level Rise: Methodology, Estimates to the Year 2100, and Research Needs*. EPA-230-09-007 121 pp.

Houghton RA, Boone RD, Fruci JR, Hobbie JE, Melillo JM, Palm CA, Peterson BJ, Shaver GR, Woodwell GM, Moore B, Skole DL, Myers N (1987) The flux of carbon from terrestrial ecosystems to the atmosphere in 1980 due to changes in land use: Geographic distribution of the global flux. Tellus 39B:122–139

Houghton RA, Schlesinger WH, Brown S, Richards JF (1985) Carbon dioxide exchange between the atmosphere and terrestrial ecosystems. In: Trabalka JR (ed) *Atmospheric Carbon Dioxide and the Global Carbon Cycle*, Vol ER-2039, US Dept of Energy, Washington, DC, pp. 113–140

Idso SB (1982) *Carbon Dioxide: Friend or Foe?* IBR Press: Tempe, Arizona, 92 pp.

Judson S (1968) Erosion of the land. Am Scientist 56:156–374

Lerman A (1979) *Geochemical Processes—Water and Sediment Environments*. New York: Wiley and Sons, 481 pp.

Lerman A, Mackenzie FT, Garrels RM (1975) Modeling of geochemical cycles—phosphorus as an example. Geol Soc Am Mem 142:205–218

Mackenzie FT (1981) Global carbon cycle: Some minor sinks for CO2. In: Likens G (ed) *Flux of Organic Carbon from the Major Rivers of the World to the Ocean*. US DOE Conf Rept 80089140, pp. 360–384

Mackenzie FT, Bischoff WD, Paterson V (1983) Biogeochemical cycles and trends in estimates of inputs of anthropogenic chemical constituents to the environment. Cornell University, Ecology Research Center Report 27

Meybeck M (1982) Carbon, nitrogen, and phosphorus transport by world rivers. Am J Sci 282:401–450

Peterson BJ, Melillo JM (1985) The potential storage of carbon caused by eutrophication of the biosphere. Tellus 37B:117–127

Redfield AC, Ketchum BH, Richard FA (1963) The influence of organisms on the composition of seawater. In: Hill MN (ed) *The Sea*. New York: Wiley and Sons, pp. 27–77

Rotty RM (1987) A look at 1983 CO$_2$ emissions from fossil fuels (with preliminary data for 1984). Tellus 39B:203–208

Siegenthaler U, Oeschger H (1987) Biospheric CO$_2$ emissions during the past 200 years reconstructed by deconvolution of ice core data. Tellus 39B:140–154

Sundquist ET (1985) Geological perspectives on carbon dioxide and the carbon cycle. Geophysical Monograph 32:5–59

Woodwell GM, Whittaker RH, Reiners WA, Likens GE, Delwiche CC, Botkin DB (1978) The biota and the world carbon budget. Science 199:141–146

Chapter 13
Biomonitoring: Closing the Loop in the Environmental Sciences

Edwin E. Herricks,[1] David J. Schaeffer,[2]
and James A. Perry[3]

For the purposes of this chapter we define biomonitoring as *the analysis of the performance of living systems structured to provide essential information for decision making*. Although some form of biomonitoring has been used for hundreds, if not thousands of years (Herricks and Schaeffer 1984), a cohesive approach to the design and use of biomonitoring programs is a recent development. This chapter discusses biomonitoring programs, emphasizing program designs that coordinate data collection and facilitate information retrieval. We highlight the role of biomonitoring in the context of environmental management and regulation. We suggest that both control theory and decision science can be used to refine biomonitoring programs to optimize the use of data in the environmental sciences, management, and regulation.

Biomonitoring can be divided into two categories (Herricks and Schaeffer 1984): bioassays and bioassessments. *Bioassays* are laboratory based tests that incorporate rigorous experimental protocols. The most common bioassays are toxicity tests, although any experimental manipulation of a biological system with appropriate controls can be considered a bioassay. *Bioassessments* are field-based analyses that lack strict experimental controls. Bioassessments range from description of organisms present in a community or ecosystem, to the measurement of a range of ecosystem properties and processes. Bioassessments may include the use

---
[1] Department of Civil Engineering, University of Illinois, 208 N. Romine, Urbana, Illinois 61801

[2] Department of Veterinary Biosciences, University of Illinois, 2001 S. Lincoln, Urbana, Illinois 61801

[3] Department of Forest Resources, University of Minnesota, 1530 N. Cleveland Avenue, St. Paul, Minnesota 55108

of experimental manipulation of contaminants, habitat, and ecological relationships in ecosystems but depend on uncontrolled reference areas to assess the consequences of the manipulation.

Selection of a bioassay or a bioassessment approach for a biomonitoring program is dependent on the complexity of the problem addressed. When it is possible to simplify environmental relationships, or when management or regulation questions only require development of dose-response information for single chemicals or simple mixtures, a bioassay approach is preferred. Where simplification is not possible (e.g., when many populations or multiple trophic levels are assessed) or confounding factors are introduced (e.g., complex contaminant mixtures or uncontrollable environmental variability), bioassessment is the preferred program design.

Bioassays and bioassessments are regularly used in applied problem solving. For example, biomonitoring is a central focus for water quality regulation and management (USEPA 1979, 1982, 1985); effluent bioassays are extensively used in discharge monitoring to determine compliance with stream standards (USEPA 1979). However, the use of bioassays in compliance monitoring has raised concern because of the limited understanding of the precision or accuracy of biomonitoring data (CMA 1984).

One of the major limitations of current applications of bioassays is the inability to accurately predict ecosystem consequences following contaminant release to the environment (Osborne 1982; USEPA 1986). Further, an emphasis on descriptive assessments and network surveillance programs has diverted attention from broader issues of environmental management to narrowly focused debates over methodological issues (GAO 1981, 1986). We feel that this concern over data issues such as precision and accuracy masks more important issues. Emphasis should be placed on sensitivity, reproducibility, and variability of the methods used (Herricks and Schaeffer 1985) and the suitability of the data collected to answer the regulatory or management questions asked (Perry et al. 1986).

Biomonitoring methods produce data, but biomonitoring *programs* provide information for decision making. Properly designed biomonitoring programs produce information using a formalized process of data collection, analysis, and interpretation. That formal process can benefit from advances in other disciplines, particularly control theory and the decision sciences. In the following sections we review methods of structuring biomonitoring programs, incorporating approaches suggested by control theory and the decision sciences.

13.1 Biomonitoring Programs for Ecosystems

If biomonitoring program information is used to support decisions regarding the management and regulation of ecosystems (Hellawell 1978), then biomonitoring methods must be improved beyond standard laboratory (e.g., enzyme function, genetic alteration) or field (e.g., enumeration of ecosystem components, impact assessment) analyses. Accurate interpretation of ecosystem manipulations requires an understanding of ecosystem complexity, the effects of homeostatic processes and natural variability, and the effects of varying temporal and spatial scales. The approach we recommend for quantitative analysis of ecosystem state condition is the use of an optimal battery of tests or test systems (Schaeffer and Janardan 1987) organized in tiers of increasing cost or complexity.

Application of tier testing to ecosystems is based on a series of assumptions: First, principles governing ecosystem structure and function are knowable and will support experimental analysis. Second, it is possible to quantify ecosystem state and condition and to define ecosystem characteristics that serve as indicators of state or condition using existing knowledge or data. Third, indicators may be measurements of a single trait or may be developed from multiple measurements made on subsets of ecosystem properties or processes that are highly correlated. Fourth, indicators of state or condition may be applicable to any ecosystem or may be applicable only in the ecosystem in which the indicator was defined and quantified. Fifth, identification of both connection between ecosystem components and interlevel relationships in an ecosystem hierarchy (Allen et al. 1984; O'Neill et al. 1986) is possible, thus making it possible to extrapolate from species and population effects to system-level consequences.

Hazard evaluation, presently used in environmental management and regulation (Cairns et al. 1978), depends on tier testing using different batteries of tests in each tier. Hazard evaluation tiers include a *Screening tier,* which employs simple and relatively inexpensive tests to make a preliminary determination of hazard and risk. Screening is followed by the *Predictive tier,* which employs more complex, longer-term testing to predict environmental and ecological effects produced by contaminant release. *Confirmatory tier* testing involves environmental and ecological analysis to determine actual effects and to assess prediction accuracy. *Monitoring tier* analyses require long-term surveillance to identify low level or unanticipated long-term effects.

Ecosystem analysis requires an expansion of tier testing. We can identify three levels of analysis within each tier: At the highest, or *Definitive level,* we quantify properties or processes that exactly define the state or condition of an ecosystem. We recognize that it is possible to quantify ecosystem characteristics, but the exact definition has eluded ecologists; indicators of state and condition are still being defined. *Definitive level* analyses are seldom obtained for any tier. The next level of analysis, *Classification,* requires quantifying properties that are being, or which could be, used to classify state or condition of an ecosystem. Classification may be based on criteria developed to meet regulatory requirements or more arbitrarily on perceptions of environmental quality. *Classification level* analyses may include measurements made through time, or the use of test systems which have appropriate experimental controls. The lowest and most common level, *Predictive,* involves analyses intended to predict Classification or Definitive level responses in an ecosystem. *Predictive level* testing is usually comprised of laboratory and field-oriented test systems that have unknown or uncertain relationships to ecosystem indicators, or are not used in developing interlevel relationships. We recognize that most currently available test systems operate at the Predictive level.

The recognition of levels of analysis within tiers simplifies the organization of data collection activities to structure biomonitoring programs. Test systems can be selected to provide experimental validation of effect (Herricks and Schaeffer 1987). Further, the applicability of laboratory experimentation to ecosystems can be clearly defined through use of tier testing and nested levels of analysis. It is also possible to develop biomonitoring programs employing bioassessment approaches that take advantage of natural or human-caused manipulations of ecosystems (Cairns 1984). Careful analysis of bioassessment data from either planned or unplanned manipulations of ecosystems can transform a simple descriptive study into a definitive analysis of ecological consequence. Experimentation on a small scale or manipulation on a large scale is the basis for hypothesis development. Although hypotheses can not always be rigorously tested through bioassessment, the outcome of similar experimental manipulations can be compared if the data collection effort adheres to defined quality assurance procedures.

In summary, tier testing that incorporates levels of analysis in each tier provides an organized approach to data collection in ecosystems. Tier testing also makes it possible to identify scientifically valid experimental or manipulative procedures for ecosystems. Tests of data sensitivity, reproducibility, and validity, using both experimental and descriptive analysis, are possible in properly organized and planned biomonitoring programs.

13.2 Improving Biomonitoring Programs

Biomonitoring has been criticized because important questions are left unanswered following program completion and because existing programs often consist of surveys undertaken only to provide a series of observations in time (GAO 1981). We feel data inadequacies in biomonitoring programs will continue unless program structure and design are improved. The remainder of this paper examines how approaches from control theory and the decision sciences can be used to improve biomonitoring programs.

13.2.1 Control Theory

Control theory provides a structure and mathematical description of the operation of complex, dynamic systems through analysis of system input, output, and feedback (Melsa and Schultz 1969; Bell 1973; Landau 1979). Input to a controlled system determines the characteristics of the output. A simple control problem improves system output to meet an established objective. An essential step in selection of the control process is the identification of operational objectives with corresponding constraints (inherent or artificial). More complex control problems are encountered when attempts are made to optimize system performance. An optimal control problem requires a model of the system, a set of inputs, a desired output, and a measure of performance (Athans and Falb 1966).

Elementary control theory distinguishes between *open-loop* (Figure 13.1a) and *closed-loop* (Figure 13.1b) systems. An open-loop system operates with a controlled output, but that output is not used as a basis for system control. In a closed-loop system, measures of output are used to modify system controls to change output characteristics. Both open-loop and closed-loop control processes are designed to meet an established output objective. An open-loop system is dependent on the capacity of prediction algorithms, operating without feedback, to meet output objectives. Closed-loop systems can have less sophisticated prediction algo-

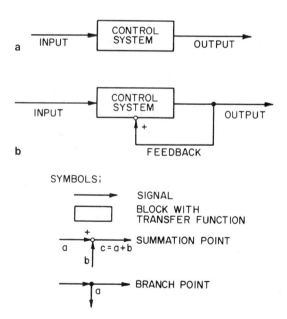

Figure 13.1.a Simple open-loop control system; **b** simple closed-loop control system.

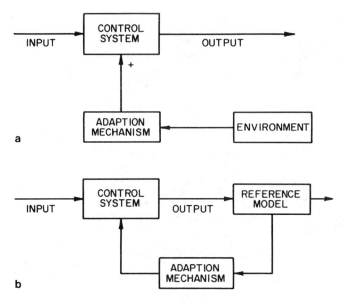

Figure 13.2.a Simple open-loop control system with an adaptive mechanism; **b** simple closed-loop control system with an adaptive mechanism.

rithms because feedback allows system adjustments, which can be tested against output objectives.

Recent advances in control theory have included adaptive control (Landau 1979), which can be used with both open-loop and closed-loop systems (Figure 13.2a and b). In adaptive control, a modification of system characteristics is made based on external measures or an index of performance. In an adaptive, open-loop system, some measure of the system environment is used to modify predictive algorithms. An example is a heating system that uses outside air temperature to determine heating cycles within a building. In an adaptive, closed-loop system, a comparison of output performance against a set of desired performances drives a control system to maintain output at a desired performance level. Adaptive control processes avoid degradation of the dynamic performance of a system in the presence of environmental variation.

Hill and Durham (1978) suggested the direct application of control theory to ecosystems is limited because ecosystem complexity makes the isolation of specific control problems difficult; further, the inability to establish a single controlling factor as a primary feedback mechanism precludes the direct application of advances in engineering control to ecosystems. Unlike ecosystem modeling, biomonitoring provides specific control problems with identifiable feedback mechanisms. Control theory is useful in biomonitoring because it improves the formal structure, and provides a mathematical formalism which assists in structuring control relationships. In addition, biomonitoring program design benefits from an understanding of factors that lead to destabilization of control systems, and the selection of control systems types to meet identified output objectives.

Examples of Control Theory in Environmental Management

An example of an open-loop system is provided by the existing National Pollution Discharge Elimination System (NPDES) permit program of the Clean Water Act. A major controlling variable in NPDES permits is the dilution flow of the receiving stream. In each permit the effluent concentration of selected parameters is set so that ambient stream standards are not exceeded at some minimum predicted flow. Although natural flow variability will produce different dilution flows, the capacity to assimilate higher contaminant loadings is not controlled by the NPDES permit. NPDES permits are open-loop systems because system control is based on some prediction of expected events that sets a single wastewater concentration that does not vary based on actual stream concentration.

Similarly, ecosystem models are often open-loop because predictive algorithms are used without continuous feedback. Although ecosystem models may incorporate new mathematical approaches to quantify causes of change (Patten 1983) or the formulation of mathematical proofs to support nodes of interaction (Levins 1981), the complexity and individual character of many models preclude regular updating of predictive equations or model coefficients. Sensitivity analyses used for model calibration is not adequate feedback to shift these models from open to closed-loop systems, since feedback is not continuous. Further, as ecosystem models are applied to more complex problems in ecosystems, the factors that must be considered in the prediction algorithm multiply. If the variables that affect system output are not completely modeled or if prediction algorithms are not constantly updated using information from output characteristics, the control system is open-loop. Open-loop control systems will not track system variability. Open-loop ecosystem models are subject to destabilization by a change in environmental conditions.

The major difference between closed-loop and open-loop systems is the use of feedback to control system output. For example, to close the loop for NPDES permits, continuous measurement of effluent parameters in the receiving stream are needed to provide feedback to modify effluent concentrations. An example of a closed-loop system used in biomonitoring is illustrated by "in-plant" biomonitoring systems (Cairns 1980). An "in-plant" approach uses a fish or other organism to detect changes in effluent characteristics that may damage the receiving stream (Predictive tier, Predictive level). When the biomonitor is stressed, that signal is used as feedback information (Predictive tier, Classification level) to modify effluent characteristics.

Herricks and Cairns (1979) introduced the concept of prediction-control (modeling) and error-control (monitoring) in relation to information flow in environmental control processes. Herricks (1984) expanded the initial concept and illustrated its utility in the use of monitoring data in river basin management (Figure 13.3). We suggest that control theory can be used to develop systems of information feedback in environmental management providing a formal basis for the structure and organization of monitoring programs that consider information input and output from decision algorithms. This change will lead to the improvement of prediction and monitoring to increase the effectiveness of biomonitoring programs.

Further improvements in open-loop systems are produced by inclusion of adaptive mechanisms (Figure 13.2a). Examples of adaptive control in river basin management and environmental impact assessment are available. In river basin management, engineers have used the Streeter–Phelps equation (Phelps 1944) to predict dissolved oxygen (Screening tier, Classification level) and develop regulation and management strategies (Thoman 1972). The use of dissolved oxygen sag analysis provides an excellent example of adaptive modification of a fundamentally open-loop system. Since 1925 engineers have regularly performed sensitivity analysis on predictions and developed an extensive empirical data base to support the use of the Streeter–Phelps type of analysis (Zison et al. 1979). This data base has contributed to the development of an adaption mechanism (selection of coefficients and exponents from reference material rather than stream sampling) for use when models are applied. Although stream monitoring results may be used to assess the sensitivity of model predictions,

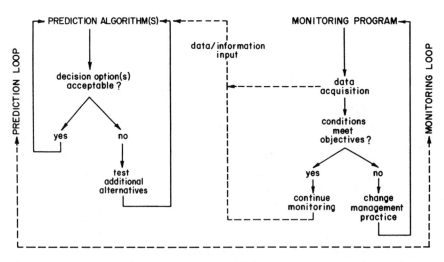

Figure 13.3. River basin management model incorporating predictive modeling and feedback using monitoring information.

the system is still open-loop because continuous monitoring is not used to modify prediction algorithms.

An example of an open-loop adaptive processes in environmental impact assessment is provided by Holling (1978). Adaptive Environmental Assessment and Management (AEAM) structures interactions among experts in analysis of environmental issues (e.g., ecologists, wildlife biologists, and geologists), agency managers, and computer modelers. The process explores a range of ecological, environmental, economic, and social factors that affect development. The goal of the process is to generate alternatives, identify effective policy, use indicators for decision-making, evaluate results, and communicate results to the public as well as to management and interested scientists. To improve impact assessment, AEAM uses a sophisticated delphi process which incorporates rapid feedback. Effects of project design or management approach are predicted using computer models. Model output is then subjected to evaluation by the group. After several iterations between scientists and managers and modelers (modeling results are the feedback in this system), an impact assessment is produced. The impact assessment is still an open-loop system because monitoring to verify predictions and change prediction algorithms is needed to close the loop. However, the algorithms used to generate predictions are adapted to the problem, assuring that prediction processes meet stringent performance criteria.

A closed-loop system with adaptive control processes is a model for river basin management (Figure 13.3). In this model, adaptive processes depend on monitoring data to modify predictive algorithms. Monitoring provides information on long-term trends in selected parameters and the natural or typical states or conditions of the system that are used as a performance index. Monitoring is also used to assess the state or condition of the ecological system, which is constantly compared with a performance index. If acceptable limits are exceeded, corrective action can be taken through a change in management strategy or regulatory response.

Unfortunately, success of the proposed river basin management model depends on observable change. In ecosystems, change is often masked by the natural variability or inadequate supporting data on the influences of temporal and spatial factors on ecosystem state or condition. When an unacceptable state or condition is detected, the processes causing that change may have sufficient momentum to cause further damage before correction to the control system is possible. Dependence on monitoring in environmental management creates a situation where failure to achieve performance requirements is likely. Response will always follow change by some finite time interval. Although it is possible to reduce the time interval between detection of an unacceptable level of performance and modification of the control system (Cairns and van der Schalie 1980), reactive systems will always allow damage and/or degradation to occur.

To prevent the ecosystem from sustaining unacceptable damage while improving the sensitivity of monitoring efforts, we suggest that biomonitoring programs must more effectively integrate modeling and monitoring. In addition, existing hazard evaluation in ecosystems can be improved if control system output is related to a tier and level. Such integration uses monitoring data to evaluate the validity and accuracy of predictions, relying on monitoring to provide the data for building models or structuring prediction algorithms.

Monitoring data are also used after models are developed to assess predictive accuracy and to assess the sensitivity of algorithms to environmental change. If models and monitoring are well integrated, prediction algorithms will be continuously updated with the availability of additional monitoring data. The dependence on appropriate models in biomonitoring can provide a cost-effective alternative to evaluation of the possible effect of changes in either input or output variables.

Better models will also improve monitoring. As prediction algorithms are improved, the focus of monitoring efforts will change, new parameters may be measured, and the control system will adapt to new performance indices. Improvement of monitoring techniques may also reduce the time delay between onset and observation of change promoting development of more efficient control strategies. Coupling prediction and monitoring loops focuses attention on critical issues while improving the probability of management and regulation success. Understanding the working level of prediction algorithms (in relation to tier and level testing) will also refine test system selection and the development of additional algorithms.

Time Considerations

Time is a critical issue for the application of control theory in biomonitoring. Two aspects of time must be considered: the time required for feedback signal processing and the inherent response time of organisms or communities. Although major advances have been made in shortening response time of the feedback signal (Cairns 1982), as discussed above, the measured response may inadequately protect ecological systems. The inherent response time of organisms and communities for important output variables is immutable and long. Because of this long response time, biomonitoring relies on trends to predict effect or models that transform state or condition data into predictions of long-term change.

Unless well designed, monitoring or surveillance programs will do little to improve the data base required for trend analysis or model sensitivity analysis. The time periods involved are also longer than the interests of the agency or organization supporting the monitoring effort. Seldom can data collection programs be sustained over a long enough time period actually to measure ecological system response or to provide conclusive evidence of long-term trends. It is rare in modern science to find a long-term environmental study that has been allowed to develop description and theory to the level where effective feedback, much less sophisticated adaptive control, can be implemented.

Space Considerations

Spatial factors are also important considerations in applying control theory in biomonitoring. One aspect of space can be dealt with as temporal, e.g., distance from a source of contamination in a stream can be assessed in terms of time-of-flow. A second aspect of space is independent of time and represents the influence of random external factors that may affect a control system. Bioassessments of communities or ecosystems may be spatially limited by ecosystem properties or processes that have different spatial scales. Movement into and out of an ecosystem should be accounted for in program design.

13.2.2 Decision Analysis

Our proposed definition of biomonitoring incorporated two elements. The first is measurement of the performance of the living system; that analysis can be improved using designs based on control theory. The second element of the definition recognizes the importance of the *use* of information in decision making. The use of information in biomonitoring can be improved through applications of decision theory, which involves making decisions under conditions of uncertainty while incorporating both rational and socio-psychological factors. Modern decision analysis ". . . studies the rational factor in order to clarify the way in which decisions should be made and does not comment, except perhaps incidentally, about how decisionmakers currently make decisions . . ." (Kaufman and Thomas 1977). The rational approach calls for a logical evaluation of alternative strategies and a systematic determination of the optimal strategy based on some criteria. A typical decision progresses through a number of stages that explore the problem and lead to an eventual decision.

Kaufman and Thomas (1977) identify the following stages of the decision analysis process: (1) structuring and decomposition of the decision problem; (2) assessing payoff (utility) values for decisions identified; (3) assessing the uncertainty of decisions identified; (4) identifying an optimum strategy; and (5) completing a sensitivity analysis to determine how robust the solution is when faced with variations in judgmental criteria. They also suggest that decision analysis is both an approach to decision making and a set of techniques that are applicable across a range of business, political, and other arenas.

An application of decision analysis in environmental monitoring is found in the Environmental Audit (Audit) as proposed by Schaeffer et al. (1985) (Figure 13.4). (The Audit in this reference should not be confused with contaminant auditing proposed by the U.S. EPA.) The Audit empha-

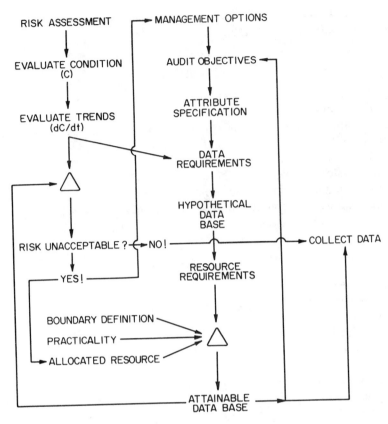

Figure 13.4. Information flow through the study design assurance process.

sizes development of clear management or regulatory objectives (structuring the decision problem), followed by the development of statistically valid sampling programs for the collection of data. The Audit also incorporates aspects of utility assessment and uncertainty analysis as the user completes several iterations through Audit elements, completing a sensitivity analysis of final monitoring design. The Audit has been modified to incorporate issues of optimality in biomonitoring (Herricks and Schaeffer 1985). The entire process, now identified as Study Design Assurance (SDA) (Herricks and Schaeffer 1986), forces a careful review of data requirements, resource requirements, and other factors that illuminate social, political, and economic issues. Incorporation of SDA principles into biomonitoring program design provides a basis for rational decision making.

The first step in the SDA process is identification of a management option or objective. Experience has demonstrated that management objectives are often broadly interpreted and inappropriate for use in biomonitoring (Perry et al. 1985). Biomonitoring program objectives must be focused. This focus is often achieved through interactions between scientists and managers (similar to initial steps in Holling's adaptive environmental assessment). Once a management objective is established, specific SDA objectives must be developed: these objectives are clear and testable, similar to hypotheses that generate from the statement of a general theory. An example of a management objective might be protection of biological integrity. One of the possible SDA objectives would be analysis of the reproductive success of a critical species to determine if the population can be sustained under existing or predicted conditions. It should be apparent that numerous SDA objectives can be developed for any single management objective.

The steps which follow identification of SDA objectives are narrowly focused to data issues. First, one or more traits, i.e., any characteristic of the organism, population, community, or ecosystem that can be expressed quantitatively (definition modified from Yablokov 1974), are selected for measurement. Following trait selection, variables are specified and the data required to meet an SDA objective are identified. At this point it is possible to identify a hypothetical data base that would meet specifications for sensitivity, reproducibility, and inherent variability (Herricks and Schaeffer 1985). The hypothetical data base is evaluated to determine if the required data can be obtained based on sampling logistics, costs, or other limits. At this time social, political, and economic factors are included. When an attainable data base is identified, the applicability of those data for assessing performance criteria and SDA objectives is assessed. The review and evaluation process at this stage of the SDA process may require several iterations through trait selection, hypothetical data base development, and attainable data base identification. When the attainable data base meets the specifications developed for the SDA, data can then be collected.

The SDA process incorporates elements of decision analysis and adaptive control approaches identified in Figure 13.2a and b. An SDA objective in this instance serves as a performance criterion, and the iterative process to define data collection procedures operates as the adaptive control process. The organization of the SDA process calls for iterative analysis at all stages in the management process, formalizing data collection to meet specific decision requirements. Adoption of SDA procedures can assure that a rigor, similar to that imposed by the scientific method in experimentation, is applied to biomonitoring.

The application of SDA principles to biomonitoring can assure success only if predictive capability matches monitoring effectiveness. Predictive capability is adequate when dealing with high-level, short-duration stress events. Changes in the numbers of species and population sizes (Confirmatory tier, Definitive level) are response measures that can be predicted for many chemicals (Kooijman and Metz 1983). Alteration of relevant ecosystem properties and processes and the importance of those alterations (NAS 1981) are much more difficult to predict because the response time of the system may be long, measurements made are of short duration, spatial issues become important, and ecosystems are very complex. If the concern is for low-level, long-duration stress effects, the reliability of predictions is low (Osborne 1982).

Where the reliability of predictions is uncertain, it is even more important that the formalized process that guides data collection and analysis also guide the formulation of realistic management and regulatory objectives. The lesson learned from use of the SDA in a number of biomonitoring designs is that standard data collection procedures are often inappropriate. We have found that it does ecotoxicologists, ecologists, and biomonitoring scientists little good to propose data collection when management objectives or SDA objectives are poorly defined (Perry et al. 1986). SDA provides a workable solution to the dilemma that was identified in Herricks and Cairns (1982). That is, decisions concerning the environment will be made with or without the benefit of ecological judgment. SDA provides a mechanism to modify objectives to fit the adequacy of the data, increasing the value of biomonitoring data in environmental decision making.

13.3 Ecotoxicological and Biomonitoring Systems

In the previous section we argued that biomonitoring can benefit from the application of control theory and decision analysis. We also suggest that environmental regulation and management can be improved if ecotoxicology can be linked more closely with biomonitoring program implementation. Applied problem solving requires integrating all efforts to produce a comprehensive understanding of chemical effects in ecosystems. This integration requires both a recognition of the utility of monitoring information and a change in experimental design for the development of models and prediction algorithms. Feedback allows prediction models to be refined by experimentation and validated by careful ecosystem analyses. From this process, practical tools for environmental management and regulation will emerge.

References

Allen TFH, O'Neill RV, Hoekstra TW (1984) Interlevel relations in ecological research and management: Some working principles from hierarchy theory. U. S. Forest Service General Technical Report RM-110

Athans M, Falb PL (1966) *Optimal Control.* New York: McGraw-Hill, 879 pp.

Bell DJ (1973) *Recent Mathematical Developments in Control.* New York: Academic Press, 446 pp.

Cairns J Jr, Dickson KL, Maki AW (1978) *Estimating the Hazard of Chemical Substances to Aquatic Life*. ASTM STP 657, 278 pp.

Cairns J Jr, van der Schalie WH (1980) Biological monitoring part I—early warning systems. Water Research 14:1179–1196

Cairns J Jr (1984) Freshwater Biological Monitoring: Keynote Address. In: Pascoe D, Edwards RW (eds) *Freshwater Biological Monitoring*. New York: Pergamon Press, pp. 1–15

CMA—Chemical Manufacturers Association (1984) Biomonitoring: A useful tool for industry, government. Chemecology, Washington, DC, 6 pp.

General Accounting Office (1981) Better monitoring techniques are needed to assess the quality of rivers and streams. Volume I. CED-81-30, United States General Accounting Office, Washington, DC, 121 pp.

General Accounting Office (1986) Key unanswered questions about the quality of rivers and streams. GAO/PEMD-86-6, United States General Accounting Office, Washington, DC, 163 pp.

Hellawell JM (1978) *Biological Surveillance of Rivers*. Water Research Centre, Stevenage, United Kingdom, 332 pp.

Herricks EE, Cairns J Jr (1979) Monitoring and mitigation of aquatic hazards. In: Fredrick ER (ed) *Control of Specific (Toxic) Pollutants*. Air Poll Cont Assoc Special Technical Publication, Washington, DC, pp. 220–231

Herricks EE, Cairns J Jr (1982) Biological monitoring Part III—Receiving system methodology based on community structure. Water Research 16:141–153

Herricks EE, Schaeffer DJ (1984) Compliance biomonitoring—standard development and regulation enforcement using biomonitoring data. In: Pascoe D, Edwards RW (eds) *Freshwater Biological Monitoring*. New York: Pergamon Press, pp. 153–166

Herricks EE (1984) Aspects of monitoring in river basin management. Water Sci Tech 16:259–274

Herricks EE, Schaeffer DJ (1985) Can we optimize biomonitoring? Environ Manage 9:487–492

Herricks EE, Schaeffer DJ (1987) *Selection of Test Systems to Evaluate the Effects of Contaminants on Ecological Systems*. Civil Engineering Studies, Environmental Engineering Series No. 71, UILU-ENG 87–2010. University of Illinois, Urbana, pp. 51

Hill J, Durham SL (1978) Input, signals, and controls in ecosystems. In: *Proceedings Conference on Acoustics, Speech and Signal Processing,* Institute of Electrical and Electronic Engineers, New York, pp. 391–397

Holling CS (1978) *Adaptive Environmental Assessment and Management*. New York: John Wiley, 377 pp.

Kooijman SALM, Metz JAJ (1983) On the dynamics of chemically stressed populations: The deduction of population consequences from effects on individuals. TNO Netherlands

Kaufmann GM, Thomas H (1977) *Modern Decision Analysis*. New York: Penguin Books, 507 pp.

Landau YD (1979) *Adaptive Control*. New York: Marcel Dekker, Inc, 406 pp.

Levins RA (1981) Ecosystem properties relevant to ecotoxicology. In: Working Papers Prepared as Background for Testing For Effects of Chemicals on Ecosystems. Washington, DC: National Academy Press, pp. 119–132

Melsa JL, Schultz DG (1969) *Linear Control Systems*. New York: McGraw-Hill, 621 pp.

National Academy of Sciences (1981) *Testing for the Effects on Ecosystems.* Washington, DC: National Academy Press, 103 pp.

O'Neill RV, DeAngelis DL, Waide JB, Allen TFH (1986) *A Hierarchical Concept of Ecosystems.* Monographs in Population Biology #23, Princeton: Princeton University Press, 253 pp.

Osborne LL (1982) EPA's recommended biomonitoring program: Assessment of potential for detecting low-level impacts on stream designated use and recommendations. Draft Report to AAAS/EPA Environmental Science and Engineering Fellowship Program, 65 pp.

Patten BC (1983) On the quantitative dominance of indirect effects in ecosystems. In: Lauenroth WK, Skogerbee GV, Flug M (eds) *Analysis of Ecological Systems: State-of-the-Art in Ecological Modeling.* New York: Elsevier, 992 pp.

Perry JA, Schaeffer DJ, Kerster HW, Herricks, EE (1985) The environmental audit. II. Application to stream network design. Environ Manage 9:199–208

Perry JA, Schaeffer DJ, Herricks EE(1986) Innovative designs for water quality monitoring: Are we asking the questions before the data are collected. ASTM STP-940, pp. 28–39

Phelps EB (1944) *Stream Sanitation.* New York: John Wiley and Sons, Inc., 276 pp.

Schaeffer DJ, Kerster HW, Perry JA, Cox DK (1985) The environmental audit I: Concepts. Environ Manage 9:191–198

Schaeffer DJ, Janardan KG (1987) Designing batteries of short-term tests with largest inter-tier correlation. Ecotox Environ Safety 13:316–323

Thoman RV (1972) *Systems Analysis and Water Quality Management.* Environmental Sciences Services Division, Environmental Research Applications, Inc, New York, 286 pp.

US Environmental Protection Agency (1979) Interim NPDES Compliance Biomonitoring Inspection Manual. MCD-62. Office of Water Enforcement, Washington, DC

US Environmental Protection Agency (1982) Water Quality Standards Handbook, Office of Water Regulations and Standards, Washington, DC

US Environmental Protection Agency (1985) Technical support document for water quality-based toxics control. Office of Water, Washington, DC, 74 pp. + appendices.

US Environmental Protection Agency (1986) Solicitation DL 86–003, Environmental Research Laboratory-Duluth, Duluth, MN

Yablokov AV (1974) *Variability of Mammals.* NTIS: Springfield, VA, TT 71–58007

Zison SW, Mills WB, Deimer D, Chen CW (1979) Rates, Constants, and Kinetics Formulations in Surface Water Quality Modeling. EPA-600/3-78-105, ORD-ERL: Athens, GA, 317 pp.

Chapter 14
The Role of Terrestrial Microcosms and Mesocosms in Ecotoxicologic Research

James W. Gillett[1]

14.1 Historical Perspective

14.1.1 Overview

Terrestrial microcosm technology is less than two decades old. Much of its efforts is based in plant growth chamber studies, soil column experiments, and environmental process measurements made in the field. Stemming from chemical, toxicological, and physiological studies begun in the latter part of the 19th century and applied widely in the first six decades of this century, microcosm technology has come to depend heavily on advanced analytical techniques, materials science, and sophisticated ecological field investigations.

Gillett and Witt (1979) reviewed terrestrial microcosms regarding the ability of selected systems to represent environmental chemistry and the fate and effects of toxic chemicals in terrestrial ecosystems. These reviewers concluded that substantial changes and improvements were needed before the laboratory model ecosystems would be as useful as initially anticipated. Subsequent review by others (Giesy 1980; Matsumura 1980; Harris 1980; Hammonds 1981a,b; Van Voris et al. 1983) presented increased evidence of accuracy, sensitivity, and reliability as the technology progressed. Although D. Goodman (1982) and Huckabee (1983) concluded that deficiencies still outweigh advantages in the application of either aquatic or terrestrial microcosm technology to the study of chemicals, a number of chemical manufacturers began using microcosm systems as management decision-making tools in the late 1970's.

[1] Institute for Comparative and Environmental Toxicology, Cornell University, Ithaca, New York 14853

They found that predictions made from microcosm test data were useful at a point in product development where choices on large expenditures of funds would have to be made. Though the naive over-expectations from early results of microcosms (Draggan 1976) have never been attained in terms of a single, generic system ubiquitously applicable to ecotoxicologic regulatory problems, terrestrial microcosm technology has achieved a decided usefulness.

Those unrealistic expectations instead prompted numerous practical approaches with more reasonable questions addressing practically needed answers. Cairns (1983), among others (e.g., Hammonds 1981b; Cairns et al. 1981; D. Goodman 1982), has often pointed out the deficiency in relying solely on single-species toxicity tests for regulation and environmental management of chemicals. Thus the need for practical systems with multiple species interactions at population and community levels of biological organization has been a strong driving force in the creation of model ecosystems.

Similarly, difficulties in developing mathematical models of chemical fate in terrestrial systems have resulted in the need for physical model systems in which the outcomes of complex and even counter-intuitive interactions of chemodynamic processes could be assessed. Model ecosystems provide a practical approach to obtaining such answers. Mathematical models of the fate of chemicals in terrestrial microcosms (e.g., Lindstrom and Piver 1984) are appearing in the scientific literature and represent an intermediate step in the complex technology of modeling and validation of chemical fate and effects. Additional interest in terrestrial microcosms lies in the need to observe both chemical fate and distribution simultaneously with effects on species and processes. In this role microcosm technology is an intermediate link between the laboratory and the field.

Microcosms and mesocosms are also intermediate links in the thread of ecological research and, as such, ought not to be considered an end in themselves. Rather they are parts of iterative and interactive steps taken to gain knowledge and form an understanding of the implications of that knowledge. This aspect is most clearly represented in development of mathematical models, where detailed information from the laboratory and the field are employed to synthesize quantitative statements, which in turn may be tested in, or used to design, model ecosystems. Data from microcosm and mesocosm studies may then lead to new questions in the lab and/or field and new design or operation of the model ecosystems.

The application of modern technologies to study the fate and effects of pesticides and other toxic chemicals with respect to environmental conditions and system composition appears both simplistic and sophisticated. On one hand, investigators may be employing state-of-the-art computerized control and data acquisition systems in carefully engineered structures, physically resembling a piece of the environment. On the other hand, the techniques and assumptions embodied in their operations may be mired in age-old problems of the heterogeneity of terrestrial environments and the complexities of interactive ecological processes. It is the confluence of attempts to simplify the many variables in environmental experimentation and the possibilities of more incisive measurements that has created the bewildering arrays of experimental systems termed model ecosystems, or microcosms and mesocosms.

14.1.2 Definitions

For the purposes of this exposition, *microcosms* are defined (Gillett and Witt 1979) as: ". . . a controlled, reproducible laboratory system which attempts to simulate the situation (i.e., processes and interaction of components) in a portion of the real world."

Thus, a microcosm has a boundary under the control of an investigator. The researcher can permit the system to be open, semi-closed, or closed to fluxes of air, water, and biota, to environmental conditions (light, temperature, humidity), and to content of soil or other components, as befits the knowledge and assumptions about the environment simulated, the experiment being executed, and, more importantly, the questions being asked.

Each of these points are important. Microcosms are not defined simply by the physical apparatus or the type of study performed. Rather, it is the comprehensive assemblage of the physical system and its operation within a scientific rationale.

Microcosms are also known as physical model ecosystems or laboratory model ecosystems, and by the name of the apparatus employed in the study. For example, Nash et al. (1977) termed their system a "microagroecosystem," and Gillett and Gile (1976) termed theirs a "terrestrial microcosm chamber" (TMC).

Terrestrial *mesocosms* will be defined for the purpose of this chapter as field engineered or constructed systems in which:

1. boundary conditions are imposed for fluxes of media (water and/or air), energy, and/or biota;
2. exposure to a chemical, pollutant, or other stressant is controlled by either the boundary condition or other apparatus within the experiment; and
3. the purpose of the experiment centers on ecosystem processes or the interactions of components structured into the test system.

Mesocosms have most of the same attributes as microcosms, but can be distinguished from microcosms by being much larger in scale, by being located outside of a laboratory setting, and by the lower degree of control exercised by the investigator. Alternative terms applied to mesocosms may be engineered field sites, enclosures, or exclosures, although each of these may not be operated as mesocosms.

14.1.3 General Considerations

Within laboratory experiments and model ecosystems there is a continuum of the scale and scope of experimental apparatus. In addition to location and degree of investigator control, one must examine the stated objectives of the experiment being performed: Are processes and their interactions in an ecological context the focus of the system? If so, then the experiment is more likely to be a study using a microcosm or mesocosm than a laboratory investigation.

Microcosm technology involves the application of many scientific disciplines and technologies. It shares in the growth of knowledge in all areas of environmental science—biology, chemistry, statistics, soil science, engineering, materials science, and applied mathematics. Many critical environmental problems cross media boundaries. Consequently, the distinctions between the technologies of aquatic and terrestrial model ecosystems are blurred, even before one approaches simulation of the interfaces (Gillett 1978; Matsumura 1980).

Generally, model ecosystems can function most effectively only in a broader context including detailed laboratory studies at the molecular level and extensive field investigations (Matsumura 1980). To set up a useful experiment, at least some knowledge of the chemical and physical properties of the test agent, and of performance of biota in their natural setting, is required to design the experiment properly. The range, variation, and type(s) of interaction of processes, both between the biota and among the physical components of the environment, should be recognized in advance or will need to be determined in preliminary experiments. The early idea, that microcosms would be a short-cut to all types of environmental information, and thus preclude the need for other laboratory and field study, is clearly false.

In this experimental context, microcosms and (to a lesser extent) mesocosms have demonstrated the comparative ease with which one may examine the outcomes of the interaction of components and processes that defy linear analysis in simpler laboratory bench studies or computer simulations. Under the appropriate circumstances, mass balance of chemical fate is readily followed. Ecological processes at the community or system level (nutrient cycling, soil respiration, population dynamics, and inter-species interactions) can be quantified. In a number of instances, microcosm experiments have shown the first quantitative view of the fate and effects of a widely used pesticide or industrial chemical. [For examples, see the work on wood preservatives (Gile et al. 1982; Gillett et al. 1983), on toxaphene (Nash et al. 1977; 1979), on 2,4-D (Gile 1983; Malanchuk and Joyce 1983), and on fly ash (Malanchuk et al. 1980; Van Voris et al. 1983).] It is in this experimental context that the ecotoxicologic value of the model systems to be discussed becomes apparent.

In the subsequent discussion of terrestrial microcosms, there will be opportunities to point out where measures of performance transcend the site-specific nature of a particular system, or where quantitative values from a generic system have some useful and even predictive power. Nevertheless, the outcomes observed may be very anecdotal. That is, within the power of the experiment to detect an effect, the specific results may only apply to that context. In this regard, microcosms are simply surrogates for the field, where identical limitations may apply, and we must place these data in an appropriate context of experience and criteria.

14.2 Microcosms as an Appropriate Technology

14.2.1 Benefits

Model laboratory ecosystems can be the most appropriate technology available to an investigator at a particular stage of research. Microcosms can be very incisive, cost-effective, and practical. The rationale in design and operation of microcosms places them between single species laboratory toxicity tests and chemical laboratory bench tests, where there is a high degree of investigator control and great freedom in simple variation of experimental conditions and components one at a time, and the field, where investigators have very limited control over components or conditions.

Safety considerations are important, and the controlled boundary of microcosms can protect investigators and the environment from potentially harmful materials. Radiolabeled chemicals or genetically engineered organisms of unknown pathogenicity are precluded commonly from field experimentation, but there are several microcosm designs readily applicable to both types of materials.

Another very practical consideration is that microcosms are the lowest level of ecological organization for which many processes or interaction of components can be observed. Physically locating a portion of the field within the laboratory affords the opportunity to make the types of measurements long sought in ecotoxicologic assessments. However, care must be taken to demonstrate the correspondence of these observations to field phenomena, where possible, as the results in the microcosm only represent a small facet of environmental fate and effects (Matsumura 1980).

Van Voris et al. (1983) demonstrated that, for evaluating effects of fly ash on phytotoxicity, macro- and micronutrient uptake, and crop yield, soil core microcosms (SCMs) are cost effective in comparison to pot studies in the greenhouse, with the additional advantage of the possibility of multiple season studies. In the second and third year, the SCMs were much more comparable to field plots than pots (Tolle et al. 1983; 1985). Ausmus et al. (1979) believe that this added realism also has at least a psychological benefit, making data from microcosms more persuasive than extrapolation of single species tests or other laboratory results to the field. In an effort to establish test protocols for ecosystem process measurements, the Office of Toxic Substances/USEPA has generated such for the SCM (Van Voris et al. 1985).

In summary, terrestrial microcosms have specific advantages over laboratory and field studies. It is an appropriate technology in the investigation of the fate and effects of a chemical or other environmental stressor, either prior to field studies or where more intensive study of particular processes is needed.

14.2.2 Limitations on Microcosms and Their Applications

Difficulties in evaluating microcosm experiments stem from the perception that either the investigator has considerable control over the system's performance or, alternatively, the investigator may be exerting inappropriate control, which leads to results different than those obtainable in the field. A particular microcosm system should be evaluated against an equivalent field system (Gillett and Witt 1979; D. Goodman 1982). Observers may anticipate achieving results fully consistent with those from the field, even though certain components had to be eliminated or significantly altered. Yet in order to bring the system into the laboratory, the consequences of such alterations may not be known to the investigators. These over-expectations lead to disappointment over, or rejection of, potentially valuable results.

What then are the limitations and reasonable expectations of a terrestrial microcosm system? The following, while not exhaustive, represent the main body of experience.

Duration

Microcosms are not self-sustaining and are too short-lived to demonstrate such ecologically significant processes as succession and other multi-generational/multi-species phenomena for higher organisms. They can, however, be operated for up to several growing seasons for plants (Van Voris et al. 1983) and are very appropriate for microbial processes (Anderson et al. 1978; Coleman et al. 1978; Cole et al. 1978; Anderson et al. 1979b; Shirazi 1980; Lighthart et al. 1982), for which species generation time is many-fold within the useful lifetime of a microcosm system. Most terrestrial model systems are directed at a single growing season (60–120 days), for which questions of duration are not significant.

Complexity

Limitations in physical size and in the extent and nature of inputs by the investigator control the selection of components and the degree of biological complexity involved in a system. Inclusion of more than three trophic levels (primary producer, consumer, and decomposer) is fraught with a host of logistical and theoretical problems. Yet elimination of even one component from the field system may significantly affect critical processes. Reduction in complexity may be highly desirable, if that were to permit analysis of a particular set of interactions and processes, even if unrealistic.

Generally, increased realism leads to decreased replicability (D. Goodman 1982). For example, a plant may not be able to complete its life cycle because it requires a pollinator (e.g., honeybee). Inclusion of that plant in a soil microcosm without the honeybee still permits study of rhizospheric interactions of chemicals, soil, soil microbiota, and plants (effects of photosynthate input, root channeling of soil, evapotranspirative flux of water, chemical fate, etc.). Soil microbial and invertebrate populations may be responsive to only a portion of the plant's life cycle (photosynthate release, litterfall, etc.), and that might be the only time when effects of a toxicant or other stressant on these seasonal processes can be observed.

In general, however, a lack of definitive knowledge of the role that biotic complexity plays in ecosystem health (Neuhold and Ruggerio 1976) severely limits interpretation of changes in complexity wrought by the experimenter. Because microcosms are typically compared within a limited range of diversity, there is little guidance as to whether deficiencies in complexity restrict usefulness in any practical manner.

14.2.3 Other Factors

Variation in Conditions

The costs of representing the natural stochastic changes in environmental conditions and biotic flux in the model system usually are very high. For example, a laboratory system achieving the complete energy spectrum of the sun, varying over the day and season, costs hundreds of thousands of dollars and is thus prohibitively expensive for most applications. Furthermore, our understanding of the effect is often limited to the average of those conditions. Representing these variations in controlled laboratory conditions may not be realistic, but reduced experimental operational complexity often simplifies interpretation of outcomes.

At times a simplified representation of the variation in natural conditions may be sufficient. For example, day length and intensity of insolation, both major determinants of plant physiologic response, appear to be adequately represented by simply controlled light sources when attention is paid to spectral balance and minimal and maximal light intensities (Mc-

Farlane 1978). However, artificial light sources for microcosms are stationary. They will shine on parts of the soil all the time they are on, and other parts of the soil will always be in shade. This contrasts with the circumstances in the field, where the angle and intensity will vary by time of day and season. An alternative therefore is to put the microcosms in a greenhouse (Nash et al. 1977; McFarlane et al. 1981; Van Voris et al. 1983). Nash et al. (1979) noted that in the greenhouse, capital investment and operating costs involved in the measurement and control of environmental conditions were 2–10 times the costs of the microcosms themselves, depending on the degree of sophistication of equipment and precision of control.

Microcosms are poorly structured for extreme excursions and cataclysmic events, although these could be represented. They are protected from incidental contamination, extreme conditions, or the many other features that typify adverse impacts in field studies. Most research—lab or field—fails to ascertain phenomena on the biogeochemical time scales and spatial scales needed for full extrapolation of results. To the extent that processes measured are truly and well represented, the data on outcomes involving them can be included in mathematical models of our understanding of impacts (Malanchuk and Kollig 1985).

As was true for deficiencies relative to biotic complexity, our lack of knowledge of the intrinsic relationships of environmental conditions and their range and rate of change to ecological outcomes is a limiting condition in the experiment. These relationships are known for only a few agronomic crop plant systems in the field, and little effort has gone into analyzing existing models for their sensitivity to anthropogenic inputs with respect to variations in conditions (Kickert 1984).

Type of Data Obtained

Just as important as any of the foregoing are considerations of the type and usefulness of data obtained from the microcosm experiment. The precise measurements of simple chemical or biological characteristics (e.g., the volatility of a pesticide or the feeding efficiency of an organism) cannot be made in a microcosm (Gillett and Witt 1979). These kinds of characterizations of chemicals or organisms are best handled in laboratory experiments designed for that purpose (Mill et al. 1980). However, if one wants to evaluate the air-borne loss of a pesticide from soil under a crop canopy in a system containing microbes and burrowing invertebrates, the microcosm, with its controllable boundary and potential for mass balance, becomes the logical place to examine the interacting phenomena.

Microcosm experiments also present problems in the generic assessment of the environments represented. The issue as related to spatial extrapolation is considered in the next section. Little effort has been devoted to such assessments over time, although they could contribute to long-term environmental assessment, as in a specimen banking program (Gillett 1985). Efforts at determining how long microcosm cores might maintain equivalency of function during storage (and the conditions thereof) and what temporal variation might be obtained in repeated tests of the same site over the years certainly could contribute to tracking environmental damage from yet unidentified sources.

The range of response for a test of a single species (e.g., EC_{50}, the concentration eliciting a given effect from half of a test population), tested in the laboratory over a wide range of environmental conditions and under circumstances of varying size, age, or physiologic status, may exceed that observed for all other species tested (Mount and Gillett 1982). This leads to the peculiar irony that the more a species is tested with a given chemical, the less confidence investigators can place in extrapolation of a specific laboratory result to a field situation where localized conditions may span the gamut of those tested. This has led to the use of large safety margins, "application factors," or similar manipulations of the test data in order to estimate outcomes in the field successfully.

Laboratory tests of a single species for intrinsic toxicity (i.e., μg of toxicant/g test organism) can attempt to assess the outcome over the range of conditions existing in a single microcosm environment. Because of the lack of control of exposure route and rate, similar testing cannot be readily performed in a terrestrial microcosm. On the other hand, the microcosm test may indicate directly for a specific treatment regime the presence or absence of a safety margin which can only be estimated as to order of magnitude from laboratory tests.

In an intact terrestrial microcosm many of the key factors which determine the outcome of competing chemodynamic processes (thus affecting exposure of a test species in the field) are on a $1:1$ scale. One may therefore conduct a bioassay for a chemical within the microcosm with a test organism for which a dosage–response relationship of the chemical has been established in the laboratory (Gillett et al. 1983) to determine the bioavailability of a toxicant to the test species. The numbers of test organisms needed for this bioassay are likely, however, to be greater than might be present at a field site. Thus, it is difficult to reach a quantifiable statement of safety or hazard in the microcosm without at least some distortion of conditions, population densities, etc., compared to the field. Similar attempts in the field, without a boundary, would be even more difficult.

The difficulty in extrapolation from the lab to either the field or the microcosm leads to a gestalt approach in development of indices of performance that may have to encompass many interacting processes. Metcalf et al. (1971) employed both an Ecological Magnification Index (EM, the ratio of the concentration of the parent material in a species to the concentration of parent as residue in the medium, e.g., soil or water) and a Biodegradation Index (BI, the ratio of the concentration of polar metabolites remaining in any organism or medium to the concentration of residual parent in that medium) to characterize the bioaccumulation and persistence, respectively, of chemicals added to his "farm-pond" system. The structure of that system precluded any use of the data to predict explicit concentrations in biota, water, soil, etc., in the real world. Nevertheless, these indices provide clear indications of the relative safety or hazard of the test chemicals.

Extrapolability

Because of limitations of scale, complexity, variability, and data specification, attempts to extrapolate directly from the microcosm to the real world can be frustrating. At the same time, much of what is included in terrestrial microcosms is on an explicit 1 : 1 scale with the real world. This begins with the heterogeneity of soil and its structure and continues through the plant–soil relationship on to the relationship of soil invertebrates in the plant rhizosphere. The dominant processes are contained within those relationships, starting with the relationship of microbiota to the soil water and organic matter in the soil crumb. This feature enables estimates of soil processes measured in microcosms to guide evaluation of energetics and nutrient dynamics (Anderson et al. 1978; Coleman et al. 1978; Cole et al. 1978).

Key issues in extrapolation are the appropriateness of the scale of the process observed (Shirazi 1980) and whether or not the process itself is linear (Wolfe et al. 1982). For example, simulation of erosion involves not only spatial scale (horizontal for flow, vertical for raindrop formation), but also the scaling factors of crop cover and soil structure. Nevertheless, a suitable microcosm (Lichtenstein et al. 1978; Lichtenstein 1980) can indicate the availability of an erosive pathway for pollutant flux from soil to surface waters. Even though the microcosm might have more or less erosion than an equivalent field, the question posed by the test was whether losses by erosion were occurring at all (Koeppe and Lichtenstein 1982, 1984).

We have observed at least two features that affect extrapolation. The most significant negative aspect is an out-of-scale process (e.g., the impact of a vole in a microcosm of 0.75 m^2 for 30 days when its territory is usually 10 m^2 for an adult for one season). The second, and more positive aspect, is the seeming intensification of certain processes, possibly as an interaction with the boundary itself. Apparent half-lives of chemicals determined in the TMC (Gillett and Gile 1976) are always shorter than anticipated from other measurements in the laboratory or reported in the literature. This difference could result from enhanced capacity of the relatively complete system, presence of additional pathways not accounted for in laboratory studies, and/or similar possibilities typical of the "attributive properties" of ecosystems (Odum 1971).

On the other hand, dieldrin residues in the TMC and in corn fields treated at the same rate and by the same method are very comparable (Gile and Gillett 1979). Tolle et al. (1983) report very similar values for plant uptake of components of fly ash and soil and for primary production in comparisons of SCMs and equivalent field plots with the same crops exposed identically and maintained as closely as possible in terms of environmental conditions. Although such comparisons may be anecdotal or fortuitous, the results reinforce the appearance of realism and thus our ability to extrapolate data of terrestrial microcosms.

Serious questions remain regarding spatial extrapolation for terrestrial environments, because the heterogeneity of the medium can lead to substantial variation even within a square meter (Gile et al. 1979). Thus both lab-to-site and site-to-site variations require increased study.

Hypothetical Construct

Selection of the hypotheses and assumptions under which a microcosm experiment may be performed are limited by our understanding of molecular events and field phenomena. When this understanding is weak, extensive side experimentation may be required to justify a simple choice, if that choice is potentially critical to the experimental outcome being studied. To the extent that assumptions are not well justified, the microcosm system will be properly criticized and results rejected.

This problem is not unique to microcosm technology per se, but appears as a consequence of the investigator control of the system. When researchers control inputs and outputs across a boundary, they are often held accountable for the outcome in a manner one would never demand of field studies.

This limitation on extrapolability has generally resulted in rejection of generic microcosms, but enhances the design and value of ad hoc or site-specific systems. As complexity is increased, the investigator must maintain a central focus, which can be rewarded by very pertinent information. The decisions that are paramount and that will control results include:

1. the physical and biotic structure of the system, including all materials used;
2. the method of treatment of the system or exposure to the stressant;
3. the experimental design and sampling schedule;
4. the endpoints selected and the exact measurements made; and
5. all of the operational variables (e.g., affecting environmental conditions) employed throughout the experiment.

Such attention also contributes to the use of microcosms of an appropriate scale for the assumptions. If several factors are to be evaluated, a very small scale system with substantial replicability must be used before higher order systems are employed.

14.3 Mesocosms as an Appropriate Technology

Microcosms and other laboratory systems at the upper limits of scale (size, complexity) have many costs and limitations. Investigators long ago turned to field plots and similar devices to evaluate anthropogenic influences on crops. The dedication of large preserves to ecological study, in this country and elsewhere, is an elaborate extension of such efforts. The boundary of the system may be a fence line, a watershed, or something wholly imaginary with respect to ecological processes.

Examples of types of terrestrial mesocosms include the "big bag" experiments in the forests of Puerto Rico (Odum and Jenkins 1970), open-top and canopied field chambers (Weinstein 1973; Lee and Weber 1983), zonal air pollution system (ZAPS) plots (Preston and Lee 1981; Lauenroth and Preston 1984), groundwater "field microcosms" (Wilson et al. 1981), and various exclosures, such as the 0.1-hectare plots used in some small rodent studies (Johnson and Barrett 1975; Anderson and Barrett 1982) and those employed in the apple orchard for the Pesticide-Orchard Ecosystem Model (POEM) (E.D. Goodman 1982; Jenkins et al. 1983; Goodman et al. 1983).

Mesocosms differ from field systems by having more substantial control of the environment and chemical or stressant exposure, but generally mesocosms rely on natural phenomena to provide environmental conditions appropriate to the study and with a greater degree of realism than attainable in the laboratory. At the extreme (where differences between field and mesocosm studies are blurred), the investigator is controlling only pollutant concentration and a very limited number of environmental variables.

A number of field chamber studies have been used to evaluate species and cultivar differences between plants exposed to air pollutants (Weinstein 1973; Heck et al. 1984a,b). The same types of systems have been employed for evaluating ecological interactions (Duchelle et al. 1982; Amundson 1983; Heggestad et al. 1985). These approaches have been compared to other studies in the field, greenhouse, or growth chamber (Lewis and Brennan 1977; Olszyk et al. 1980; Laurence et al. 1982; Davis et al. 1983). Improvements in design continue to be sought (Hogsett et al. 1985).

At the borderline between microcosm and mesocosm are the "field microcosms" of Wilson et al. (1981), 10-m deep soil microcosms which are operated outside of the laboratory for study of groundwater contamination. The technology is identical to that of smaller laboratory soil microcosm systems; they are principally process-oriented at microbial and physico-chemical levels. The conditions of experimentation in the field microcosms are practically identical to those of the surrounding soil zones (Wilson and Wilson 1985; Rees et al. 1985). However, no laboratory system includes the vadose zone (the unsaturated soil between the root zone and the water table), an area of critical interest in microbial ecology of groundwater.

14.3.1 Comparison of Terrestrial and Aquatic Systems

Aquatic and terrestrial microcosms share a number of features or concerns in common (such as those limitations discussed above), but terrestrial mesocosms differ radically from similarly scaled aquatic systems. The aquatic mesocosms generally must contend with operational problems similar to those of microcosms (i.e., epiphytic growth on the sides, mixing rate within the system, means of developing a representation of the benthos, and so forth). For these features, the scaling factors may be critical (Perez et al. 1977).

Terrestrial mesocosms are freed from many of the constraints of the laboratory systems. For example, if one wishes to examine inputs and effects of rainfall, wind, or sunlight, these are available without applying scaling factors. Exclosures can include or exclude even large herbivores. However, the costs of controlling energy inputs or fluxes of air and biota for terrestrial mesocosms can be very high. Specific intervention is required, and the systems may be subject to extreme situations, which is not altogether undesirable for realism.

To date, terrestrial mesocosms have concentrated on outcomes of controls on either air flux (and contaminants), water flux, and/or biotic flux, although there have been attempts at controlling all three. Such limited use of control reduces cost and complexity of operation and maintains realism relative to the field, so that replication of the test units is feasible. Thus, only a few mesocosm systems closely approximate test designs and operations in laboratory microcosms; many more resemble field studies. The increase in the number, density, and extent of measurements that may be required for statistical significance outside of laboratory control (depending in part on the nature of questions being asked) economically inhibits extensive proliferation of test units.

14.3.2 Benefits and Limitations

The appropriateness of the mesocosm is in the degree of control differentiating it from being simply another field plot. Construction of similar systems in different parts of the country or world permits analysis of effects of environmental conditions (soil type, weather and climate, water quality and quantity) in relation to a particular stressant. Typical of this approach are open-top or canopied field chambers (into which controlled amounts of air pollutants are introduced) and biological exclosures of various types. Because of close approximation to the field, most parameters are scaled 1 : 1, and the dependency on a rigid set of assumptions is greatly reduced. In certain cases, it is possible to use remote sensing data in conjunction with direct measurement ("ground truth") in the mesocosm to add further credibility to the mesocosm approach (Lauenroth and Preston 1984).

Mesocosms also have detractions, one of the most important being the cost of control of the boundary. Terrestrial mesocosm studies are either very expensive or suffer from some serious lack of control at the boundary, which might then permit toxic materials and hazardous organisms to emigrate from (as well as contaminate) the system.

Most terrestrial mesocosm systems tend to be more like biological "islands" than microcosms. They can be self-sustaining and demonstrate highly significant processes requiring long time periods or broader spatial scales. Variability and extrapolability appear close to field values (Mandl et al. 1980; Laurence et al. 1982), although the extrapolation may be hindered by the same lack of predictability as field studies when comparing one crop to another under identical conditions, or one place or set of conditions to those of other places for the same crop (Heck et al. 1984b). Mesocosms are relatively complete in terms of numbers and complexity of processes, even where there is some biological exclosure. Undoubtedly, the boundary introduces artifacts in and of itself (Lewis and Brennan 1977; Davis et al. 1983), but at times this can be discounted through experimental design (including physical structure, sampling, and endpoint selection).

Terrestrial mesocosm systems are relatively sparse, possibly due to the costs, including dedication of land resources. Only open-topped chambers enjoy wide-spread use. These are largely operated in a mode resembling greenhouse or growth chamber studies, rather than model ecosystems. The limited control and experimental design of the open-topped chambers provides, as a consequence, limited insight into factors affecting the outcomes observed. Lack of attention to ecological processes at lower levels of biological organization and to physico-chemical processes and their complications relative to the experiment is common, because terrestrial mesocosms are usually operated to examine effects on higher organisms and higher level processes. Most microcosm systems are limited to examination of the lower levels only. Failure of mesocosm experiments to incorporate studies at these lower levels, which may be viewed by experimenters as removal of a constraint in comparison to microcosms, may restrict ecological realism and extrapolability.

14.3.3 System Vulnerability

Mesocosms are subject to external catastrophe, including volcanic eruption, severe weather such as hail, high winds, and blizzards, and extremes of heat and cold. Although these factors can be anticipated in the design, the costs of physically securing a terrestrial mesocosm system are very high. Investigators therefore appear to construct systems that (a) have a low replacement cost, minimizing economic losses if destroyed and permitting increased replication, and/or (b) attempt to exert little control over natural forces.

The experience with the ZAPS (Zonal Air Pollution System) studies on the Colestrip coal-fired power plant complex in eastern Montana gives anecdotal evidence of these types of problems (Lauenroth and Preston 1984). These mesocosms were tracts of high prairie fenced off to prevent cattle, deer, and elk from stepping on the network of aluminum tubing delivering a pre-determined amount of SO_x to the test sites. In a five-year study, the sites were subjected to: (1) the coldest winter recorded in Montana, resulting in the extirpation of at least one vole species on the site and probably many terrestrial invertebrates; (2) one of the hottest summers on record, with a very protracted drought that severely affected reproductive success of birds studied on the sites and adjacent areas; (3) the importation of substantial quantities of DDT-R residues in migrating birds (especially insectivores, such as meadowlarks); and (4) the eruption of Mt. St. Helens, which deposited several inches of ash on the project and terminated it.

14.3.4 Alternatives to Mesocosms

Field studies of the ecological impact of toxic chemicals and, presumably, genetically engineered organisms require some sort of boundary. In large preserves, such as the National Laboratories at Oak Ridge and Savannah River, radioisotopes have been used in the field. The boundary is merely a patrolled fence line and a large buffer zone. Most investigators do not have access to such sites and must create more stringent containment.

One option is to employ stable isotopes; the other is to work on naturally occurring pollutants (O_3, SO_2, NO_x), unregulated stressants (heavy metals, acid deposition, anthropogenic wastes of one form or another), or registered pesticides applied at approved rates. Each of these can serve as surrogates for classes of restricted substances, while not unnecessarily constraining the investigator.

Similarly, for genetically engineered organisms (GEOs) criteria may be developed which define the boundary conditions and provide "containment" (Gillett 1986). The degree of control or management required in the field would be determined by assessment of the potential risks (hazard identification) and by criteria established for potential exposure to the organism from the test site. Many studies may eventually take place in the field with no substantial boundaries with respect to fluxes of air, water, energy, and biota. Other organisms will require considerable management, suggesting mesocosm investigations after greenhouse and/or laboratory microcosm studies. The equivalent of domed stadia in which to test

GEOs to be deliberately released in the field has also been suggested. That is an economically unattractive alternative for any system where many different types of organisms might have to be tested.

Containment or control may also involve additional manipulations of the organism. For example, genetic determinants of ecological interaction may be introduced into the genome, making the test organism dependant on specific conditions (nutrients, other organisms, environmental conditions) maintained in the test area—so-called biological containment. Alternatively, chemical controls (bactericide, herbicide) at the boundary may be used to create zones of containment. The usefulness of biological containment and the appropriateness of various means of control, mitigation, and remediation need immediate investigation (Gillett et al. 1985).

14.4 Relationship to Mathematical Modeling

Only one mathematical model of toxicant fate for a terrestrial model ecosystem (TMC) has been published (Lindstrom and Piver 1984). Mathematical models coupling fate and effects within microcosms have not appeared, but the relationships have encouraged investigators to make such links, at least in the planning and evaluation phases of both types of models (Malanchuk and Kollig 1985). Mesocosm studies or field engineered sites have contributed heavily to the modeling of surface runoff [Universal Transport Model (UTM; Patterson et al. 1974); Agricultural Runoff Management model (ARM; Donigian et al. 1977); POEM (E.D. Goodman 1982)] and inputs to groundwater [PESTicide ANalysis System (PESTANS; Enfield et al. 1982); Pesticide Root Zone Model (PRZM; Carsel et al. 1984)]. "Boundary conditions" ranged from fences and surficial water containment (UTM, ARM, POEM) to extensive well and lysimeter systems (PESTANS).

Mathematical models make important contributions to the design of microcosm experiments. They can be used to estimate exposure concentrations in portions of the system, help form test hypotheses, or evaluate and analyze data. In a few cases terrestrial microcosms and mesocosms have been used to confirm model estimates and act as an intermediate level of validation in the modeling process (Wilson et al. 1981; Gillett 1983; Lindstrom and Piver 1984). To date, POEM is the only physical system to be designed and operated by the mathematical modeler. Frequently, efforts have been thwarted by the complexity of interactions, by limitations on the ability of the experimenters to provide accurate parameters to the modeler, by demands created by the modeler for detail and accuracy beyond the ability of current methods and laboratory resources, and finally by simple lack of understanding of the processes involved. "Supercomputer" facilities may overcome the earlier limitations on model structure and process, which were imposed by deficiencies in computational speed and data storage capacity. This may radically affect both physical and mathematical modeling and result in stronger linkages.

For example, PESTANS and PRZM are relatively simple, one-dimensional models of pesticide movement to groundwater (vertical flux only). The processes involved in a chemical's movement in soil have been examined both individually and in combination using laboratory studies. Even when scientists know how all of these processes interact, the computational resources required for an explicit numerical solution to the simultaneous partial differential equations making up a three-dimensional model over the time periods of interest (years) at the computational increment significant to the processes (seconds to minutes) are enormous. Lindstrom (unpublished report, 1981) estimated that 10^{12} computations would be required per run. Use of the supercomputer with models with numerical solutions of this order would free experimental design from its present simplistic one-dimensional outlook (and thus permit tests for interactions), while allowing the experimenter to use more readily obtained, explicit measurements (time, space). Presently, two spatial dimensions may be modeled on minicomputers (Lindstrom and Piver 1986).

14.5 Ecotoxicological Applications of Microcosms and Mesocosms

14.5.1 Microcosm Studies

Chemical Fate

Most researchers agree that chemical mass balance in microcosms is an important feature both in qualifying a microcosm system for use in ecotoxicologic studies and in supporting direct studies of fate itself. When used in conjunction with radioactive test materials, microcosms provide the opportunity to track the chemical and its transformation products accurately over time and space.

"Bound" residues (those not extracted by normal methods) can be quantified by direct measurement, radioautography, or sample destruction (ashing, hydrolysis, etc.). Volatile residues can be distinguished from those undergoing mineralization to gases. In earlier field and laboratory studies, these losses were assumed to be due to degradation and/or volatilization and were ignored. With the discovery of latent mobilization of bound residues and transport of toxic materials out of the soil to other systems, such losses could no longer be neglected. By employing chemical mass balance, all major and numerous minor pathways of chemical fate could be observed and quantified over time. Factors affecting chemical fate (soil type, environmental conditions, biota present) could then be evaluated more readily.

The role of mass balance in microcosm technology was central to criticism of terrestrial microcosms in 1977 (Gillett and Witt 1979). Although data were available from variants of five systems, differing radically in size and scope, most of the criticism was of the Metcalf-type microcosm (Metcalf et al. 1971). This system had been employed, in various configurations and modifications, to examine the fate and effects of about 200 chemicals. As an assembled or synthetic system, ecologists were righteously outraged that it was termed a "laboratory ecosystem." Chemists were only slightly more charitable, pointing out deficiencies in mass balance. The systems were fully open to the air, and the mode of application of the chemical (spotted on to foliage or soil by micropipette) was unrealistic.

In spite of the validity of these criticisms, it was quite apparent that the Metcalf model systems served a very useful purpose through measurement of the two indices noted earlier. The Biodegradation Index (BI) compared the relative extent and ease of transformations. Exposure of vulnerable ecosystem components or processes in the field to a chemical which had been found to have a high BI in a 30-day microcosm experiment would be for only a part of a growing season (although there might be short term or acute injury). Chemicals with very low BI's (<0.003), such as DDT and dieldrin, would remain essentially unaltered and provide continuous exposure to evoke chronic effects.

The more important index, Ecological Magnification (EM), integrates bioaccumulation and biodegradation effects on fate. Chemicals which might be bioaccumulated but are easily degraded show a low EM. This index sums bioconcentration effects from the medium, bioaccumulation by trophic uptake, and excretion/metabolism in one step. When soil is the medium, EM's of greater than 0.05 are of serious concern, equivalent to an EM of about 10^3 for the same chemical in water. This difference is attributable to soil sorption, reflecting a significantly lower soil water concentration of the chemical available for interaction with the test species and its food or environment.

In applying these model systems to a host of pesticides, feed additives, and industrial chemicals, Metcalf and his coworkers provided a critical link between the octanol–water partition coefficient (K_{ow}) and estimates of bioaccumulation in whole organisms. Extensive studies by Leo et al. (1971) had demonstrated that K_{ow} (or its expression as a logarithm, Log P) could be estimated from chemical structure. Structure–activity relationships (SAR) based on laboratory feeding or exposure of single species can be employed in estimation of bioaccumulation potential of neutral organic molecules (Kenaga 1980; Trabalka and Garten 1982). Metcalf (1977) demonstrated that EM could be predicted safely from Log P in the relatively more complete "farm-pond" microcosm system. Thus, microcosm technology provided a link in the chain of evidence establishing Log P as one axis of the three-dimensional array of chemical properties used in the Pre-Biologic Screen (PBS) for ecotoxicologic effects (Gillett 1983). The PBS permits ecotoxicological testing needs to be evaluated early in the development of a chemical, before much biological information is available.

Still, the bioaccumulation potential of covalently reacting materials (e.g., methyl mercury) and highly ionized substances is not readily or accurately predicted by SAR considerations. For these, testing in the model ecosystem permits the direct evaluation of this potential in a relatively concise manner.

Biodegradation in accumulating organisms and by the community of microbes present in microcosms is another important facet of chemical fate. This can permit tracking of transformation pathways (see Metcalf and Sanborn 1975 for examples) and a closer look at heretofore difficulty followed processes. Bengtsson (1985) and Wilson et al. (1985) have used small microcosms of aquifer material transferred antiseptically and anaerobically to evaluate microbial degradation of groundwater pollutants. Such systems can be used in fate and transport studies or can contribute to understanding of impacts on the relatively sparse microbial communities in such material (Bengtsson 1985).

Table 14.1 (modified from Gillett 1981) lists the chemicals tested in model ecosystems (terrestrial or terrestrial–aquatic) for which chemical fate data were obtained. There are a number of questions that remain regarding the nature, extent, and validity of microcosm fate studies that require further laboratory bench research, including detailed chemical analysis of mechanisms and pathways. This is particularly true of the bioavailability of bound residues in soils and plants, which have not been well studied in the field.

Use of Bioassay. Because field studies and consequent chemical analyses are so expensive, most measures of the fate of pesticides and toxic substances in the environment have been on a very limited basis. Typically, the chemical is followed over some progressively increasing time interval after application, frequently with only a few randomly selected and composited samples. The data analysis takes the form of plots of log (chemical) versus time, from which a rate constant is derived yielding the environmental half-life (the time it takes for half of the parent chemical to dissipate).

Most microcosm studies have tended to use the example of field sampling as the model for experimental design. Destructive or invasive sampling of microcosms is a costly part of operations, because it increases the number of replicates required and decreases the realism or functioning of a unit. Consequently, the intimate time course of a chemical's fate may not be tracked any better than in the field.

Before the advent of chromatographic analyses of environmental resi-

Table 14.1. Chemicals studied in assembled model ecosystems

Chemicals tested[1]	System type
Acephate[a] alachlor[a] aldicarb[a] aldrin[ab] anthracene[c] atrazine[a] Banamite(TM)[a] bentazon[ad] benzidine[e] benzo(a)pyrene[e] bifenox[a] CdCl$_2$[f] captan[a] carbaryl[a] carbofuran[a] carbazole[c] cartap[h] chlordane[ah] chlordimeform[a] chlorpyrifos[a] chlorpyrifos-methyl[a] Counter(TM)[a] cyanazine[a] DDD[ag] DDE[bj] DDT[abklm] desbromoleptophos and 5 other phenylphosphonothioates[n] dibenzofuran[c] dibenzothiophene[c] dicamba[a] 2,4-D[a] dieldrin[ao] diethylstilbestrol[g] dimilin[a] disulfaton[h] edifenphos[h] endosulfan isomers[p] endosulfan sulfate[p] endrin[ab] EPN[n] 2-ethylhexyl phthalate[q] fenitrothion[a] fluorene[d] formetanate[a] heptachlor[a] hexachlorobenzene(HCB)[ab] α,β,γ-isomers of hexachlorocyclohexane[h] kepone[g] Kitazin P(TM)[h] Pb(NO$_3$)$_2$[f] leptophos[an] lindane[ab] malathion[a] metalkamate[a] methoprene[a] methoxychlor[aklm] 14 halo-alkyl substituted DDT analogs[klmrst] methyl parathion[u] metrabuzin[a] mirex[ab] monosodium methanearsenate[v] parathion[au] 2,4,5-2',5'-pentachlorobiphenyl[j] pentachlorophenol(PCP)[ah] phenmediphan[a] propachlor[a] propoxur[a] pyridaphenthion[h] pyrazon[a] sewage sludge[f] sodium selenite[w] temephos[a] 2,5-2',5'-tetrachlorobiphenyl[j] toxaphene[ai] trifluralin[a] 2,5,2'-trichlorobiphenyl[j] 2,4,6-trichlorophenyl-4'-nitrophenyl ether(CNP)[h] triphenyl phosphate[n] 3,5-xylyl methyl carbamate[x]	"Farm pond" [sand or soil; plants; water fauna]
Clopidol(TM)[y] DDT[z] diethylstilbestrol[y] phenothiazine[y] Robinine-HCl[z] sulfamethazine[y]	"Feed lot" [animals over soil; plants; water; fauna]
Atrazine[aabb] carbaryl[aabb] mephosfolan[aa] methyl parathion[aabb] pentachloronitrobenzene[bb] propanil[aabb]	"Rice paddy" [water over soil; plants; fauna]
Aldrin[ccdd] DDT[dd] fonofos[dd] methoxychlor[dd]	"Soil-corn microcosm" [soil; plants; fauna]
Captan dieldrin HCB methyl parathion parathion PCNB PCP phorate simazine 2,4,5-T trifluralin[dd]	"Physical model ecosystem" [soil; plants; fauna; water; aquatic fauna]
Bromacil[ee] captan[ff] dieldrin[eeffgghh] HCB[ff] methyl parathion[hh] p-nitrophenol[hh] parathion[hh] PCNB[ff] PCP[ff] simazine[ee] 2,4,5-T[ee] trifluralin[ee]	"Terrestrial microcosm chamber (TMC)" [synthetic soil; plants; fauna]

Table 14.1 (continued)

Chemicals tested[1]	System type
Creosote(phenanthrene, acenaphthene)[iiij] 2,4-D[kk] dieldrin[iiij] PCP[iiij] bis-tributyltin oxide[iiij]	"TMC II" [soil; plants; fauna]
Chlordane[ll] chlordene[ll] DDD[mm] DDT(and related compounds)[mm] endrin[ll] heptachlor[ll] heptachlor epoxide[ll] lindane[ll] maneb[ll] PCBs[ll] phenthoate[ll] silvex[nn] 2,3,5-T[nn] tetrachlorodibenzodioxin[nn] TH-6040[ll] toxaphene[mm] TPTH[ll] trifluralin and 8 other substituted nitroaniline herbicides[ll] zineb[ll]	"Microagroecosystem" [soil, plants]
Phorate phorate sulfone phorate sulfoxide[oo]	"Soil-Lake Mud-Water" [soil; lake mud; water coupled]
Carbofuran[pp] phorate[qq] Stauffer N-2596[rr]	"Soil-Plant-Water" [soil; plants; water coupled; plants; fauna]
Carbaryl dimethoate fenitrothion malathion methomyl[tt]	Terrestrial/Quail [modular]

[1] Where all chemicals are reported in a single reference, superscript for reference is given at the end of the list.

[a] Metcalf and Sanborn 1975
[b] Metcalf et al. 1973b
[c] Lu et al. 1978
[d] Booth et al. 1973
[e] Lu et al. 1977
[f] Lu et al. 1975
[g] Metcalf 1979
[h] Tomizawa and Kazano 1979
[i] Sanborn et al. 1976
[j] Metcalf et al. 1975
[k] Kapoor et al. 1970
[l] Kapoor et al. 1972
[m] Kapoor et al. 1973
[n] Francis et al. 1980
[o] Sanborn and Yu 1973
[p] Ali 1978
[q] Metcalf et al. 1973a
[r] Coats et al. 1974
[s] Coats et al. 1979
[t] Hirwe et al. 1975
[u] Metcalf 1976
[v] Anderson et al. 1979a
[w] Nassos et al. 1980
[x] Kazano et al. 1975
[y] Coats et al. 1976
[z] Cabella et al.1979

[aa] Tomizawa 1980
[bb] Au 1979
[cc] Cole et al. 1976b
[dd] Metcalf et al. 1979
[ee] Gile et al. 1980a
[ff] Gile et al. 1980b
[gg] Gile and Gillett 1979
[hh] Gile and Gillett 1981
[ii] Gile et al. 1982
[jj] Gillett et al. 1983
[kk] Gile 1983
[ll] Nash et al. 1979
[mm] Nash et al. 1977
[nn] Nash and Beall 1980
[oo] Walter-Echols and Lichtenstein 1977, 1978a,b
[pp] Koeppe and Lichtenstein 1982, 1984
[qq] Lichtenstein et al. 1974
[rr] Lichtenstein et al. 1977
[ss] Liang and Lichtenstein 1979
[tt] Howe 1977

dues, the only technique sensitive enough to detect biologically active residues of pesticides was bioassay using a particularly sensitive species. The subject chemical had to be extracted from the medium (soil, plant material, biota) and, for example, applied to newly hatched female houseflies of a highly inbred susceptible strain. Standard curves of dilutions of the known parent material were constructed from the mortalities observed at fixed times. Residues were then calculated by comparison of test sample mortalities to those of the standard curve. Although fraught with numerous problems, this bioassay technique was cheap, reasonably reliable in trained hands, and fairly precise.

In studies of the fate and effects of wood preservatives in a ryegrass microcosm in the TMC, Gile et al. (1982) employed a bioassay of the "available" dieldrin by introducing a fixed number of crickets (*Acheta domestica*) and counting their numbers several times a day. By this approach, Gillett et al. (1983) were able to obtain a rate of cricket loss which could be compared to the same rate resulting from exposure of crickets to contaminated food in the laboratory. This comparison revealed an apparent cyclic process, with a period of about 16 days, in the bioavailability of dieldrin to crickets in the TMC. Moreover, an large, unexplained peak of dieldrin trapped on exit filters in the air system was accompanied by a similar extreme peak in dieldrin bioavailability, albeit offset by several days. Unless the chemical were monitored on such a frequent basis, this apparent and unexplained cycle might never have been revealed. What microbial or soil processes are involved? How is bioavailability related to vapor-phase losses? What is the immediate source of dieldrin to which the crickets respond? How does the existence of unknown processes affect the assessment of risk of this and other chemicals?

These and other intriguing questions cannot be examined in either microcosm or field systems sampled on a quarterly basis looking for gross disposition. The mass balance operations and opportunity for intimate analysis, particularly by non-invasive and biologically accurate techniques available in the laboratory, encourage attempts at better understanding of the complete set of processes operating in the soil-plant-invertebrate ecosystem. Moreover, bioassay provides an opportunity for detecting active transformation products generated abiotically or biologically within the microcosm system. If all metabolites could be detected by chemical analysis, and if the biological activity of each metabolite were known, then environmental studies could be carried out solely with chemical analyses. It is unfortunate that chromatographic systems and radioactive tracers have made bioassay analysis passé.

Chemical Uptake and Bioaccumulation. As a part of chemical fate analyses, bioaccumulation in various organisms can be considered a trophic pathway to a terminal residue pool in a target organism. This process is particularly important for lipophilic substances and heavy metals. The analyses of chemicals in microcosms can also identify locations of residues in soil with depth, revealing potential for leaching and groundwater contamination. Residues in plants followed over time can reveal bioavailability to consumers or direct toxic threats to the plant. However, simple residue analyses alone do not distinguish between different types of movement from sources to sinks—they merely reveal outcomes. Such questions regarding pathways and rates are important in mathematical modeling of uptake on either an a priori or an empirical basis.

Van Voris et al. (1983) determined the concentrations of 26 elements in the plants and soil in the SCMs and field plots. Initially, they used the ratios of elements as measures of comparability, to demonstrate that the three species of plants in the SCMs performed as did their counterparts in the field plots. The authors were also very much interested in the specific relationship between exposure to fly ash and uptake of nutrients and toxic elements. The ratio of concentrations in the plant to concentrations in the soil was found to be dependent on the plant and the element, rather than exposure rate, for a number of elements toxic to animals. The ratio could be used to calculate exposure of herbivores and their consumers, where feeding efficiency or uptake were known, as a direct measure for exposure assessment in assessing risk of using fly ash amendment for land reclamation.

The recent development of the highly computerized plant uptake chamber by McFarlane et al. (1987) is an excellent example of application of modern techniques to chemical fate issues, which may lead to further advances in microcosm technology. Generally, data on plant uptake of nutrients, toxic chemicals, etc., have been obtained on an empirical basis–grow a plant in a given soil with a set concentration of test chemical, then measure the residues over time in different portions of the plant. Pesticide tolerance data may require 5 crops to be grown in 30 different soils under a variety of treatments and environmental conditions. By separation of the aerial and root portions of the test plants and monitoring of conditions in both the hydroponic and atmospheric portions of their apparatus, MacFarlane and coworkers are attempting to demonstrate the relationship of organic chemical uptake to evapotranspirative water flux and plant energetics (photosynthesis and respiration), thus permitting a physiologically based model of uptake (Lindstrom et al. 1986) that may be applicable to other species and conditions.

More importantly, such systems may be useful in evaluating phytotoxicity and indirect influences of chemicals acting through plants in terrestrial ecosystems. The bases of the apparatus are the use of (a) separated plant root and aerial chambers, (b) transistorized circuit boards to read a variety of environmental conditions from sensors as direct (analog) outputs, then convert these analog values and store the resulting digitized data in a microcomputer, and (c) computer software and firmware to control data acquisition and to sort, store, and analyze the data in an efficient manner. Such systems form a useful adjunct to microcosm studies, whereas the application of some of the technical advances (b,c) could improve capabilities of terrestrial and aquatic microcosms.

Problems. The outcomes observed in microcosms do not always lend themselves to ready interpretation. Manipulations may affect biotic interactions which in turn affect physical conditions. In preliminary experiments using a dwarf corn variety (*Zea mays*) in the TMC, voles (*Microtus canicaudus*) usually cut down the corn after feeding, in order, on alfalfa (*Medicago sativa*), soybeans (*Phaseolus vulgaris*), and ryegrass (*Hordendum vulgaris*). However, 4 of 16 individual voles cut down or girdled the corn stalks first, thus radically altering soil temperature and moisture levels. Had the systems with these corn plants been treated with a test chemical, the outcomes would have been quite different between these two behaviors.

The chemical in the TMC without corn stalks would have been more readily photolyzed, while the increased soil temperature would have favored higher volatilization of the test chemical from the soil. Reduced soil moisture would probably reduce microbial degradation, but the outcome would depend on the capabilities of the altered microbial population responding to the change in nutrients and environmental conditions. Although the differences in results would depend on the characteristics of the test chemical and which chemodynamic processes most affected its fate, for a chemical such as dieldrin, the vole that cut down the corn plant first would have been much more heavily exposed to dieldrin and its more highly toxic photodecomposition products.

These early terrestrial microcosm chamber (TMC) experiments were highly criticized (Gillett and Witt 1979) in the same light as Metcalf's "farm pond" system (Kapoor et al. 1970) and the "laboratory physical model ecosystem" (Metcalf et al. 1979). They had shallow soil depth and used artificial substrates which restricted plant growth and did not permit earthworms to survive well. Furthermore, the air system was inefficient in trapping volatiles. A substantial improvement in this regard was made with the microagroecosystem of Nash et al. (1977, 1979), which employed polyurethane foam (PUF) plugs as absorbants. Based on these constructive criticisms, the TMC (Gillett and Gile 1976) was modified substantially for work with wood preservatives (Gile et al. 1982; Gillett et al. 1983). Increased depth (from 20 to 50 cm) of native soils supported vastly improved earthworm survival, permitting intermediate sampling of this species as a bioaccumulator, and improved plant growth. At the same time, increased head space (from 30 to 80 cm) and use of the PUF plugs permitted higher airflow rates (approaching field diffusion rates), greater trapping efficiency, and enhanced accountability.

Inclusion of voles as representative mammals in the TMC and physical model ecosystem has also been criticized because of the lack of consistency with scale, for their voracious appetites and burrowing activities radically alter the system. Nevertheless, the *M. canicaudus* in the TMCs are able to raise a litter (4–6 pups) to a normal weaning weight (10 g). This can be compared to litters of 5–7 pups in both the laboratory rearing rooms and the field.

Not all of the plant material is consumed. Some is also used for nesting material and may be stored in underground "hay mows," where the chemical may be subjected to fermentative metabolism and cometabolism. In many cases it was possible to demonstrate placental or lactational transfer to pups of females exposed while pregnant or nursing. The voles and invertebrate species included in these studies act as integrative samplers of the system, particularly in regard to bioavailable residues of a parent chemical and its metabolites.

Effects

Ecotoxicologic effects of toxics and stressants observed in terrestrial microcosms have included the following:

Death or debilitation of individual species (Metcalf et al. 1971; Cole et al. 1976a; Gile and Gillett 1979; Gillett et al. 1983)

Changes in population numbers, age/size, class structure, or recruitment (Anderson et al. 1978; Anderson et al. 1979b; Gillett et al. 1983)

Altered genotypic distributions (Lighthart 1979)

Altered plant growth, primary productivity, and crop yield (Cole et al. 1976b; Van Voris et al. 1983)

Altered rate and extent of soil respiration (Ausmus et al. 1978; Van Voris et al. 1980; Lighthart et al. 1982; Mitchell et al. 1982; Arthur et al. 1984)

Enhanced leaching of macro- and micro-nutrients (Jackson et al. 1977; Draggan 1979; Van Voris et al. 1980; Malanchuk et al. 1980; Gile et al. 1979; Tolle et al. 1983)

Altered organismic behavior and inter-species interactions (Anderson et al. 1979b; Gillett et al. 1983)

Not all effects can be observed in any one system or with every chemical. In a number of instances, these effects can be tied to treatment rate or route and to ambient concentrations of test chemical. This important connection between fate and effects is a vital part of most microcosm studies. The observed effects cover the gamut of significant ecotoxicologic responses anticipated in the field and have been obtained on a variety of agents and materials. The results provide strong encouragement to the use of microcosms in ecotoxicology. Unfortunately, many studies of chemicals in excised terrestrial microcosms (Table 14.2) have concentrated largely on the ecotoxicologic effect and not on chemical fate.

In addition to application in ecotoxicologic assessment, these systems

Table 14.2. Chemicals studied in excised soil systems

Chemical	Core size (depth × height in cm)	Components
Ampicillin, cadmium chloride, chloromycetin, gentamicin, streptomycin[a]	1.5 × 10	Fir litter
Dieldrin,[b] hexachlorobenzene,[c] methyl parathion,[b] sodium arsenate[c,d] 2,4,5-T[b]	5 × 10	Intact soil
Cadmium chloride,[e] 2,4-D,[f] copper sulfate,[g] fly ash,[h] lead nitrate,[g] zinc chloride[g]	15 × 10	Intact soil; plants; litter; invertebrates
Sewage sludge[i]	5 × 20	Intact soil; invertebrates
Fly ash[j]	17 × 60	Soil; crops
Lead nitrate[g]	100 × 60	Intact soil; red maple; litter; invertebrates

[a] Lighthart 1979
[b] Gile et al. 1979
[c] Harris 1980
[d] Draggan 1976
[e] Van Voris et al. 1980

[f] Malanchuk and Joyce 1983
[g] Jackson et al. 1977; Ausmus et al. 1978
[h] Malanchuk et al. 1980
[i] Mitchell et al. 1982
[j] Van Voris et al. 1983

have also contributed to a better understanding of ecological principles. Much as in the use of toxicants to study the connections in biochemistry and physiology, toxicants used in microcosms reveal connections within ecosystems. Attention to processes in unexposed model laboratory systems forms the basis for enhanced understanding of ecosystems.

This principle is best represented by the experiments of Van Voris et al. (1980), using $CdCl_2$ as a stressant. Intact grassland microcosms (15 cm diameter × 10 cm deep) were held for several months, during which time the CO_2 concentration in the canopy was monitored hourly and leaching of calcium was examined weekly. A power spectrum analysis of the CO_2 concentration was used to obtain a measure of the extent and intensity of interactions of components in the system, as postulated by the authors. Microcosms with greater numbers of intense peaks in the power spectrum corresponding to frequencies with periods >24 hrs were hypothesized to represent greater complexity.

The post-treatment, integrated Ca export was similarly hypothesized to represent ecosystem stability. Close examination of the temporal loss rate revealed that this stability might arise from greater resistance (onset or extent of loss) or greater resilience (time to return to pretreatment norms). Nevertheless, the postulated correlation ($P < 0.05$) between increased complexity and high ecosystem stability was obtained. Other measures of species abundance and diversity, both before and at the end of the test, were statistically less consistent than the experimental observations. Although the experiments did not fully elucidate the relationships between complexity, diversity, and species richness, they are a start in the quest for experimental verification of ecological theories.

Organismic Toxicology. Metcalf et al. (1971 1979) observed acute mortality of, and chronic toxicity to, various plants and animals caused by chemicals introduced into their "farm pond" and "physical model ecosystems." It is possible to set up treatment–outcome relationships, both as to mortality and as to residues associated with mortality in a situation comparable to the field. Gile and Gillett (1979) reported that six voles exposed in TMCs treated with dieldrin at 1.12 kg/hectare all died within a few days with brain dieldrin concentrations averaging 60 ± 5 (S.E.) $\mu g/g$ (fresh wt.). Time-to-death and hours of exposure were directly related ($r^2 = 0.98$) to the time between treatment and when the vole was added to the system. From this it was estimated that the "re-entry time" for voles in a ryegrass/alfalfa pasture so treated would be about 42 ± 3 days. Conversely, estimates of exposure could be made from cricket mortality measured in both the TMCs and laboratory systems (Gillett et al. 1983).

The presence or absence of species before and after treatment provides additional qualitative estimates of toxicity to those species. Microcosms are inappropriate for obtaining statistically valid estimates of intrinsic toxicity (μg of toxicant/g body weight of organism), due to limitations on the numbers of test organisms and the heterogeneous nature of exposure.

Comparative Toxicity. Various microcosms have given good comparisons of the toxicity between classes of chemicals and classes of organisms, but they are weak tools for assaying between specific species. A given microcosm system provides excellent comparisons between chemicals within a given type of exposure and between types of exposures for the same total environmental burden. This permits verification of laboratory estimates of selective toxicity relationships (the dose–response relationships compared between species for a given chemical or between chemicals for a given species) under more realistic environmental conditions which involve numerous interactions of the chemical and biota with the environment.

Van Voris et al. (1983) reported differential dosage effects on three plant species in the SCMs evaluating fly ash. Since the soil was boron-deficient and the fly ash contained high levels of boron, the responses were bi-phasic. Low rates of fly ash stimulated growth more effectively for timothy, but high rates were more toxic to red clover.

Chronic and Non-Lethal Effects. For persistent organics and heavy metals a number of effects have been observed, in particular reduction in growth or crop yield of plants (Van Voris et al. 1983) with heavy applications of fly ash. Acute toxicity of aldrin/dieldrin to corn seedlings (Metcalf et al. 1979) was overcome by later growth, even though the dieldrin persisted in the soil and was transferred to the corn leaves via volatilization. Generally, such effects are measured by destructive sampling at termination of the microcosm and require the whole plant biomass for statistical significance. Until passive or nondestructive measures are available, plant growth inhibition (or consumptive herbivory) appears to be too costly or insensitive a quantitative measure of toxicant effect (Gillett and Gile 1976).

Certain types of interactions (e.g., predator–prey relationships, some behavioral phenomena) may only be relevant to the more inclusive types of microcosms and the field. That is, the tested phenomenon is only elicited in a realistic situation. With large microcosms and high populations of visible species (Gillett et al. 1983), and by appropriate sampling techniques with microscopic invertebrates and microorganisms (Anderson et al. 1978; Anderson et al. 1979b; Mitchell et al. 1982), reproductive and related population effects can be assayed.

Anecdotal behavioral changes observed between voles exposed to various chemicals included reduced number of nest entrances, a smaller or more shallow nest, reduced numbers of storage sites for uneaten, but harvested "hay," fewer tunnels and more surficial trails, and more time spent in the open, all in comparison to unexposed controls. A typical dieldrin-exposed vole seen trembling on the surface was usually found dead within a few days. In the wild, it would likely have been consumed as prey before that time, passing its dieldrin burden on to a hawk or a fox. These behavioral changes suggest microcosms could be usefully applied in greater depth than is presently the case.

System-Level Effects

By far the best and most numerous use of microcosms is to assess the impact of a stressant or chemical on critical ecosystem processes, especially those integrating the actions of many species. Two processes, soil respiration and micronutrient cycling, have been investigated extensively.

Soil Respiration. Numerous devices have been developed to measure release of CO_2 and/or uptake of O_2 as measures of system function. Most of these measurements can be automated. Although attractive because of the explicit involvement of all aerobic organisms and relative simplicity of measurement, inhibition of respiration is a relatively insensitive indication of a chemical's toxicity to a terrestrial system. It may be made more sensitive when coupled to measurements of productivity to ascertain productivity/respiration (P/R) ratios. Since measurement of respiration can be readily made at all levels of scale and complexity, it potentially can be used in any system. Nevertheless, it is best suited to small, easily replicated units, such as soil or soil-litter microcosms (Anderson et al. 1978; Anderson et al. 1979b; Lighthart 1979; Lighthart et al. 1982) or small SCMs (Jackson et al. 1977; Draggan 1979; Gile et al. 1979).

Shirazi et al. (1984) demonstrated how data on the toxicity of metals in various soils (Lighthart et al. 1982) could be connected to soil properties (porosity, bulk density, organic matter content) through appropriate scaling factors representing diffusivity of O_2/CO_2 and growth of the endemic organisms. As noted earlier, Van Voris et al. (1980) had shown how the intensity of complex patterns of CO_2 concentration increased in proportion to system diversity in SCMs with intact vegetation. However, Gile et al. (1979) found insignificant changes in respiration by concentrations of hexachlorobenzene, pentachlorophenol, and parathion which affected micronutrient loss by leaching (see below). Apparently, respiratory effects may be highly buffered by sorption of organic chemicals or complexation of inorganic cations to soil.

Micronutrient Leaching. Nutrient losses in watersheds exposed to pesticides (Likens et al. 1970) or heavy metals (O'Neill et al. 1977) prompted microcosm studies of heavy metals (Hg, Cd, Pb) by various workers at Oak Ridge National Laboratory. Jackson et al. (1977) and Van Voris et al. (1980) then demonstrated that leaching of cations (especially Ca^{+2}, Mg^{+2}, Mn^{+2}, and Fe^{+3}), dissolved organic carbon (DOC), and some nitrogen forms (especially NO_2^- and NO_3^-) provided an excellent measure of disruption of the soil system. The extent of leaching was inversely proportional to the frequency of intense peaks estimated by the spectral analysis of CO_2 respiration (noted above). In particular, Ca^{+2} leaching greatly exceeded possible displacement by added Cd^{+2}, but ^{45}Ca added later in the experiment (after respiration returned to normal values) was totally retained. These results suggest that Cd cannot eliminate binding sites completely nor displace Ca from those sites available.

Later, Tolle et al. (1983) employed Ca^{+2} loss in assessing the impacts of fly ash on soil microcosms and field plots. Again, disruption of soil micronutrient cycling was revealed. However, an amount of fly ash deposited from stacks over a period of decades would be required to produce a measurable effect. Gile et al. (1979) found, however, that there could be significant release of leachable cations with environmentally realistic levels of the organics tested. Malanchuk et al. (1980) also reported forest litter effects at concentrations of metals found in the environment. Overall, excessive leaching of soil nutrients appears to be a more sensitive measure of impact than inhibition of respiration or cellular growth.

The exact nature of the processes interrupted by these treatments, the reason for reversibility over time, and the extent of sensitivity under a range of conditions are unknown. Furthermore, the lack of a mechanistic interpretation of this process inhibits characterization of the results as more than a "suggestion" of a relationship (Van Voris et al. 1980) between complexity and stability. Interestingly, in a early experiment on fly ash, Van Voris and colleagues (1983) found that naturally acidic precipitation ("acid rain") collected for treatment of the microcosms was more effective at releasing indigenous soil cations than the fly ash itself.

It is unfortunate that the nature of the processes by which micronutrients are lost is so poorly understood. Free anions (NO_2^-, NO_3^-) may be immeasurably low in untreated controls and in the recovered system (Gile et al. 1979). The recovery process itself, which appears sensitive to ecosystem complexity (Van Voris et al. 1980), suggests that responses of soil microcosms are related to the time course of speciation of toxic metals, forms of which become unavailable biologically (Lighthart et al. 1982), or to adaptation of the community of microorganisms present to surviving, unaffected ionic species (Lighthart 1979).

The results of these soil microcosm studies have implications for such areas as groundwater quality protection and the questions of long-term forest debilitation observed in Europe and elsewhere. Damage of the surface and root-zone microbial systems can deprive plants of valuable nutrients which can become groundwater pollutants. There may also be direct toxic effects on plant growth and increased vulnerability to disease and other stresses. Reduced plant growth favors erosion and reduced soil water-holding capacity. Thus, both surface and groundwater quality can suffer, and one would predict reduced water quantity as the long-term consequence of toxicant-induced nutrient loss. With emergence of the linkage of soil and water quality issues as important national priorities, microcosms and mesocosms appear as potentially useful tools in analyzing both the affected processes and the fate of the stressants perturbing the systems.

14.5.2 Mesocosm Studies

Generally, mesocosm studies are neither as numerous nor as wide-ranging as microcosm studies, particularly for organics. Most of the effort has been directed at air pollutant research (acid deposition and effects of NO_x, SO_x, and O_3). Moreover, except for field chambers, only a few system are currently in operation. Because most terrestrial mesocosms do not differ from field studies in practical terms, an extensive discussion of their applications would not be useful. However, certain systems show promise.

Chemical Fate

Wilson et al. (1981) reported difficulties in achieving mass balance, even in their closed system, with various volatile organic solvents. In all other terrestrial mesocosm systems to date, mass balance and chemical fate have been largely ignored. Lauenroth and Preston (1984) reported SO_x air concentrations and loading from stochastically delivered concentration patterns. However, the boundary conditions were not intended to control the toxicant. Interestingly, the resultant dispersal of SO_x from the test sites provided excellent correlation with infrared remote sensing photographs. The differences in infrared absorbance between treatments is clearly visible, with a plume of pollution leaving the ZAPS plot up a small ravine.

Failure of terrestrial mesocosms to have adequate boundary controls can create a serious problem regarding safety (an Environmental Impact Statement may be needed), but the degree of "openness" is at the heart of the rationale for terrestrial mesocosms. Open-topped chambers are operated to permit exposure of test plants to air pollutants, insects, and pathogenic agents other than the test chemical or stressant. The results are then readily compared to 'control' field plots of the crop grown under similar conditions without added stressant. Since efforts to retain the test agent in the mesocosm may impede these other inputs and conditions, the safety considerations which are part of the justification for microcosms (i.e., little or no uncontrolled release of stressant) may not be applicable to terrestrial mesocosms. Control of the pollutant exposure concentration, however, can be very good, allowing the investigator to manipulate exposure to simulate patterns in the environment (Male et al. 1978) and control releases within acceptable limits.

Ecological Effects

The various terrestrial mesocosm systems in which system-level ecological effects were investigated, excepting simple exclosures, have not included organic toxicants. Exclosures have tended to focus more narrowly on impacts at the species population level, rather than at the system level. Although effects on processes (respiration, primary productivity, nutrient cycling) can be measured, there is little direct evidence of such applications. Thus, demonstration of direct utility and cost effectiveness of terrestrial mesocosms in ecotoxicologic assessment has been limited.

Much of the effort in the ZAPS plots (Lauenroth and Preston 1984) was an extension of the grassland biome studies of the 1970's. In general, the measures applied could distinguish statistically significant effects between the highest level of treatment and control (unexposed) plots, but neither of these were distinguishable from intermediate levels of SO_2 exposure. These measures included standing crop, net annual primary production, root carbohydrate storage, and insect diversity. Remote sensing was more sensitive and precise, demonstrating differences in infrared absorption quantifiable with respect to SO_2 exposure. These results pose a typical methodologic dilemma—how to bring together measures that can be used to estimate economic or aesthetic losses with those that are accurate, sensitive, precise, and comprehensive, but are poorly linked to decision making. The solution would appear to be (a) development of "ground truth" for remote sensing, and/or (b) modification of remote sensing technology to link it more closely to traditional measures. Both approaches are currently under investigation in ecological research.

Use of a field engineered system for acid deposition studies permitted demonstration of impacts on nutrient leaching (Lee and Weber 1980) and litter decomposition rate (Lee and Weber 1980; 1983) of material in mesh leaf bags, and on nutrient (NO_3^-, Zn^+) availability in soil lysimeters (Lee 1985).

Open-topped chambers have been useful in evaluating tree seedling growth under ambient (polluted) conditions in native soils (Duchelle et al. 1982). Using the chambers to supply charcoal-filtered (O_3-free) air, they found increased tree growth and lower incidence of visible injury in comparison to either field-grown plants or those in which unfiltered air was supplied in the chamber. In most other open-topped chamber studies, ozone is added at different levels and pot-grown plants are used. Heck et al. (1984a) describes the use of this system in crop loss assessment research. Improvements (Hogsett et al. 1985) in chamber design and operation portend greater usefulness for air pollutants.

Simple exclosures have been very useful in a variety of ecological studies. Because these systems are at the borderline between "mesocosms" and "field studies," one must focus on those in which chemical fate and effects are considered.

POEM is an excellent example, for mathematical modeling of pesticide fate was integrated with modeling of a "non-target" organism's population dynamics as an impact of the pesticide. An old orchard was divided by galvanized sheets buried halfway into the ground, largely to facilitate the collection of runoff surface water. The orchard was then treated with azinophos-methyl (Guthion®). Using life tables developed in the laboratory under controlled conditions (Snider 1979; Snider and Shaddy 1980) and field calibration of a mathematical model of azinophos-methyl fate in an orchard (Goodman et al. 1983), E. Goodman (1982) demonstrated that they could predict monitored populations of the isopod *Trachelipus rathkei* within a factor of about 3. Probably the largest, single contributing factor in their underestimation was the assumption in the model that the barriers between plots in the orchard would result in "closed" populations; differences were attributed to reinvasion, rather than recruitment.

Barrett and coworkers have employed similarly structured field exclosures to study the effects of various chemicals on small rodents using a "capture-mark-release-recapture" methodology. Johnson and Barrett (1975) evaluated field population changes engendered by exposure of feral mice (*Mus musculus* L.) to diethylstilbestrol in what constituted a pioneering study. More recently Anderson and Barrett (1982) examined the impact of repeated sewage sludge disposal on meadow vole (*Microtus pennsylvanicus* L.) in grasslands. In these and other similar studies they demonstrated both direct and indirect population effects at levels of stressant without acute or chronic mortality.

The USEPA currently uses a similar type of study ('large-scale field pen') for evaluation of chronic avian toxicity of pesticides. Either mallard (*Anas platyrhynchos*) or bobwhite (*Colinus virginianus*) may be used in the test, which concentrates on reproductive endpoints (USEPA 1978). Other protocols for controlled or open field studies have been suggested, but not defined, and a manufacturer can be required to perform a field study, on an ad hoc basis, prior to registration. With the emerging interest in microbial pesticides and genetically engineered organisms in pest control, renewed emphasis in mesocosm approaches to effects assessment appears at hand. The degree and nature of containment needed are under study and debate, particularly regarding requirements for mesocosm studies before field evaluation. Based upon a single example (Watrud et al. 1985), it would appear that small scale-field testing of genetically engineered pesticides will be significantly more rigorous as to boundary conditions and release of test agents than most mesocosm systems to date.

14.6 Conclusions

Terrestrial microcosm and mesocosm technology can provide an incisive method of attack on environmental problems, especially where there are questions of environmental fate and effects of toxic chemicals. The available range in size and complexity of microcosms is encouraging, although much work is still needed to clarify various aspects of the limitations of this technology. For most toxic organic chemicals, terrestrial mesocosms offer little more generically than do nominal field trials. However, the technology will continue to be of use in critical air pollutants and for specific questions requiring the large-scale setting of the mesocosm. If there are significant questions of environmental safety remaining after laboratory and microcosm tests, substantially improved control of the boundary conditions will be needed for terrestrial mesocosms to be able to answer such questions safely and adequately. Testing of genetically engineered organisms and other potential environmental stressors of unknown safety may also be well served by containment within model ecosystems. Both microcosms and mesocosms will benefit from an improved understanding of the effects and implications of that boundary control on the responses of the test ecosystem. Both technologies will also gain from advances in chemodynamic and ecological effects modeling, which in turn may be served by laboratory and field model ecosystems in validation steps.

Ecotoxicologic effects of natural and synthetic organic chemicals have only begun to be revealed by model ecosystem research. It seems likely that terrestrial microcosms will play a critical role in methodologic development for process-level studies and regulatory testing. Although meso-cosm applications may remain more limited, much will depend upon recognition of the need for increased emphasis on ecotoxicologic concerns in environmental management. Development of criteria for ascertaining the most appropriate level of testing (i.e., lab bench, microcosm, mesocosm, or field) for a given problem, and commensurate research of greater breadth (i.e., chemicals, components, and measures made), to provide a validated, experiential data base to support these criteria across the range of anthropogenic pollutants, are both needed. As other areas of toxicology progress toward *in vitro* assays and similar simplistic systems, the need for the realism and complexity of model ecosystems in ecotoxicologic research will not simply be enhanced, but indeed become urgent.

14.7 Summary

Microcosms and mesocosms are model ecosystems subject to investigator control of conditions at a boundary between the model ecosystem and the "real world." They have been applied to pragmatic studies of a variety of problems involving the fate and distribution of synthetic and natural chemicals and other stressants and the consequent effects of these stressants on ecological processes.

Microcosms are practical compromises between laboratory bench tests and field studies. They afford a measure of safety for the investigator and the environment, can be operated to provide mass balance of introduced pollutants, and therefore have been judged to be particularly good for fate and transport studies of toxic chemicals. However, there are limitations in their use and interpretation of data therefrom. In particular, the duration, complexity, extrapolability, and relationship to ecological theory have been questioned or found to be deficient. Nevertheless, microcosms can be more cost-effective, comprehensive, and incisive than individual laboratory or greenhouse tests.

Mesocosms differ from microcosms by being located outside of laboratories, so that they are often difficult to distinguish from field studies. Their advantage is in the high degree of realism, direct extrapolability of data, long duration, and scale and scope of processes examined. They are, however, vulnerable to catastrophic impacts of environmental phenomena, costly in terms of maintenance of the boundary, and presently lack a broad experiential base regarding synthetic organic chemicals. For certain questions mesocosm are the only practical methodology for environmental safety.

Mathematical modeling of environmental fate and effects is closely linked to model ecosystem design and application. Mesocosms may be particularly appropriate in testing the mathematical models, whereas microcosms may be useful in developing a better understanding of processes included in the models. The models in turn influence model ecosystem structure and operation.

Microcosms played a key role in the development of the use of physicochemical data to predict bioaccumulation and of comprehensive indices of chemical fate. Although there remain methodologic problems, laboratory model ecosystems have been used to detect effects of toxicants on species mortality, growth, and behavior and on ecological processes such as respiration, decomposition, nutrient cycling, and species interactions.

Studies of fate of chemicals moving to groundwater and the effects of air pollutants and other chemicals on plant growth, animal populations, and ecosystem processes are among the accomplishments of mesocosm research. Mesocosms have also supported mathematical modeling, pollutant regulation, and specific safety questions.

References

Ali S (1978) Degradation and environmental fate of endosulfan isomers and endosulfan sulfate in mouse, insects, and laboratory model ecosystem. Ph. D. Thesis. University of Illinois - Champaign/Urbana. 101 pp. [Diss. Abstr. Int. B 39:2117]

Amundson RG (1983) Yield reduction of soybeans due to exposure to sulfur dioxide and nitrogen dioxide in combination. J Environ Qual 12:454–458

Anderson AC, Abdelghani AA, McDonell D (1979a) Fate of the herbicide MSMA in microcosms. Trace Substances Environ Health 13:274–284

Anderson RV, Elliot ET, McClellan JF, Coleman DC, Cole CV, Hunt HW (1978) Trophic interactions in soils as they affect energy and nutrient dynamics. III. Biotic interactions of bacteria, amoebae, and nematodes. Microbiol Ecol 4:381–387

Anderson RV, Coleman DC, Cole CV, Elliott ET, McClellan JF (1979b) The use of soil microcosms in evaluating bacteriophagic nematode response to other organisms and effects on nutrient cycling. Int J Environ Stud 13:175–182

Anderson TJ, Barrett GW (1982) Effects of dried sewage sludge on meadow vole (*Microtus pennsylvanicus*) populations in two grassland communities. J Appl Ecol 19:759–772

Arthur MF, Zwick TC, Tolle DA, Van Voris P (1984) Effects of fly ash on microbial carbon dioxide evolution from an agricultural soil. Water Air Soil Pollut 22:209–216

Au LA (1979) Pesticide interactions in the laboratory rice paddy model ecosystem. Ph. D. Thesis. University of Illinois - Champaign/Urbana. 156 pp. [Diss. Abstr. Int. B 40:3567–3568]

Ausmus BS, Dodson GJ, Jackson DR (1978) Behavior of heavy metals in forest microcosms: III. Effects on litter-soil carbon metabolism. Water Air Soil Pollut 10:19–26

Ausmus BS, Jackson DR, Van Voris P (1979) The accuracy of screening techniques. In: Witt JM, Gillett JW, Wyatt CJ (eds) *Terrestrial Microcosms and Environmental Chemistry*, Proceedings of the Colloquia held at Corvallis, OR,

June, 1977. NSF/RA 79–0026. Washington DC: National Science Foundation, pp. 123–130

Bengtsson G (1985) Microcosm for groundwater research. In: Ward CH, Giger W, McCarty PL (eds) *Ground Water Quality*. New York: John Wiley & Sons, pp. 330–341

Booth GM, Yu C-C, Hansen DJ (1973) Fate, metabolism, and toxicity of 3-isopropyl-1H-2,1,3-benzothiadiazin-4(3H)-1–2,2-dioxide in a model ecosystem. J Environ Qual 2:408–411

Caballa SH, Patterson M, Kapoor IP (1979) A terrestrial-aquatic model ecosystem for evaluating the environmental fate of drugs and related residues in animal excreta. In: Khan MAQ, Lech JJ, Menn JJ (eds) *Pesticide and Xenobiotic Metabolism in Aquatic Organisms*. ACS Symposium Series 99: 183–194

Cairns J Jr (1983) Are single species toxicity tests alone adequate for estimating environmental hazard? Hydrobiology 100:47–57

Cairns J Jr, Alexander M, Cummins KW, Edmondson WT, Goldman R, Harte J, Isensee AR, Levin R, McCormick JF, Peterle TJ, Zar JH (1981) *Testing Effects of Chemicals on Ecosystems*. Washington DC: National Academy Press, 103 pp.

Carsell RF, Smith CM, Mulkey LA, Dean JD, Jowise P (1984) Users Manual for Pesticide Root Zone Model (PRZM), Release 1. US/EPA 600/3–84–109. Washington, DC: US Environmental Protection Agency

Coats JR, Metcalf RL, Kapoor IP (1974) Metabolism of the methoxychlor isostere, dianisyl neopentane, in mouse, insects, and a model ecosystem. Pestic Biochem Physiol 4:201–211

Coats JR, Metcalf RL, Lu P-Y, Brown DD, Williams JF, Hansen LG (1976) Model ecosystem evaluation of the environmental impacts of the veterinary drugs phenothiazine, sulfamethazine, Clopidol, and diethylstilbestrol. Environ Health Perspect 18:167–179

Coats JR, Metcalf RL, Kapoor IP, Chio L-C, Boyle PA (1979) Physical-chemical and biological degradation studies of DDT analogs with altered aliphatic moieties. J Agric Food Chem 27:1016–1022

Cole CV, Elliot ET, Hunt HW, Coleman DC (1978) Trophic interactions in soils as they affect energy and nutrient dynamics. V. Phosphorus transformations. Microbiol Ecol 4:381–387

Cole LK, Metcalf RL, Sanborn JR (1976a) Environmental fate of insecticides in terrestrial model ecosystems. Int J Environ Stud 10:7–14

Cole LK, Sanborn JR, Metcalf RL (1976b) Inhibition of corn growth by aldrin and the insecticide's fate in the soil, air, and wildlife of a terrestrial model ecosystem. Environ Entomol 5:583–589

Coleman DC, Anderson RV, Cole CV, Elliot ET, Woods L, Campion MK (1978) Trophic interactions in soils as they affect energy and nutrient dynamics. IV. Flows of metabolic and biomass carbon. Microbiol Ecol 4:373–380

Davis JM, Riodan AJ, Lawson RE Jr (1983) Wind tunnel study of the flow field within and around open-top chambers used for air pollution studies. Boundary-Layer Meteorol 25:193–214

Donigian AS Jr, Beyerlein DC, Davis HH, Crawford MH (1977) *Agricultural Runoff Management (ARM) Model, Version II Refinement and Testing*. EPA-600/2–77–098. Athens, GA: US Environmental Protection Agency, Environmental Research Laboratory, 293 pp.

Draggan S (1976) The microcosm as a tool for estimation of environmental transport of toxic materials. Int J Environ Stud 10:1–7

Draggan S (1979) Effects of substrate type and arsenic dosage level on arsenic behavior in grassland microcosms. Part I: Preliminary results on ^{74}As as transport. In: Witt JM, Gillett JW, Wyatt CJ (eds) *Terrestrial Microcosms and Environmental Chemistry*. Proceedings of the Colloquia held at Corvallis, OR, June, 1977. NSF/RA 79–0026. Washington DC: National Science Foundation, pp. 102–110

Duchelle SF, Skelly JM, Chevone BI (1982) Oxidant effects on forest tree seedling growth in the Appalachian mountains. Water Air Soil Pollut 18:363–373

Enfield CG, Carsel RF, Cohen SZ, Phan T, Walters DM (1982) Approximation of pollutant transport to ground water. Ground Water 20:711–722

Francis BM, Hansen LG, Fukuto TR, Lu P-Y, Metcalf RL (1980) Ecotoxicology of phenylphosphonothioates. Environ Health Perspect 36:187–195

Giesy JP Jr (ed) (1980) *Microcosms in Ecological Research*. Proceedings of the Symposium in Augusta, GA, August, 1979. U.S. Department of Energy Symposium Series No. 52. (CONF-781101). Springfield, VA: U.S. Technical Information Service, 1034 pp.

Gile JD (1983) 2,4-D: Its distribution and effects in a rye grass (*Lolium perenne*) ecosystem. J Environ Qual 12:406–412

Gile JD, Gillett JW (1979) The fate of ^{14}C-dieldrin in a simulated terrestrial ecosystem. Arch Environ Contam Toxicol 8:107–124

Gile JD, Collins JC, Gillett JW (1979) *The Soil Core Microcosm—A Potential Screening Tool* EPA-600/3–79–089. Corvallis, OR: U.S. Environmental Protection Agency, 40 pp.

Gile JD, Gillett JW, Collins JC (1980a) Fate of selected herbicides in a terrestrial laboratory microcosm. Environ Sci Technol 14:1124–1128

Gile JD, Collins JC, Gillett JW (1980b) Fate of selected fungicides in a terrestrial laboratory ecosystem. J Agric Food Chem 27:1159–1164

Gile JD, Gillett JW (1981) Transport and fate of organophosphate insecticides in a terrestrial laboratory ecosystem. J Agric Food Chem 29:616–621

Gile JD, Collins JC, Gillett JW (1982) Fate and impact of selected wood preservatives in a terrestrial model ecosystem. J Agric Food Chem 30:295–301

Gillett JW (1978) Terrestrial laboratory microcosms: Relationship to water resources. In: Klingman P (ed) *Toxic Materials in the Aquatic Environment* WR-024–078. Water Resources Research Institute, Corvallis, OR: Oregon State University, pp. 17–42

Gillett JW (1981) Model ecosystems in fate and movement of toxicants. In: Beroza M (ed) *Test Protocols for Environmental Fate and Movement of Toxicants*. Proceedings of the Symposium, 94th annual meeting in Arlington, VA. Arlington VA: Association of Official Analytical Chemists, pp. 214–232

Gillett JW (1983) A comprehensive pre-biologic screen for ecotoxicologic effects. Environ Toxicol Chem 2:463–476

Gillett JW (1985) Annex A: The role of microcosms in specimen banking and monitoring, In: Lewis RA, Gillett J, Van Loon JC, Hushon JM, Ludke JL, Watson AP (eds) *"Richtlinien für den Einsatz von Umweltprobenbanken in der Bundesrepublik Deutschland auf ökologischer Grundlage. Umweltforschungsplan des Bundesministers des Innern."* [Guidelines for the Role of Environmental Specimen Banking in the Federal Republic of Germany on an Ecologic

Basis. Environmental Research Plan of the Ministry of the Interior.],
Saarbrücken, W. Germany: Universität des Saarlandes, pp. A-1–A-9

Gillett JW (ed) (1986) *Prospects for Physical and Biological Containment of Genetically Engineered Organisms*. Proceedings of the Shackelton Point Workshop on Biotechnology Impact Assessment, October, 1985, Bridgeport, NY.
ERC-114. Ecosystems Research Center, Cornell University, Ithaca, NY. 137
pp.

Gillett JW, Gile JD (1976) Pesticide fate in terrestrial laboratory ecosystems. Int J
Environ Stud 10:15–22

Gillett JW, Witt JM (1979) *Terrestrial Microcosms*. Proceedings of the workshops
at Otter Rock and Portland, OR, June and December, 1977. NSF/RA-790034.
Washington DC: National Science Foundation, 34 pp.

Gillett JW, Russell LK, Gile JD (1983) Predator–prey (vole–cricket) interactions:
The effects of wood preservatives. Environ Toxicol Chem 2:83–93

Gillett JW, Levin SA, Harwell M, Alexander M, Andow DA, Stern AM (eds)
(1985) *Potential Impacts of Environmental Release of Biotechnology Products:
Assessment, Regulation, and Research Needs*. ERC-075. Ecosystems Research Center, Cornell University, Ithaca, NY. 241 pp. See also, (1986) Environ. Manage. 10:433–563 .

Goodman D (1982) *The Limits of Microcosms: Problems in the Interpretation of
Toxicity Results from Laboratory Multispecies Systems*. ERC-13. Ecosystems
Research Center, Cornell University, Ithaca, NY. 15pp.

Goodman ED (1982) Modeling the effects of pesticides on populations of soil/litter
invertebrates in an orchard ecosystem. Environ Toxicol Chem 1:45–60

Goodman ED, Jenkins JJ, Zabik MJ (1983) A model for azinophosmethyl attenuation and movement in a Michigan orchard ecosystem. II. Parameterization of
the field-based model. Arch Environ Contam Toxicol 12:111–119

Hammonds A (ed) (1981a) *Methods for Ecological Toxicology,* ORNL-5708 (EPA
560/11–80–026) Oak Ridge, TN: Oak Ridge National Laboratory, 307 pp.

Hammonds A (ed) (1981b) *Ecological Test Systems* (Proceedings of a series of
workshops, Oak Ridge, TN) ORNL-5709 (EPA 560/6-81-004) Oak Ridge, TN:
Oak Ridge National Laboratory, 179 pp.

Harris WF (ed) (1980) *Microcosms as Potential Screening Tools for Evaluating
Transport and Effects of Toxic Substances,* Environmental Sciences Div. Publication No. 1506 (EPA 600/3-80-042). Oak Ridge, TN: Oak Ridge National Laboratory, 379 pp.

Heck WW, Cure WW, Rawlings JO, Zaragoza LJ, Heagle AS, Heggestad HE,
Kohut RJ, Kress LW, Temple PJ (1984a). Assessing impacts of ozone on agricultural crops: I. Overview. J Air Pollut Control Assoc 34:729–735

Heck WW, Cure WW, Rawlings JO, Zaragoza LJ, Heagle AS, Heggestad HE,
Kohut RJ, Kress LW, Temple PJ (1984b) Assessing impacts of ozone on agri-
cultural crops. II. Crop yield functions and alternative exposure statistics. J Air
Pollut Control Assoc 34: 810–817

Heggestad HE, Gish TJ, Lee EH, Bennett JH, Douglass LW (1985) Interaction of
soil moisture stress and ambient ozone on growth and yields of soybeans.
Phytopathology 75:472–477

Hirwe AS, Metcalf RL, Lu P-Y, Chio LC (1975) Comparative metabolism of 1,1-
bis-(*p*-chlorophenyl)-2-nitropropane (Prolan) in mouse, insects and a model
ecosystem. Pestic Biochem Physiol 5:65–72

Hogsett WE, Tingey DT, Holman SR (1985) A programmable exposure control system for determination of the effects of pollutant exposure regimes on plant growth. Atmos Environ 19:1135–1145

Howe GJ (1977) The effects of various insecticides applied to a terrestrial model ecosystem or fed in the diet on the serum cholinesterase level and reproductive potential of Coturnix quail. Ph.D. Thesis. University of Manitoba. [Diss. Abstr. Int. B 38:4785]

Huckabee J (1983) Evaluation of tests to predict chemical injury to ecosystems: Microcosms. Palo Alto, CA: Electric Power Research Institute, 43 pp. [draft ms]

Jackson DR, Washburne CD, Ausmus BS (1977) Loss of calcium and nitrate nitrogen from terrestrial microcosms as in indicator of soil pollution. Water Air Soil Pollut 8:279–284

Jenkins JJ, Zabik MJ, Kon R, Goodman ED (1983) A model for azinophos-methyl attenuation and movement in a Michigan orchard ecosystem. I. Development and presentation of the experimental data base. Arch Environ Contam Toxicol 12:99–110

Johnson RF, Barrett GW (1975) Effects of diethylstilbestrol on feral house mouse (*Mus musculus* L.) population dynamics under experimental field conditions. J Appl Ecol 12:741–747

Kapoor IP, Metcalf RL, Nystrom RF, Sangha GK (1970) Comparative metabolism of methoxychlor, methiochlor, and DDT in the mouse, insects, and in a model ecosystem. J Agric Food Chem 18:1145–1152

Kapoor IP, Metcalf RL, Hirwe AS, Lu P-Y, Coats JR, Nystrom RF (1972) Comparative metabolism of DDT, methoxychlor, and ethoxychlor in mouse, insects and in a model ecosystem. J Agric Food Chem 20:1–6

Kapoor IP, Metcalf RL, Hirwe AS, Coats JR, Khalsa MS (1973) Structure-activity correlations of biodegradability of DDT analogs. J Agric Food Chem 21:310–315

Kazano H, Asakawa M, Tomizawa C (1975) Fate of 3,5-xylyl methylcarbamate insecticide (XMC) in a model ecosystem. Appl Ent Zool 10:108–115

Kenaga EE (1980) Correlation of bioconcentration factors of chemicals in aquatic and terrestrial organisms with their physical and chemical properties. Environ Sci Technol 14:553–556

Kickert RN (1984) Sensitivity of agricultural ecological system models and implications for vulnerability to toxic chemicals. Environ Toxicol Chem 3:309–324

Koeppe MK, Lichtenstein EP (1982) Effects of percolating water, captafol, and S-ethyl-N,N-dipropylthiocarbamate on the movement and metabolism of soil-applied carbon-14-labeled carbofuran in an agromicrocosm. J Agric Food Chem 30:116–121

Koeppe MK, Lichtenstein EP (1984) Effects of organic fertilizers on the fate of 14C carbofuran in an agro-microcosm under soil run-off conditions. J Econ Entomol 77:1116–1122

Lauenroth WK, Preston EM (1984) *Effects of SO₂ on a Grassland: a Case Study in the Northern Great Plains of the United States*. EPA-600/3-84-107. Corvallis, OR: U.S. Environmental Protection Agency

Laurence JA, Maclean DC, Mandl RH, Schneider RE, Hansen KS (1982) Field tests of a linear gradient system for exposure of row crops to SO₂ and HF. Water Air Soil Pollut 17:399–407

Lee JJ (1985) Effects of simulated sulfuric acid rain on the chemistry of a sulfate-adsorbing forest soil. EPA-600/J-85-015. Corvallis, OR: U.S. Environmental Protection Agency

Lee JJ, Weber DE (1980) Effects of sulfuric acid rain on two model hardwood forests: throughfall, litter leachate, and soil solution. EPA-600/3-80-014. Corvallis, OR: U.S. Environmental Protection Agency

Lee JJ, Weber DE (1983) Effects of sulfuric acid rain on decomposition rate and chemical element content of hardwood leaf litter. EPA-600/J-83-038. Corvallis, OR: U.S. Environmental Protection Agency

Leo AJ, Hansch C, Elkins D (1971) Partition coefficients and their uses. Chem Rev 71:525–616

Lewis E, Brennan E (1977) A disparity in the ozone response of bean plants grown in a greenhouse, growth chamber or open-top chamber. J Air Pollut Control Assoc 27:889–891

Liang TT, Lichtenstein EP (1979). Effects of cover crops on the movement and fate of soil-applied (14C)-fonofos in a soil-plant-water microcosm. J Econ Entomol 73:204–210

Lichtenstein EP (1980) Fate and behavior of pesticides in a compartmentalized microcosm, In: Giesy JP Jr (ed) *Microcosms in Ecological Research.* (Augusta, GA, Nov. 8–10, 1978). DOE Symposium Series, Vol. 52. Department of Energy CONF-781101. Springfield, VA: National Technical Information Service, pp. 954–970

Lichtenstein EP, Fuhremann TW, Schulz KR (1974) Translocation and metabolism of [^{14}C]phorate as affected by percolating water in a model soil-plant ecosystem. J Agric Food Chem 22:991–996

Lichtenstein EP, Liang TT, Fuhremann TW (1978) A compartmentalized microcosm for studying the fate of chemicals in the environment. J Agric Food Chem 26:948–953

Lichtenstein EP, Schulz KR, Liang TT (1977) Fate of fresh and aged soil residues of the insecticide [^{14}C]-N-2596 in a soil-corn-water ecosystem. J Econ Entomol 70:169–175

Lighthart B (1979) Enrichment of cadmium-mediated antibiotic-resistant bacteria in a Douglas fir (*Pseudotsuga menziesii*) litter microcosm. Appl Environ Microbiol 37:859–861

Lighthart B, Baham J, Volk VV (1982) Microbial respiration and chemical speciation in metal-amended soils. J Environ Microbiol 46:1073–1079

Likens GE, Bormann FH, Johnson NM, Fisher DW, Pierce RS (1970) Effects of forest cutting and herbicide treatment on nutrient budgets in the Hubbard Brook watershed ecosystem. Ecol Monogr 40:23–47

Lindstrom FT, Piver WT (1984) *A Mathematical Model of the Vertical Transport and Fate of Toxic Chemicals in a Terrestrial Microcosm System.* Tech. Rep. No. 51, Dept. of Mathematics. Corvallis, OR: Oregon State University

Lindstrom FT, Piver WT (1986) Vertical-horizontal transport and fate of low water solubility chemicals in unsaturated soil. J Hydrol 86:93–131

Lindstrom FT, McCoy EL, McFarlane JC, Boersma L (1986) Uptake of organic chemicals by plants: A theoretical model. Corvallis OR: U.S. Environmental Protection Agency, 54 pp. [draft]

Lu P-Y, Metcalf RL, Furman R, Vogel R, Hassett J (1975) Model ecosystem studies of lead and cadmium and of urban sewage sludge containing these elements. J Environ Qual 4:505–509

Lu P-Y, Metcalf RL, Plummer N, Mandel D (1977) The environmental fate of three carcinogens: benzo(a)pyrene, benzidine, and vinyl chloride evaluated in laboratory model ecosystems. Arch Environ Contam Toxicol 6:129–142

Lu P-Y, Metcalf RL, Carlson EM (1978) Environmental fate of five radiolabeled coal conversion by-products evaluated in a laboratory model ecosystem. Environ Health Perspect 24:201–208

Malanchuk JL, Mueller CA, Pomerantz SM (1980) Microcosm evaluation of the agricultural potential of fly ash amended soils, In: Giesy JP Jr (ed) *Microcosms in Ecological Research*. (Augusta, GA, Nov. 8–10, 1978). DOE Symposium Series, Vol. 52. Department of Energy CONF-781101. Springfield, VA: National Technical Information Service, pp. 1034–1049

Malanchuk JL, Joyce K (1983) Effects of 2,4-D on nitrogen fixation and carbon dioxide evolution in a soil microcosm. Water Air Soil Pollut 20:181–190

Malanchuk JL, Kollig HP (1985). Integrated use of physical and mathematical models to evaluate ecological effects. Water Air Soil Pollut 24:267–282

Male LM, Van Sickle J, Wilhour R (1978) *Time Series Experiments for Studying Plant Growth Response to Pollution*. EPA 600/3-78-038. Corvallis, OR: US Environmental Protection Agency

Mandl RH, Weinstein L, Dean M, Wheeler M (1980) Response of sweet corn to HF and SO_2 under field conditions. Environ Exp Bot 20: 359–365

Matsumura F (1980) Use of a microcosm approach to assess pesticide biodegradability. In: *Biotransformation and Fate of Chemical in the Aquatic Environment*. Proceedings of workshop, Aug. 14–18, 1979. Washington DC: American Society of Microbiology, pp. 126–135.

McFarlane JC (1978) Light. In: Langhans RW (ed) *A Growth Chamber Manual*. Ithaca, NY: Cornell University Press, pp. 1–44

McFarlane JC, Cross A, Frank C, Rogers RD (1981) Atmospheric benzene depletion by soil microorganisms. Environ Monit Assess 1:75–81

McFarlane JC, Nolt C, Wickliff C, Pfleeger T, Shimabuku R, McDowell M (1987) The uptake, distribution, and metabolism of four organic chemicals by soybean plants and barley roots. Environ Toxicol Chem 6:847–856

Metcalf RL (1976)Model ecosystem studies of bioconcentration and biodegradation of pesticides, In: Khan MAQ (ed) *Pesticides in Aquatic Environments*. New York: Plenum Press, pp. 127–144

Metcalf RL (1977) Model ecosystem approach to insecticide degradation. Ann Rev Entomol 22:241–261

Metcalf RL (1979) Model ecosystems for environmental studies of estrogens. In: McLachlan JA (ed) *Estrogens in the Environment*. New York: Elsevier North Holland, Inc, pp. 203–211

Metcalf RL, Sangha GK, Kapoor IP (1971) Model ecosystem for the evaluation of pesticide biodegradability and ecological magnification. Environ Sci Technol 5:709–713

Metcalf RL, Booth GM, Schuth CK, Hansen DJ, Lu P-Y (1973) Uptake and fate of di-2-ethylhexyl phthalate in aquatic organisms and in a model ecosystem. Environ Health Perspect 4:27–34

Metcalf RL, Kapoor IP, Lu P-Y, Schuth CK, Sherman P (1973). Model ecosystem studies of the environmental fate of six organochlorine pesticides. Environ Health Perspect 4:35–44

Metcalf RL, Sanborn JR (1975) Pesticides and environmental quality in Illinois. Bull Ill Nat Hist Surv 31:381–436

Metcalf RL, Sanborn JR, Lu P-Y, Nye D (1975). Laboratory model ecosystem studies of the degradation and fate of radiolabeled tri-, tetra-, and pentachlorobiphenyl compared with DDE. Arch Environ Contam Toxicol 3:151–165

Metcalf RL, Cole LK, Wood SG, Mandel DJ, Milbrath ML (1979) *Design and Evaluation of a Terrestrial Model Ecosystem for Evaluation of Substitute Pesticide Chemicals*. EPA 600/3-79-004. Corvallis, OR: U.S. Environmental Protection Agency, 200 pp.

Mill T, Mabey WR, Bomberger DC, Chou T-W, Hendry DG, Smith JH (1980) *Laboratory Protocols for Evaluating the Fate of Organic Chemicals in Air and Water*. SRI Project PYU-4396. Palo Alto, CA: SRI International, 329 pp.

Mitchell MJ, Parkinson CM, Hamilton WE, Dindal DL (1982) Role of the earthworm, *Eisenia foetida,* in affecting organic matter decomposition in microcosms of sludge-amended soil. J Appl Ecol 19:805–812

Mount D, Gillett JW (1982) Impact of pollution on wildlife and habitat resources. In: Mason WT (ed) *Research on Fish and Wildlife Habitat,* EPA-600/8-82-022. Washington DC: U.S. Environmental Protection Agency, pp. 143–164

Nash RG, Beall ML Jr (1980) Distribution of silvex, 2,4-D, and TCDD applied to turf in chambers and field plots. J Agric Food Chem 28:614–623

Nash RG, Beall ML Jr, Harris WG (1977) Toxaphene and 1,1,1- trichloro-2,2-bis(p-chlorophenyl)-ethane (DDT) losses from cotton in an agroecosystem chamber. J Agric Food Chem 25:336–341

Nash RG, Beall ML Jr, Kearney PC (1979) A microagroecosystem to monitor environmental fate of pesticides. In: Witt JM, Gillett JW, Wyatt CJ (eds), *Terrestrial Microcosms and Environmental Chemistry,* (Proceedings of the Colloquia held at Corvallis, OR, June, 1977) NSF/RA 79–0026. Washington DC: National Science Foundation, pp. 86–94

Nassos PA, Coats JR, Metcalf RL, Brown DD, Hansen LG (1980) Model ecosystem, toxicity, and uptake evaluation of ^{75}Se-selenite. Bull Environ Contam Toxicol 24:752–758

Neuhold J, Ruggerio L (1976) *Ecosystem processes and organic contaminants*. NSF/RA 76-0008. Washington DC: National Science Foundation, 44 pp.

Odum EP (1971) *Fundamentals of Ecology*. Philadelphia: W.B. Saunders, 574 pp.

Odum HT, Jenkins CF (1970) Metabolism and evapotranspiration of the lower forest in a giant plastic cylinder. In: Odum HT (ed) *A Tropical Rain Forest*. TID 24270. Washington DC: Div. of Technical Information, US Atomic Energy Commission, pp 165–189

Olszyk DM, Tibbits TW, Hertzberg WM (1980) Environment in open-top field chambers utilized for air pollution studies. J Environ Qual 9:610–615

O'Neill RV, Ausmus BS, Jackson DR, Van Hook RI, Van Voris P, Washburne C, Watson AP (1977) Monitoring terrestrial ecosystems by analysis of nutrient export. Water Air Soil Pollut 8:271–277

Patterson MR, Mankin JB, Brooks AA (1974) Overview of a unified transport model. Proc Ann NSF Trace Contam Conf 1:12–23

Perez KT, Morrison GM, Lackie NK, Oviatt CA, Nixon SW, Buckley BA, Heltshe JF (1977)The importance of physical and biological scaling to the experimental simulation of a coastal marine ecosystem. Helgol Wiss Meeresunters 30:144–162

Preston EM, Lee JJ (1981) *Design and Performance of a Field Exposure System for Evaluation of the Ecological Effects of SO$_2$ on Native Grassland*. EPA-600/J-82-221. Corvallis, OR: U.S. Environmental Protection Agency

Rees JF, Wilson BH, Wilson JT (1985) Biotransformation of toluene in methanogenic subsurface material. 85th Annual Meeting of the American Society for Microbiology, Las Vegas, NV, Mar. 3–7, 1985. Abstracts Ann Meet Amer Soc Microbiol 85:258

Sanborn JR, Yu C-C (1973) The fate of dieldrin in a model ecosystem. Bull Environ Contam Toxicol 10:340–346

Sanborn JR, Metcalf RL, Bruce WN, Lu P-Y (1976) The fate of chlordane and toxaphene in a terrestrial-aquatic model ecosystem. Environ Entomol 5:533–538

Shirazi MA (1980) Development of scaling criteria for terrestrial microcosms. ISEM Journal 2:97–116

Shirazi MA, Lighthart B, Gillett JW (1984) A method for scaling biological response of soil microcosms. Ecol Model 13:203–26

Snider RM (1979) The effects of azinophos-methyl (Guthion^R) on a population *Trachelipus rathkei* (Isopoda) in a Michigan orchard. Pedobiology 19:99–105

Snider RM, Shaddy JW (1980)The ecobiology of *Trachelipus rathkei* (Isopoda). Pedobiology 20:394–410

Tolle DA, Arthur MF, Van Voris P (1983) Microcosm and field comparison of trace element uptake in crops grown in fly ash amended soil. Sci Total Environ 31:243–261

Tolle DA, Arthur MF, Chesson J, Van Voris P (1985) Comparison of pots vs. microcosms for predicting agroecosystem effects due to waste amendment. Environ Toxicol Chem 4:501–510

Tomizawa C (1980) Biological accumulation of pesticides in an ecosystem – Evaluation of biodegradability and ecological magnification of rice pesticides by a model ecosystem. Japan Agric Research Quart 14:143–149

Tomizawa C, Kazano H (1979) Environmental fate of rice paddy pesticides in a model ecosystem. J Environ Sci Health B 14:121–152

Trabalka JR, Garten CT (1982) *Development of Predictive Models for Xenobiotic Bioaccumulation in Terrestrial Ecosystems*. Environmental Sciences Div. Publication No. 2037 Oak Ridge, TN: Oak Ridge National Laboratory, 256 pp.

U.S. Environmental Protection Agency (1978) Proposed guidelines for registration of pesticides in the United States [40 CFR Parts 162, 163, 181]. Federal Reg. 43:29697–29741

Van Voris P, O'Neill RV, Emanuel WR, Shugart HH Jr (1980) Functional complexity and ecosystem stability. Ecology 61:1352–1360

Van Voris P, Arthur MF, Tolle DA (1983) *Field and Laboratory Evaluation of Terrestrial Microcosms for Assessing Ecological Effects of Utility Wastes*. Project report No. 1224–5. Palo Alto, CA: Electric Power Research Institute, 183 pp.

Van Voris P, Tolle DA, Arthur MF (1985) *Experimental Terrestrial Soil-Core Microcosm Test Protocol*. EPA 600/3-85-047. Corvallis, OR: US Environmental Protection Agency, Environmental Research Laboratory, 71 pp.

Walter-Echols G, Lichtenstein EP (1977) Microbial reduction of phorate sulfoxide to phorate in a soil-lake mud-water microcosm. J Econ Entomol 70:505–509

Walter-Echols G, Lichtenstein EP (1978a) Movement and metabolism of [^{14}C]Phorate in a flooded soil system. J Agric Food Chem 26:599–604

Walter-Echols G, Lichtenstein EP (1978b) Effects of lake bottom mud on the movement and metabolism of ^{14}C-phorate in a flooded soil-plant system. J Environ Sci Health B 13:149–168

Watrud LS, Perlak FJ, Tran MT, Kusano K, Mayer EJ, Miller-Wideman MA, Obukowicz MG, Nelson DR, Kreitinger JP, Kaufman RJ (1985) Cloning of the *Bacillus thurengiensis* sub. sp. kurstaki delta-endotoxin gene into *Psuedomonas fluorescens:* Molecular biology and ecology of an engineered microbial pesticide. In: Halverson HO, Pramer D, Rogul M (eds) *Engineered Organisms in the Environment: Scientific Issues.* Washington DC: American Society of Microbiology

Weinstein LH (1973) A cylindrical, open-top chamber for the exposure of plants to air pollutants in the field. J Environ Qual 2:371–376

Wilson JT, Enfield CG, Dunlop WJ, Cosby RL, Foster DA, Baskin LB (1981) Transport and fate of selected organic pollutants in a sandy soil. J Environ Qual 10:501–506

Wilson JT, Noonan MJ, McNabb JF (1985) Biodegradation of contaminants in the subsurface. In: Ward CH, Giger W, McCarty PL (eds) *Ground Water Quality.* New York: John Wiley & Sons, pp 483–492

Wilson JT, Wilson BH (1985) Biotransformation of trichloroethylene in soil. Appl Environ Microbiol 49:242–243

Wolfe NL, Burns LA, Stern WC (1982) Use of linear free energy relationships and an evaluative model to assess the fate and transport of phthalate esters in the aquatic environment. Chemosphere 9:393–402

Chapter 15
The Role of Aquatic Microcosms in Ecotoxicologic Research as Illustrated by Large Marine Systems

Juanita N. Gearing[1]

Microcosms, in theory, should be one of the most valuable research tools available to ecotoxicology. They provide the opportunity to study entire ecosystems under controlled conditions where the level and form of toxicants as well as many other environmental parameters can be varied by the experimenter. They make possible experimental controls and replicates.

The strength of microcosms lies in the ability to examine the end result of many organisms and processes interacting in a manner similar to nature. That is to say, microcosms can be used as empirical models of nature. This is useful because ecosystems are "something other than the sum of their parts" (Oviatt et al. 1977). "Scientists have not been able to discover all the variables" (Parsons 1978a) necessary to understand an ecosystem, making it impossible to predict precisely the ecological effects of stresses from laboratory experiments on single species or mathematical modeling alone (for example see Steele 1979; Oiestad 1982; Franklin 1983; Sabatini and Marcotte 1983).

Microcosms attempt to bridge the gap between laboratory experiments and field observations. They can be thought of as biological models for measuring the integrated result of a stress (or combination of stresses) on a living ecosystem. For example, suppose we need to know what a certain chemical will do if released into the marine environment at certain levels. Rather than trying to make a prediction based on knowledge of the chemical properties of the compound and its toxicity to individual biological species (a calculation fraught with many assumptions) or than adding the

---

[1] Maurice Lamontagne Institute, Fisheries and Oceans, Mont-Joli, Quebec G5H 3Z4, Canada

chemical directly to the environment (a dangerous and often legally difficult experiment), we can measure results directly after adding the chemical to a microcosm. These can then be extrapolated to natural ecosystems, making the assumption that the processes in the microcosm act as in nature. Moreover, with microcosms, one can have experimental controls and replication for better statistical interpretation of the data as well as control over many environmental parameters.

This approach differs from laboratory experiments in that it attempts to examine an entire, naturally functioning system, the microcosm. It differs from field experiments in that some experimental control is maintained.

Such quantitative measurements of the effect of pollutants added to or eliminated from ecosystems can be directly useful to regulators for decision making. For example, replicated additions of No. 2 fuel oil at 90 ppb were followed by a statistically significant decline in the benthos but no verifiable change in the plankton over four months. The benthos had not recovered after over a year. A higher level of oil in the water column (190 ppb) caused changes in both the plankton and benthos (Oviatt et al. 1982).

For microcosms to be most useful to ecotoxicology, their limitations as well as their advantages must be clearly understood. All types of investigations–laboratory experiments, microcosm work, field studies, and mathematical models–have their particular uses. This chapter will attempt to delineate the place of microcosms by:

1. Illustrating the variety of microcosms which have been used. The variety and flexibility of microcosms make them useful for a wide range of ecotoxicologic experiments.
2. Examining the applicability of results from microcosms to nature. How do microcosms compare with the natural ecosystems which they attempt to mimic? Can their results be extrapolated to the real world better than those of laboratory experiments? What properties of microcosms are most important for enhancing their usefulness?
3. Setting forth the advantages and disadvantages of microcosm experiments relative to other types of studies.
4. Illustrating some particular experiments in which microcosms have been successfully used for ecotoxicology.

Definitions

The first priority is to define exactly what is meant by a microcosm because it has no generally accepted definition for science. The name has been given to systems of diverse sizes (milliliters to millions of liters), kinds and numbers of organisms (two picked species to natural assemblages), and operational time scales (hours to years). Using the broadest interpretation, any system from an in vitro experiment to a field study can be considered "a world in miniature, a small representation of the whole," the dictionary definition of microcosm. The National Research Council (1981) has defined microcosms as "samples from natural ecosystems housed in artificial containers and kept in a laboratory environment." Some scientists require microcosms to contain only *natural* assemblages, for example Davies and Gamble (1979), while others include gnotobiotic systems containing only certain defined species chosen by the experimenter (Harte et al. 1980). Draggan and Reisa (1980) state that microcosms are "functionally similar to but may differ in origin or in structure from the natural ecosystem that is simulated." The term mesocosm has also been used to refer to large-scale microcosms (size limit undefined), but since there is no planned functional difference between the two, only the term microcosm will be used herein.

The focus here will be on the larger, seawater microcosms. Laboratory scale systems (up to about 10 liters) are the backbone of microcosm work, but it is not possible in this short review to do more than mention a few representative systems. Freshwater systems are also not covered in great detail. The engineering and conceptual problems presented here for large marine microcosms are common to all microcosms: marine, freshwater, and terrestrial.

Sources of Additional Information

There are so many studies that might be considered to have used microcosms, that one chapter cannot hope to cover them all. A selection of papers covering different kinds of microcosms is presented here, and further information on other studies can be obtained by consulting the bibliographies of the individual papers. Nontechnical presentations of the rationale of using microcosms are given by Gibson and Grice (1980) and Strickland (1967). Several papers cover the theory behind the use of microcosms (e.g., Donaghay 1984; Dudzik et al. 1979; Kemp et al. 1980; Mullin 1982; Parsons 1978b; Schindler et al. 1980b; Steele 1979; Steele and Henderson 1981). Historical reviews of large microcosms are also available (Banse 1982; Boyd 1981; Davies and Gamble 1979; Gamble and Davies 1982; Giddings 1980; Giesy and Odum 1980; Hill and Wiegert 1980; Levandowsky 1977; Menzel and Steele 1978; Parsons 1978a; Parsons 1982; Pilson and Nixon 1980). Adams and Giddings (1982), Draggan and Reisa (1980), Kuiper (1984), Santschi (1982), and Santschi et al. (1984) review the uses of microcosms in pollution research.

There are, in addition, published papers summarizing particular microcosms or types of microcosms: Warren and Davis (1971), laboratory streams; Cooke (1971), freshwater laboratory microcosms; Balch et al. (1978), Scripps Deep Tank and Dalhousie Aquatron; Hood (1978), upwelling impoundments; Kelly (1984) and Zeitzschel and Davies (1978), benthic chambers; and Muschenheim et al. (1986), laboratory flumes. Two books (Giesy 1978; Grice and Reeve 1982) present papers dealing exclusively with microcosm research. Other groups of papers on the use of microcosms have arisen from symposia, particularly: The Role of Microcosms in Ecological Research (Draggon 1976), International Helgoland Symposium on Ecosystem Research (Kinne 1977), and Controlled Ecosystems Experiments—Joint Oceanographic Assembly (Parsons 1978c).

15.1 Types of Microcosms

15.1.1 Freshwater Systems

Freshwater microcosms can mimic streams, ponds, and lakes. Some examples representative of these types are presented here to illustrate the variety and flexibility of microcosms available.

Model streams present a unique set of experimental difficulties, being directed toward short-term effects studies. Warren and Davis (1971) have reviewed much of the work in this field. Table 15.1 summarizes a variety of studies in artificial streams.

There have also been many pond and lake microcosms. Considerable numbers of these studies have been carried out in the laboratory, for example, see Table 15.2 and the review by Cooke (1971). These laboratory systems have a greater possibility of deviation from nature than systems *in situ*—artificial light, changes in turbulent regime, altered assemblage of organisms, etc. Balanced against this is the greater experimental control possible and relatively lower cost. With smaller systems the problem of adequate sample needed for chemical analysis has often been addressed by using radiolabeled compounds. The smaller systems, however, are limited in the size and number of organisms that can be maintained. It is particularly noteworthy that some researchers (e.g., Medine et al. 1980) have also enclosed the column of air above the water to determine evaporation and condensation. Air currents across open tanks were regulated in a Toronto experiment (Oil and Gas Working Group 1978). The importance of wind was also noted by Bower and McCorkle (1980). The control and sampling of air for understanding air–water exchange is a very important yet undeveloped area in microcosm research.

In general, the larger microcosms have been employed in an attempt to achieve controlled ecosystems as similar as possible to naturally occurring ones in their range of organisms, physical environment, and chemical processes. These large containers tend to be situated outdoors where they enjoy the advantage of natural environmental conditions (light, temperature, rain, etc.). Biological realism has been attempted in some by enclosing an actual body of water. Table 15.3 shows the multiformity of these experiments. Of particular interest are the large (18,000,000 liters) limno-

Table 15.1. Some representative studies in artificial streams

Reference	Length (meters)	Description	Studies
Lamberti and Resh 1983	<1	four outdoor plexiglass troughs, tiles colonized in natural streams. CA	effects of thermal stress
McIntire et al. 1964	3	indoor streams (6), enamel over wood. OR and ID	effects of nutrients
Rogers and Harvey 1976 Rogers et al. 1980	4	outdoor streams (6). Glen Lyn, VA	effects of nutrients, copper, chromium, chlorine
Hoffman and Horne 1980	5	outdoor wooden flumes (12), plastic lined. Truckee River, CA	effects of wastewaters
Warren et al. 1964 Davis and Warren 1965 Hansen and Garton 1982	6	indoor streams (10).	effects of sucrose and insecticides
Maki and Johnson 1976 Maki 1980	8	indoor channels (6).	effects of lampreycide
Sigmon et al 1977 Williams and Giesy 1978 Bowling et al. 1980	90	outdoor concrete streams and pools lined with sand. Savannah River Facility	effects of mercury, cadmium, nitrilotriacetic acid, and aromatic hydrocarbons
Irvine 1985	100	outdoor streams (2) with riffles and pools. Waitaki River, NZ	effects of flow perturbations
Armitage 1980 Rogers 1980	114	outdoor channels (12). Browns Ferry, AL	thermal effects
Phaup and Gannon 1967	200	outdoor aluminum trough lined with PVC. Huron River, MI	studies of attached plants
Stout and Cooper 1983	1000	outdoor stream. Monticello, MN	effects of p-cresol

corrals originally constructed and tested in Blelham Tarn, Great Britain. One of these reinforced, butyl-rubber cylinders, 45 m in diameter and about 11 m deep, was set up in 1970 and was still in good condition seven years later. Five others of the same design have been successfully used at Grote Rug in the Netherlands and at Farmoor Reservoir in Great Britain. These tubes are anchored into the lake sediments. Over periods of up to

Table 15.2. Some representative laboratory freshwater microcosms

Reference	Volume (liters)	Description	Controlled parameters	Studies
Maguire 1980	—	sealed flasks		mortality
Ringelberg 1976	—	aquaria	gnotobiotic	2 years, observations
Isensee 1976	—			bioaccumulation of pesticides
Metcalf et al. 1971	—			fate & effect of DDT
Hattori et al. 1980 Heath 1979	.1	flasks	gnotobiotic	2 months, nutrients
Nixon 1969	.25	flasks	gnotobiotic	microcosm operation
Kitchens and Copeland 1980	.5	flasks (252)	temperature, light	3 months, algal mats
Lichtenstein 1980	2	compartments, aquaria	soil	1 month, pesticides
Taub 1969 Taub 1976 Taub and Crow 1980 Taub et al. 1980	3	aquaria	gnotobiotic, sediment, light	microcosm operation
Beyers 1962 Beyers 1963	4	aquaria (12)	temperature, light, sediment, turbulence	primary production
Weinberger et al. 1982	4		gnotobiotic, light, sediment	
Coats 1980	8	aquaria	gnotobiotic, light, turbulence	insecticides
Brockway et al. 1979	14	30 aquaria	temperature, light, sediment, turnover	methoxychlor
Medine et al. 1980	14	sealed cylinders	sediment, pH	fate of Zn, Cr, Cd, Pb, Hg
Cooper 1973	15	aquaria (15)	temperature, light, sediment	3 weeks, herbicides
dePinto et al. 1980	15		sediment	
Raymont and Adams 1958	15	battery jars		phytoplankton
Cole and Metcalf 1980	19	compartments: glass carboys	sediment	organics

Table 15.2 (continued)

Reference	Volume (liters)	Description	Controlled parameters	Studies
Bretthauer 1980	20+	aquaria & channel	gnotobiotic, temperature, light	sewage, herbicides, pesticides, metals, temperature
King 1980	28	boxes	light	sewage
King 1980	85	boxes	light	sewage
Giddings 1980	80	aquaria (12)		arsenic
Giddings and Eddlemon 1978				
Cooper and Copeland 1973	170	6 groups of 5 aquaria	salinity, turbulence	1 year, microcosm operation
Nadeau and Roush 1973	2,000	aquaria with salt marsh blocks		4 months, oil
Raymont and Adams 1958	2,400	concrete ponds	gnotobiotic	phytoplankton
Harte et al. 1980	7,000	aquaria (2)		
Oil and Gas Working Group 1978	8,400	cylinders (2)	light, turbulence, air currents	hydrocarbons

22 months, the water columns thus enclosed have been shown to be qualitatively similar in temperature, phytoplankton, and zooplankton to the water in unenclosed portions of the tarn. One significant difference is the nutrient flux, the tarn being supplied by runoff not available within the enclosure. These microcosms have been used for seasonal studies of the effects of eutrophication; compounds labeled with nitrogen-15 have aided in the elucidation of natural nutrient cycling. Although large, these systems have not been overburdened by engineering difficulties and have proven cost-effective due to years of successful experimentation in the same facility.

15.1.2 Seawater Systems

Benthic Marine Microcosms

These include seawater ecosystems which focus on the sediment and sediment–water exchanges. Zeitzschel and Davies (1978) and Kelly (1984) have reviewed some of the literature on benthic microcosms. One

Table 15.3. Some representative outdoor freshwater microcosms

Reference	Volume (liters)	Description	Studies
McLaren 1969	—	plastic tubes under ice	zooplankton
Hellebust et al. 1975	—	metal cylinders (6)	oil
Miller et al. 1978	—	Arctic ponds	oil
Snow and Scott 1975	—	lakes divided with plastic sheeting	oil
Marshall and Mellinger 1980	8	carboys in Lake Michigan	effects of cadmium
Giddings 1980 Giddings and Eddlemon 1978	700	pond enclosures	arsenic
Jackson et al. 1980	860	cylinders (5)	fate of metals at different pH
Weinberger et al. 1982	1,000	lake enclosure, with sediment	1 month, pesticides
Brunskill et al. 1980	1,000	cylinders (8), with sediments	effects of sucrose and arsenic
Bender and Jordan 1970	2,000	plastic bag in lake	primary production
Goldman 1962	3,000	plastic tube	
Hellebust et al. 1975	5,700	plastic cylinders (6)	3 months, oil
Goldman 1962	11,000	plastic tubes	
Boyd 1981	42,000	steel tower	effects of cold and pressure on fish
Hesslein and Quay 1973	74,000	triangular, plastic limnocorral	turbulence
Bower and McCorkle 1980	77,000	plastic cylinders	primary production
Dutka and Kwan 1984 Nagy et al. 1984 Scott and Glooschenko 1984 Scott et al. 1984a Scott et al. 1984b Sherry 1984	81,000	ponds (5), Lake Huron	fate and effects of oil and dispersant
Kaushik et al. 1985 Solomon et al. 1980 Solomon et al. 1985 Stephenson et al. 1984	125,000	enclosures (3), with sediments	effects of pesticides
Rudd and Turner 1983a Rudd and Turner 1983b Salki et al. 1984 Turner and Rudd 1983	130,000	enclosures (5), some with sediment	effects of mercury and selenium
Jackson et al. 1980 Marshall and Mellinger 1980 Muller 1980 Schindler et al. 1980a	150,000	plastic cylinders (4) in sediments	fate and effects of metals and acidification

Table 15.3 (continued)

Reference	Volume (liters)	Description	Studies
Hall et al. 1970	650,000	20 ponds	effects of nutrients and predators
Stephenson et al. 1984	1,000,000	enclosures with sediments (3)	effects of size
Jones 1973	18,000,000	rubber limnocorrals UK, Lund tubes	effects of enclosure, nutrients, and mixing
Jones 1975			
Jones 1976			
Lack and Lund 1974			
Lund 1972			
Lund 1978			
Lund and Reynolds 1982			
Reynolds 1983			
Reynolds 1985			
Reynolds et al. 1983			
Reynolds et al. 1985			
Smyly 1976			
Thompson et al. 1982			
Baccini et al. 1979	—	limnocorral, MELIMEX	heavy metals
Bachter 1979			
Imboden et al. 1979			
Lang and Lang-Dobler 1979			

inexpensive procedure is to take intact portions of sediment (generally cores) into the laboratory. With flowing seawater on top, such cores have been held for up to three months without extensive deviation from normal chemical and biological activity (Kelly and Nixon 1984). Using such techniques, benthic fluxes and denitrification rates can be measured directly (Kelly and Nixon 1984; Nixon 1969; Seitzinger et al. 1984). Cores from the field can also be held in laboratory flumes and exposed to different water and flow regimes (see review by Muschenheim et al. 1986).

Sediment chambers have also been used to enclose patches of benthos in the field. Gust (1977) used both static and flow-through boxes to monitor the behavior of *Fucus*. Two chambers were used for nine months in Norway to examine the effects of oil in situ (Bakke and Johnson 1979; Bakke et al. 1982). The Loch Ewe facility has deployed three fiberglass cylinders for eleven months to examine the effects of sewage sludge (Eleftheriou et al. 1982). Rectangular microcosms containing phytoplankton and individual species of bivalve and fish were also used there to study the effects of nutrients, cadmium, copper, mercury, and lead (McIntyre 1977; Saward et al. 1975; Steele 1979).

Pelagic Marine Microcosms

Table 15.4 summarizes the best known pelagic microcosms, or "big bags." These consist of plastic containers (generally in the form of inverted silos) which are anchored to a floating metal framework or raft. They usually contain no sediments. The enclosed water columns may be studied for several days or weeks before diverging significantly from the surrounding waters. Case (1978) has presented a good summary of the engineering necessary to maintain such large-scale microcosms. Two similar, highly publicized microcosm facilities were begun in Canada

Table 15.4. Some representative pelagic marine microcosms

Reference	Volume (liters)	Location/ description	Studies
Oiestad 1982	—	Norway	fish larvae
Lacaze 1974	—	Rance, France 15 bags in lagoon, no mixing	effects of oil & dispersants
Koike et al. 1982 Kuiper 1977a Kuiper 1977b Kuiper 1981 Kuiper 1982 Kuiper et al. 1982	1,500	Den Helder, Netherlands 8 plastic bags	4–6 weeks, replication, plankton growth, mercury, phenol, chlorophenol, dichlorophenol, dichloroaniline
Brockmann et al. 1977a Brockmann et al. 1977b Brockmann et al. 1979 Brockmann et al. 1974	4,000	Helgoland, W. Germany 3 bags, 5m deep mixed continuously	phytoplankton succession
Boyd 1981	10,000	Marsaille, France	—
Barth 1984 Sjkoldal et al. 1982 Sjkoldal et al. 1983 Throndsen 1982 Tjessem et al. 1984	15,500	Lindaspollene, Norway 2 bags, to 20m deep, stratified water	2 weeks, microcosm operation 2 months, fate & effect of oil and dispersant
von Bodungen et al. 1976 Smetacek et al. 1976 Smetacek et al. 1982	30,000	Kiel Bight, 4 bags, anchored in sediment, mixed by bubbling	effects of light/dark and different sediments on water column
Brockmann et al. 1979 Kattner et al. 1982	31,000	POSER (Norway) 3 bags, 40m deep	1 mo., comparison with sea, mercury, oil
Antia et al. 1963 McAllister et al. 1961 Strickland and Trehune 1961	113,000	Nanaimo, B.C., Canada 6m diam. plastic sphere, filtered water	22 days, photosynthesis
see Table 5	100,000 300,000	Loch Ewe, Scotland bags, 17m deep	see Table 5
see Table 6	68,000	CEPEX, Saanich Inlet	see Table 6

Table 15.5. Some Loch Ewe experiments

Experiment	Reference
Comparison with bay, circulation & temperature	Steele et al. 1977
Comparison with bay, 46 days, 1 big bag	Davies and Williams 1984
Trophic structure	Steele and Baird 1968
Growth of herring larvae	Gatten et al. 1983
Predation effects on fish larvae	Steele and Gamble 1982
Mercury, fate and effect, 3 big bags, control, 1 μg/l, 10 μg/l	Davies and Gamble 1979
Mercury, comparison with bay	Topping et al 1982
Copper & nutrients, 1 month, small bags	Gamble et al. 1977
Copper, comparison with CEPEX	Topping and Windom 1977
Oil, fate & effects, 2 months, 2 big bags	Davies et al. 1980
Glucose, effects on production	Irvine et al. 1985
	Ducklow et al. 1986
Methods test, primary productivity	Davies et al. 1975

(CEPEX) and Scotland (Loch Ewe) during the 1970's. Tables 15.5 and 15.6 summarize some of the results from them.

Land-Based Marine Microcosms

Table 15.7 shows the wide range of sizes and experimental designs used for land-based microcosms. These systems present fewer experimental difficulties than bags, being easier to control. Although, they are usually considered less realistic and more likely to deviate from conditions obtained in the ocean, some land-based systems have operated for several years without large-scale deviations from nature, much longer than pelagic systems. The effects of enclosure over time in ecosystems have been detailed by Odum et al. (1963), deWilde and Kuipers (1977), and Polk (1978).

The best-known land-based microcosm now operating is the Marine Ecosystem Research Laboratory (MERL), where 14 tanks have been used to study how such systems operate, their basic ecology and geochemistry, and the ecotoxicology of a variety of pollutants. Table 15.8 summarizes the results of experiments at MERL.

15.2 Applicability of Microcosm Results to Nature

15.2.1 Properties Controlling Applicability

In order that results from microcosms be applicable to nature, the processes relevant to the parameters being studied should operate similarly in the microcosms as in naturally operating ecosystems. Studies of differ-

Table 15.6. Some CEPEX experiments

Experiment	References
Description, comparison with nature	
General description	Menzel and Steele 1978
Description	Parsons 1978a
Comparison with bay	Takahashi and Whitney 1977
Comparison with warm core rings	Menzel 1980
Zooplankton comparison, CEPEX vs lab vs bay	King 1982
Comparison, plankton ecology	King et al. 1980
Summary 1977	Menzel 1977
Limitations	Menzel 1980
Replication, 4 bags, 30 days	Takahashi et al. 1975
Testing of methods	
Zooplankton sampling	Williams et al. 1977
Sampling comparison	Lawson and Grice 1977
Cadmium I & II (2 bags, 68,000 l, 1 month)	
Effects	von Brockel 1982
Speciation and fate	Kremling et al. 1978
Copper (1 control, 5 μg/l, 10 μg/l)	
(2 controls, 10 μg/l, 50 μg/l)	
Effect on bacteria	Vaccaro et al. 1977
Effects on phytoplankton	Goering et al. 1977
Effect on phytoplankton	Harrison et al. 1977
Effect on primary production	Thomas et al. 1977a
Effect on algal species	Thomas and Seibert 1977
Effects on microzooplankton	Beers et al. 1977b
Effect on zooplankton feeding	Reeve et al. 1977a
Effects on ctenophores	Gibson and Grice, 1977
Effect on fish larvae	Koeller and Parsons 1977
Effect on nitrogen cycling	Harrison and Davies 1977
Biological mediation of fate	Topping and Windom 1977
Mercury (3 bags, 1,300,000 l, 1 control, 1 μg/l, 5 μg/l, 72 days)	
Comparison with Kiel and Loch Ewe	von Brockel 1982
Effects on bacteria	Azam et al. 1977
Effects on zooplankton	Beers et al. 1977a
Summary	Grice and Menzel 1978
Summary	Grice et al. 1977
Effect on nutrient cycling (lab)	Harrison et al. 1978
Effect on salmon chum	Koeller and Wallace 1977
Effect on bacteria (lab)	Koike et al. 1978
Effects	Takahashi et al. 1977
Effect on algal species and production	Thomas et al. 1977b
Biological mediation on fate	Wallace et al. 1982

Table 15.6 (continued)

Experiment	References
Petroleum — No. 2 fuel oil (2 exp, 6 bags, 68,000 l 2 control, 10 μg/l, 20 μg/l, 2 at 40 μg/l, 1 month, water-soluble fraction)	
Effects on bacteria	Hodson et al. 1977
Effects on phytoplankton	Parsons et al. 1976
Fate and effect	Lee and Takahashi 1977
Fate and effect	Lee et al. 1977
Other organics	
Fate of Prudhoe Bay crude oil, polycyclic aromatic hydrocarbons, and radiolabeled benzo(a)pyrene	Lee et al 1978
Fate and effect of naphthalenes (2 bags, 68,000 l, 20 days)	Lee and Anderson 1977
Effects of polychlorinated biphenyls (2 bags, 68,000 l, 20 days)	Iseki et al. 1981
Effects of pentachlorophenol (3 bags, 68,000 l, 25 days)	Whitney et al. 1981
Effects of glucose (3 bags, 68,000 l, 25 days)	Parsons et al. 1981
Nutrients	
Nitrogen-15 tracer, effects of Cu on cycling	Hattori et al. 1980
Nitrogen-15 tracer study of cycling	Koike et al. 1982
Effects of added nutrients	Parsons et al. 1977a
Effects of added nutrients	Parsons et al. 1977b
Effects (phytoplankton) of added nutrients	Takahashi et al. 1982
Nutrient effects (lab)	Reeve et al. 1977b
Foodweb (3 bags, 1,300,000 l, 111 days)	
Effects of manipulation on phytoplankton species	Davis 1982
Effects of stirring on phytoplankton species	Eppley et al. 1978
Summary of experiment	Grice et al. 1980
Summary	Harris et al. 1982
Growth of herring larvae	Houde and Berkeley 1982
Factors causing phytoplankton blooms	Parsons et al. 1978
Effects of different size phytoplankton on secondary production	Reeve et al. 1982
Effects of mixing on phytoplankton	Sonntag and Parsons 1979
Seaflux (3 bags, 68,000 l, 2–3 weeks)	
Effects of copper and glucose on bacteria	Seki et al. 1983
Effects of oil & dispersant on bacteria	Lee et al. 1984
Fate and microbial effect of oil & dispersant	Wong et al. 1984

Table 15.7. Some representative land-based marine microcosms

Reference	Volume (liters)	Description	Controlled parameters T,L,S,M,W	Studies/comments
Pratt 1949	—	natural pond	n,n,y,n,n	eutrophication
Nadeau and Roush 1973	—	salt marsh	u,n,y,y,y	3 months, fate of oil
de la Cruz 1982	—	4 ponds	n,n,y,n,n	oil effects
Bourquin et al. 1979 Tagatz et al. 1983 Tagatz et al. 1981 Wilkes 1978	.125 to 125	five separate systems		fate & effects of organics
Raymont and Adams 1958	15	battery jars	n,y,n,n,n	effects of light and CO
Spies and Parsons 1985	20	12 flasks	y,y,n,y,n	effects of salinity
Nixon et al. 1979 Oviatt et al. 1979 Oviatt et al. 1977 Perez et al. 1977	150	12 cylinders	y,y,y,y,y	1 month, replication, effects of temperature, light, turbulence, relative benth area, predation, turnover rate, sewag effects at different seasons
Henderson and Smith 1980	500	boxes	n,n,y,y,y	naturally colonized benthos, 4 months, eutrophication
Evans 1977 Henderson and Smith 1980	500	12 boxes	n,n,y,y,y	naturally colonized benthos, eutrophication, copper, effect (contaminated sediment on overlying water
Kemp et al. 1980	55 700	aquaria 6 tanks	u,u,y,u,u	effects of herbicides o macrophytes
Fulton 1984a Fulton 1984b	600	cylinders, 40 cm diam.	n,n,n,y,n	eutrophication
Heinle et al. 1979	750	6 cylinders	n,y,y,n,n	sewage effects
Legendre et al. 1985	1,200	fiberglass tank	y,y,n,n,n	effects of tides
Malone et al. 1975	2,000	4 concrete tanks	n,n,n,y,y	nutrients
Raymont and Adams 1958	2,400	concrete ponds	n,y,n,n,n	effects of light and CC
Wilson 1978 Brown et al. 1978	4,000	5 tanks, .9 m diam.	n,n,n,n,y	fate of oil & dispersan
Hagstrom 1977	4,200	5 plastic ponds	n,n,y,y,y	oil in littoral zone
Edmondson 1955	5,000	concrete tank	n,n,n,n,n	eutrophication
Gordon et al. 1976	8,000	2 tanks, 4 m diam.	n,n,y,n,n	3 months, fate of oil
Linden et al. 1987	8,000	6 tanks	n,n,y,y,y	4 months, effects of o and dispersant
Adey 1983 Walton 1980 Tangley 1985	10,500	reef & lagoon intertidal tank	y,y,y,y,n	demonstrations, fish added
see Table 15.8	13,000	insulated tank, fiberglass	y,n,e,y,e	see Table 15.8

Table 15.7 (continued)

Reference	Volume (liters)	Description	Controlled parameters T,L,S,M,W	Studies/comments
		and resin, sediment tray, 1.8 m diam., 5.5 m high		
Edmondson and Edmondson 1947	18,000	concrete tank	n,n,n,n,n	eutrophication
Raymont and Miller 1962	20,000	2 concrete ponds	n,y,n,n,n	11 weeks, enclosure effects
Kuiper et al. 1984	25,000	4 tidal flats	n,n,y,y,y	14 months, oil effects
de Wilde and Kuipers 1977	25,000	2 tidal flats	y,y,y,y,n	2 years, observations
Petterson et al. 1939	37,000	tower, 2 m diam.	y,y,u,u,n	sterilized water
Fanuko 1984		2 basins in a lagoon	n,n,y,n,n	2 years, sewage, 63 m^2
Odum et al. 1963	to 57,000	concrete ponds	n,n,y,n,y	enclosure effects
Balch et al. 1978 Eppley et al. 1968 Hirota 1974 Kamykowski and Zentara 1977 Mullin and Evans 1974 Paffenhofer 1976 Strickland and Trehune 1961	73,000	Scripps Deep Tank insulated steel, viewing ports, 3 m diam, 10 m high black plastic-lined	n,e,n,n,e	plankton behavior & trophic dynamics
Balch et al. 1978 Conover and Paranjape 1977 Kamykowski and Zentara 1977 Lane and Collins 1985	108,000	Dalhousie Aquatron concrete tank viewing ports 3.7 m diam., 10.5 m high	y,y,n,n,e	4 months, zooplankton behavior, species succession, nutrients
Brown and Parsons 1972	450,000	tank, 24 m diam.	n,n,n,y,y	upwelling production
Hood 1978	852,000	2 ponds	n,n,y,y,y	upwelling production
Oiestad 1982	24,000 2,500,000 4,400,000	concrete basins	n,n,n,n,n	enclosure effects, fish larvae
Paul et al. 1979	1,000,000	pond	n,n,y,y,y	upwelling production
Polk 1978	1,000,000,000	sluice-dock	n,n,y,n,n	enclosure effects

T = temperature: *yes* = controlled, *no* = not controlled, e = either possible, u = unknown
L = light: *yes* = controlled (artificial), *no* = not controlled (natural sunlight), e = either possible
S = sediment: *yes* = present, *no* = not present, e = either possible, u = unknown
M = mixing: *yes* = mixed, *no* = not mixed, e = either possible, u = unknown
W̄ = water turnover: *yes* = turnover, *no* = no turnover, e = either possible, u = unknown

Table 15.8. Some MERL experiments

Experiment	References
Microcosm operation	
Replication, 9 tanks, 4 months	Pilson et al. 1979
Variability	Smith et al. 1979
Replication	Smith et al. 1982
Comparison with bay, metabolism	Oviatt et al. 1981a
Comparison with bay, multivariate technique	Oviatt et al. 1980
Comparison with bay, nutrients, temperature, benthic fluxes, chlorophyll-a	Pilson et al. 1980
Nutrient cycling	Nixon and Pilson 1983
Particle rates (settling, removal, reworking) measurement of air-sea gas exchange (radon)	Bopp et al. 1981
Effectiveness of wall cleaning, stagnant boundary film thickness	Caron and Sieburth 1981a Santschi et al. 1982
Measurement of turbulence	Nixon et al. 1980
Effects of mixing, stratification	Oviatt 1981
Effects of "storm" mixing	Oviatt et al. 1981b
Radiolabeled metals	
Fate: Cd, Co, Cr, Fe, Hg	Adler et al. 1980
Fate	Amdurer et al. 1983
Fate, winter vs. summer, Am, As, Ba, Be, Cd, Co, Cr, Cs, Fe, Hg, Mn, Pa, Pb, Po, Pu	Santschi et al. 1982
Fate, summary	Amdurer et al. 1982
Fate comparison, Th, 1501, MERL, bay	Santschi et al. 1980
Removal Rates, Cs, Pb, Pu, Th, MERL vs. bay	Santschi et al. 1983b
Bioaccumulation, benthic organisms	Frithsen 1984
Metals	
Fate, Mn, biological mediation	Hunt 1983a
Fate, Mn, benthic flux, seasonal variation	Hunt 1983b
Fate, Mn, during diatom bloom	Hunt and Smith 1980
Speciation, Cu	Johnson 1978
Speciation, As, during diatom bloom	Johnson and Burke 1978
Fate of copper	
Petroleum	
Preliminary results	Pilson et al. 1977
Effects of fouling communities	Caron and Sieburth 1981b
Effects on bacterioplankton & nanoplankton	Davies et al. 1979
Effects on hydrocarbon-degrading bacteria	Lee and Ryan 1983
Effects on phytoplankton species	Vargo et al. 1982
Effects on zooplankton physiology	Vargo 1981
Effects on zooplankton feeding	Berman and Heinle 1980
Effects on phytoplankton, zooplankton, macrobenthos, and meiobenthos	Elmgren et al. 1980
Effects on meiofauna	Frithsen et al. 1985

Table 15.8 (continued)

Experiment	References
Effects on benthic organisms	Grassle et al. 1981
Effect on metal cycling	Hunt and Smith 1982
Comparison of effects with *Tsesis* spill	Elmgren and Frithsen 1982
Fate, preliminary results	Gearing et al. 1979
Effects on benthic organisms	Grassle et al. 1981
Effect on metal cycling	Hunt and Smith 1982
Comparison of effects with *Tsesis* spill	Elmgren and Frithsen 1982
Fate, preliminary results	Gearing et al. 1979
Fate, effect of turbulence & suspended load	Gearing and Gearing 1983
Fate, variation with temperature and biological activity	Gearing and Gearing 1982a
Fate, volatilization	Gearing and Gearing 1982b
Fate, in sediments	Gearing et al. 1980
Fate of saturated hydrocarbons in sediments	Wade and Quinn 1980
Bioaccumulation in macrobenthos	Farrington et al. 1982
Summary & comparison with published results	Olsen et al. 1982
Summary & synthesis	Oviatt et al. 1982
Other organics	
Fate, C1 to C4 hydrocarbons	Bopp et al. 1981
Fate, toluene	Wakeham et al. 1985
Fate, 2- to 4-ring PAH	Gearing and Gearing 1983
Fate, winter, benzanthracene	Hinga 1984
Fate, summer, benzanthracene	Hinga et al. 1980
Fate, dimethylbenzanthracene	Hinga et al. 1986
Fate, benzanthracene, dimethylbenzanthracene, and pentachlorophenol	Lee et al. 1982
Fate, C6 to C17 hydrocarbons and halogenated hydrocarbons	Wakeham et al. 1982
Fate, C6 to C17 organics, variation with season and biological activity	Wakeham et al. 1983
Recovery of polluted sediments	
Fate, Cd, Cu, Mn, Pb	Hunt and Smith 1983
Fate, aromatic hydrocarbons	Pruell and Quinn 1985
Effects, summary	Oviatt et al. 1984
Eutrophication	
Effects, preliminary	Nixon et al. 1984
Effects, nutrients	Kelly et al. 1985
Effects, productivity	Oviatt et al. 1986a
Effects, zooplankton	Sullivan and Ritacco 1985a
	Sullivan and Ritacco 1985b
Effects, benthic organisms	Grassle et al. 1986
Other experiments	
Fate, acid iron wastes, stratified tanks	Brown et al. 1983
	Fox et al. 1986

Table 15.8 (continued)

Experiment	References
Test of sampling methods	Frithsen et al. 1983
Test of methods, hydrocarbons	Gearing et al. 1978
Variability in methods	Gearing et al. 1979
	Oviatt et al. 1986b
Stable isotope distribution	Gearing et al. 1984
Fate and effect of sewage sludge	Oviatt et al. 1987

ent microcosms have shown parameters summarized in Table 15.9 to be particularly important in achieving this goal. Depth and size in general determine the number and size of organisms which can be grown. Larger size reduces effects caused by the presence of walls, changes in currents, and horizontal exchange, etc. and allows the experiment to continue longer (more water and organisms available for sampling without greatly affecting the overall population). Increasing size also necessitates higher costs and usually more intricate engineering. Primarily because of the size-cost relationship, much thought has gone into the effect of size. Boyd (1981) concluded that the "relative value of several studies employing microcosms seems to be almost inversely related to the volume chosen for manipulation." This fact has resulted from an overemphasis on engineering in some microcosms. The technical problems associated with maintaining large microcosms are important and interesting engineering exercises which perhaps should be funded separately from scientific studies in microcosms. The initial set-up and maintenance costs for a facility decline as time goes by, a point in favor of stable, long-term (5–10 year) funding for the larger microcosm complexes. On the other hand, it is possible to run intermediate-sized microcosms without expensive engineering, as is demonstrated by several cost-effective studies, for example the fate of oil in 8,000 liter tanks (Gordon et al. 1976) and the effects of eutrophication in 600 liter cylinders (Fulton 1984a; b). Much of the pioneering work on microcosms (e.g., Raymont and Miller 1962; Odum et al. 1963) was relatively inexpensive.

Factors other than size are also important, if not usually as costly. Natural levels of light and nutrients are needed to maintain primary production. Sediments provide a natural source and sink for nutrients as well as a habitat for benthic animals where shallow environments are the prototypes. A natural assemblage of organisms representing several trophic levels is necessary for measuring indirect effects of pollutants caused by species interactions such as competition and predator–prey relationships (Cairns 1980). This is a fundamental property which distinguishes microcosms from laboratory experiments on individual species, and which makes microcosms particularly useful for ecotoxicology.

Table 15.9. Controlling properties of microcosms

Property	Reference
Depth	
should not restrict zooplankton migration	Dudzik et al. 1979
Lifetime of possible experiments	
longer than generation time of organisms	Menzel and Case 1977
Light	
quantity and quality affect production	Adams and Giddings 1982
	de la Cruz 1982
	Nixon et al. 1979
	Oviatt et al. 1979
	Parsons et al. 1978
	Perez et al. 1977
	Pilson and Nixon 1980
	Smetacek et al. 1982
Nutrients	
	Davis 1982
	Parsons et al. 1978
Sediment	
source of nutrients and importance for seasonal cycling	Adams and Giddings 1982
	Donaghay 1984
	Mullin 1982
	Parsons 1982
	Perez et al. 1977
	Pilson and Nixon 1980
	Schindler et al. 1977
	Smetacek et al. 1976
	Smyly 1976
Size	
necessity for several trophic levels and for repeated sampling—must trade-off with expense	Adams and Giddings 1982
	Adler et al. 1980
	Draggan and Reisa 1980
	Menzel and Case 1977
	Pilson and Nixon 1980
necessity for natural behavior of organisms	Boyd 1981
reduces artifacts (wall effects, etc.) and increases extrapolatability	Gamble and Davies 1982
	Giesy and Odum 1980
	Kemp et al. 1980
	Nadeau and Roush 1973
not always important, especially with radioisotopes	Giddings 1980
	Harte et al. 1980
Temperature	
especially important for chemical and biological rates	Adams and Giddings 1982
	Gearing and Gearing 1982a
	Hoffman and Horne 1980

Table 15.9 (continued)

Property	Reference
	Oviatt et al. 1979
	Wakeham et al. 1983
Trophic levels	
several needed for ecosystem functioning	Menzel and Case 1977
presence of predators important	Oviatt et al. 1979
	Pilson and Nixon 1980
	Santschi et al. 1982
	Steele and Gamble 1982
Turbulence	
	Davis 1982
	Donaghay and Klos 1985
	Eppley et al. 1978
	Gearing and Gearing 1983
	Harte et al. 1980
	Mullin 1982
	Nixon et al. 1980
	Oviatt 1981
	Oviatt et al. 1981b
	Parsons et al. 1978
	Perez et al. 1977
	Pilson and Nixon 1980
	Polk 1978
	Reynolds 1983
	Reynolds et al. 1983
	Sonntag and Parsons 1979
Turnover	
to provide nutrients	Paul et al. 1979
for recruitment and nutrients	Perez et al. 1977
	Pilson and Nixon 1980
Wind	
for mixing and evaporation	de Wilde and Kuipers 1977
	Oil and Gas Working Group 1978
	Bower and McCorkle 1980

Physical parameters such as temperature, turbulence, and wind speed have also been shown to have important effects on the functioning of microcosms. The physical environment is as much a part of an ecosystem as the biological environment, as recent studies have shown (Legendre and Demers 1984). Physical parameters affect not only the fate and availability of pollutants (rates of evaporation and dissolution, type of particles present, rates of chemical and biological degradation) but also the behavior of exposed organisms. Detailed examinations of the extents and mechanisms of effects for all these parameters are given in the references listed in Table 15.9.

Ideal conditions can never be obtained, especially when trying to maintain experimental control. For each individual experiment, it is necessary to make compromises based on money available, manpower, purpose of experiment, etc. It is, however, very important to bear in mind that each deviation from reality imposes certain limitations on the applicability of the final results. For example, Santschi and co-workers (Adler et al. 1980; Santschi et al. 1980) found similar results from short-term experiments conducted both in 150 liter and in 13,000 liter microcosms, with the larger systems, however, being necessary for long-term (>1 month) experiments. In papers describing microcosm work, the limitations of applicability of results (if known) should be clearly stated.

15.2.2 Tests for Applicability

The bugbear of researchers working in microcosms is that their results will not be believed because they were obtained in microcosms. Verduin (1969) speaks for some scientists by saying that microcosms have not been proven to be valid or even superior to laboratory experiments. This is especially worrisome when considering the large amount of time and money that has been spent on designing, maintaining, and conducting experiments in microcosms. It is a measure of our lack of knowledge of any ecosystem that there is "no generally agreed upon criteria for evaluating the credibility of living models or for comparing the behavior of microcosms with the 'real' systems they represent" (Oviatt et al. 1977). Microcosms must simply be compared with natural systems and judgments made of their similarity based on a limited number of observations. Results from microcosms will not always be applicable to nature because ecosystems are so variable. However, comparisons of microcosm and field results have shown, in general, good applicability from well-designed microcosms. Microcosms have a valuable place in an overall scheme of research including all types of experiments.

Although the question of microcosm validity cannot be answered unambiguously, many comparisons have now been made in many different microcosms over different time periods. Table 15.10 summarizes some of these comparisons, the general conclusion being that newly isolated microcosms are not greatly different from natural ecosystems. Some of the microcosms have functioned successfully (with physical parameters and seasonal cycles of nutrients and plankton similar to those in nature) for several years—particularly the British limnocorrals (Lack and Lund 1974; Lund 1972). Some MERL microcosms have been run for over two years without great deviations from behavior predicted from studies of Narragansett Bay (e.g., Oviatt et al. 1981a). There appears to be a time constraint on the use of some of the pelagic microcosms. Unmixed bag

Table 15.10. Comparisons of microcosms with natural ecosystems

Microcosm	Duration	Observations made[a]	Reference
General			
Loch Ewe	—	overall not the same as surrounding waters	Steele and Gamble 1982
aquaria	1 year	qualitatively similar	de la Cruz 1982
Loch Ewe	—		Davies et al. 1979
100 ml flasks	—	qualitatively similar to lakes	Heath 1980
boxes	—	similar to sewage lagoons	King 1980
4 salt marsh boxes	—	similar	Kitchens et al. 1979
3 tanks	—	generally similar, some differences	Odum et al. 1963
12, 150 l, EPA	—	similar to Narragansett Bay	Oviatt et al. 1977
Lindespollene	2 weeks	similar to bay	Skjoldal et al. 1983
McGill bags	1 week	similar in temperature, salinity, oxygen, chl-a	de Lafontaine and Leggett 1987
Particulates			
MERL	—	particulate load & particulate alumina	Hunt and Smith 1982
bags	—	sedimentation rate (+)	von Brockel 1982
MERL	—	particle removal rates, sedimentary rates of reworking, & air-sea film thickness (+)	Santschi 1982
Physical properties			
limnocorral	—	mixing energy (−)	Boyce 1974
limnocorral	3 months	light	Muller 1980
benthic chamber	—	turbulence	Gust 1977
MERL	—	vertical (+) & horizontal (−) diffusivity	Nixon et al. 1980
CEPEX	—	sinking rate of fecal pellets (−)	Bienfang 1982
Loch Ewe	—	turbulence (−) and temperature	Steele et al. 1977
CEPEX	—	temperature, light (−), & salinity	Takahashi and Whitney 1977
limnocorral	2.5 years	temperature & plankton (qualitatively the same)	Lack and Lund 1974; Lund 1972
MERL, 9 tanks	—	temperature, chlorophyll-a, benthic fluxes of nutrients	Pilson et al. 1980 Pilson 1985
700 l aquaria	6 months	nutrient amounts (−) & fluxes (−)	Kemp et al. 1980
MERL	—	air-sea gas exchange	Santschi et al. 1982
Organisms			
POSER	20 days	phytoplankton succession	Kattner et al. 1982

Table 15.10 (continued)

Microcosm	Duration	Observations made[a]	Reference
Den Helder bags	8 weeks	plankton behavior	Kuiper et al. 1982
limnocorral	3 months	biomass, species composition, diversity index	Muller 1980
Scripps Deep Tank	—	production ratios (primary, secondary, tertiary)	Mullin and Evans 1974
MERL	1 year	phytoplankton P/R, yearly averages	Oviatt et al. 1981a
limnocorral	22 months	zooplankton (qualitatively the same)	Smith et al. 1979
Scripps Deep Tank	—	zooplankton feeding behavior	Paffenhofer 1976
CEPEX	—	zooplankton growth rates	King 1982
Aquatron	4 months	zooplankton community	Conover and Paranjape 1977
CEPEX	—	larval fish growth (−)	Houde and Berkeley 1982
MERL	—	benthic community structure	Grassle et al. 1981
Other			
bags	—	weathering of oil	Barth 1984
MERL	—	oil fate & effects, versus *Tsesis*	Elmgren and Frithsen 1982
MERL	—	fate of thorium	Santschi et al. 1980
MERL	—	metal removal rates	Santschi et al. 1983b
700 l enclosure	—	response to arsenic	Giddings and Eddlemon 1979
model stream	—	effects of insecticide	Maki 1980

[a] (+) = higher in microcosm; (−) = lower in microcosm; no indication is given when the microcosm is similar to nature.

microcosms often start to deviate significantly from the surrounding water column after a few weeks (Takahashi and Whitney 1977; Kattner et al. 1982; Kuiper et al. 1982). In their present state, the big bags may be more suitable for short-term (<1 month) experiments.

15.3 Comparison of Microcosms with Other Experimental Approaches

15.3.1 Disadvantages of Microcosms

Some of the drawbacks attributed to microcosms are listed in Table 15.11. Principally, it can be argued that: many systems are too large to have

Table 15.11. Disadvantages of microcosms

General	
costly	Davies and Gamble 1979
	Parsons 1978a
	Steele 1979
cannot model *all* significant environmental	Bourquin et al. 1979
processes	Manuel and Minshall 1980
loss of natural refuges	Menzel 1977
problems with new engineering	Santschi 1982
difficult to fill replicably without stressing	Grice et al. 1977
organisms	
difficult to fund and manage because of large	
scale	
Wall effects	
attenuate light	Goldman 1962
wall material may be biologically and	Dudzik et al. 1979
chemically active	
substrate for organisms not natural to	Balch et al. 1978
ecosystem	
walls must be cleaned (brushed usually)	Caron and Sieburth 1981a
periodically	Caron and Sieburth 1981b
	Giddings 1980
	Grice et al. 1980
	Harte et al. 1980
	Hellebust et al. 1975
	Odum et al. 1963
	Santschi 1982
no horizontal exchange, less or no	Giddings 1980
recruitment	Menzel 1980
	Odum et al. 1963
	Parsons 1978a
Light	
not natural, affected by walls, super-	Balch et al. 1978
structure, etc.	Menzel 1977
Turbulence	
often less than in nature, turbulent processes	Boyce 1974
may differ, turbulence affects many other	Davies and Gamble 1979
processes	Davis 1982
	Harte et al. 1980
	Imboden et al. 1979
	Menzel 1977
	Menzel 1980
	Mullin 1982
	Nixon et al. 1979
	Odum et al. 1963
	Parsons 1978a
	Santschi 1982

Table 15.11 (continued)

Replicates/controls	
not possible as defined for laboratory	Hill and Wiegert 1980
experiments	Steele 1979
Samples	
limited amount of sample available	Balch et al. 1978
difficult to sample representatively	Santschi 1982
Size	
limits number of trophic levels	Dudzik et al. 1979
larger size means more work, more expense,	Lund 1978
fewer replicates, and less control	Steele 1979
Complexity	
harder to interpret results, less information	Bowling et al. 1980
on individual processes, less isolation of	Bretthauer 1980
cause and effect	Giddings 1980
	Grice et al. 1980
	Kuiper 1984
	Parsons et al 1976
	Steele 1979
Validity	
not proven valid or superior to laboratory	Donaghay 1984
results, needs field verification	Heath 1979
	Hill and Wiegert 1980
	King 1980
	Verduin 1969

good experimental control and replication, such as is possible for laboratory experiments; they are, in general, too costly, with the emphasis often on engineering to the detriment of science; they are too complex, making it more difficult to interpret results and to isolate causes and effects; and their results cannot be used to predict effects in nature because they differ from natural systems in significant ways.

These arguments have merit. Microcosms are far from perfect analogs of nature. However, the principal objection is one of cost versus result. As microcosms are used more and understood better, the ratio of their cost to result is decreasing. Microcosms make possible the experimental manipulation of entire ecosystems; as such they should be further developed and used. Their defects should not be overlooked but examined closely in an effort to better understand both microcosms and nature.

15.3.2 Advantages of Microcosms

Just as they have real disadvantages, microcosms have distinct advantages for investigating some types of problems. Their unique advantage is in providing a means of studying a complete and functional ecosystem under controlled conditions. With them, one can investigate the overall results of pollutants on an ecosystem and extrapolate the results quantitatively to nature with reasonable confidence.

However, as detailed in the previous section, microcosms are neither completely natural nor completely controlled. Tests of their "naturalness" indicate small but definite differences between microcosms and nature (Table 15.10). As with naturalness, there is no consensus on what constitutes control. Many studies have been done to measure the similarity of different operating microcosms (Table 15.12). In general, microcosms are more similar to each other than any one microcosm may be to nature. Isensee (1976) has quantified the extent of variability to be expected among microcosms. In MERL, for example, nine separate microcosms were begun and run identically for four months (Pilson et al. 1979); later three tanks were run together and studied for over two years (Pilson et al. 1980). These studies showed that replication of microcosms is necessary to quantify subtle changes in population structure, because each

Table 15.12. Replicability studies of microcosms

Number of replicates	Duration of study	Type of microcosm	Reference
—	—	—	Isensee 1976
—	—	—	Smith et al. 1982
4	—	salt marsh	Kitchens et al. 1979
4	10 weeks	benthic chambers	Tagatz et al. 1983
4	10 weeks	benthic chambers	Tagatz et al. 1981
12	30 days	carboys	Abbott 1966
12	—	aquaria	Beyers 1963
10	10 weeks	aquaria	Brockway et al. 1979
6	18 months	aquaria	Cooper and Copeland 1973
12	15 days	150 l (EPA)	Oviatt et al. 1977
4	6 weeks	1,400 l	Kuiper 1977a
8	6 weeks	1,500 l bags	de Kock and Kuiper 1981
3	—	MERL (fate of metals)	Amdurer et al. 1983
9	4 months	MERL	Pilson et al. 1979
6	6 months	MERL	Pilson et al. 1980
3	1 year	MERL (natural levels of 5 metals)	Hunt and Smith, 1982
3	2+ years	MERL	Pilson et al. 1980
2	—	CEPEX (copper effects)	Beers et al. 1977b
3	—	CEPEX (sedimentation rates)	von Brockel 1982
3	10 days	CEPEX (1,300,000 l)	Grice et al. 1977
4	30 days	CEPEX (68,000 l)	Takahashi et al. 1975
2	—	isolated Arctic lakes	Snow and Scott 1975

microcosm operates independently. The tanks do diverge from each other, but only within limits of variability similar to that found spatially and temporally in a natural environment. One can conclude that individual microcosms show some of the different behaviors potential in the natural environment. All of the microcosms show seasonal patterns typical of nature. Thus, while different microcosms are not replicates as strictly defined for laboratory experiments (see Table 15.11), they replicate within the limits of natural variability. Such replication is necessary for statistically differentiating small, chronic effects of pollutants.

One other advantage of microcosms which has not been previously emphasized is their flexibility. Because of their size, they have not been considered as flexible as laboratory experiments. However, microcosms have been operated in a great number of different configurations, depending on the needs of a particular experiment. Their potential flexibility is just as great as that of laboratory experiments, although the costs are in some cases greater. Table 15.13 lists some of the different ways in which microcosms have been run. The only real restraints on microcosms are the controlling properties (Table 15.9) necessary to provide a realistic simulation of nature. Within these limits (and they are as wide as natural variation), microcosms can be and have been used to simulate a wide variety of environments and environmental conditions. Physical conditions such as light, wind speed, salinity, and turbulence have been varied. The effects of sediments have been examined: for example, how the presence of benthic predators or of contaminated sediments affects the overlying water. Pelagic predators have been added in some experiments and screened out for others. With such control, different possible sets of natural conditions can be studied scientifically without depending on chance storms, winds, etc.

In summary, experiments in microcosms can have advantages over field observations, laboratory experiments, and mathematical models for some studies. Each type of study provides different sorts of information. Laboratory experiments are cost effective for studying the mechanisms of individual processes and effects on particular organisms, but they are hard to extrapolate directly to a complex, natural ecosystem. Field observations provide the invaluable direct studies of nature, yet they are costly, impossible to control, and difficult to interpret. Results from one area at one time do not necessarily apply to other areas at other times. Mathematical models attempt to mesh the data from laboratory and field into a comprehensive, predictive tool, but are limited by the amount of data available. Microcosms act as biological models, making possible the direct observation and quantification of fates and effects of substances in replicable environments as close to nature as possible while still being under experimental control. Their specific advantages are catalogued in Tables 15.14, 15.15, and 15.16.

Table 15.13. Flexibility of microcosms

Variables altered	Reference
Air currents above water	Medine et al. 1980
	Oil and Gas Working Group 1978
	Parsons 1978a
Light	Balch et al. 1978
	Conover and Paranjape 1977
Range of organisms present	Davis 1982
	Tagatz et al. 1983
	Tagatz et al. 1981
Pressure	Boyd 1981
Salinity	Cooper and Copeland 1973
Sediment: presence and type	Bopp et al. 1981
	Gearing and Gearing 1983
	Oviatt et al. 1979
Size	Hellebust et al. 1975
Temperature	Gearing and Gearing 1982a
	Hinga 1984
	Hinga et al. 1980
	Kitchens and Copeland 1980
	Wakeham et al. 1983
Tidal mixing	Kitchens et al. 1979
	Kuiper et al. 1984
Turbulence (from unmixed to storm)	Gearing and Gearing 1983
	Oviatt 1981
	Oviatt et al. 1981b
Water: filtered (different mesh) or unfiltered	Balch et al. 1978
	Lawrence et al. 1979
	Strickland et al. 1969
Water: poisoned	Gearing and Gearing 1982b
	Wakeham et al. 1983
Water: stratified	Balch et al. 1978
	Brockway et al. 1979
	Brown and Kester 1984
	Donaghay 1984
	Donaghay and Klos 1985
	Fox et al. 1986
	Goldman 1962
	Reynolds 1983
	Strickland et al. 1969

Table 15.14. Advantages of microcosms over nature

Can use radiotracers	Amdurer et al. 1983
	Hattori et al. 1980
Less costly, especially compared to open	Elmgren et al. 1980
ocean	Harte et al. 1980
Sampling is easier, less time consuming, and	Balch et al. 1978
possibly more accurate	Elmgren et al. 1980
	Ringelberg 1976
More experimental control (see Table 15.13)	Balch et al. 1978
	Davis 1982
	Donaghay 1984
	Fulton 1984b
	Gibson and Grice 1980
	Giddings 1980
	Harte et al. 1980
	Leffler 1980
	Menzel 1977
	Menzel and Steele 1978
	Oviatt et al. 1984
	Ringelberg 1976
Restricted environment more easily quantified,	Amdurer et al. 1983
mass balances possible	Donaghay 1984
	Harrison and Davies 1977
	Harte et al. 1980
	Oiestad 1982
Can observe and sample same water column	Balch et al. 1978
over time—important for plankton work and	Davies and Gamble 1979
for measuring rate processes under semi-	Gibson and Grice 1980
natural conditions	Giddings 1980
	Grice et al. 1977
	Menzel 1977
	Menzel and Steele 1978
	Mullin 1982
	Oiestad 1982
Replicates and controls may be possible	Davies and Gamble 1979
	Donaghay 1984
	Giddings 1980
	Harte et al. 1980
	Menzel and Steele 1978
	Nadeau and Roush 1973
	Parsons 1982
Because flexible, results may be extrapolated	Nadeau and Roush 1973
to more than one specific area in nature	

Table 15.15. Advantages of microcosms over laboratory studies

Can grow and study more different organisms	Balch et al. 1978
	Donaghay 1984
	Elmgren and Frithsen 1982
	Parsons 1978a
Plankton grown at low concentrations, more natural	Grice et al. 1977
	Oiestad 1982
	Smith et al 1982
Behavior of organisms more natural	Boyd 1981
Can do experiments on behavior	Conover and Paranjape 1977
Can study several trophic levels living together	Kuiper 1982
	Parsons 1978a
	Saward et al. 1975
Long-term and seasonal studies possible	Gamble and Davies 1982
	Grice et al. 1977
	Lund 1978
	Maguire 1980
Artifacts minimized	Nadeau and Roush 1973
More comparable to nature in complexity	Kuiper 1982
	Steele 1979
Better chance of pollutant being in a natural physical and chemical form	Kuiper 1984
Results more directly applicable to nature	Farrington et al. 1982
	Hansen and Garton 1982
	Sabatini and Marcotte 1983
	Saward et al. 1975
	Steele 1979

Table 15.16. Advantages of microcosms over mathematical models

Results sometimes unpredicted by theory and therefore by models	Boyd 1981
	Nixon et al. 1984
Automatically integrates processes, no assumptions needed	Giddings 1980
All parameters included, not just those thought to be important	Mann 1979

15.4 Representative Results from Microcosms

The ecotoxicologic information obtained from microcosm experiments is proof of their usefulness. Such experiments have quantified effects not predicted by laboratory studies or mathematical models. Moreover, many of these effects have now been validated by field observations. This section contains a few representative examples of the value of using microcosms, in conjunction with other methods.

Table 15.17 lists references to microcosm work on various trace metals. Over twenty elements have been examined, with particular emphasis having been placed on arsenic, cadmium, copper, and mercury. Results from different microcosms in widely separated locations on these toxic metals have been reproducible within analytical variability (Kuiper 1984). Effects of metals on the ecosystem have been demonstrated at lower concentrations than those predicted from laboratory experiments on single species. Details of these studies have been well summarized by Santschi (1982) and Kuiper (1984).

Similar results have been obtained for organic pollutants. Table 15.18 lists some of the microcosm studies that have been reported to date. The fate and effects of petroleum and chemicals related to its exploration and control (drilling muds and chemical dispersants) have been extensively studied and well illustrate the type of results which can and cannot be obtained from microcosm experiments. Elmgren and Frithsen (1982) have compared the effects of an actual oil spill in the Baltic with the results predicted from a series of experiments at MERL and found very good agreement. The same groups of organisms were adversely affected in both cases and the temporal extents of measurable effects were similar. The effects were not, however, always those predicted from laboratory bioassays. For example, initially the oil slightly inhibited primary production by the phytoplankton, in agreement with laboratory studies. Over the course of several weeks, however, the deleterious effect of the oil on zooplankton and other phytoplankton grazers was so great that the overall result was an enhancement of primary productivity. This result was not predicted from mathematical modeling based on bioassays and has been valuable for interpreting field results.

The chemical fate of the oil added in MERL microcosm experiments has also been compared with results obtained in the field and predictions based on laboratory experiments (Gearing and Gearing 1982a, b; 1983; Gearing et al. 1980; Lee and Ryan 1983). The microcosms were able to give general ideas about which processes were the most important for particular compounds under natural conditions, but gave little idea of the mechanisms of individual processes. Laboratory experiments were necessary to elucidate each process. On the other hand, microcosms were able to give rates of transfer within the ecosystem (total loss, sedimenta-

Table 15.17. Microcosm studies of metals

Aluminum	Schindler et al. 1980a	Copper	Beers et al. 1977b
Americium	Santschi et al. 1983a		Evans 1977
	Santschi et al. 1980		Gamble et al. 1977
Antimony	Amdurer et al. 1982		Gibson and Grice 1977
	Santschi et al. 1983a		Harrison and Davies 1977
Arsenic	Amdurer et al. 1982		Harrison et al. 1977
	Brunskill et al. 1980		Hedtke 1984
	Giddings 1980		Hunt and Smith 1983
	Johnson and Burke 1978		Hunt and Smith 1982
	Santschi et al. 1983a		Johnson 1978
Barium	Amdurer et al. 1982		Koeller and Parsons 1977
	Jackson et al. 1980		Reeve et al. 1977a
	Santschi et al. 1983a		Saward et al. 1975
	Schindler et al. 1980a		Thomas et al. 1977a
Beryllium	Frithsen 1984		Thomas and Seibert 1977
Cadmium	Adler et al. 1980	Iron	Adler et al. 1980
	Amdurer et al. 1982		Amdurer et al. 1982
	Frithsen 1984		Brown et al. 1983
	Hunt and Smith 1982		Fox et al. 1986
	Hunt and Smith 1983		Frithsen 1984
	Kremling et al. 1978		Hunt and Smith 1982
	Kuiper 1982		Hunt and Smith 1983
	Marshall and Mellinger		Jackson et al. 1980
	1980		Santschi et al. 1980
	Medine et al. 1980		Santschi et al. 1983a
	Santschi et al. 1983a		Schindler et al. 1980a
	Williams and Giesy 1978	Lead	Hunt and Smith 1982
Cesium	Amdurer et al. 1982		Hunt and Smith 1983
	Jackson et al. 1980		Santschi et al. 1983a
	Frithsen 1984		Santschi et al. 1980
	Santschi et al. 1983a		Santschi et al. 1983b
	Santschi et al. 1983b	Manganese	Amdurer et al. 1982
	Schindler et al. 1980a		Frithsen 1984
Chromium	Adler et al. 1980		Hunt 1983a
	Amdurer et al. 1982		Hunt 1983b
	Frithsen 1984		Hunt and Smith 1980
	Jackson et al. 1980		Hunt and Smith 1982
	Medine et al. 1980		Hunt and Smith 1983
	Santschi et al. 1980		Jackson et al. 1980
	Santschi et al. 1983a		Santschi et al. 1983a
Cobalt	Adler et al. 1980		Schindler et al. 1980a
	Amdurer et al. 1982	Mercury	Adler et al. 1980
	Frithsen 1984		Amdurer et al. 1982
	Jackson et al. 1980		Azam et al. 1977
	Santschi et al. 1983a		Beers et al. 1977a
	Schindler et al. 1980a		Frithsen 1984

Table 15.17 (continued)

Mercury		Radium	Amdurer et al. 1982
(continued)			Santschi et al. 1983a
	Jackson et al. 1980	Selenium	Amdurer et al. 1982
	de Kock and Kuiper 1981		Jackson et al. 1980
	Kuiper 1977b		Rudd and Turner 1983a
	Kuiper 1981		and b
	Kuiper 1982		Salki et al. 1984
	Medine et al. 1980		Santschi et al. 1983a
	Rudd and Turner 1983a		Schindler et al. 1980a
	and b	Thorium	Jackson et al. 1980
	Santschi et al. 1983a		Santschi et al. 1983a
	Santschi et al. 1980		Santschi et al. 1980
	Schindler et al. 1980a		Santschi et al. 1983b
	Sigmon et al. 1977	Tin	Amdurer et al. 1982
	Steele 1979		Frithsen 1984
	Topping et al. 1982		Santschi et al. 1983a
	Turner and Rudd 1983	Vanadium	Amdurer et al. 1982
	Wallace et al. 1982		Jackson et al. 1980
Polonium	Santschi et al. 1983a		Schindler et al. 1980a
	Santschi et al. 1980	Zinc	Amdurer et al. 1982
	Santschi et al. 1983b		Frithsen 1984
Proactinium	Amdurer et al. 1982		Jackson et al. 1980
	Frithsen 1984		Medine et al. 1980
	Santschi et al. 1983a		Santschi et al. 1983a
			Schindler et al. 1980a

tion, degradation, etc.) which, often unlike those from laboratory experiments, agreed with those measured in the field.

The same group of microcosm experiments has also shown the value of microcosms for testing methods. Careful, detailed sampling of the sediment and analyses of oil with depth showed that the hydrocarbons were initially present in a mobile, flocculent layer at the sediment–water interface. This layer was not collected by the usual apparati used in the field. Because of these methods studies in a microcosm (Gearing et al. 1980; Frithsen et al. 1983), it was recognized that calculations of amounts of oil in sediments following a spill were generally too low (Boehm et al. 1979) and that field sampling methods for sediments needed to be modified.

Another example of how microcosms can be used to verify and calibrate methods is the discrepancy between the rates of microbial degradation of petroleum observed in the field and those measured in short-term

Table 15.18. Microcosm studies of organic chemicals

Oil effects
  General, Arctic lakes   Hellebust et al. 1975
  General, MOTIF tidal flats   Kuiper et al. 1984
  General, salt marsh microcosms   Nadeau and Roush 1973
  General, MERL, no. 2 fuel   Elmgren and Frithsen 1982
                                 Elmgren et al. 1980
                                 Olsen et al. 1982
                                 Oviatt et al. 1982
                                 Pilson et al. 1977
  General, CEPEX, no. 2 fuel   Lee and Takahashi 1977
                                 Lee et al. 1977
  General, Lindaspollen bags, Ekofisk   Skjoldal et al. 1982
    crude
  General, Arctic ponds, Norman Wells   Snow and Scott 1975
    & Pembina crudes
  General, Rance bags, Kuwait crude   Lacaze 1974
  General, Loch Ewe, North Sea crude   Davies et al. 1980
  General, CEPEX, Prudhoe Bay crude   Lee et al. 1978
  On primary producers, ponds   de la Cruz 1982
  On algae, ponds   Scott and Gloosckenko 1984
  On phytoplankton, MERL, no. 2 fuel   Vargo et al. 1982
    oil
  On plankton (flagellates), Lindaspollen   Throndsen 1982
    bags, Ekofisk crude
  On plankton (phyto- and zoo-), Arctic   Miller et al. 1978
    ponds
  On zooplankton, MERL, no. 2 fuel   Berman and Heinle 1980
                                   Vargo 1981
  On benthos, benthic chambers   Bakke et al. 1982
                                   Bakke and Johnson 1979
  On benthos, MERL, no. 2 fuel   Grassle et al. 1981
  On meiobenthos, MERL, no. 2 fuel   Frithsen et al. 1985
  On fauna, ponds   Scott et al. 1984a
  On wall growth, MERL, no. 2 fuel   Caron and Sieburth 1981b
  On bacteria, MERL, no. 2 fuel   Davies et al. 1979
  On bacteria, MERL, no. 2 fuel   Lee and Ryan 1983
  On bacteria, SEAFLUX, Prudhoe Bay   Lee et al. 1984
    crude
  On microbes, ponds, Norman Wells   Dutka and Kwan 1984
    crude
  On fungi, ponds   Sherry 1984
Behavior and fate of oil
  Overall   Barth 1984
  Overall, Baltic ponds   Hagstrom 1977
  Overall, Bedford tanks   Gordon et al. 1976
  Overall, MERL, no. 2 fuel   Gearing et al. 1978
                                   Olsen et al. 1982
                                 Oviatt et al. 1982
                                 Pilson et al. 1977

Table 15.18 (continued)

Behavior and fate of oil (continued)	
Overall, CEPEX, no. 2 fuel	Lee and Takahashi 1977
	Lee et al. 1977
Overall, ERDA tanks, no. 2 fuel oil	Wilson 1978
Overall, Loch Ewe, North Sea crude	Brown et al. 1978
Overall, CEPEX, Prudhoe Bay crude	Davies et al. 1980
	Lee et al. 1978
	Wong et al. 1984
Overall, Lindaspollen bags, Ekofisk crude	Skjoldal et al. 1982
Overall, Arctic ponds, Norman Wells & Pembina crude	Snow and Scott 1975
Overall, ponds, Norman Wells crude	Nagy et al. 1984
Photooxidation, Lindaspollen bags, Ekofisk crude	Tjessem et al. 1984
Evaporation, MERL, no. 2 fuel	Gearing and Gearing 1982b
In water column, MERL, no. 2 fuel	Gearing and Gearing 1982a
In sediments, MERL, no. 2 fuel	Gearing et al. 1979
	Wade and Quinn 1980
Individual hydrocarbons	
Individual hydrocarbons	Oil and Gas Working Group 1978
Individual hydrocarbons, fate in different seasons, MERL	Wakeham et al. 1982
	Wakeham et al. 1983
	Wakeham et al. 1985
c1 to c4 alkanes, fate, MERL	Bopp et al. 1981
Aromatics, fate, MERL	Lee et al. 1982
Aromatics, fate, MERL	Pruell and Quinn 1985
Aromatics including benzopyrene, CEPEX	Lee et al. 1978
Aromatics, fate in streams	Bowling et al. 1980
Naphthalenes, fate & effect, CEPEX	Lee and Anderson 1977
Benzanthracene, fate in winter, MERL	Hinga 1984
Benzanthracene, fate in summer, MERL	Hinga et al. 1980
Dimethylbenanthracene, fate, MERL	Hinga et al. 1986
Sewage	
Effects on microorganisms, Konstanz	Bretthauer 1980
Effects & fate, Loch Ewe	Eleftheriou et al. 1982
Effects & fate, MERL	Oviatt et al. 1987
Effects, Yugoslavia lagoon	Fanuko 1984
Effects, Maryland tanks	Heinle et al. 1979
Effects, Michigan lagoon	King 1980
Effects, EPA tanks	Oviatt et al. 1979
	Oviatt et al. 1977

Table 15.18 (continued)

Others	
Wastewaters, effects on stream	Hoffman and Horne 1980
Contaminated sediments, effects on overlying water	Oviatt et al. 1984
Added organic carbon, effect on sediments	Kelly and Nixon 1984
Added organic carbon, effect on benthic organisms	Grassle et al. 1986
Dispersants for oil spills, effects, Rance	Lacaze 1974
Dispersants for oil spills, effects, SEAFLUX	Lee et al. 1984
Dispersants for oil spills, effects, Lindaspollen	Throndsen 1982
Dispersants for oil spills, effects, ERDA	Wilson 1978
PCBs, effects, CEPEX	Imboden et al. 1979
DDT, fate & effects	Metcalf et al. 1971
Pesticides, fate & effect	Coats 1980
	Cole and Metcalf 1980
Pesticides	Solomon et al. 1980
Pesticides	Lichtenstein 1980
Pesticides, effects in lakes	Weinberger et al. 1982
Insecticides, effects in streams	Hansen and Garton 1982
Pesticides & herbicides	Bretthauer 1980
Herbicides, effects, Maryland	Kemp et al. 1980
NTA (nitrilotriacetic acid), effects in stream	Bowling et al. 1980
Methoxychlor	Brockway et al. 1979
Permethrin, fate and effects	Kaushik et al. 1985
	Solomon et al. 1985
Phenol, fate, Den Helder	de Kock and Kuiper 1981
Four phenols & anilines, fate, Den Helder	Kuiper 1982
PCP (pentachlorophenol), fate, MERL	Lee et al. 1982
PCP, effects	Tagatz et al. 1981
PCP, effects, CEPEX	Whitney et al. 1981
TFM (trifluoromethylnitrophenol), effects in streams	Maki 1980
	Maki and Johnson 1976
p-Cresol, effects, streams	Stout and Cooper 1983
Glucose, effects, CEPEX	Parsons et al. 1981
Glucose & glucose plus copper, effects, SEAFLUX	Seki et al. 1983
Glucose, Loch Ewe	Ducklow et al. 1986
Sucrose, effects in streams	Warren et al. 1964
Sucrose, effect in lakes	Brunskill et al. 1980

bottle experiments with radiolabeled compounds. These laboratory tests are a quick and effective way of determining a compound's potential for biodegradation in different waters and sediments, yet only the relative rates can be directly applied to the field. For example, Kuiper (1983) reported significant differences between the degradation of 4-nitrophenol in laboratory experiments and in three microcosm experiments (pelagic bags), and hypothesized that they were due to a lack of sunlight in laboratory experiments. Much lower rates of degradation of 2-methylnaphthalene have been reported from bottle experiments using water from microcosms (Lee and Ryan 1983) than those measured when the same compound was added to the whole microcosms (Gearing and Gearing 1982a; 1983; Wakeham et al. 1983). Both methods gave similar trends with temperature, but whole microcosm rates were 30 to 40 times higher. If enough such comparisons were made, it might be possible to find factors allowing extrapolation of rates from bottle experiments to actual field conditions.

Finally, because microcosm experiments can be carried out under different conditions, these results can be extrapolated to a wide variety of environmental conditions and used to predict the residence times of different chemical components of oil in different parts of the marine ecosystem (water, surface microlayer, suspended particulates, sediments, etc.). For example, the chemical behavior of oil and petroleum hydrocarbons has been examined in the MERL microcosms with various loadings of suspended particles, different kinds and levels of biological activity, and temperatures ranging from 0 to 25°C. Sediments had little or no effect on chemical cycling; particles of high organic content did. The relative importance of biodegradation and photo-oxidation could be measured by poisoning the tanks. Temperature had the most dramatic effect on the behavior of oil, particularly on the one- to three-ring aromatics. These toxic chemicals are rapidly biodegraded at high temperatures but can remain in the water column for weeks or months during cold periods, greatly increasing the possible effects of oil.

Other examples of how the flexibility of microcosms can make possible the investigation of pollution in a wide variety of areas are the series of experiments at MERL on eutrophication (Oviatt et al. 1986a; Sullivan and Ritacco 1985b). Six different levels of inorganic nutrients were added to microcosms and the effects compared with the dose. Later experiments compared the same levels of nutrients added in the inorganic form with organic nutrients present in sewage sludge (Oviatt et al. 1987). Comparison of effects with dosage and form makes deciding on an acceptable permissible level in nature much easier. These kinds of studies measuring ecotoxicologic results under different environmental conditions can only be done by taking advantage of the flexibility and "naturalness" of microcosms.

15.5 Conclusion

In summary, aquatic microcosms are being and will continue to be used as one of several techniques for examining the environment. The results of microcosm studies can be directly and quantitatively extrapolated to nature. They are a tool capable of providing valuable and unique data, especially for ecotoxicology.

The work done to date in microcosms well illustrates their range. These experimental results stand on their own merits, not those of the method. Microcosms are flexible systems that have been used to study a great variety of ecosystem types under many different environmental conditions. They have been compared with natural ecosystems and been found similar in biological composition, nutrient cycling, and physical parameters. They are replicable within the limits of natural variability. Microcosms are valuable tools for understanding the ecological effects of pollutants when used in conjunction with other types of studies.

References

Abbott W (1966) Microcosm studies in estuarine waters. I. The replicability of microcosms. J Water Pollut Control Fed 39:258–270

Adams SM, Giddings JM (1982) Review and evaluation of microcosms for assessing effects of stress in marine ecosystems. Environ Intern 7:409–418

Adey WH (1983) The microcosm: a new tool for reef research. Coral Reefs 1:193–201

Adler D, Amdurer M, Santschi PH (1980) Metal tracers in two marine microcosms: sensitivity to scale and configuration. In: Giesy JP Jr (ed) *Microcosms in Ecological Research*. Washington DC: U.S. Dept. of Energy, Symposium Series 52 (Conf-781101), pp. 348–368

Amdurer M, Adler D, Santschi PH (1982) Radiotracers in studies of trace metal behavior in mesocosms: advantages and limitations. In: Grice GD, Reeve MR (eds) *Marine Mesocosms, Biological and Chemical Research in Experimental Ecosystems*. New York: Springer-Verlag, pp. 81–95

Amdurer M, Adler D, Santschi PH (1983) Studies of the chemical forms of trace metals in sea water using radiotracers. In: Wong CS, Boyle E, Bruland KW, Burton JD, Goldberg ED (eds) Trace Metals in Sea Water. *New York: Plenum Press, pp. 537–562*

Antia NJ, McAllister CD, Parsons TR, Stephens K, Strickland JDH (1963) Further measurements on primary production using a large volume plastic sphere. Limnol Oceanogr 8:166–184

Armitage BJ (1980) Effects of temperature on periphyton biomass and community composition in the Browns Ferry experimental channels. In: Giesy JP Jr (ed) *Microcosms in Ecological Research*. Washington DC: U.S. Dept. of Energy, Symposium Series 52 (Conf-781101), pp. 668–683

Azam F, Vaccaro RF, Gillespie PA, Moussalli EI, Hudson RE (1977) Controlled ecosystem pollution experiment: effect of mercury on enclosed water columns. II. Marine bacterioplankton. Mar Sci Comm 3:313–329

Baccini P, Ruchti J, Warner O, Grieder E (1979) MELIMEX, an experimental heavy metal pollution study: regulation of trace metal concentrations in limnocorrals. Schweiz Z Hydrol 41:202–227

Bakke T, Dale T, Thingstad TF (1982) Structural and metabolic responses of a subtidal sediment community to water extracts of oil. Neth J Sea Research 16:524–537

Bakke T, Johnson TM (1979) Response of a subtidal sediment community to low levels of oil hydrocarbons in a Norwegian fjord. In: 1979 Oil Spill Conference (Prevention, Behavior, Control, Cleanup).Washington DC: American Petroleum Institute, pp. 633–639

Balch N, Boyd CM, Mullin M (1978) Large-scale tower tank systems. Rapp PV Reun Cons Int Explor Mer 173:13–21

Banse K (1982) Experimental marine ecosystem enclosures in a historical perspective. In: Grice GD, Reeve MR (eds) *Marine Mesocosms, Biological and Chemical Research in Experimental Ecosystems*. New York: Springer-Verlag, pp. 11–24

Barth T (1984) Weathering of crude oil in natural marine environments: the concentration of polar degradation products in water under oil as measured in several field studies. Chemosphere 13:67–86

Beers JR, Reeve MR, Grice GD (1977a) Controlled ecosystem pollution experiment: effect of mercury on enclosed water columns. Mar Sci Comm 3:355–394

Beers JR, Stewart GL, Hoskins KD (1977b) Dynamics of micro-zooplankton populations treated with copper: controlled ecosystem pollution experiment. Bull Mar Sci 27: 66–79

Bender ME, Jordan RA (1970) Plastic enclosure versus open lake productivity measurements. Trans Amer Fish Soc 99:607–610

Berman M, Heinle DR (1980) Modification of the feeding behavior of marine copepods by sublethal concentrations of water-accommodated fuel oil. Mar Biol 56:59–64

Beyers RJ (1962) Relationship between temperature and the metabolism of experimental ecosystems. Science 136:980–982

Beyers RJ (1963) The metabolism of twelve laboratory microecosystems. Ecol Monogr 33:281–306

Bienfang PK (1982) Phytoplankton sinking-rate dynamics in enclosed experimental ecosystems. In: Grice GD, Reeve MR (eds) *Marine Mesocosms, Biological and Chemical Research in Experimental Ecosystems*. New York: Springer-Verlag, pp. 261–274

vonBodungen B, vonBrockel K, Smetacek V, Zeitzschel B (1976) The plankton tower. I. A structure to study water/sediment interactions in enclosed water columns. Mar Biol 34:369–372

Boehm P, Barak J, Fiest D, Elskus A (1979) The analytical chemistry of *Mytilus edulis, Macoma balthica,* sediment trap, and surface sediment samples—a one-year study. In: The *Tsesis* Oil Spill. Asko Laboratory, Stockholm, Sweden

Bopp RF, Santschi PH, Deck BL (1981) Biodegradation and gas exchange of gaseous alkanes in model estuarine ecosystems. Org Geochem 3:9–14

Bourquin AW, Garmas RL, Pritchard PH, Wilkes FG, Cripe CR, Rubinstein NI (1979) Interdependent microcosms for the assessment of pollutants in the marine environment. Intern J Environ Studies 13:131–140

Bower P, McCorkle D (1980) Gas exchange, photosynthetic uptake, and carbon budget for a radiocarbon addition to a small enclosure in a stratified lake. Can J Fish Aquat Sci 37:464–471

Bowling JW, Giesy JP Jr, Kania HJ, Knight RL (1980) Large-scale microcosms for assessing fates and effects of trace contaminants. In: Giesy JP Jr (ed) *Microcosms in Ecological Research*. Washington DC: U.S. Dept. of Energy, Symposium Series 52 (Conf-781101), pp. 224–247

Boyce FM (1974) Mixing within experimental enclosures: a cautionary note on the limnocorral. J Fish Res Board Can 31:1400–1405

Boyd CM (1981) Microcosms and experimental planktonic food chains. In: Longhurst AR (ed) *Analysis of Marine Ecosystems*. New York: Academic Press, pp. 627–649

Bretthauer R (1980) Laboratory aquatic microcosms. In: Giesy JP Jr (ed) *Microcosms in Ecological Research*. Washington DC: U.S. Dept. of Energy, Symposium Series 52 (Conf-781101), pp. 416–445

vonBrockel K (1982) Sedimentation of phytoplankton cells within controlled experimental ecosystems following launching and implications for further enclosure studies. In: Grice GD, MR Reeve (eds) *Marine Mesocosms, Biological and Chemical Research in Experimental Ecosystems*. New York: Springer-Verlag, pp. 251–259

Brockmann UH, Eberlein K, Hentzschel G, Schone HK, Siebers D, Wandschneider K, Weber A (1977a) Parallel plastic tank experiments with cultures of marine diatoms. Helgol Wiss Meeresunters 30:201–216

Brockmann UH, Eberlein K, Hosumbec P, Trageser H, Maier-Reimer E, Schone HK, Junge HD (1977b) The development of a natural plankton population in an outdoor tank with nutrient-poor sea water. I. Phytoplankton succession. Mar Biol 43:1–17

Brockmann UH, Eberlein K, Junge HD, Maier-Reimer E, Siebers D (1979) The development of a natural plankton population in an outdoor tank with nutrient-poor sea water. II. Changes in dissolved carbohydrates and amino acids. Mar Ecol Prog Ser 1:283–291

Brockmann UH, Eberlein K, Junge HD, Trageser H, Trahms KJ (1974) Einfache Folientanks zur Planktonuntersuchung *in situ*. Mar Biol 24:163–166

Brockmann UH, Kattner G, Dahl E (1982) Plankton spring development in a south Norwegian fjord. In: Grice GD, Reeve MR (eds) *Marine Mesocosms, Biological and Chemical Research in Experimental Ecosystems*. New York: Springer-Verlag, pp. 195–204

Brockway DL, Hill J, Maudsley JR, Lassiter RR (1979) Development, replicability and modeling of naturally derived microcosms. Intern J Environ Studies 13:149–158

Brown CW, Lynch PF, Ahmadjian M (1978) Chemical analysis of dispersed oil in the water column. In: McCarthy LT, Lindblom GP, Walter HF (eds) *Chemical Dispersants for the Control of Oil Spills*. Philadelphia: American Society for Testing and Materials, pp. 188–202

Brown MF, Kester DR, Dowd JM (1983) Fate of ocean dumped acid-iron waste in a MERL stratified microcosm. In: Duedall IW, Ketchum BH, Park PK, Kester DR (eds) *Wastes in the Ocean: Chemical and Sewage Wastes*. New York: Wiley-Interscience, pp. 157–169

Brown PS, Parsons TR (1972) The effect of simulated upwelling on the maximization of primary productivity and the formation of photodetritus. Mem Inst Ital Idrobiol, 29 suppl:169–183

Brunskill GJ, Graham BW, Rudd JWM (1980) Experimental studies on the effect of arsenic on microbial degradation of organic matter and algal growth. Can J Fish Aquat Sci 37:415–423

Cairns J (1980) Beyond single species toxicity testing. Mar Environ Research 3:157–159

Caron DA, Sieburth JM (1981a) Disruption of the primary fouling sequence on fiberglass-reinforced plastic submerged in the marine environment. Appl Environ Microbiol 41:268–273

Caron DA, Sieburth JM (1981b) Response of peritrichous ciliates in fouling communities to seawater-accommodated hydrocarbons. Trans Amer Microsc Soc 100:183–203

Case JN (1978) The engineering aspects of capturing a marine environment, CEPEX and others. Rapp PV Reun Cons Int Explor Mer 173:49–58

Coats JR (1980) A stream microcosm for environmental assessment of pesticides. In: Giesy JP Jr (ed) *Microcosms in Ecological Research*. Washington DC: U.S. Dept. of Energy, Symposium Series 52 (Conf-781101), pp. 715–723

Cole LK, Metcalf RL (1980) Environmental destinies of insecticides, herbicides, and fungicides in the plants, animals, soil, air, and water of homologous microcosms. In: Giesy JP, Jr (ed) *Microcosms in Ecological Research*. Washington DC: U.S. Dept. of Energy, Symposium Series 52 (Conf-781101), pp. 971–1007

Conover RJ, Paranjape MA (1977) Comments on the use of a deep tank in planktological research. Helgol Wiss Meeresunters 30:105–117

Cooke GD (1971) Aquatic laboratory microsystems and communities. In: Cairns J Jr (ed) *The Structure and Function of Microbial Communities*. Research Division Monograph 3. American Microbial Society Symposium. Blacksburg, VA: Virginia Polytechnic Institute and State University,pp. 47–85

Cooper DC (1973) Enhancement of net primary productivity by herbivore grazing in aquatic laboratory microcosms. Limnol Oceanogr 18:31–37

Cooper DC, Copeland BJ (1973) Responses of continuous-series estuarine microecosystems to point-source input variations. Ecol Monogr 43:213–236

de la Cruz AA (1982) Effects of oil on phytoplankton metabolism in natural and experimental estuarine ponds. Mar Environ Research 7:257–263

Davies JM, Baird IE, Massie LC, Hay SJ, Ward AP (1980) Some effects of oil derived hydrocarbons on a pelagic food web from observations in an enclosed ecosystem and a consideration of their implications for monitoring. Rapp PV Reun Cons Int Explor Mer 179:201–211

Davies JM, Gamble JC (1979) Experiments with large enclosed ecosystems. Phil Trans R Soc Lond B 286:523–544

Davies JM, Gamble JC, Steele JH (1975) Preliminary studies with a large plastic enclosure. In: Cronin LE (ed) *Estuarine Research*. New York: Academic Press, pp. 251–264

Davies JM, Williams PJ leB (1984) Verification of ^{14}C and O_2 derived primary organic production measurements using an enclosed ecosystem. J Plankton Research 6:457–474

Davies PG, Hefferman RF, Sieburth JM (1979) Heterotrophic microbial populations in estuarine microcosms: influence of season and water accommodated hydrocarbons. Trans Amer Microsc Soc 98:152 (abst.)

Davis CO (1982) The importance of understanding phytoplankton life strategies in the design of enclosure experiments. In: Grice GD, Reeve MR (eds) *Marine*

Mesocosms, Biological and Chemical Research in Experimental Ecosystems. New York: Springer-Verlag, pp. 323–332

Davis GE, Warren CE (1965) Trophic relations of a sculpin in laboratory stream communities. J Wildlife Manage 29:846–871

Donaghay PL (1984) Utility of mesocosms to assess marine pollution. In: White H (ed) *Concepts in Marine Pollution Measurements*. College Park, MD: Maryland Sea Grant Program, pp. 589–620

Donaghay PL, Klos E (1985) Physical, chemical and biological responses to simulated wind and tidal mixing in experimental marine ecosystems. Mar Ecol Prog Ser 26:35–45

Draggan S (ed) (1976) The Role of Microcosms in Ecological Research. Inter J Environ Studies, Volume 10

Draggan S, Reisa JJ (1980) Controlling toxic substances: historical perspective and future research needs. In: Giesy JP Jr (ed) *Microcosms in Ecological Research*. Washington DC: U.S. Dept. of Energy, Symposium Series 52 (Conf-781101), pp. iii–xii

Ducklow HW, Purdie DA, Williams PJLeB, Davies JM (1986) Bacterioplankton: a sink for carbon in a coastal marine plankton community. Science 232:865–867

Dudzik M, Harte J, Jassby A, Lapan E, Levy D, Rees J (1979) Some considerations in the design of aquatic microcosms for plankton research. Intern J Environ Studies 13:125–130

Dutka BJ, Kwan KK (1984) Study of long term effects of oil and oil-dispersant mixtures on freshwater microbial populations in man made ponds. Sci Total Environ 35:135–148

Edmondson WT (1955) Factors affecting productivity in fertilized salt water. Pap Mar Biol Oceanogr, Deep Sea Research (Suppl) 3:451–463

Edmondson WT, Edmondson YH (1947) Measurements of production in fertilized salt-water. J Mar Research 6:228–246

Eleftheriou A, Moore DC, Basford DJ, Robertson MR (1982) Underwater experiments on the effects of sewage sludge on a marine ecosystem. Neth J Sea Research 16:465–473

Elmgren R, Frithsen JB (1982) The use of experimental ecosystems for evaluating the environmental impact of pollutants: a comparison of an oil spill in the Baltic Sea and two long-term, low-level oil addition experiments in mesocosms. In: Grice GD, Reeve MR (eds) *Marine Mesocosms, Biological and Chemical Research in Experimental Ecosystems*. New York: Springer-Verlag, pp. 153–165

Elmgren R, Vargo GA, Grassle GF, Grassle JP, Heinle DR, Langlois G, Vargo SL (1980) Trophic interactions in experimental marine ecosystems perturbed by oil. In: Giesy JP Jr (ed) *Microcosms in Ecological Research*. Washington DC: U.S. Dept. of Energy, Symposium Series 52 (Conf-781101), pp. 779–800

Eppley RW, Holm-Hansen O, Strickland JDH (1968) Some observations on the vertical migration of dinoflagellates. J Phycol 4:333–340

Eppley RW, Koeller P, Wallace GT (1978) Stirring influences the phytoplankton species composition within enclosed columns of coastal sea water. J Exp Mar Biol Ecol 32:219–239

Evans EC (1977) Microcosm responses to environmental perturbants. An extension of baseline field survey. Helgol Wiss Meeresunters 30:178–191

Fanuko N (1984) The influence of experimental sewage pollution on lagoon phytoplankton. Mar Pollut Bull 15:195–198

Farrington JW, Tripp BW, Teal JM, Mills G, Tjessem K, Davis AC, Livramento J, Hayward NA, Frew NM (1982) Biogeochemistry of aromatic hydrocarbons in the benthos of microcosms. Toxicol Environ Chem 5:331–346

Fox MF, Kester DR, Hunt CD (1986) Vertical transport processes of an acid-iron waste in a MERL stratified mesocosm. Environ Sci Technol 20:62–68

Franklin FL (1983) Laboratory tests as a basis for the control of sewage sludge dumping at sea. Mar Pollut Bull 14:217–223

Frithsen JB (1984) Metal incorporation by benthic fauna; relationship to sediment inventory. Est Coastal Shelf Sci 19:523–539

Frithsen JB, Elmgren R, Rudnick DT (1985) Responses of benthic meiofauna to long-term, low-level additions of No. 2 fuel oil. Mar Ecol Prog Ser 23:1–14

Frithsen JB, Rudnick DT, Elmgren R (1983) A new, flow-through corer for the quantitative sampling of surface sediments. Hydrobiol 99:75–79

Fulton RS (1984a) Effects of chaetognath predation and nutrient enrichment on enclosed estuarine copepod communities. Oecologia (Berl) 62:97–101

Fulton RS (1984b) Predation, production and the organization of an estuarine copepod community. J Plankton Research 6:399–415

Gachter R (1979) MELIMEX, an experimental heavy metal pollution study: goals, experimental design and major findings. Schweiz Z Hydrol 41:169–176

Gamble JC, Davies JM (1982) Application of enclosures to the study of marine pelagic systems. In: Grice GD, Reeve MR (eds) *Marine Mesocosms, Biological and Chemical Research in Experimental Ecosystems.* New York: Springer-Verlag, pp. 25–48

Gamble JC, Davies JM, Steele JH (1977) Loch Ewe bag experiment, 1974. Bull Mar Sci 27:146–175

Gatten RR Jr, Sargent JR, Gamble JC (1983) Diet-induced changes in fatty acid composition of herring larvae reared in enclosed ecosystems. J Mar Biol Assoc UK 63:575–584

Gearing JN, Gearing PJ (1983) The effects of suspended load and solubility on sedimentation of petroleum hydrocarbons in controlled estuarine ecosystems. Can J Fish Aquat Sci 40:54–62

Gearing JN, Gearing PJ, Lytle TF, Lytle JS (1978) Comparison of thin layer and column chromatography for separation of sedimentary hydrocarbons. Anal Chem 50:1833–1836

Gearing JN, Gearing PJ, Rudnick DT, Requejo AG, Hutchins MJ (1984) Isotopic variability of organic carbon in a phytoplankton-based, temperate estuary. Geochim Cosmochim Acta 48:1089–1098

Gearing JN, Gearing PJ, Wade T, Quinn JG, McCarty HB, Farrington J, Lee RF (1979) The rates of transport and fates of petroleum hydrocarbons in a controlled marine ecosystem, and a note on analytical variability. In: *Proceedings of the 1979 Oil Spill Conference.* Washington DC: American Petroleum Institute, pp. 555–564

Gearing PJ, Gearing JN (1982a) Behavior of no. 2 fuel oil in the water column of controlled ecosystems. Mar Environ Research 6:115–132

Gearing PJ, Gearing JN (1982b) Transport of no. 2 fuel oil between water column, surface microlayer, and atmosphere in controlled ecosystems. Mar Environ Research 6:133–143

Gearing PJ, Gearing JN, Pruell RJ, Wade TL, Quinn JG (1980) Partitioning of no.

2 fuel oil in controlled estuarine ecosystems: sediments and suspended particulate matter. Environ Sci Technol 14:1129–1136

Gibson VR, Grice GD (1977) Response of macro-zooplankton populations to copper: controlled ecosystem pollution experiment. Bull Mar Sci 27:85–91

Gibson VR, Grice GD (1980) A comparison between a beaker and a bay: the big bag. Oceans 13:21–25

Giddings JM (1980) Types of aquatic microcosms and their research applications. In: Giesy JP Jr (ed) Microcosms in Ecological Research. Washington DC: U.S. Dept. of Energy, Symposium Series 52 (Conf-781101), pp. 248–266

Giddings JM, Eddlemon GK (1978) Photosynthesis/ respiration ratios in aquatic microcosms under arsenic stress. Water Air Soil Pollut 9:207–212

Giddings JM, Eddlemon GK (1979) Some ecological and experimental properties of complex aquatic microcosms. Intern J Environ Studies 13:119–123

Giesy JP Jr (ed) (1978) Microcosms in Ecological Research. Washington DC: U.S. Dept. of Energy, Symposium Series 52. Conf-781101

Giesy JP Jr, Odum EP (1980) Microcosmology: introductory comments. In: Giesy JP Jr (ed) Microcosms in Ecological Research. Washington DC: U.S. Dept. of Energy, Symposium Series 52 (Conf-781101), pp. 1–13

Goering JJ, Boisseau D, Hattori A (1977) Effects of copper on silicic acid uptake by a marine phytoplankton population: controlled ecosystem pollution experiment. Bull Mar Sci. 27:58–65

Goldman CR (1962) A method of studying nutrient limiting factors in situ in waters isolated by polyethylene film. Limnol Oceanogr 7:99–101

Gordon DC Jr, Keizer PD, Hardstaff WR, Aldous DG (1976) Fate of crude oil spilled on seawater contained in outdoor tanks. Environ Sci Technol 10:580–585

Grassle JF, Elmgren R, Grassle JP (1981) Response of benthic communities in MERL experimental ecosystems to low-level chronic additions of #2 fuel oil. Mar Environ Research 4:279–297

Grassle JF, Grassle JP, Brown-Leger LS, Petreccal RF, Copley NJ (1986) Subtidal macrobenthos of Narragansett Bay. Field and mesocosm studies of the effects of eutrophication and organic input on benthic populations. In: Gray JS, Christianson ME (eds) Marine Biology of Polar Regions and Effects of Stress on Marine Organisms. New York: Wiley, pp. 421–434

Grice GD, Harris RP, Reeve MR, Heinbokel JF, Davis CO (1980) Large scale enclosed water column ecosystem. An overview of Foodweb. I. The final CEPEX experiment. J Mar Biol Assoc U K 60:401–414

Grice GD, Menzel DW (1978) Controlled ecosystem pollution experiment: effect of mercury on enclosed water columns. VIII. Summary of results. Mar Sci Comm 4:23–31

Grice GD, Reeve MR (eds) (1982) Marine Mesocosms, Biological and Chemical Research in Experimental Ecosystems. New York: Springer-Verlag

Grice GD, Reeve MR, Koeller P, Menzel DW (1977) The use of large volume, transparent, enclosed sea-surface water columns in the study of stress on plankton ecosystems. Helgol Wiss Meeresunters 30:118–133

Gust G (1977) Turbulence and waves inside flexible-wall systems designed for biological studies. Mar Biol 42:47–53

Hagstrom A (1977) The fate of oil in a model ecosystem. Ambio 6:229–231

Hall DJ, Cooper WE, Werner EE (1970) An experimental approach to the production dynamics and structure of freshwater animal communities. Limnol Oceanogr 15:839–928

Hansen SR, Garton RR (1982) Ability of standard toxicity tests to predict the effects of the insecticide diflubezuron on laboratory stream communities. Can J Fish Aquat Sci 39:1273–1288

Harris RP, Reeve MR, Grice GD, Evans GT, Gibson VR, Beers JR, Sullivan BK (1982) Trophic interactions and production processes in natural zooplankton communities in enclosed water columns. In: Grice GD, Reeve MR (eds) *Marine Mesocosms, Biological and Chemical Research in Experimental Ecosystems.* New York: Springer-Verlag, pp. 353–387

Harrison WG, Davies JM (1977) Nitrogen cycling in a marine planktonic food chain: nitrogen fluxes through the principal components and the effects of adding copper. Mar Biol 43:299–306

Harrison WG, Eppley RW, Renger EH (1977) Phytoplankton nitrogen metabolism, nitrogen budgets and observations on copper toxicity: controlled ecosystem pollution experiment. Bull Mar Sci 27:44–57

Harrison WG, Renger EH, Eppley RW (1978) Controlled ecosystem pollution experiment: effect of mercury on enclosed water columns. VII. Inhibition of nitrogen assimilation and ammonia regeneration by plankton in seawater samples. Mar Sci Comm 4:13–22

Harte J, Levy D, Rees J, Saegebarth E (1980) Making microcosms an effective assessment tool. In: Giesy JP Jr (ed) *Microcosms in Ecological Research.* Washington DC: U.S. Dept. of Energy, Symposium Series 52 (Conf-781101), pp. 105–137

Hattori A, Koike I, Ohtsu M, Goering JJ, Boisseau D (1980) Uptake and regeneration of nitrogen in controlled aquatic ecosystems and the effects of copper on these processes. Bull Mar. Sci 30:431–443

Heath RT (1979) Holistic study of an aquatic microcosm: theoretical and practical implications. Intern J Environ Studies 13:87–93

Heath RT (1980) Are microcosms useful for ecosystem analysis? In: Giesy JP Jr (ed) *Microcosms in Ecological Research.* Washington DC: U.S. Dept. of Energy, Symposium Series 52 (Conf-781101), pp. 333–347

Hedtke SF (1984) Structure and function of copper-stressed aquatic microcosms. Aquat Toxicol 5:227–244

Heinle DR, Flemer DA, Huff RT, Sulkin ST, Ulanowicz RE (1979) Effects of perturbations on estuarine microcosms. In: Dame RF (ed) *Marsh-Estuarine Systems Simulations.* Columbia, SC: University of South Carolina Press, pp. 119–141

Hellebust JA, Hanna B, Sheath RG, Gergis M, Hutchinson TC (1975) Experimental crude oil spills on a small subarctic lake in the Mackenzie Valley, N.W.T.: Effects on phytoplankton, periphyton, and attached aquatic vegetation. In: *1975 Conference on Prevention and Control of Oil Pollution.* Washington DC: American Petroleum Institute, pp. 509–515

Henderson RS, Smith SV (1980) Semitropical marine microcosms: facility design and an elevated-nutrient-effects experiment. In: Giesy JP Jr (ed) *Microcosms in Ecological Research.* Washington DC: U.S. Dept. of Energy, Symposium Series 52 (Conf-781101), pp. 869–910

Hesslein RH, Quay P (1973) Vertical eddy diffusion studies in the thermocline of a small stratified lake. J Fish Res Board Can 30:1491–1500

Hill J IV, Wiegert RG (1980) Microcosms in ecological modeling. In: Giesy JP Jr
    (ed) *Microcosms in Ecological Research.* Washington DC: U.S. Dept. of En-
    ergy, Symposium Series 52 (Conf-781101), pp. 138–163
Hinga KR (1984) The fate of polycyclic aromatic hydrocarbons in enclosed ma-
    rine ecosystems. Narragansett, RI: Univ. of Rhode Island. Ph.D. Dissertation
Hinga KR, Pilson MEQ, Almquist G, Lee RF (1986) The degradation of 7, 12-
    dimethylbenz(a)anthracene in an enclosed marine ecosystem. Mar Environ Re-
    search 18:79–91
Hinga KR, Pilson MEQ, Lee RF, Farrington JW, Tjessem K, Davis AC (1980)
    Biogeochemistry of benzanthracene in an enclosed marine ecosystem. Environ
    Sci Technol 14:1136–1143
Hirota J (1974) Quantitative natural history of *Pleurobrachia bachei* in LaJolla
    Bight. Fish Bull US 72:295–335
Hodson RE, Azam F, Lee RF (1977) Effects of four oils on marine bacterial
    populations: controlled ecosystem pollution experiment. Bull Mar Sci 27:119–
    126
Hoffman RW, Horne AJ (1980) On-site flume studies for assessment of effluent
    impacts on stream aufwuchs communities. In: Giesy JP Jr (ed) *Microcosms in
    Ecological Research.* Washington DC: U.S. Dept. of Energy, Symposium Se-
    ries 52 (Conf-781101), pp. 610–624
Hood DW (1978) Upwelled impoundments as a means of enhancing primary
    productivity. Rapp PV Reun Cons Int Explor Mer 173:22–30
Houde ED, Berkeley SA (1982) Food and growth of juvenile herring, *Clupea
    harengus pallasi,* in CEPEX enclosures. In: Grice GD, Reeve MR (eds) *Marine
    Mesocosms, Biological and Chemical Research in Experimental Ecosystems.*
    New York: Springer-Verlag, pp. 239–249
Hunt CD (1983a) Incorporation and deposition of Mn and other trace metals by
    flocculent organic matter in a controlled marine ecosystem. Limnol Oceanogr
    28:302–308
Hunt CD (1983b) Variability in the benthic Mn flux in coastal marine ecosystems
    resulting from temperature and primary production. Limnol Oceanogr 28:913–
    923
Hunt CD, Smith DL (1980) Conversion of dissolved Mn to particulate Mn during a
    diatom bloom: Effects on the Mn cycle in the MERL microcosms. In: Giesy JP
    Jr (ed) *Microcosms in Ecological Research.* Washington DC: U.S. Dept. of
    Energy, Symposium Series 52 (Conf-781101), pp. 850–868
Hunt CD, Smith DL (1982) Controlled marine ecosystems - a tool for studying
    stable trace metal cycles: long term response and variability. In: Grice GD,
    Reeve MR (eds) *Marine Mesocosms, Biological and Chemical Research in
    Experimental Ecosystems.* New York: Springer-Verlag, pp. 111–122
Hunt CD, Smith DL (1983) Remobilization of metals from polluted marine sedi-
    ments. Can J Fish Aquat Sci. 40:132–142
Imboden DM, Eid BSF, Joller T, Schurter M, Wetzel J (1979) MELIMEX, an
    experimental heavy metal pollution study: vertical mixing in a large limnocor-
    ral. Schweiz Z Hydrol 41:177–189
Irvine JR (1985) Effects of successive flow perturbations on stream invertebrates.
    Can J Fish Aquat Sci 42:1922–1927
Iseki K, Takahashi M, Bauerfeind E, Wong CS (1981) Effects of polychlorinated

biphenyls (PCBs) on a marine plankton population and sedimentation in controlled ecosystem enclosures. Mar Ecol Prog Ser 5:207–214

Isensee AR (1976) Variability of aquatic model ecosystem-derived data. Intern J Environ Studies 10:35–41

Jackson TA, Kipphut G, Hesslein RH, Schindler DW (1980) Experimental study of trace metal chemistry in soft-water lakes at different pH levels. Can J Fish Aquat Sci 37:387–402

Johnson DL (1978) Biological mediation of chemical speciation. I. Microcosm studies of the diurnal pattern of copper species in seawater. Chemosphere 7:641–644

Johnson DL, Burke RM (1978) Biological mediation of chemical speciation. II. Arsenate reduction during marine phytoplankton blooms. Chemosphere 7:645–648

Jones JG (1973) Studies on freshwater bacteria; the effect of enclosure in large experimental tubes. J Appl Bact 36:445–456

Jones JG (1975) Some observations on the occurrence of the iron bacterium *Leptothrix ochracea* in fresh water, including reference to large experimental enclosures . J Appl Bact 39:63–72

Jones JG (1976) The microbiology and decomposition of seston in open water and experimental enclosures in a productive lake. J Ecol 65:241–278

Kamykowski D, Zentara S-J (1977) The diurnal vertical migration of motile phytoplankton through temperature gradients. Limnol Oceanogr 22:148–151

Kattner G, Brockmann UH, Eberlein K, Hammer KD (1982) Enclosed planktonic ecosystems during different stages of a spring bloom in south Norway. Neth J Sea Research 16:353–361

Kaushik NK, Stephenson GL, Solomon KR, Day KE (1985) Impact of permethrin on zooplankton communities in limnocorrals. Can J Fish Aquat Sci 42:77–85

Kelly JR (1984) Microcosms for studies of sediment-water interactions. In: Persoone G, Jaspers E, Claus C (eds) *Ecotoxicological Testing for the Marine Environment,* vol. 2. Bredene, Belgium: State Univ. Ghent and Inst. Mar. Sci. Res., pp. 315–330

Kelly JR, Nixon SW (1984) Experimental studies of the effect of organic decomposition on the metabolism of a coastal marine bottom community. Mar Ecol Prog Ser 17:157–169

Kelly JR, Berounsky VM, Nixon SW, Oviatt CA (1985) Benthic pelagic coupling and nutrient cycling across an experimental eutrophication gradient. Mar Ecol Prog Ser 26:207–219

Kemp WM, Lewis MR, Cunningham JJ, Stevenson JC, Boynton WR (1980) Microcosms, macrophytes, and hierarchies: environmental research in the Chesapeake Bay. In: Giesy JP Jr (ed) *Microcosms in Ecological Research.* Washington DC: U.S. Dept. of Energy, Symposium Series 52 (Conf-781101), pp. 911–936

King DL (1980) Some cautions in applying results from aquatic microcosms. In: Giesy JP Jr (ed) *Microcosms in Ecological Research.* Washington DC: U.S. Dept. of Energy, Symposium Series 52 (Conf-781101), pp. 164–191

King KR (1982) The population biology of the larvacean *Oikopleura dioica* in enclosed water columns. In: Grice GD, Reeve MR (eds) *Marine Mesocosms,*

Biological and Chemical Research in Experimental Ecosystems. New York: Springer-Verlag, pp. 341–351

King KR, Hollobaugh JT, Azam F (1980) Predator–prey interactions between the larvacean *Oikopleura dioica* and bacterioplankton in enclosed water columns. Mar Biol 56:49–57

Kinne O (ed) (1977) International Helgoland Symposium on Ecosystem Research. Helgoländer wissenschatliche Meeresuntersuchungen, Vol. 30

Kitchens WM, Copeland BJ (1980) Succession in laboratory microecosystems subjected to thermal and nutrient-addition stress. In: Giesy JP Jr (ed) *Microcosms in Ecological Research.* Washington DC: U.S. Dept. of Energy, Symposium Series 52 (Conf-781101), pp. 536–561

Kitchens WM, Edwards RT, Johnson WV (1979) Development of a "living" salt marsh ecosystem model: A microecosystem approach. In: Dame RF (ed) *Marsh-Estuarine Systems Simulation.* Columbia, SC: University of South Carolina Press, pp. 107–117

de Kock WC, Kuiper J (1981) Possibilities for marine pollution research at the ecosystem level. Chemosphere 10:575–603

Koeller P, Parsons TS (1977) The growth of young salmonids (*Oncorhynchus keta*): controlled ecosystem pollution experiment. Bull Mar Sci 27:114–118

Koeller P, Wallace GT (1977) Controlled ecosystem pollution experiment: effect of mercury on enclosed water columns. V. Growth of juvenile chum salmon (*Oncorhynchus keta*). Mar Sci Comm 3:395–406

Koike I, Hattori A, Goering JJ (1978) Controlled ecosystem pollution experiment: effect of mercury on enclosed water columns. VI. Denitrification by marine bacteria. Mar Sci Comm 4:1–12

Koike I, Hattori A, Takahashi M, Goering JJ (1982) The use of enclosed experimental ecosystems to study nitrogen dynamics in coastal water. In: Grice GD, Reeve MR (eds) *Marine Mesocosms, Biological and Chemical Research in Experimental Ecosystems.* New York: Springer-Verlag, pp. 291–303

Kremling K, Pinze J, vonBrockel K, Wong CS (1978) Studies on the pathways and effects of cadmium in controlled ecosystem enclosures. Mar Biol 48:1–10

Kuiper J (1977a) Development of North Sea coastal plankton communities in separate plastic bags under identical conditions. Mar Biol 44:97–107

Kuiper J (1977b) An experimental approach in studying the influence of mercury on a North Sea coastal plankton community. Helgol Wiss Meeresunters 30:652–665

Kuiper J (1981) Fate and effects of mercury in marine plankton communities in experimental enclosures. Ecotox Environ Safety 5:106–134

Kuiper J (1982) Ecotoxicological experiments with marine plankton communities in plastic bags. In: Grice GD, Reeve MR (eds) *Marine Mesocosms, Biological and Chemical Research in Experimental Ecosystems.* New York: Springer-Verlag, pp. 181–193

Kuiper J (1984) Marine ecotoxicological tests: Multispecies and model ecosystem experiments. In: Persoone G, Jaspers E, Claus C (eds) International Symposium on Ecotoxicological Testing for the Marine Environment, vol. 2. Bredene, Belgium: State Univ. Ghent and Inst. Mar. Sci. Res, pp. 527–588

Kuiper J, Van Het Groenewoud H, Hoornsman G (1982) Diurnal variations in some plankton parameters in an enclosed marine community. Neth J Sea Research 16:345–352

Kuiper J, de Wilde P, Wolff W (1984) Effects of an oil spill in outdoor model tidal flat ecosystems. Mar Pollut Bull 15:102–106

Lacaze JC (1974) Ecotoxicology of crude oils and the use of experimental marine ecosystems. Mar Pollut Bull 5:153–156

Lack TJ, Lund JW (1974) Observations and experiments on the phytoplankton of Blelham Tarn, English Lake District. I. The experimental tubes. Freshwater Biol 4:399–415

de Lafontaine Y, Leggett WC (1987) Evaluation of in situ enclosures for larval fish studies. Can J Fish Aquat Sci 44:54–65

Lamberti GA, Resh VH (1983) Geothermal effects on stream benthos: separate influences of thermal and chemical components on periphyton and macroinvertebrates. Can J Fish Aquat Sci 40:1995–2009

Lane PA, Collins TM (1985) Food web models of a marine plankton community network: an experimental mesocosm approach. J Exp Mar Biol Ecol 94:41–70

Lang C, Lang-Dobler B (1979) MELIMEX, an experimental heavy metal pollution study: oligochaetes and chironomid larvae in heavy metal loaded and control limnocorrals. Schweiz Z Hydrol 41:271–276

Lawrence GC, Halavik TA, Burns BR, Smigielski AS (1979) An environmental chamber for monitoring "in situ" growth and survival of larval fish. Trans Amer Fish Soc 108:197–203

Lawson TJ, Grice GD (1977) Zooplankton sampling variability: controlled ecosystem pollution experiment. Bull Mar Sci 27:80–84

Lee K, Wong CS, Wu JP (1984) Microbial response to Prudhoe Bay crude oil and Corexit 9527: Seaflux enclosure study. Abstracts of Papers for the 47th Annual Meeting, Amer Soc Limnol Oceanogr, June 11–14, 1984, University of British Columbia, Vancouver, B.C., p. 48

Lee RF, Anderson JW (1977) Fate and effect of naphthalenes: controlled ecosystem pollution experiment. Bull Mar Sci 27:127–134

Lee RF, Gardner WS, Anderson JW, Blaylock JW, Barwell-Clarke J (1978) Fate of polycyclic aromatic hydrocarbons in controlled ecosystem enclosures. Environ Sci Technol 12:832–838

Lee RF, Hinga K, Almquist G (1982) Fate of radiolabeled polycyclic aromatic hydrocarbons and pentachlorophenol in enclosed marine ecosystems. In: Grice GD, Reeve MR (eds) Marine Mesocosms, Biological and Chemical Research in Experimental Ecosystems. New York: Springer-Verlag, pp. 123–135

Lee RF, Ryan C (1983) Microbial and photochemical degradation of polycyclic aromatic hydrocarbons in estuarine waters and sediments. Can J Fish Aquat Sci 40:86–94

Lee RF, Takahashi M (1977) The fate and effect of petroleum in controlled ecosystem enclosures. Rapp PV Reun Cons Int Explor Mer 171:150–156

Lee RF, Takahashi M, Beers JR, Thomas WH, Seibert DLR, Koeller P, Green DR (1977) Controlled ecosystems: Their use in the study of the effects of petroleum hydrocarbons in plankton. In: Vernberg FJ, Calabrese A, Thurberg FP, Vernberg WB (eds) Physiological Responses of Marine Biota to Pollutants. New York: Academic Press, pp. 323–342

Leffler JW (1980) Microcosmology: Theoretical applications of biological models. In: Giesy JP Jr (ed) Microcosms in Ecological Research. Washington DC: U.S. Dept. of Energy, Symposium Series 52 (Conf-781101), pp. 14–29

Legendre L, Demers S (1984) Towards dynamic biological oceanography and limnology. Can J Fish Aquatic Sci 41:2–19

Legendre L, Demers S, Therriault J-C, Boudreau C-A (1985) Tidal variations in the photosynthesis of estuarine phytoplankton isolated in a tank. Mar Biol 88:301–309

Levandowsky M (1977) Multispecies cultures and microcosms. In: Kinne O (ed) *Marine Ecology*. London: Wiley-Interscience, pp. 1399–1458

Lichtenstein EP (1980) Fate and behavior of pesticides in a compartmentalized microcosm. In: Giesy JP Jr (ed) *Microcosms in Ecological Research*. Washington DC: U.S. Dept. of Energy, Symposium Series 52 (Conf-781101), pp. 954–970

Linden O, Rosemarin A, Lindskog A, Hoglund C, Johansson S (1987) Effects of oil and oil dispersant on an enclosed marine ecosystem. Environ Sci Technol 21:374–382

Lund JWG (1972) Preliminary observations on the use of large experimental tubes in lakes. Verh internat Verein Limnol 18:71–77

Lund JWG (1978) Experiments with lake phytoplankton in large enclosures. Rep Freshw Biol Assn 46:32–39

Lund JWG, Reynolds CS (1982) The development and operation of large limnetic enclosures in Blelham Tarn, English Lake District, and their contribution of phytoplankton ecology. Prog Phycol Res 1:1–65

McAllister CD, Parsons TR, Stephens K, Strickland JDH (1961) Measurements of primary production in coastal sea water using a large-volume plastic sphere. Limnol Oceanogr 6:237–258

McIntire CD, Garrison RL, Phinney HK, Warren CE (1964) Primary production in laboratory streams. Limnol Oceanogr 9:92–102

McIntyre AD (1977) Effects of pollution of inshore benthos. In: Coull BC (ed) *Ecology of Marine Benthos*. Columbia, SC: University of South Carolina Press, pp. 301–318

McLaren IA (1969) Population and production ecology of zooplankton in Ogae Lake, a landlocked fjord on Baffin Island. J Fish Res Board Can 26:1485–1559

Maguire B Jr (1980) Some patterns in post-closure ecosystem dynamics (failure). In: Giesy JP Jr (ed) *Microcosms in Ecological Research*. Washington DC: U.S. Dept. of Energy, Symposium Series 52 (Conf-781101), pp. 319–332

Maki AW (1980) Evaluation of toxicant effects on structure and function of model stream communities: Correlations with natural stream effects. In: Giesy JP Jr (ed) *Microcosms in Ecological Research*. Washington DC: U.S. Dept. of Energy, Symposium Series 52 (Conf-781101), pp. 583–609

Maki AW, Johnson HE (1976) Evaluation of a toxicant on the metabolism of model stream communities. J Fish Res Board Can 33:2740–2746

Malone TC, Garside C, Haines KC, Roals OA (1975) Nitrate uptake and growth of *Chaetoceros* sp. in large outdoor continuous cultures. Limnol Oceanogr 20:9–19

Mann KH (1979) Qualitative aspects of estuarine modeling. In: Dane RF (ed) *Marsh-Estuarine Systems Simulations*. Columbia, SC: University of South Carolina Press, pp. 207–220

Manuel CY, Minshall GW (1980) Limitations on the use of microcosms for predicting algal response to nutrient enrichment in lotic systems. In: Giesy JP Jr (ed) *Microcosms in Ecological Research*. Washington DC: U.S. Dept. of Energy, Symposium Series 52 (Conf-781101), pp. 645–667

Marshall JS, Mellinger DL (1980) Dynamics of cadmium-stressed plankton communities. Can J Fish Aquat Sci 37:403–414

Medine AJ, Porcella DB, Adams VD (1980) Heavy-metal and nutrient effects on sediment oxygen demand in three-phase aquatic microcosms. In: Giesy JP Jr (ed) *Microcosms in Ecological Research*. Washington DC: U.S. Dept. of Energy, Symposium Series 52 (Conf-781101), pp. 279–303

Menzel DW (1977) Summary of experimental results: controlled ecosystem pollution experiment. Bull Mar Sci 27:142–145

Menzel DW (1980) Applying results derived from experimental microcosms to the study of natural pelagic marine ecosystems. In: Giesy JP Jr (ed) *Microcosms in Ecological Research*. Washington DC: U.S. Dept. of Energy, Symposium Series 52 (Conf-781101), pp. 742–752

Menzel DW, Case J (1977) Concept and design: controlled ecosystem pollution experiment. Bull Mar Sci 27:1–7

Menzel DW, Steele JH (1978) The application of plastic enclosures to the study of pelagic marine biota. Rapp PV Reun Cons Int Explor Mer 173:7–12

Metcalf RL, Sangha GK, Kapoor IP (1971) Model ecosystem for the evaluation of pesticide biodegradability and ecological magnification. Environ Sci Technol 5:709–713

Miller MC, Alexander V, Barsdate RJ (1978) The effects of oil spills on phytoplankton in an Arctic lake and ponds. Arctic 31:192–198

Muller P (1980) Effects of artificial acidification on the growth of periphyton. Can J Fish Aquat Sci 37:355–363

Mullin MM (1982) How can enclosing seawater liberate biological oceanographers? In: Grice GD, Reeve MR (eds) *Marine Mesocosms, Biological and Chemical Research in Experimental Ecosystems*. New York: Springer-Verlag, pp. 399–410

Mullin MM, Evans PM (1974) The use of a deep tank in plankton ecology. 2. Efficiency of a planktonic food chain. Limnol Oceanogr 19:902–911

Muschenheim DK, Grant J, Mills EL (1986) Flumes for benthic ecologists: theory, construction and practice. Mar Ecol Prog Ser 28:185–196

Nadeau NJ, Roush TH (1973) A salt marsh microcosm: An experimental unit for marine pollution studies. In: *Proceedings of the 1973 Joint Conference on Prevention and Control of Oil Spills*. Washington DC: American Petroleum Institute, pp. 671–683

Nagy E, Scott BF, Hart J (1984) The fate of oil and oil-dispersant mixtures in freshwater ponds. Sci Total Environ 35:115–133

National Research Council (1981) *Testing for Effects of Chemicals on Ecosystems (A Report by the Committee to Review Methods for Ecotoxicology)*. Washington DC: National Academy Press

Nixon SW (1969) A synthetic microcosm. Limnol Oceanogr 14:142–145

Nixon SW, Alonso D, Pilson MEQ, Buckley BA (1980) Turbulent mixing in aquatic microcosms. In: Giesy JP Jr (ed) *Microcosms in Ecological Research*. Washington DC: U.S. Dept. of Energy, Symposium Series 52 (Conf-781101), pp. 818–849

Nixon SW, Oviatt CA, Kremer JN, Perez K (1979) The use of numerical models and laboratory microcosms in estuarine ecosystem analysis—simulations of a winter phytoplankton bloom. In: Dane RF (ed) *Marsh-Estuarine Systems Simulations*. Columbia, SC: University of South Carolina Press, pp. 165–188

Nixon SW, Pilson MEQ (1983) Nitrogen in estuarine and coastal marine ecosystems. In: Carpenter EJ, Capone DG (eds) *Nitrogen in the Marine Environment.* New York: Academic Press, pp. 565–648

Nixon SW, Pilson MEQ, Oviatt CA, Donaghay P, Sullivan B, Seitzinger S, Rudnick D, Frithsen J (1984) Eutrophication of a coastal marine ecosystem—an experimental study using the MERL microcosms. In: Fasham MJR (ed) *Flows of Energy and Materials in Marine Ecosystems.* New York: Plenum Press, pp. 105–135

Odum HT, Siler WL, Beyers RJ, Armstrong N (1963) Experiments with engineering of marine ecosystems. Contrib Mar Sci 9:373–403

Oiestad V (1982) Application of enclosures to studies on the early life history of fishes. In: Grice GD, Reeve MR (eds) *Marine Mesocosms, Biological and Chemical Research in Experimental Ecosystems.* New York: Springer-Verlag, pp. 49–62

Oil and Gas Working Group (1978) *Studies of the Effects of Hydrocarbons on Laboratory Aquatic Ecosystems.* Report of the Institute for Environmental Studies, University of Toronto

Olsen S, Pilson MEQ, Oviatt CA, Gearing JN (1982) Ecological Consequences of Low, Sustained Concentrations of Petroleum Hydrocarbons in Temperate Estuaries. Univ. of Rhode Island, Narragansett, RI, 30pp.

Oviatt CA (1981) Effects of different mixing schedules on phytoplankton, zooplankton, and nutrients in marine microcosms. Mar Ecol Prog Ser 4:57–67

Oviatt CA, Buckley BA, Nixon SW (1981a) Annual phytoplankton metabolism in Narragansett Bay calculated from survey field measurements and microcosm observations. Estuaries 4:167–175

Oviatt CA, Frithsen J, Gearing J, Gearing P (1982) Low chronic additions of No. 2 fuel oil: Chemical behavior, biological impact and recovery in a simulated estuarine environment. Mar Ecol Prog Ser 9:121–136

Oviatt CA, Hunt CD, Vargo GA, Kopchynski KW (1981b) Simulation of a storm event in marine microcosms. J Mar Research 39:605–626

Oviatt CA, Keller A, Sampou P, Beatty LL (1986a) Patterns of productivity during eutrophication: a mesocosm experiment. Mar Ecol Prog Ser 28:69–80

Oviatt CA, Nixon SW, Perez KT, Buckley B (1979) On the season and nature of perturbations in microcosm experiments. In: Dane RF (ed) *Marsh-Estuarine Systems Simulations.* Columbia, SC: University of South Carolina Press, pp. 143–164

Oviatt CA, Perez KT, Nixon SW (1977) Multivariate analysis of experimental marine ecosystems. Helgol Wiss Meeresunters 30:30–46

Oviatt CA, Pilson MEQ, Nixon SW, Frithsen JB, Rudnick DT, Kelly JR, Grassle JF, Grassle JP (1984) Recovery of a polluted estuarine system: A mesocosm experiment. Mar Biol Prog Ser 16:203–217

Oviatt CA, Quinn JG, Maughan JT, Ellis JT, Sullivan BK, Gearing JN, Gearing PJ, Hunt CD, Sampou PA, Latimer JS (1987) Fate and effects of sewage sludge in the coastal marine environment: a mesocosm experiment. Mar Biol Prog Ser 41:187–203.

Oviatt CA, Rudnick D, Keller A, Sampou P, Almquist G (1986b) A comparison of system (O_2 and CO_2) and C-14 measurements of metabolism in estuarine mesocosms. Mar Ecol Prog Ser 28:57–67

Oviatt CA, Walker H, Pilson MEQ (1980) An exploratory analysis of microcosm and ecosystem behavior using multivariate techniques. Mar Ecol Prog Ser 2:179–191

Paffenhofer G-A (1976) Continuous and nocturnal feeding of the marine plank-tonic copepod *Calanus helgolandicus*. Bull Mar Sci 26:49–58

Parsons TR (1978a) Controlled aquatic ecosystem experiments in ocean ecology research. Mar Pollut Bull 9:203–205

Parsons TR (1978b) Controlled ecosystem experiments. Rapp PV Reun Cons Int Explor Mer 173:5–6

Parsons TR (ed) (1978c) Controlled Ecosystem Experiments - Joint Oceano-graphic Assembly. Rapp PV Reun Cons Int Explor Mer, Vol. 173

Parsons TR (1982) The future of controlled ecosystem enclosure experiments. In: Grice GD, Reeve MR (eds) *Marine Mesocosms, Biological and Chemical Research in Experimental Ecosystems*. New York: Springer-Verlag, pp. 411–418

Parsons TR, Albright LJ, Whitnew F, Wong CS, Williams PJLeB (1981) The effect of glucose on the productivity of seawater: An experimental approach using controlled aquatic ecosystems. Mar Environ Research 4:229–242

Parsons TR, von Brockel K, Koeller K, Takahashi M, Reeve MR, Holm-Hansen O (1977a) The distribution of organic carbon in a marine planktonic food web following nutrient enrichment. J Exp Mar Biol Ecol 26:235–247

Parsons TR, Harrison PH, Waters R (1978) An experimental simulation of changes in diatom and flagellate blooms. J Exp Mar Biol Ecol 32:285–294

Parsons TR, Li WK, Waters R (1976) Some preliminary observations on the enhancement of phytoplankton growth by low levels of mineral hydrocarbons. Hydrobiology 51:85–89

Parsons TR, Thomas WH, Siebert D, Beers JR, Gillespie P, Bawden C (1977b) The effect of nutrient enrichment on the plankton community in enclosed water columns. Int Rev Ges Hydrobiol 62:565–572

Paul AJ, Paul JM, Shoemaker PA (1979) Artificial upwelling and phytoplankton production in Alaska. Mar Sci Comm 5:79–89

Perez KT, Morrison GM, Lackie NF, Oviatt CA, Nixon SW, Buckley BA, Helt-she JF (1977) The importance of physical and biotic scaling to the experimental simulation of a coastal marine ecosystem. Helgol Wiss Meeresunters 30:144–162

Petterson H, Gross F, Koczy FF (1939) Large scale plankton culture. Nature 144:332–333

Phaup JD, Gannon J (1967) Ecology of *Sphaerotilus* in an experimental outdoor channel. Water Research 1:523–541

Pilson MEQ (1985) Annual cycles of nutrients and chlorophyll in Narragansett Bay, Rhode Island. J Mar Research 43:849–873

Pilson MEQ, Nixon SW (1980) Marine microcosms in ecological research. In: Giesy JP Jr (ed) *Microcosms in Ecological Research*. Washington DC: U.S. Dept. of Energy, Symposium Series 52 (Conf-781101), pp. 724–741

Pilson MEQ, Oviatt CA, Nixon SW (1980) Annual nutrient cycles in a marine microcosm. In: Giesy JP Jr (ed) *Microcosms in Ecological Research*. Washing-ton DC: U.S. Dept. of Energy, Symposium Series 52 (Conf-781101), pp. 735–778

Pilson MEQ, Oviatt CA, Vargo GA, Vargo SL (1979) Replicability of MERL microcosms: Initial observations. In: Jacoff FS (ed) *Advances in Marine Environmental Research*. Narragansett RI: Environmental Protection Agency (EPA-600/9-79-035), pp. 359–381

Pilson MEQ, Vargo GA, Gearing PJ, Gearing JN (1977) Investigations of effects and fates of pollutants. In: Energy/Environment II, Proceedings 2nd National

Conference Interagency R&D Program. Narragansett, RI: Environmental Protection Agency (EPA-600/9–77–012), pp. 513–516

dePinto JV, Guminiak RF, Howell RS, Edzwald JK (1980) Use of microcosms to evaluate acid lake recovery techniques. In: Giesy JP Jr (ed) *Microcosms in Ecological Research*. Washington DC: U.S. Dept. of Energy, Symposium Series 52 (Conf-781101), pp. 562–582

Polk P (1978) The sluice-dock at Ostend. Rapp PV Reun Cons Int Explor Mer 173:43–48

Pratt DM (1949) Experiments in the fertilization of a salt water pond. J Mar Research 8:36–59

Pruell RJ, Quinn JG (1985) Polycyclic aromatic hydrocarbons in surface sediments held in experimental mesocosms. Toxicol Environ Chem 10:183–200

Raymont JEG, Adams MME (1958) Studies on the mass culture of *Phaeodactylum*. Limnol Oceanogr 3:119–136

Raymont JEG, Miller RS (1962) Production of marine zooplankton with fertilization in an enclosed body of sea water. Int Rev Ges Hydrobiol 47:169–209

Reeve MR, Gamble JC, Walter MA (1977a) Experimental observations on the effects of copper on copepods and other zooplankton: controlled ecosystem pollution experiment. Bull Mar Sci 27:92–104

Reeve MR, Grice GD, Harris RP (1982) The CEPEX approach and its implications. In: Grice GD, Reeve MR (eds) *Marine Mesocosms, Biological and Chemical Research in Experimental Ecosystems*. New York: Springer-Verlag, pp. 389–398

Reeve MR, Walter MA, Darcy K, Ikeda T (1977b) Evaluation of potential indicators of sub-lethal toxic stress on marine zooplankton (feeding, fecundity, respiration, and excretion): controlled ecosystem pollution experiment. Bull Mar Sci 27:105–113

Reynolds CS (1983) Growth-rate responses of *Volvox aureus* Ehrenb (Chlorophyta, Volvocales) to variability in the physical environment. British Phycol J 18:422–433

Reynolds CS (1985) Experimental manipulations of the phytoplankton periodicity in large limnetic enclosures in Blelham Tarn, English Lake District. Hydrobiologia, in press

Reynolds CS, Graham GP, Harris P, Gouldney DN (1985) Comparison of carbon-specific growth rates and rates of cellular increase in phytoplankton in large limnetic enclosures. J Plankton Research 7:791–820

Reynolds CS, Wiseman SW, Godfrey BM, Butterwick C (1983) Some effects of artificial mixing on the dynamics of phytoplankton populations in large limnetic enclosures. J Plankton Research 5:203–234

Ringelberg J (1976) The possibilities of a new kind of micro-ecosystem in the aquatic ecosystem research. Hydrobiol Bull 10:17–18

Rogers EB (1980) Effects of elevated temperature on macroinvertebrate populations in the Browns Ferry experimental enclosures. In: Giesy JP Jr (ed) *Microcosms in Ecological Research*. Washington DC: U.S. Dept. of Energy, Symposium Series 52 (Conf-781101), pp. 684–702

Rogers JH Jr, Harvey RS (1976) The effect of current on periphytic productivity as determined using carbon-14. Water Research Bull 12:1109–1118

Rogers JH Jr, Clark JR, Dickson KL, Cairns J Jr (1980) Nontaxonomic analyses of structure and function of *aufwuchs* communities in lotic microcosms. In:

Giesy JP Jr (ed) *Microcosms in Ecological Research*. Washington DC: U.S. Dept. of Energy, Symposium Series 52 (Conf-781101), pp. 625–644

Rudd JWM, Turner MA (1983a) The English-Wabigoon River System: II. Suppression of mercury and selenium bioaccumulation by suspended and bottom sediments. Can J Fish Aquat Sci 40:2218–2227

Rudd JWM, Turner MA (1983b) The English-Wabigoon River System: V. Mercury and selenium bioaccumulation as a function of aquatic primary productivity. Can J Fish Aquat Sci 40:2251–2259

Sabatini G, Marcotte BM (1983) Water pollution: a view from ecology. Mar Pollut Bull 14:254–256

Salki A, Turner M, Patalas K, Rudd J, Findlay D (1984) The influence of fish-zooplankton-phytoplankton interactions on the results of selenium toxicity experiments within large enclosures. Can J Fish Aquat Sci 42:1132–1143

Santschi PH (1982) Application of enclosures to the study of ocean chemistry. In: Grice GD, Reeve MR (eds) *Marine Mesocosms, Biological and Chemical Research in Experimental Ecosystems*. New York: Springer-Verlag, pp. 63–80

Santschi PH, Adler D, Amdurer M (1983a) The fate of particle-reactive trace metals in coastal waters: Radioisotope studies in microcosms. In: Wong CS, Boyle E, Bruland KW, Burton JD, Goldberg ED (eds) *Trace Metals in Sea Water*. New York: Plenum Press, pp. 331–349

Santschi PH, Adler D, Amdurer M, Li Y-H, Bell J (1980) Thorium isotopes as analogs for "particle-reactive" pollutants in coastal marine environments. Earth Planet Sci Lett 47:327–335

Santschi PH, Carson S, Li Y-H (1982) Natural radionuclides as tracers for geochemical processes in MERL mesocosms and Narragansett Bay. In: Grice GD, Reeve MR (eds) *Marine Mesocosms, Biological and Chemical Research in Experimental Ecosystems*. New York: Springer-Verlag, pp. 97–109

Santschi PH, Li Y-H, Bell J, Adler D, Amdurer M, Nyfeller UP (1983b) The relative mobility of natural (Th, Pb, Po) and fallout (Pu, Cs, Am) radionuclides in the coastal marine environment: Results from model ecosystems (MERL) and Narragansett Bay studies. Geochim Cosmochim Acta 47:201–210

Santschi PH, Nyfeller UP, Anderson R, Schiff S (1984) The enclosure as a tool for the assessment of transport and effects of pollutants in lakes. In: White HH (ed) *Concepts in Marine Pollution Measurements*. College Park, MD: Maryland Sea Grant College, pp. 549–562

Saward D, Stirling A, Topping G (1975) Experimental studies on the effects of copper on a marine food chain. Mar Biol 29:351–361

Schindler DW, Hesslein R, Kipphot G (1977) Interactions between sediments and overlying waters in an experimentally eutrophied Precambrian Shield Lake. In: Golterman HL (ed) *Interactions Between Sediments and Fresh Water*. The Hague: Dr W Junk Publishers, pp. 235–243

Schindler DW, Hesslein RH, Wagemann R, Broecker WS (1980a) Effects of acidification on mobilization of heavy metals and radionuclides from the sediments of a freshwater lake. Can J Fish Aquat Sci 37:373–377

Schindler JE, Waide JB, Waldron MC, Hains JJ, Schreiner SP, Freeman ML, Benz SL, Pettigrew DR, Schissel LA, Clark PJ (1980b) A microcosm approach to the study of biogeochemical systems. I. Theoretical rationale. In: Giesy JP Jr (ed) *Microcosms in Ecological Research*. Washington DC: U.S. Dept. of Energy, Symposium Series 52 (Conf-781101), pp. 192–203

Scott BF, Glooschenko V (1984) Impact of oil and oil-dispersant mixtures on flora and water chemistry parameters in freshwater ponds. Sci Total Environ 35:169–190

Scott BF, Wade PJ, Taylor WD (1984a) Impact of oil and oil-dispersant mixtures on the fauna of freshwater ponds. Sci Total Environ 35:191–206

Scott BF, Nagy E, Dutka BJ, Sherry JP, Hart J, Taylor WD, Glooschenko V, Wade PJ (1984b) The fate and impact of oil and oil-dispersant mixtures in freshwater pond ecosystems: introduction. Sci Total Environ 35:105–113

Seitzinger SP, Nixon SW, Pilson MEQ (1984) Denitrification and nitrous oxide production in a coastal marine ecosystem. Limnol Oceanogr 29:73–83

Seki H, Whitney FA, Wong CS (1983) Copper effect on dynamics of organic materials in marine controlled ecosystems. Arch Hydrobiol 96:176–189

Sherry JP (1984) The impact of oil and oil-dispersant mixtures on fungi in freshwater ponds. Sci Total Environ 35:149–167

Sigmon CF, Kania HF, Beyers RJ (1977) Reductions in biomass and diversity resulting from exposure to mercury in artificial streams. J Fish Res Bd Can 34:493–500

Skjoldal HR, Dale T, Haldorsen H, Pengerud B, Thingstad TF, Tjessem K, Aaberg A (1982) Oil pollution and plankton dynamics. I. Controlled ecosystem experiment during the 1980 spring bloom in Lindaspollene, Norway. Neth J Sea Research 16:511–523

Skjoldal HR, Johannessen P, Klinken J, Haldorsen H (1983) Controlled ecosystem experiment in Lindaspollene, western Norway, June 1979: Comparisons between the natural and two enclosed water columns. Sarsia 68:47–64

Smetacek V, von Bodungen B, von Brockel K, Zeitzschel B (1976) The plankton tower. II. Release of nutrients from sediments due to changes in the density of bottom water. Mar Biol 34:373–378

Smetacek V, vonBodungen B, Knoppers B, Pollehne F, Zeitzschel B (1982) The plankton tower. IV. Interactions between water column and sediment in enclosure experiments in Kiel Bight. In: Grice GD, Reeve MR (eds) *Marine Mesocosms, Biological and Chemical Research in Experimental Ecosystems*. New York: Springer-Verlag, pp. 205–216

Smith W, Gibson VR, Brown-Leger LS, Grassle JF (1979) Diversity as an indicator of pollution: cautionary results from microcosm experiments. In: Grassle JF, Patel GP, Smith WK, Taillie D (eds) *Ecological Diversity in Theory and Practice*. Fairland, MD: International Cooperative Publishing House, pp. 269–277

Smith W, Gibson VR, Grassle JF (1982) Replication in controlled marine systems: presenting the evidence. In: Grice GD, Reeve MR (eds) *Marine Mesocosms, Biological and Chemical Research in Experimental Ecosystems*. New York: Springer-Verlag, pp. 217–225

Smyly WJP (1976) Some effects of enclosure on the zooplankton in a small lake. Freshwater Biol 6:241–251

Snow NB, Scott BF (1975) The effect and fate of crude oil spilt on two arctic lakes. In: *1975 Conference on Prevention and Control of Oil Spills*. Washington DC: American Petroleum Institute, pp. 527–534

Solomon KR, Smith K, Guest G, Yoo JY, Kaushik NK (1980) Use of limnocorrals in studying the effects of pesticide in the aquatic ecosystem. Can Tech Rep Fish Aquat Sci 975:1–9

Solomon KR, Yoo JY, Lean D, Kaushik NK, Day KE, Stephenson GL (1985) Dissipation of permethrin in limnocorrals. Can J Fish Aquat Sci 42:70–76

Sonntag NC, Parsons TR (1979) Mixing an enclosed, 1,300 m³ water column: effects on the planktonic food web. J Plankton Research 1:85–102

Spies A, Parsons TR (1985) Estuarine microplankton: an experimental approach in combination with field studies. J Exp Mar Biol Ecol 92:63–81

Steele JH (1979) The uses of experimental ecosystems. Phil Trans R Soc Lond B 286:583–595

Steele JH, Baird IE (1968) The ecology of O-group plaice and common dabs at Loch Ewe. I. Population and food. J Exp Mar Biol Ecol 2:215–238

Steele JH, Farmer DM, Henderson EW (1977) Circulation and temperature structure in large marine enclosures. J Fish Research Board Can 34:1095–1104

Steele JH, Gamble JC (1982) Predator control in enclosures. In: Grice GD, Reeve MR (eds) *Marine Mesocosms, Biological and Chemical Research in Experimental Ecosystems*. New York: Springer-Verlag, pp. 228–237

Steele JH, Henderson EW (1981) A simple plankton model. Amer Nat 117:676–691

Stephenson GL, Hamilton P, Kaushik NK, Robinson JB, Solomon KR (1984) Spatial distribution of plankton in enclosures of three sizes. Can J Fish Aquat Sci 41:1048–1054

Stout RJ, Cooper WE (1983) Effect of p-cresol on leaf decomposition and invertebrate colonization in experimental outdoor streams. Can J Fish Aquat Sci 40:1647–1657

Strickland JDH (1967) Between beakers and bays. New Scientist (Feb. 2, 1967):276–278

Strickland JDH, Holm-Hansen O, Eppley RW, Linn RJ (1969) The use of a deep tank in plankton ecology. I. Studies of the growth and composition of phytoplankton crops at low nutrient levels. Limnol Oceanogr 14:23–34

Strickland JDH, Trehune LDB (1961) The study of *in situ* marine photosynthesis using a large plastic bag. Limnol Oceanogr 6:93–96

Sullivan BK, Ritacco PJ (1985a) The response of dominant copepod species to food limitation in a coastal marine ecosystem. Arch Hydrobiol Beih Ergebn Limnol 21:407–418

Sullivan BK, Ritacco PJ (1985b) Ammonia toxicity to larval copepods in eutrophic marine ecosystems: a comparison of results from bioassays and enclosed experimental ecosystems. Aquat Toxicol 7:205–217

Tagatz ME, Deans CH, Moore JC, Plaia GR (1983) Alterations in composition of field- and laboratory-derived estuarine benthic communities exposed to di-n-butyl phthalate. Aquat Toxicol 3:239–248

Tagatz ME, Ivey JM, Gregory NR, Oglesby JL (1981) Effects of pentachlorophenol on field- and laboratory-derived estuarine benthic communities. Bull Environ Contam Toxicol 26:137–143

Takahashi M, Koike I, Iseki K, Bienfang PK, Hattori A (1982) Phytoplankton species responses to nutrient changes in experimental enclosures and coastal waters. In: Grice GD, Reeve MR (eds) *Marine Mesocosms, Biological and*

Chemical Research in Experimental Ecosystems. New York: Springer-Verlag, pp. 333–340

Takahashi M, Thomas WH, Seibert DLR, Beers J, Koeller P, Parsons TR (1975) The replication of biological events in enclosed water columns. Arch Hydrobiol 76:5–23

Takahashi M, Wallace GT, Whitney FA, Menzel DW (1977) Controlled ecosystem pollution experiment: effect of mercury on enclosed water columns. I. Manipulation of experimental enclosures. Mar Sci Comm 3:313–329

Takahashi M, Whitney FA (1977) Physical features of controlled experimental ecosystems (CEE) with special reference to their temperature, salinity and light penetration structures. Bull Mar Sci 27:8–16

Tangley L (1985) And live from the East Coast—a miniature Maine ecosystem. BioScience 35:618–619

Taub FB (1969) A biological model of a freshwater community: a gnotobiotic ecosystem. Limnol Oceanogr 14:136–142

Taub FB (1976) Demonstration of pollution effects in aquatic microcosms. Intern J Environ Studies 10:23–33

Taub FB, Crow ME (1980) Synthesizing aquatic microcosms. In: Giesy JP Jr (ed) *Microcosms in Ecological Research*. Washington DC: U.S. Dept. of Energy, Symposium Series 52 (Conf-781101), pp. 69–104

Taub FB, Crow ME, Hartmann HJ (1980) Responses of aquatic microcosms to acute mortality. In: Giesy JP Jr (ed) *Microcosms in Ecological Research*. Washington DC: U.S. Dept. of Energy, Symposium Series 52 (Conf-781101), pp. 513–535

Thomas WH, Seibert DLR (1977) Effects of copper on the dominance and the diversity of algae: controlled ecosystem pollution experiment. Bull Mar Sci 27:23–33

Thomas WH, Holm-Hansen O, Seibert DLR, Azam F, Hodson R, Takahashi M (1977a) Effects of copper on phytoplankton standing crop and productivity: controlled ecosystem pollution experiment. Bull Mar Sci 27:34–43

Thomas WH, Seibert DLR, Takahashi M (1977b) Controlled ecosystem pollution experiment: effect of mercury on enclosed water columns. III. Phytoplankton population dynamics and production. Mar Sci Comm 3:331–354

Thompson JM, Ferguson ADJ, Reynolds CS (1982) Natural filtration rates of zooplankton in a closed system: the derivation of a community grazing index. J Plankton Research 4:545–560

Throndsen J (1982) Oil pollution and plankton dynamics. III. Effects on flagellate communities in controlled ecosystem experiments in Lindaspollene, Norway, June 1980 and 1981. Sarsia 67:163–170

Tjessem K, Pedersen D, Aaberg A (1984) On the environmental fate of a dispersed Ecofisk crude oil in sea-immersed plastic columns. Water Research 18:1129–1136

Topping G, Davies IM, Pirie JM (1982) Processes affecting the movement and speciation of mercury in the marine environment. In: Grice GD, Reeve MR (eds) *Marine Mesocosms, Biological and Chemical Research in Experimental Ecosystems*. New York: Springer-Verlag, pp. 167–179

Topping G, Windom H (1977) Biological transport of copper at Loch Ewe and Saanich Inlet: controlled ecosystem pollution experiment. Bull Mar Sci 27:134–141

Turner MA, Rudd JWM (1983) The English-Wabigoon River System: III. Its geochemistry, bioaccumulation, and ability to reduce mercury bioaccumulation. Can J Fish Aquat Sci 40:2228–2240

Vaccaro RF, Azam F, Hodson RE (1977) Response of natural marine bacterial populations to copper: controlled ecosystem pollution experiment. Bull Mar Sci 27:17–22

Vargo GA, Hutchins M, Almquist G (1982) The effect of low, chronic levels of No.2 fuel oil on natural phytoplankton assemblages in microcosms: I. Species composition and seasonal succession. Mar Environ Research 6:245–264

Vargo SL (1981) The effects of chronic low concentrations of No. 2 fuel oil on the physiology of a temperate estuarine zooplankton community in the MERL microcosms. In: Vernberg FJ, Calabrese A, Thurberg FP, Vernberg WB (eds) *Biological Monitoring of Marine Pollutants*. New York: Academic Press, pp. 295–322

Verduin J (1969) Critique of research methods involving plastic bags in aquatic environments. Trans Amer Fish Soc 98:335–336

Wade TL, Quinn JG (1980) Incorporation, distribution, and fate of saturated petroleum hydrocarbons in sediments from a controlled marine ecosystem. Mar Environ Research 3:15–33

Wakeham SG, Canuel EA, Doering PH, Hobbie JE, Helfrich JVK (1985) The biogeochemistry of toluene in coastal water: radiotracer experiments in controlled ecosystems. Biogeochemistry 1:307–328

Wakeham SG, Davis AC, Goodwin JT (1982) Biogeochemistry of volatile organic compounds in marine experimental ecosystems and the estuarine environment—initial results. In: Grice GD, Reeve MR (eds) *Marine Mesocosms, Biological and Chemical Research in Experimental Ecosystems*. New York: Springer-Verlag, pp. 137–151

Wakeham SG, Davis AC, Karas JA (1983) Mesocosm experiments to determine the fate and persistence of volatile organic compounds in coastal seawater. Environ Sci Technol 17:611–617

Wallace GT Jr, Seibert DL, Holzknecht SM, Thomas WH (1982) The biogeochemical fate and toxicity of mercury in controlled experimental ecosystems. Estuar Coastal Shelf Sci 15:151–182

Walton S (1980) The reef's tale. Bioscience 30:805

Warren CE, Davis GE (1971) Laboratory stream research: objectives, possibilities and constraints. Ann Rev Ecol Syst 2:111–144

Warren CE, Wales JH, Davis GE, Doudoroff P (1964) Trout production in an experimental stream enriched with sucrose. J Wildlife Manage 28:617–660

Weinberger P, Greenhalgh R, Moody RP, Boulton B (1982) Fate of fenitrothion in aquatic microcosms and the role of aquatic plants. Environ Sci Technol 16:470–473

Whitney FA, Perry K, Philpott C, Ramey A, Wong CS (1981) Pentachlorophenol in a pelagic marine ecosystem: effects on the ecosystem. Pacific Science Report 81-3, Institute of Ocean Sciences. Sidney, B.C., Canada

deWilde PAWJ, Kuipers BR (1977) A large indoor tidal mud-flat ecosystem. Helgol Wiss Meeresunters 30:334–342

Wilkes FG (1978) Laboratory microcosms for use in determining pollutant stress. In: Hutzinger O, vanLelyveld IH, Zoeteman BCJ (eds) *Aquatic Pollutants:*

Transformation and Biological Effects. New York: Pergamon Press, pp. 309–321

Williams DR, Giesy JP Jr (1978) Relative importance of food and water sources to cadmium uptake by *Gambusia affinis* (Poeciliidae). Environ Research 16:326–332

Williams IP, Gibson VR, Smith WK (1977) Horizontal distribution of pumped zooplankton during a controlled ecosystem pollution experiment: implications for sampling strategy in large volume enclosed water columns. Mar Sci Comm 3:239–253

Wilson MP (1978) Assessment problems of whether or not to treat oil spills. In: McCarthy LT, Lindblom GP, Walter HF (eds) *Chemical Dispersants for the Control of Oil Spills.* Philadelphia: American Society for Testing and Materials, pp. 119–126

Wong CS, Whitney FA, Cretney WJ, Lee K, McLaughlin F, Wu J, Fu T, Zhuang D (1984) An experimental marine ecosystem response to crude oil and Corexit 9527: Part 1. Fate of a chemically dispersed crude oil. Mar Environ Research 13:247–263

Zeitzschel B, Davies JM (1978) Benthic growth chambers. Rapp PV Reun Cons Int Explor Mer 173:31– 42

Part IV
Ecotoxicological Decision Making

This final section further considers ecological indicators and ecosystem responses to chemical stress, but here the issues extend to ecotoxicological decision making in the context of environmental laws and regulations. The first chapter broaches some of the issues via a case study approach, which illustrates the challenge to provide a scientific basis for ecological risk assessment of chemical stresses. The study relates to protection against ''unreasonable degradation of the marine environment'' from a complex discharge with multiple potential stresses, and focuses on a particular ''area of biological concern''—shallow water seagrass ecosystems that fringe the shoreline of the Gulf of Mexico. Here, the need is to describe and evaluate the potential for ecological change in response to human activities, as is required for any environmental risk assessment. A theme, in particular, is to highlight how this task fundamentally differs from past efforts at environmental protection that have been charged only to set rough guidelines on levels of permissible discharges based on information from single-species laboratory tests.

The seagrass case is not unique. Societal concern for maintaining the health of ecosystems in general has been great; indeed, there are legislative mandates covering many different situations where protecting against significant ecological change from chemicals in ecosystems is required. The second chapter, by C. Harwell, provides an overview of these situations by extracting ecotoxicologically relevant facets of the major federal environmental laws and associated regulations promulgated by the U.S. Environmental Protection Agency (EPA). Key to these discussions are the regulatory endpoints of concern, i.e., those phrases in the laws and/or regulations that ultimately describe the factors to be considered as the intended decision-making point for environmental regulators. These include, for example, the phrase "unreasonable degradation" considered in the first chapter. By highlighting the ecologically relevant aspects of U.S. environmental laws and regulations, this review provides a sense of the current basis, or intended basis, for ecotoxicological management for many types of stresses and ecosystems.

The final chapter, by Harwell and Harwell, illustrates the divergence between these current rules and specified processes for regulation and the actual types of information and considerations required for a comprehensive understanding of toxic chemical fate and effects in the environment. By first indicating those factors necessary to characterize reliably ecological effects from anthropogenic chemicals, the limited scientific scope of the current, or any, environmental regulatory scheme becomes evident. Consequently, regulatory evaluations must fundamentally rely on extrapolations from data bases limited in both quantity and quality. Concomitant with such extrapolations are uncertainties, some reducible by enhanced empirical studies, others effectively irreducible. Nevertheless, environmental decision making must proceed even in the presence of uncertainties, and suggestions are presented to facilitate that process.

Chapter 16

Ecotoxicology Beyond Sensitivity: A Case Study Involving "Unreasonableness" of Environmental Change

John R. Kelly[1]

Aquatic ecotoxicology has been devoted to research that would allow for protection of the environment by proxy. For example, single-species tests for toxicity often are conducted to determine, with great precision, levels of permissible discharge; often then a "safety" factor is applied in hopes that natural ecosystems are no more than some arbitrary factor more sensitive than the most sensitive species tested. On the ecological side, many studies focus on the search for early-warning indicators within nature; these indicators then would serve as surrogates, signaling the upcoming ecological changes of actual concern. With either the traditional toxicological or ecological approach, there are obvious problems with extrapolation across the range of natural ecosystems and across the range of potential stresses. Since both approaches rely on uncertain extrapolation, and this is an uncomfortable situation, a concern has been to conduct tests with the "most sensitive species" and to evaluate response in ecosystems by the "most sensitive indicator." The philosophy of protection by proxy can be simply stated: in ignorance of the response of most natural ecosystems to most chemical stresses, we had better try to hedge our bets as much as possible.

Ignorance of potential or likely ecological response is unacceptable in cases where ecotoxicology has tools to enlighten us. In these cases, reliance on the philosophy of protection by proxy should be lessened; indeed, the scientific challenge does not end with the identification of sensitive indicators, nor even sensitive responses. Where some understanding of ecological response can be developed and where certain sensitivities can be estimated, a central issue for ecotoxicology becomes not whether

---

[1] Ecosystems Research Center, Cornell University, Ithaca, New York 14853

changes will occur (i.e., not whether the ecosystem is sensitive), but whether the ecological changes will be acceptable to society. Clearly, at some point, refining understanding of sensitivities, to the virtual exclusion of addressing these other relevant concerns, is unwarranted.

The need for expansion of research past simple questions on sensitivity and into the very difficult realm of characterizing the *significance* of an ecological change was heightened for me by studies on the potential impact of complex discharges on shallow coastal marine ecosystems dominated by seagrasses. This chapter draws on a recent synthesis of the seagrass problem (Kelly et al. 1987) to examine further the relevance of various ecological measurements in assessing the acceptability of environmental change. Some risk assessment issues related to ecological indicators of stress, raised in Chapter 2, are addressed explicitly using this case study approach. The case itself is not unique; many of the regulatory and scientific concerns illustrated here are common to impacts upon a wider variety of ecosystems (see following Chapters 17 and 18).

16.1 Potential Impacts on Seagrasses as an Ecotoxicological Case Study

16.1.1 Areas of Biological Concern

Specifically identified in environmental legislation on protection of the marine and coastal environment are "areas of biological concern." Included in these areas are *Thalassia* ecosystems. *Thalassia testudinum* (turtlegrass) beds, a prevalent feature of the shallow coastal waters along Florida's Gulf Coast (Figure 16.1), are singled out for many reasons: 1. they are highly productive ecosystems, where the grass is a direct or indirect source of food for a diverse group of fauna; 2. they support a complex biotic structure by offering habitat and substrate for many invertebrate and epiphytic algal species that either grow attached to the leaves or live associated with the subsurface roots and rhizomes; 3. for both of these reasons, they function as a refuge and nursery area for commercially valuable fish and shellfish; and 4. the presence of grass blades extending up into the water column also enlarges the role of *Thalassia* and other seagrasses beyond biological significance, as beds are important in the trapping, binding, and stabilization of sediments.

Seagrass beds are subjected to natural perturbations (e.g., hurricanes, salinity, and thermal stress), but they also can experience a range of disturbances from human activities (e.g., coastal development involving dredge and fill, modifications to land runoff, cultural eutrophication, and effects of other waste discharges). Seagrass ecosystems are highly vulnerable partly by virtue of their proximity to the shoreline, where many anthropogenic impacts can focus. Depending on both the nature and de-

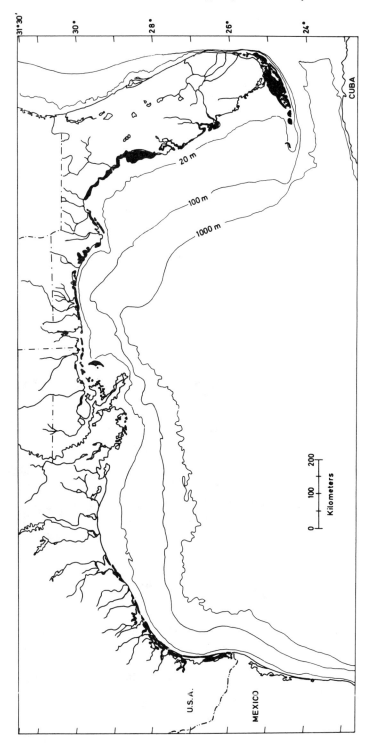

Figure 16.1. Estimated distribution of seagrasses in the Gulf of Mexico. Darkened areas in nearshore regions show the best known and comprehensive estimate. No data are available outside the U.S. Note the interruption of a continuous distribution that occurs between Mississippi Sound and Galveston Bay, associated with low salinity and turbidity of the Mississippi River discharge. (From NOAA 1986, with permission.)

gree of damage, recovery from disruption can take decades or more without intervention of restoration techniques (Thorhaug 1986).

For the reasons of high probability of exposure to stress and potentially slow recovery rates, seagrasses thus are viewed as potentially sensitive, as well as important, ecosystems. These recognitions are reflected in the designation as areas "of concern"; moreover, of particular interest is additional language of environmental legislation, as this is where the societal concerns regarding acceptability of environmental change directly challenge science to provide a more informed basis for ecological risk assessment.

16.1.2 Acceptability of Impact

Relating to the potential impacts of any discharge to the marine environment, regulations (under Clean Water Act §403(c); see Harwell 1984 and Chapter 17) call for determination of whether or not the discharge will cause "unreasonable degradation of the marine environment." "Unreasonable degradation" is not specified, but purposely is defined broadly to include:

- significant adverse changes in ecosystem diversity, productivity, and stability of the biological community within the area of discharge and surrounding biological communities;
- threats to human health through direct exposure to pollutants or consumption of exposed aquatic organisms; or
- loss of aesthetic, recreational, scientific, or economic values which are unreasonable in relation to the benefits derived from the discharge.

Thus, direct from the regulatory language, there are two relevant notions for discussion. Whether there is an *undesired* change, rather than just change per se, is a key concern for discharges that may impact *Thalassia*. Surely, ultimate decisions on acceptability are extrascientific; the challenge to scientists is to develop relevant information to evaluate the nature of changes likely to occur in response to a certain type of discharge.

Secondly, the criteria suggested to judge ecological changes speak to various levels of biological organization, from populations and communities to the ecosystem level, thus indicating concern for the health of the whole ecological system as well as for selected important components. This concern is critical in cases such as presented here, where both indirect and direct effects are likely; an ecosystem perspective thus is required. But the concern with impacts at different levels relates also to the possibility of long-term, undesired change involving linkages between structural and functional aspects of seagrass ecosystems; the importance of such linkages relates to the issues of sensitivity and relevance of ecological measurements and indicators, issues which form the theme for this chapter.

16.1.3 Effects of Drilling Fluid Discharges: Results and a Framework for Assessment

Seagrass Microcosm Experiments

Significant ecological impacts may arise from drilling fluids used in proposed oil drilling operations if discharges released into shallow-water, nearshore areas are transported to the seagrass beds fringing the Florida Gulf shoreline. Drilling fluids, used to perform various functions in drilling, may contain various mixtures of barite, bentonite, lignite, and lignosulfonate, as well as additives including bacteriocides, lubricants, or agents for pH control (NRC 1983, Duke and Parrish 1984). Potential effects of exposure of *Thalassia* beds to discharges of these complex mixtures have been investigated in microcosm experiments at the U.S. EPA Environmental Research Laboratory, Gulf Breeze, Florida, USA (Morton et al. 1986; Price et al. 1986); these references should be consulted for details of the experimental methodology.

Briefly, microcosm research has used intact cores ($\sim$ 200 cm^2) of sediment ($\sim$ 15 cm deep) and overlying water ($\sim$35 cm column) taken from a nearby natural seagrass bed, thereby containing *Thalassia* and its associated epiphytic and benthic biota. The cores are brought quickly into the laboratory, where they are held, with ambient Santa Rosa Sound seawater flow-through to maintain temperature and water quality, and with lighting to simulate, if not duplicate, the natural situation. Ecotoxicological experiments have involved dosing replicate cores ($n = 16$) of a treatment either with the suspended particulate phase (SPP) of a "used" drilling fluid (i.e., taken from a drilling operation) or with Milgel®, a common additive consisting of sodium montmorillonite clay, mixed to the appropriate density and delivered as a suspended solids load similar to the drilling fluid SPP treatment; drilling fluid (also called drilling mud) and clay treatments were compared with untreated controls. Lacking adequate fate models or direct studies of the influence of different physical turbulence regimes in the field on drilling fluid exposures in shallow areas vegetated by seagrasses (Figure 16.1), the extent to which these microcosm studies simulated the range of real world exposure scenarios is unknown. That difficult and important topic is broached elsewhere (Morton et al. 1986; Kelly et al. 1987; Morton and Montgomery 1988), but furthermore, lies apart from the generic considerations discussed in this paper. Experiments have lasted up to several months and have tested for effects on biotic composition, on *Thalassia,* on epiphtyes, and on important processes in the seagrass system.

Biological Effects

Several 10-day and 6-week microcosm tests were conducted with SSP delivered at nominal exposures (i.e., in seawater inflow to the microcosms) at concentrations within the range of acute LC_{50} or EC_{50} values as indicated from shorter-term (48 or 96 hour) single-species tests. In a summary of results, Kelly et al. (1987) concluded the experimental evidence established that significant effects could occur both by physical disturbance (i.e., the clay and drilling mud exposure both produced changes, some of which were similar between those two treatments) and from some other, presumably toxicological, feature of the drilling mud (i.e., where only the drilling mud exposure caused significant deviation from the control). Both direct and indirect effects were suggested and both structural and functional features of the seagrass system were impacted, either by disturbance, ecotoxicological response, or both (Table 16.1). Specifically

Table 16.1. Aspects of the *Thalassia* ecosystem shown to be significantly altered by exposure to clay or drilling mud in seagrass microcosms[a]

Component	Parameter (nature of)	Nature of effect (treatment effect)[b]
Autotrophs		
Thalassia[c]	Carbon fixation (Primary production **process**)	**Ecotoxicological** (drilling mud only)
Thalassia[d]	Chlorophyll a (g *Thalassia*)$^{-1}$ (Species physiological **index**)	**Mixed** (clay *and/or* drilling mud)
Epiphytes[d]	Chlorophyll a (cm^2 *Thalassia*)$^{-1}$) (Community **index**)	**Mixed** (clay *and/or* drilling mud)
Epiphytes[d]	Chlorophyll a (g epiphyte)$^{-1}$ (Community physiological **index**)	**Mixed** (either clay *or* drilling mud)
Heterotrophs[e]		
Decomposers	Loss of dried *Thalassia* (Decomposition **process**)	**Ecotoxicological** (drilling mud only)
Benthic macrofauna	Individual species (Top 10 in population **abundance**)	3 cases **Disturbance** (clay *and* drilling mud)
		1 case **Ecotoxicological** (drilling mud only)
Benthic macrofauna	Adundances of all species (Community **composition**)	**Disturbance and Ecotoxicological** (clay and drilling mud communities separable)
Benthic macrofauna	All species (3 community **diversity/biotic indices**)	**Disturbance** (clay *and* drilling mud)

[a] Includes results detailed by Price et al. (1986) and Morton et al. (1986). Includes only tests at nominal 200 ppm or 190 ppm suspended particulate phase, other tests and lists of parameters measured are given in Kelly et al. (1987) from which this Table is taken and modified.

[b] Classes are based on similarity of response in drilling mud and clay treatments; further classification of mixed effects are given in Kelly et al. (1987). Mixed classes for autotrophs arise from different results seen in tests at different seasons.

[c] Same effect shown in 2 tests, measured after 10-day exposure.

[d] Included are results from 6-week tests for parameters that showed an effect in at least 2 of 3 different tests.

[e] Results from one 6-week test (Morton *et al* 1986), further data analyses of Kelly et al. (1987).

affected were both autotrophic and heterotrophic individuals and populations. There were changes in the benthic invertebrate and epiphytic algal community attributes, and there were also effects upon major processes involving energy flow in the seagrass system.

Pathways for Ecological Change

There are a number of ways for complex chemical discharges to the environment to affect individual biota; in particular, the sometimes overlooked effects by indirect pathways have been discussed by Levin et al. (1984) and Kimball and Levin (1985). A perspective encompassing both direct and indirect effects is useful to summarize drilling muds' effects on seagrass (Figure 16.2). For example, directly toxic effects (path 1 in Figure 16.2) were expressed upon individuals of the decomposer community and upon *Thalassia* in terms of reduced primary production (carbon fixation rate). Additionally, indirect effects were seen via a disturbance mechanism, where modification of the environment by both clay and drilling mud treatments lead to reduction in the numbers of several dominant benthic macroinvertebrates. Whether increased particle deposition created, for example, conditions of sediment anoxia sufficient to cause mortality is not known; thus, indirect *toxicity* is not necessarily suggested, but indirect *changes* caused by the environmental modification are, and these are considered indirect ecotoxicological effects.

One type of indirect effect (via path 2 in Figure 16.2) was not explicitly considered, as it was presumed to be of lesser concern. This path is relevant where a chemical becomes modified (e.g., as with methylation of mercury) to result in, for example, greater toxicity at lower concentrations and/or toxic effects upon a previously unaffected component of the biota. With the drilling fluids, it seems as plausible (based on knowledge of the constituents) that environmental modification might reduce toxicity. In any event, temporal studies on chemical fate of drilling fluid constituents would be required to suggest path 2 effects; neither the chemical fate of drilling fluids nor their ecological effects were followed as a time course in these initial microcosm studies.

Some Prospects for Long-Term Ecological Consequences

Relating to temporal phenomena, however, two indirect paths (Figure 16.2 4a,4c) were identified as potential agents for long-term ecological change. Both must be regarded in our case as speculative since they involve extrapolation.

The first effect involved change in community composition. The number of certain species of macrobenthos decreased, and the resultant community composition, distinct in each treatment (see Kelly et al. 1987), was lacking strongly dominant species. While perturbations in the field might have compensating mechanisms to lessen the decrease in numbers that

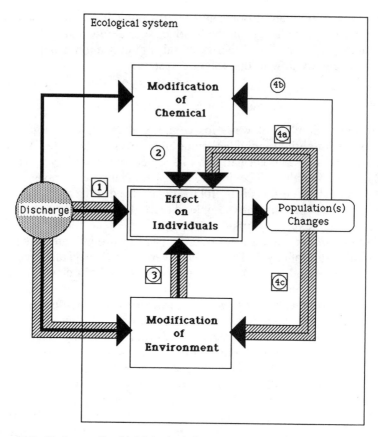

Figure 16.2. Pathways for the biological changes observed after exposure to drilling fluids (suspended particulate phase), as suggested from seagrass core microcosm experiments. Bold lines for arrow shafts indicate paths that may produce toxic (or non-toxic) effects. Thin lines indicate paths producing changes as a result of previous biotic change, and thus do not by themselves produce toxic stress. Pathway 1 is the only direct toxic effect; other paths are considered indirect, producing change by affecting individuals via toxic or non-toxic mechanisms. Microcosms dosed either with drilling fluids or with non-toxic addition of a clay suspension were used to suggest paths by which observed effects (Table 1) occurred; these paths are indicated by diagonally-striped highlighting of arrow shafts. (Modified from Kelly et al. 1987.)

were observed in the microcosm experiments, Kelly et al. (1987) argued that if similar reductions occurred in the field they might lead to invasion by certain opportunists, with the result being (path 4a, Figure 16.2) continuing change in benthic composition over some period of time. Such change could be caused by niche space being made available by a general decrease in numbers of organisms of many species. At a finer level, with decreased numbers of certain prey, predator, or competitor species, interacting populations could be affected quite specifically. Indirect, interspecific effects are plausible with effects upon some critical species; furthermore, the possibility of intraspecific effects upon such species, involving decreased fecundity and/or recruitment, also offers potential for long-term effects (path 4a). Thus, biotic–biotic linkage effects that started either by direct or indirect paths, by toxic or disturbance mechanisms, could induce significant, lasting change in species composition.

Two key processes were affected by the drilling muds, and these represent the second type of mechanism for producing lasting ecological changes. If certain organisms controlling key processes are affected in a way that alters the physico-chemical environment, this can produce an indirect effect by a biotic–abiotic linkage, thereby involving ecosystem functional change (path 4c). This was just the case in these experiments: the productivity and growth of *Thalassia* itself, the species which, by its physical presence offers habitat and protection from sediment erosion, and by its carbon production offers food resource, were significantly reduced at times. Perhaps no other single effect, depending of course on the severity and length of the reduction, could have as serious long-term consequences.

Secondly, possible blockage of the activity of the detrital decomposer community was indicated by reduction of the rate at which *Thalassia* blades were decomposed (Morton et al. 1986; Kelly et al. 1987). The possible consequences are wide-ranging, from disruption of fragile biological associations that are timed to detrital processing by season (e.g., Imai et al. 1950), to reduction in nutrient recycling rates and pathways that may influence primary production levels. The latter scenario relates to modification of the biogeochemical environment (path 4c), which normally has consequence for species composition, if not always for the level of productivity and carbon flow.

Further Perspective on Ecosystem Effects

Related to effects as represented by path 4c, the significance of some of the effects of Table 16.1 and Figure 16.2 can be illustrated by another conceptualization of the ecological system particularly relevant to the seagrass/drilling muds case. Figure 16.3 also depicts the direct and indirect paths of disturbance and toxic response, but additionally emphasizes linkages between various structural and functional features in ecosystems. For example, ecological change of individual biota could lead to further individual and population changes, some effected through processes that, themselves being affected, modify the environment in an unusual way. For our case, a specific example is the one just mentioned where the effect upon certain *individuals*, in part responsible for the decomposition *process*, could slow nutrient recycling, thereby modifying

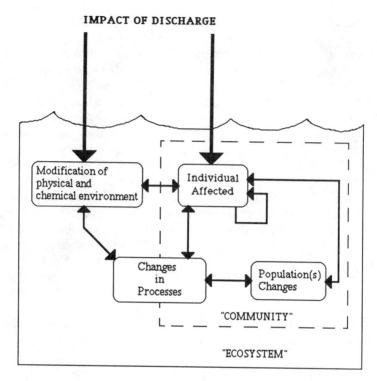

Figure 16.3. Potential ecological effects produced by discharges to natural eco-systems. Direct and indirect impacts upon individuals (see Figure 2) can be felt across levels of biological organization. Ecological change at the level of individuals, populations, communities, or ecosystems can arise from effects on biotic–biotic interactions or on biotic–abiotic interactions, the latter of which usually involve effects upon processes, seen as interface between organisms and their chemical and physical environment. (Modified from Kelly 1986.)

the chemical nutrient environment available for plant growth, and restrict growth of *Thalassia*.

In the above paragraph "modify the environment in an *unusual* way" is stated because processes, in particular those involving nutrient trans-formations, continuously modify the chemical environment and are a ma-jor contributor to "normal" (i.e., unperturbed) daily and seasonal ecolog-ical change in both structure and function. Consequently, the problem of demonstrating biotic change as directly linked to a process change, even if it occurs, could be confounded often by a small "signal-to-noise ratio" and a changed process could be ascribed incorrectly to the "natural vari-ability" associated with the process. This particular problem, in our case, was circumvented to some extent by using controlled microcosm experi-ments to examine changes in process rates.

Processes can be defined broadly to incorporate not only biogeochemical reactions, but also any biologically mediated material transfers that involve abiotic components in the environment. Processes, thus defined, include activities that are often thought of as strictly "biological" only because the abiotic interaction may be slightly less obvious or less the focal point of study (e.g., these might include activities involving food gathering that cause redistribution of inorganic particles). In the broad use of the term, and as depicted in Figure 16.3, processes stand at one interface between community and ecosystem. Importantly, however, they are also shown as links between individuals and populations and their physical environment—links which, like others in the figure, are depicted by double-headed arrows to emphasize that changes *can,* although not necessarily will, run in either direction.

Surely, ecosystem change can result from changes in processes. However, ecosystem change comes from the total interacting system of changes, which cuts across all "levels" (Figure 16.3), and includes both biotic–biotic and biotic–abiotic interactions.

This perspective constructs a framework within which ecological impacts occur in nature. The framework fully considers biotic structure, but suggests also that consideration of processes may be required. Unfortunately, the latter sometimes are dismissed as less sensitive than species or other biotic structure changes, and for that (often unsubstantiated) reason, among others, the more functional attributes of ecosystems are sometimes prematurely (I believe) excluded from consideration in ecological risk assessments.

In sum, recent experimental studies using a stress of drilling mud exposure indicate potential effects on most aspects of the *Thalassia* ecosystem are possible, identify certain pathways of concern, and suggest where some long-term consequences may lie. There are many unresolved issues relating further to extrapolating from seagrass core microcosms and laboratory exposures to a field situation (e.g., Kelly et al. 1987). Here, the case itself, with initial findings, is used to provide backdrop for focus on how concerns beyond sensitivity issues might shape a scientific process where the ultimate needs are to address the acceptability of impacts.

16.2 Beyond Sensitivity: Raising the Acceptability Issue

As brought out in the introduction, a prominent theme in ecotoxicology has been the sensitivity concern, involving either that of looking for sensitive species to test or looking for sensitive indicators of stress in ecosystems. A secondary concern, centered on the latter topic, has provoked argument as to which aspect of ecosystems, biotic structure or function, is generally the more sensitive. The *Thalassia* case brings additional perspective, illustrating that our scientific vision must not be limited to these concerns.

Recent papers elaborate current limitations of testing for the effects of chemicals on natural ecosystems. For example, in arguing the necessity of broader perspective for ecotoxicology extending beyond, but not excluding, single-species laboratory toxicity tests, Levin et al. (1984) and Kimball and Levin (1985) accurately reflected the consensus of many ecologists. Yet, surprisingly, at least to this author, some resistance was still voiced by Weiss (1986), who wrote a letter that concluded: "It seems to me, therefore, that in looking for the effects of toxicants, single-species tests may be the most sensitive if one chooses the species and parameters carefully." This letter, in part, prompted Cairns (1986a) to expound on the "myth of the most sensitive species."

Cairns argued that, in employing a most sensitive species approach to protection of the environment, there were inherent, though not always explicit, assumptions and that four assumptions, which he considered, had serious flaws. His arguments have great merit, as do his (Cairns 1986a,1986b) and many earlier urgings (see Levin et al. 1984, Odum 1984, Kimball and Levin 1985) to include multispecies testing, with micro- and mesocosm tests for toxicity, in order to enhance our ability to protect natural ecosystems.

The comments of Weiss and Cairns, however, largely relate to testing. As considerations go beyond the initial phases of setting environmental concentration guidelines by proxy, and of *testing for effects on individual organisms,* to pursue questions that would provide a basis for *evaluating ecological change,* surely the realm of ecological risk assessment is entered. Ecological risk assessment must employ greater understanding of the natural ecosystem than is required for testing that originally was intended to act only as surrogate, in absence of sufficient understanding of nature.

For the *Thalassia* case, three topics often discussed with regard to species' tests will be considered here. The objective is to suggest additional concerns that may arise, in particular showing how the questions asked of ecotoxicology change markedly:

1. Can one select a most sensitive species whose response corresponds to those organisms exposed in the natural environment? Furthermore, even if one could, would the corresponding response in nature suggest the undesirability of the change?
2. Will a species shown to be most sensitive to a limited array of tested chemicals also be so to a larger array, including multiple stresses as often experienced in nature? If not, what is the likelihood that a response, unindicated by the test species, will prove to be more unacceptable than a response indicated by this most sensitive species?
3. Are the endpoints chosen for a most sensitive species toxicity test at

least as sensitive as responses, at any level, in nature? Even if so, using such a species test, could the actual response in nature be suggested such that its acceptability could be judged?

The answer to the first question for each of these topics is, of course, "not always;" it is not clear under what restricted circumstances the answer to all three would be "yes." Fundamentally, our interests often should lie more with the second question than with the first for each topic; this is where the question of "unreasonable degradation" enters. The first two topics relate most directly to the use of a most sensitive test or indicator species, and thus are considered apart from the third, which pertains to sensitive indicators in general, but relates to the structural/functional schism in ecotoxicology.

16.2.1 Species Sensitivity and the Significance of Response to Perturbations

Beyond Problems with Sensitivity Correspondence of Tests and Nature

It is possible that "overprotection" of the biota may result from the approach of the most sensitive species, i.e., where no biological response in the natural ecosystem occurs at an exposure lower than that indicated by response of a most sensitive species. "Underprotection" seems as likely in cases where indirect effects (Figure 16.2) predominate. In general, matching the sensitivity from among the many organisms within a given ecosystem to that of a laboratory tested species for a given stress may be possible, but long-term study of that ecosystem's response will be necessary to corroborate correspondence. Moreover, since aquatic test species generally are not chosen as representative of different types of aquatic ecosystems, but are chosen on the basis of sensitivity to (only) direct exposure to chemicals, the probability of appropriate usage in extrapolation across an array of ecosystems seems very low. Unfortunately, sensitivity correspondence is not simple to address, especially due to the difficulty of establishing actual exposure regimes in nature, which hinders direct comparison of sensitivities.

The exposure characterization problem is acute indeed with ecosystems like seagrasses—where biota may live free-swimming or suspended in the water, within the sediment to various depths, in both the water and the sediments (e.g., *Thalassia*), or attached to substrate extending into the water column. Each mode of living carries its own microexposure regime, a regime varying among biota for a given chemical, but also varying from chemical to chemical in terms of the relative exposure among microhabitats.

Because species in seagrass ecosystems can be exposed differentially, in a heterogeneous fashion, a multispecies approach to sensitivity is reasonable, and the use of species from the actual ecosystem is warranted. Additionally, as the possibility of environmental modification exists (Figure 16.2), toxicity of a discharged chemical may be either reduced or increased. Because of this, matching of sensitivities between a most sensitive test species and the biota of the ecosystem could be simply fortuitous—if the chemical or the environment (Figure 16.2) is modified sufficiently, the simple single-species test system and nature (or microcosms) may examine sensitivity to different chemicals! Therefore, a microcosm technique such as with the seagrass cores, which includes the possibility of modification, ensuing indirect effects, and a simulated heterogeneous exposure, is a reasonable approach to assess sensitivity as potentially corresponding to some patch of nature.

This is not to argue that a microcosm, or "mesocosm," approach solves all. In particular, relating to this topic of correspondence in sensitivities, there remain many uncertainties with the use of microcosms (Kelly et al. 1987). Extrapolation across scales, to other seagrass beds with different biota, to different physical regimes shaping different exposure regimes, are major concerns and there are others. For example, in areas already perturbed by similar or different chemicals than are being tested, are the very sensitive species already lost so that the system tested is not what it could be? Do the "most sensitive species" survive the removal from the field and continue to function normally in the laboratory under the simulated, not actual, conditions?

In the initial seagrass/drilling fluids studies, the sensitivity correspondence really has not been examined directly. The nominal *chronic* SPP concentrations for 6-week microcosm tests were similar to the lowest *acute* 48-hour EC_{50} value from species tests (see Duke and Parrish 1984; Kelly et al. 1987). It is suspected that chronic exposure at still lower levels in microcosm tests would result in effects, but the sensitivity in terms of lowest discharge concentration yielding an effect has yet to be examined in the laboratory and the likely exposure regime of different components (e.g., benthos versus epiphytes) has not been fully characterized in the field. Even supposing such exposure characterization for a given situation could be satisfactorily achieved, however, if all species from a natural bed were considered in search for most sensitive ones, the task would be large because the species associated with *Thalassia* are so numerous, yet so diverse. Should this path be taken when the real problem is to address when ecosystem change (Figure 16.3) arising from individual effects is of concern?

As an example of size of the task this search for a most sensitive species would be, consider the following: In one microcosm experiment, the benthic invertebrates of one size class (> 0.5 mm) had 74 or 75 species in each treatment (clay, drilling mud) and the control, for a total of 95 species across the total 48 cores used for the experiment. Characterizing the tails of the species abundance curves, i.e., quantifying rare species past the top 20 or so that are common enough to identify readily, requires great effort and significant taxonomic expertise. Additionally, rare species generally are spotty in their distribution across replicate cores. Therefore, ascribing treatment effects other than for fairly common and relatively uniformly distributed species is problematic; even if possible this requires unwieldy numbers of replicates or larger-scale enclosures. Alternatively, the rare species effects assessment problem could be approached in a different way, through experimental manipulation, such as by addition of constant numbers of a given species to cores. That technique would decrease realism of the test in the sense of aggregating numbers in smaller space than normally occupied; more critically, if required for all species, it certainly expands the task enormously. Consider doing such standard addition tests for each of these 95 species, singly, let alone in combination!

Beyond questions of whether sensitivities can be matched across all species and between *any* laboratory tests and the field, lies the fundamental concern regarding acceptability of the ecological change. At issue, particularly in the seagrass case with emphasis on an evaluation of "unreasonable degradation" from discharges, are "so what" questions. For example, even if species #87 were identifiable as the most sensitive to a certain stress, and even if it were affected by a discharge to a certain area, so what? Why should anyone care? Relevant are several topics: Is this species a *useful* indicator? Would an effect upon it carry concern regarding impacts on important aspects of the seagrass ecosystem, such as the four major functions listed in discussing *Thalassia* as an "area of concern"? Could an effect on this species translate into changes classifiable as "unreasonable?" Obviously, these do not represent trivial questions, nor do they dissolve easily into a set of hypotheses to be tested.

To be sure, sensitivity and acceptability sometimes go hand in hand. For one example, if an endangered species were seen as sensitive to a particular stress, a discharge causing such stress would normally be regarded as an unacceptable risk. Additionally, the importance of maintaining biotic diversity should not be undervalued. Related to diversity in the natural world, the sensitivity of each and every species *is* key, though these sometimes may be estimated in the aggregate. The maintenance of diversity, in fact, relates strongly to one issue involving acceptability, the spatial extent of damage—a topic discussed more later.

At present, very little is known about the roles of the 95 or more species of large macrofauna or the many others (e.g., microbes, epiphytes, meiofauna, fish, grasses other than *Thalassia*) in this and other seagrass systems, but knowing more precisely all their individual sensitivities would not address the more fundamental acceptability concern. As a start towards acceptability questions, individual species, or perhaps groups, need to be identified that play roles critical to the major functions of seagrasses in coastal ecology. Using the few species or groups already identified (Table 16.1) as initial focus for examining roles would seem efficient. Once deemed to have critical roles *and* be sensitive, there are further steps in the assessment.

For example, if *Thalassia* itself seemed sensitive, as it was in the effects studies, the concern for undesired change is raised immediately, for the role of this species is central to others. From this concern emerges considerations of many factors: the level of damage to growth, including spatial and temporal extent, and the prospect and time-frame for recovery would be key general considerations on a decision on acceptability. Thus, the need to address acceptability of change, and not just sensitivity of species or whether laboratory tested species match those of organisms in nature, suggests a principal need for improved ecological understanding. The need to evaluate both the immediate response to stress and future consequence dictates that we develop an understanding of species roles in shaping biotic changes in, and surrounding, seagrass beds.

In brief, then, it can be argued that with a very pressing concern to evaluate ecosystem change, it would be unwise only to focus continuing research efforts on comparative sensitivity evaluation of all species within the system. There comes a point in assessment where the efforts should turn to look squarely at species roles offering the potential for ecological change, yet this rarely happens early enough. As we move to this stage, I do not advocate dropping all examinations of species sensitivity, nor will I next advocate dropping species tests for tests of processes; but to concentrate effort on sensitivity search among single species tested totally out of context from their environmental setting, where correspondence in sensitivity to nature is doubtful (or only fortuitous) because of mismatch of exposure regimes and exclusion of indirect effects, surely seems folly.

Beyond the Sensitive Species or Indicator: Problems with Extrapolation of Sensitivity Across Stresses

A species shown to be the most sensitive to one chemical will not invariably be so for others, singly nor in combination. Thus, a single "most sensitive species" acts as poor surrogate indicator across different chemicals.

The drilling muds example offers a case in point. In single-species testing, the overall most sensitive species and endpoint and the particular constituent evoking that response could be found for a given mud; however, like sewage sludges, the composition of drilling muds varies widely. Subsequently, the most sensitive organism or endpoint likely will not be a constant across muds. Secondly, since a single drilling fluid discharge is a complex mixture, exposed organisms may respond to a number of the constituents.

This latter suggestion may be a small concern in direct laboratory exposure for toxicity, but it becomes more critical in the environmental setting. There, differential sensitivities of species to different constituents, expressed by different mechanisms and pathways, can produce an array of effects; additionally, different constituents of a mixed discharge may be transported very differently, creating unpredictable exposures to any particular individual constituent. Since *Thalassia* beds tend to stabilize and trap material, any particle-reactive constituent may tend to be "focused" into a bed to develop concentrations well above what one would calculate using a simple dilution model that would be more appropriate for dissolved components.

Because of these types of concerns, when focus turns towards the acceptability of effects, questions on which effects will lead to ecological changes of *significance* and which may not, are of major concern. Thus, particularly if there is the issue to protect against certain levels and kinds of ecological change, a suite of organisms and indicators responsive, in toto, to the discharge, with all its potential stresses and exposures in nature, is required.

In our example, where testing was carried beyond single species and furthermore included the environmental framework offered by intact seagrass cores, not only toxicity, but also disturbance effects were recognized (Table 16.1). As nominal toxicity of a drilling mud decreases, the disturbance effects, and perhaps indirect effects in general, take on greater relative importance. If we take the results in Table 16.1 at face value we may expect some discharges to affect, more noticeably, ecosystem processes (i.e., if certain toxic components were present), whereas other discharges, lower in those constituents, might be detected more easily as changes in abundance of those particular invertebrates sensitive to increased particle deposition (i.e., the ecological "disturbance" possible from the discharge). Focus on only one or the other of these taxonomic or functional groups of species, for all discharges, might miss some of the early, important, or unacceptable effects of some discharges.

Drilling fluids, with their complex and variable composition, are not an atypical example of chemical stresses to which seagrasses and other shallow water marine ecosystems may be exposed. In nature, exposure to multiple stresses from overlapping sources of different chemicals and/or from heavy loads of a single type of complex discharges (e.g., municipal sewage effluents would be another) is relatively common. However, coastal ecosystems, in particular, by virtue of position downstream from other ecosystems, yet often being proximal to high concentrations of the human population, can act as receptacles for many of these anthropogenic wastes. Thus, the lesson from the drilling fluids example—to look for multiple ecological endpoints that are sensitive to different categories of perturbation from multiple chemical stress and are perhaps indicative of different types of ecological effects—has generic relevance to assessing acceptability of impacts in coastal waters.

16.2.2 Functional and Compositional Indicators
for Ecotoxicology

Beyond Indicators of Sensitive Responses in Nature

Recent discussions, initiated both by ecologists and toxicologists, have focused on the relative efficacy of indicators of toxicological effects on ecosystem structure versus effects on ecosystem function (cf. Schindler 1982; Schindler et al. 1985; Schindler 1987; Cairns and Pratt 1986; Chapter 6). On the basis of few actual studies, it nevertheless seems generally perceived by aquatic scientists that structural changes, involving effects on biotic populations, are more sensitive than functional changes, such as alterations in rates of productivity or some biogeochemical processes.

Generalizations often are developed too quickly, which are largely untested or for which are attempted extrapolations beyond the bounds of the studies from which they emerge. In the case of Schindler's studies at the Experimental Lakes Area (ELA) in Canada on eutrophication and acidification, for example, the ecosystems considered were quite pristine, geologically young, oligotrophic lakes in which the dominant autotrophs were the plankton. However, there exist a wide array of ecosystems, equally aquatic, which differ from the ELA lakes in many ways, including: trophic status; chemistry as influenced by watershed and geology; relative change already induced by human activities; biology in terms of the endemic species and in terms of the structure of the food chain; as well as many others. Estuaries in general and seagrass ecosystems in particular, as another type of aquatic ecosystem, are far removed from those in Schindler's classic whole-lake perturbation studies, and perhaps from the paradigms to emerge from them as well.

A number of factors in estuaries and with seagrasses may contribute to a different response in species relative to processes. As mentioned, coastal marine areas are heavily populated and the load, for example, of nutrients, relative to pristine lakes, usually is several orders of magnitude higher (Kelly and Levin 1986; Nixon et al. 1986). In fact, a "pristine" estuary comparable to a "pristine" lake may not exist, not just for reasons of human impingement, but because of the peculiar position of estuaries in the ecological landscape that makes them the more integrated receiver of all upstream exports. With the last several decades of increasing impingement by humans, most coastal areas recently studied intensively may include many more "resistant" species that have already replaced some more sensitive ones; the present species responses to additional stress may be less dramatic than in a pristine situation. Additionally, for seagrasses, we expect the response of autotrophs to vary substantially relative to fast-growing, high turnover planktonic species' changes with a given stress. Dominated by attached benthic macrophytes, the seagrass ecosystem response might even more resemble the often-generalized terrestrial response, where the primary producers are relatively long-lived. In such cases, autotrophic species changes with stress therefore can be slow, and changes in biogeochemical processes (mediated by microbial and detrital activity) seem sometimes earlier to forecast an ecosystem change such as long-term productivity. The point is we do not know the relative efficacy of various classes of indicators, either of structure or of function, for all aquatic ecosystems, or for different stresses, and we should not presume that we do.

Aside from these disclaimers, one can find arguments in the literature to support the generalization that aquatic species composition and biotic structure effects may be seen before detecting changes in processes. One argument is that effects from toxic chemicals first are manifested as effects on individuals (e.g., as in presented in Figure 16.2) and on populations; consequently, changes in functional attributes in ecosystems would imply prior changes in populations of the organisms performing those functions. A second argument is that there often exists a considerable degree of functional redundancy in ecosystems; i.e., a particular function typically can be effected by more than one species. Because of this redundancy, there is often a buffering of ecosystem effects by species, such that toxicological reduction in one species may at least in part be compensated for by another species, which has a similar functional role, but which is not directly affected by the toxic stress. As a rule, changes in species composition may or may not be associated with changes in ecosystem functioning; however, the outcome of the converse change seems less flexible—changes in processes almost always have concomitant changes in species composition. A further argument suggests that recovery from disturbance for functional attributes may be often more rapid and more complete than is recovery of the structural attributes of an ecosystem. Thus, while being perhaps more resistant to change, functional aspects may be often more resilient (see Chapter 2).

Arguments, uncertain extrapolations, and presumptions aside, there is too much suggestion of a dichotomy that inadvertently casts structure against function; several issues need further examination in the case of most stresses and most ecosystems. For example, functional indicators may be as sensitive, or even more sensitive, than compositional indicators *for certain ecosystems or particular stresses*. Additionally, functional indicators may be more readily measured, again *under specified conditions*. Finally, and most critical to the concern with the acceptability of ecological change, functional indicators sometimes *may be more relevant* than compositional measures, with respect to ecosystem attributes that are of importance to humans.

Considering first the relative sensitivity issue, the example of drilling effects on *Thalassia* ecosystems is a case in point. As seen in Table 16.1, several functional attributes of the seagrass systems illustrated statistically significant effects at the lowest dose level tested (190 μl l^{-1} SPP), a level comparable to the minimal level for demonstrable effects as indicated in acute single-species toxicity tests. Tellingly, the functional indicators seemed most responsive to the toxicity aspects of the drilling fluid exposures (Table 16.1). More functional experiments, including a suite of doses at lower levels, might even show a greater sensitivity than the toxicity tests. In the case of primary production or chlorophyll content (a measure of functional potential), this could occur by the plants photosynthetic mechanisms and biochemical pathways being responsive to the toxic stress, but not leading immediately to plant mortality. In the case of decomposition, greater sensitivity could occur by the toxic stress affecting all microorganisms that perform decomposition processes; thus, whereas there may indeed be functional redundancy of the many species involved in decomposition of *Thalassia* detritus, the toxic stress could cut across that redundancy, so that effectively the *suite* of microbial species respond as one.

The second issue, of relative ease of measurement of functional versus compositional attributes, is again illustrated by the drilling fluid studies. The same microcosm experiments used for functional testing included examination of the benthic invertebrate community (Table 16.1), as characterized by the abundances of all invertebrate species within a particular sieve size range (Morton et al. 1986). The problems associated with physical scaling and rare species as identified earlier and by Kelly et al. (1987) are particularly relevant on this issue. For example, if the more sensitive species are among the rarer ones, a much larger sample size is needed to discern statistically where effects occur, i.e., to distinguish the response signal from the natural noise (heterogeneity) that exists in all ecosystems. Exacerbating this problem is the real possibility that the species that are more sensitive to toxic stress may also be more vulnerable to effects of core handling; consequently, there may be a disproportionate loss of those species from even the control treatment, requiring a larger sample size. This latter point also belies the argument for identifying the "most sensitive species." As has been suggested, it is quite possible that the most sensitive species are inadvertently precluded from testing or may be locally missing from the sampled environment because of other adverse conditions. Thus, researchers may be deceived by assuming the actual most sensitive species in an ecosystem would be protected based on toxicity data derived for the species found to be most sensitive *under any laboratory conditions*.

This leads discussion once again beyond sensitivity, to the third issue listed above, the relevancy of ecotoxicological indicators. There is no single criterion by which to select an indicator of ecosystem effects from chemical discharges; rather, different characteristics of ecosystems are most effectively evaluated by different indicators. Thus, looking for the "most sensitive species" is only one objective of ecotoxicological testing; there are other objectives, often which are more relevant to ecological risk assessment.

On this point, refer back to the brief description of the regulatory framework for drilling fluid effects on *Thalassia* ecosystems, i.e., §403(c) of the Clean Water Act with its primary regulatory endpoint emphasizing the phrase "unreasonable degradation of the marine environment." The characterization of the degree of degradation of the ecosystem itself must involve different measures, which *may* include individual species measures, whether seen as population- or community-level, but *must* consider all of the interacting attributes of the ecosystem (Figure 16.3). As a part of the overall judgment of degradation, the *relevancy* of an observed impact must be considered, and scientific assessments must reflect this need. For example, a species level change, such as reduction in numbers of a particular population of detritus-feeding polychaete annelid worms, may have little importance to qualities of the ecosystem that are seen of value to humans. On the other hand, a functional change, such as the demonstrated effects on decomposition, may have considerable implications for the rest of the ecosystem in general, which in this case is substantially supported by detritus-based energetics, and for things of direct concern to humans, such as harvestable fish productivity. Consequently, having ecotoxicological indicators that relate to those attributes of concern (i.e., *relevant* indicators) is appropriate in order to forewarn of future ecological changes of significance to regulatory endpoints.

In conclusion, it seems that arguments for using compositional indicators in nature preferentially because of a purported high degree of sensitivity are inappropriate for ecotoxicological evaluations in the context of providing the scientific basis for regulatory decision making. There is no scientific basis for casting structural and functional indicators against each other as an either/or decision; aspects of both structure and function within ecosystems have purpose for evaluating ecological change independent of their relative sensitivities. Many aspects of structure and function sensitivity itself remain unstudied for most ecosystems and most stresses; for example, indeed it may be found that the longer lasting and more irreparable harm to seagrasses relates sometimes more directly to functional perturbations than species replacements. Furthermore, one cannot presume that compositional indicators are necessarily more sensitive (as in this example, they have not proven so yet), nor should sensitivity be the sole criterion for selecting indicators and establishing testing schemes. Rather, there is a stronger case for the argument that functional as well as compositional indicators should be sought, with a suite of indicators selected to meet a suite of criteria, including relating to the importance of particular effects, the sensitivity and early-warning roles of the indicators, and the ease and economy of measurement and monitoring. Currently, a reliance of ecotoxicological evaluations on compositional measures, including some that are clearly unresponsive or responsive only to a proportion of the constituents of a multiple-stress discharge, is neither warranted nor efficacious in meeting decision-making needs.

16.3 Conclusion and Prospectus

As ecotoxicology moves from toxicity testing to ecological risk assessment, concerns emerge that extend past issues associated with identifying sensitive species for testing or choosing the most sensitive indicators of response. Some are these are illustrated by the case of regulation of discharges to coastal ecosystems such as *Thalassia* seagrass ecosystems, where the issues revolve around determination, not of sensitivity alone, but of degradation of the ecosystem that is "unreasonable." Experimental microcosm studies on potential impacts on the *Thalassia* ecosystem from exposure to one type of discharge, that of drilling fluids which could be generated from shallow energy exploration along Florida's Gulf Coast, suggest that effects will often cut across biological "levels" of organization; in this case, effects included impacts on heterotrophic and autotrophic components, and perturbations of processes as well as species. Mechanisms of effects may include toxicity as well as the physical disturbance of higher suspended loads and more particle settling; additionally, both direct and indirect effects were suggested by the microcosm studies.

One of the most difficult and frustrating tasks facing the ecological sciences is the translation of such effects seen in experiments (whether in the laboratory, in microcosms, mesocosms, or even in a unique and isolated field situation) to those ecological responses that should be of concern to humans and are to be expected for natural ecosystems . Yet, in particular for the seagrass case, the question of acceptability of the change is key, and this must shape the scientific process. Because of this need, a narrow-vision approach that focuses on finding sensitive species indicators should be argued against, and instead the broader approach of including multiple indicators and searching for indicators relevant to the ecological change itself should be argued for. This is the approach by which the issue of acceptability may be more feasibly addressed. Related to this, the extrapolation of effects, for *Thalassia* as for most ecosystems, is currently most hindered by a serious deficiency in a fundamental area— understanding what structures natural ecosystems and what makes them work. To address "so what" questions and to develop the best scientific basis for risk assessment and eventual judgement on regulatory endpoints such as "unreasonable degradation", very fundamental ecological issues need to be looked at. This is a very large topic; indeed, the requirement for understanding of ecological uncertainty is addressed more fully in Chapter 18 (Harwell and Harwell) and touched upon only briefly here. Three major concerns arise: issues involving both (1) spatial and (2) temporal aspects of ecological changes, and (3) issues relating to the importance/relevance of an ecological change. Many spatial concerns center on the relative proportion of impact to a seagrass bed that may be tolerable within a given region or locally. For example, one facet of scale research would examine the distribution and size of *Thalassia* beds in relation to fisheries yield or the presence and abundance of a particularly important species. Many temporal issues center around the rate of recovery of *Thalassia* after perturbation, as the plant itself is central to the other biota. The acceptability of impact will relate to long-term growth and survival of the grass. Lastly, many ecosystem effects arise from cascading chains of ecological events—individuals, populations, communities, and processes all interact to produce change. An important, and relevant, ecotoxicological research question is: what species or processes are critical to the structure and function of *Thalassia*?

Acknowledgments. I thank Mark Harwell and Tom Duke, with whom it has been a pleasure to work on the seagrass research, and who graciously commented on this paper.

This publication is ERC-152 of the Ecosystems Research Center, Cornell University, and was supported by the U.S. Environmental Protection Agency Cooperative Agreement Number CR812685–01. Additional funding was provided by Cornell University.

The work and conclusions herein represent the views of the author and do not necessarily represent the opinions, policies, or recommendations of the Environmental Protection Agency.

References

Cairns J Jr (1986a) The myth of the most sensitive species. Bioscience 36: 670–672

Cairns J Jr (1986b) Overview. In: Cairns J (ed) Community toxicity testing. ASTM STP 920, Philadelphia: American Society for Testing and Materials, pp. 1–5

Cairns J Jr, Pratt JR (1986) On the relation between structural and functional analyses of ecosystems. Environ Toxicol Chem 5(9): 785–786

Duke TW, Parrish PR (1984) Results of the drilling fluids research program sponsored by the Gulf Breeze Environmental Research Laboratory, 1976–1984, and their application to hazard assessment. EPA-600/4-84-055. Gulf Breeze, FL: U.S. Environmental Protection Agency

Harwell CC (1984) Analysis of Clean Water Act Section 403—Ocean Discharge Criteria. ERC-026. Ecosystems Research Center, Cornell University, Ithaca, NY, 14+ pp.

Imai T, Hatanaka M, Sato R, Sakai S (1950) Ecology of Mangoku-Ura inlet with special reference to the seed-oyster production. Rep Instit Agric Res, Tohoku Univ 1–2:137–155

Kelly JR (1986) How might enhanced levels of UV-B radiation affect marine ecosystems? In: Titus G (ed) *Effects of Changes in Stratospheric Ozone and Global Climate, Vol. 2: Stratospheric Ozone*. Proceedings of EPA/UNEP International Conference on Health and Environmental Effects of Ozone Modification and Climate Change, Washington, DC, pp. 237–251

Kelly JR, Levin SA (1986) A comparison of aquatic and terrestrial nutrient cycling and production processes in natural ecosystems, with reference to ecological concepts of relevance to some waste disposal issues. In: Kullenberg G (ed) *The Role of the Oceans as a Waste Disposal Option*. NATO Advanced Research Workshop Series. Dordrecht, Holland: D. Reidel, pp. 165–203

Kelly JR, Duke TW, Harwell MA, Harwell CC (1987) An ecosystem perspective on potential impacts of drilling fluid discharges on seagrasses. Environ Manage 11:537–562

Kimball KD, Levin SA (1985) Limitations of laboratory bioassays: the need for ecosystem-level testing. Bioscience 35:165–171

Levin SA, Kimball KD, McDowell WH, Kimball SF (1984) New perspectives in ecotoxicology. Environ Manage 8:375–442

Morton RD, Montgomery SO (1988) Microcosm studies on the effects of drilling fluids on seagrass communities. Proceedings of 1988 International Conference on Drilling Wastes held April 1988, Calgary, Alberta, Canada. London: Elsevier Publishers (in press)

Morton RD, Duke TW, Macauley JM, Clark JR, Price WA, Hendricks SJ, Owsley-Montgomery SL, Plaia GR (1986) Impact of drilling fluids on seagrasses: An experimental community approach. In: Cairns J (ed) *Community Toxicity Testing*. ASTM STP 920, Philadelphia: American Society for Testing and Materials, pp. 199–212

National Oceanic and Atmospheric Administration (1986) Gulf of Mexico coastal and ocean zones strategic assessment: Data atlas. Strategic Assessment Branch and Southeast Fisheries Center. Rockville, MD 20852, 190 pp.

National Research Council (U.S.) (1983) Drilling discharges in the marine environment. Panel on assessment of fates and effects of drilling fluids and cuttings in the marine environment. Washington DC: National Academy Press, 192 pp.

Nixon SW, Hunt CD, Nowiki B (1986) The retention of nutrients (C,N,P), heavy metals (Mn, Cd,Pb,Cu) and petroleum hydrocarbons in Narragansett Bay, RI (USA). In: Lasserre P, Martin JM (eds), *Biogeochemical Processes at the Land–Sea Boundary*. Amsterdam: Elsevier, pp. 99–122

Odum EP (1984) The mesocosm. Bioscience 34:558–562

Price WA II, Macauley JM, Clark JR (1986) Effects of drilling fluids on *Thalassia testudinum* and its epiphytic algae. Environ Exp Bot 26(4):321–330

Schindler DW (1982) Vulnerability of noncalcareous softwater lakes to eutrophication and acidification. ERC-022. Ecosystems Research Center, Cornell University, Ithaca, NY. 24 pp.

Schindler DW, Mills KH, Malley DF, Findlay DL, Shearer JA, Davies IJ, Turner MA, Linsey GA, Cruikshank DR (1985) Long-term ecosystem stress: the effects of years of experimental acidification on a small lake. Science 228:1395–1401

Schindler DW (1987) Detecting ecosystem responses to anthropogenic stress. Can J Fish Aquat Sci 44, Supplement No. 1:6–25

Thorhaug A (1986) Review of seagrass restoration efforts. Ambio 15(2): 110–117

Weiss JS (1985) Letters to the editor: Species in ecosystems. Bioscience 35: 330

Chapter 17

Regulatory Framework for Ecotoxicology

Christine C. Harwell[1]

The U.S. Environmental Protection Agency (EPA) has primary responsibility for much of environmental decision making in the United States. This responsibility is derived from a number of statutes and Executive Orders. In this chapter are reviewed seven such pieces of legislation with relevancy to ecotoxicology; specifically, the Toxic Substances Control Act (TSCA); the Clean Water Act (CWA); the Clean Air Act (CAA); the Resource Conservation and Recovery Act (RCRA); the Comprehensive Environmental Response, Compensation and Liability Act (CERCLA or Superfund); the Safe Drinking Water Act (SDWA); and the Federal Insecticide, Fungicide, and Rodenticide Act (FIFRA). Much of each of these pieces of legislation relates to administrative or procedural details or other considerations not central to the science of ecotoxicology; consequently, the review here extracts only those elements of relevance to ecotoxicology.

17.1 Toxic Substances Control Act (TSCA)

The Toxic Substances Control Act (TSCA; 15 USC Sections 2601 *et seq.*) is intended to allow federal government oversight and control of the introduction of certain chemical substances and mixtures into the environment. Before TSCA was enacted in 1976, no central federal requirement existed to test or regulate the use of all of the many new chemicals entering the environment, either for human health effects or environmen-

[1] Center for Environmental Research, Cornell University, Ithaca, New York 14853

tal impact. Experiences with human health and environmental injuries from toxic compounds of organic mercury, polychlorinated biphenyls, vinyl chloride, and Kepone led to passage of TSCA, with emphasis on regulation of potential toxins prior to environmental exposure and damage.

The Environmental Protection Agency has authority to require testing of manufactured and processed chemical substances and mixtures that may present an unreasonable risk of injury to human health or the environment and to regulate these substances and mixtures where necessary. Regulations implementing this legislative intent are to be designed to consider the environmental, economic, and social impact of any actions taken by EPA. Actions available to EPA include prohibiting or limiting the manufacture, processing, distribution, use, or disposal of chemicals which test data, or the lack of adequate data, indicate may present unreasonable health or environmental risk. Chemical manufacturers and processors are responsible for developing the data base on existing chemicals upon which decisions are made by EPA about disposition of these chemicals (TSCA Section 2[b][1]).

The central area of scientific concern, then, is in the formulation of adequate testing methodologies to produce a data base upon which environmentally sound decisions can be made by EPA.

17.1.1 General Testing Requirements

Testing under TSCA is primarily focused on new chemical substances that have not yet been introduced in commerce or on chemicals already in commerce for which a *significant new use* is proposed that increases human or environmental exposure. TSCA Section 5 controls the mechanisms by which a manufacturer must notify EPA of the intent to manufacture or make new use of such a substance. The manufacturer submits available test data to support its premanufacture notice in order for EPA to determine the reasonableness of risk of manufacturing/processing the chemical (TSCA Section 5[b]). Thus, the manufacturer can conceive toxicological tests for new chemicals or new-use chemicals; EPA then would review the adequacy of such test data. The testing data are to demonstrate the endpoint of regulatory concern, i.e., that there will be no "unreasonable risk of injury to health or the environment" from the manufacture, processing, distribution in commerce, use, and disposal of a new chemical substance or from the intended significant new use of a chemical substance already in commerce (TSCA Section 5[b]).

In contrast, TSCA Section 4 authorizes EPA to promulgate a test rule requiring specified tests be conducted by a manufacturer for chemicals or categories of chemicals *already in commerce* which, in the opinion of EPA, may present an unreasonable risk of injury to health or the environment and for which insufficient data and experience exist to determine or predict the effects of such chemical. New chemicals which would fall under TSCA Section 5 control, but which are in a category covered by a TSCA Section 4 test rule, must be tested under TSCA Section 4 requirements. At this time, EPA is assessing the applicability of categorical approaches to regulating new and existing chemicals. Categorical approaches to testing are being considered under Section 4 test rules, under Sections 5, 6, and 9 limiting manufacture and use, and under Section 8 to provide more information on manufacture and use (50 FR 44658, 29 October 1985).

17.1.2 TSCA Section 5 Testing

Under TSCA Section 5 and regulations contained in 40 CFR 720, manufacturers and importers of new chemical substances to be used for commercial purposes are required to notify EPA at least 90 days in advance. New chemicals are those not already included on an inventory list prepared by EPA under Section 8(b), which inventory is regularly updated to include chemicals that have already undergone Section 5 review and have entered commercial production. Certain chemicals are exempt from the Section 5(a) notice requirements. Once a substance is, in effect, approved by being added to the inventory, any person may manufacture, import, or process the substance for any use.

Section 5(d)(1) lists information that manufacturers/importers must provide in Section 5(a) notices to EPA: test data in their possession or control; descriptions of health and environmental effects data that they know or can reasonably ascertain; and known or reasonably ascertainable information on chemical identity, proposed categories of use, proposed volume of production, byproducts resulting from manufacture, processing, use, or disposal, workplace exposure, and manner or method of disposal. While development of any particular test data before submitting a notice to EPA is not required, EPA encourages consultation before selecting testing protocols.

Section 5(b) imposes additional data requirements for chemicals already subject to test rules under Section 4 (see discussion below) and for chemicals that EPA has determined, by rule under Section 5(b)(4), may present unreasonable risks of injury to health or the environment.

EPA has 90 days (extendable for a further 90 days for good cause) to review a Section 5 notice, designated a Premanufacture Notice (PMN), and during the review period may act under Sections 5(e) or 5(f) to regulate production or use. If EPA has not prohibited manufacture or import during the review period, these activities can begin when the review period expires, subject to restrictions or testing requirements imposed during the review period.

In conducting its review, EPA assesses the potential risks during all phases of the life cycle of the substance, from manufacture or importation to disposal. The review is based on information provided by the PMN submitter and on information EPA may obtain elsewhere, e.g., in literature searches on new substances and structural analogues conducted as part of a new chemical review. However, the PMN submitter is responsible in principle for providing enough information for an adequate risk assessment.

During review, EPA considers, among many factors, potential toxicity determined from test data submitted or from elsewhere, data on structural analogues, and structure-activity analyses. EPA also reviews the potential for human and environmental exposure during proposed uses of the substance. In addition, in determining the regulatory endpoint of "reasonableness" of any risk, EPA evaluates such nonrisk factors as possible economic benefits and availability of substitutes, both more and less toxic. Also included for evaluation are impurities, byproducts, degradation products, unintended reaction products, and other chemical substances associated with the life cycle of the new substance.

EPA issued premanufacture testing guidelines in 1981 (46 FR 8985, 27 January 1981). It was recommended that PMN submitters use the Organization for Economic Cooperation and Development (OECD) Minimum Premarket Data set as an initial point in setting up a testing program for new chemicals (see discussion in 17.1.3, below). EPA does not require these data, however, and a submission not containing them is not *de facto* incomplete, if it otherwise meets TSCA requirements. It is the stated intention of EPA that "submitters use the testing guidelines flexibly, tailoring their evaluations of new chemicals to the specific new chemicals in question" (48 FR 21740, 13 May 1983).

In addition to the normal PMN process for new chemicals not yet evaluated by EPA, TSCA Section 5(a)(2) authorizes EPA to determine that a use of a chemical substance is a "significant new use." Once this determination is made by issuance of a Significant New Use Rule (SNUR), persons must submit a notice to EPA at least 90 days before they manufacture, import, or process the substance for that use (40 CFR 721). This notice is generally subject to the same statutory requirements and procedures as a PMN (49 FR 35011, 5 September 1984), and submitters must use the PMN form and follow the PMN procedures which have been codified in 40 CFR 720 (see discussion above). EPA may regulate the substance during the review period. Environmental effects testing may follow the OECD guidelines recommended by EPA for the PMN process.

17.1.3 TSCA Section 4 Testing Guidelines

Section 4(b)(1) of TSCA, which governs the formulation of the required testing for specified chemicals or categories of chemicals already in commerce, requires that such prescribed testing, termed "test rules," include standards for the development of test data. EPA has created regulations, codified as 40 CFR Parts 796, 797, and 798, which incorporate guidelines that may be used by EPA in establishing test standards in new TSCA Section 4 test rules (50 FR 39252, 27 September 1985). The guidelines, covering chemical fate (Part 796), environmental effects (Part 797), and health effects (Part 798), present "generally formulated procedures for laboratory testing of an effect or characteristic deemed important for the evaluation of health and environmental hazards of a chemical" (50 FR 39252, 27 September 1985). All of these guidelines have been extensively peer reviewed and have been subject to previous publication and public comment; they are reviewed and updated yearly.

The guidelines for Section 4 testing are not now automatically mandatory for persons subject to a Section 4 test rule; they become effective when they are specified in an individual Section 4 proceeding producing a chemical-specific test rule. Even then, the particular guideline becomes a test standard only for that test rule and does not become a generic test standard. The guidelines may be modified by EPA as they are incorporated into a chemical-specific test rule.

However, after EPA review of the test guidelines which have been incorporated into 40 CFR Parts 796, 797, and 798, it was determined that some of the suggested or recommended elements of each of those guidelines would often be made minimum requirements for individual test rules. Accordingly, EPA has proposed a rule that these elements of each guideline be explicitly designated (51 FR 1522, 14 January 1986). This proposed rule is set forth in the belief that this designation would provide more guidance on the minimum requirements for each study and thus avoid repetitive chemical-specific changes to the guidelines when adopted as test standards for specific test rules under Section 4. The proposed rule would then specify, as *minimum requirements* for the test standards for Section 4 test rules, the elements in the guidelines which pertain to: test procedures; analytical measurements; test conditions; animal selection (number of species to be tested, age of animals, sex, number of animals per dose level); use of controls; number of dose levels; facilities (test apparatus, dilution water, test substance delivery system, test parameters); observation period; administration of the substance; observation of animals; clinical exams; gross necropsy; histopathology; and reporting of data. As can be seen from this list, most of these elements of the guidelines affect primarily the testing for human health effects as opposed to environmental effects.

Discussed below is a brief description of the specific types of information sought under the guidelines which may be incorporated into a TSCA Section 4 test rule.

40 CFR Part 796 contains chemical fate testing guidelines. Subpart B of Part 796 describes testing guidelines for determining physical and chemical properties of toxic substances. Regulatory sections within this Subpart describe standard and alternate procedures for determining the octanol–water partition coefficient (K_{ow}) of solid or liquid organic chemicals in order to correlate it to water solubility, soil–sediment adsorption coefficient, and bioconcentration, and to incorporate it in the formulation of structure-activity relationships. Methods are described for determining water solubility, in order to define the extent and rate of chemical transformations via hydrolysis, photolysis, oxidation, reduction, and biodegradation in water. Specific procedures are incorporated in the guideline to measure the water solubility of very hydrophobic compounds and to mitigate problems of colloid and emulsion formation. Methodologies are described for measurement of vapor pressure, which, when combined with water solubility data, permits determination of volatility from water and, thus, potential for airborne transport and possible atmospheric oxidation and photolysis.

Subpart C of Part 796 describes testing guidelines for determining transport processes of toxic substances in the environment. Soil thin-layer chromatography is used as a qualitative screening tool to estimate leaching potential in assessing the fate of chemicals in the environment. An adsorption isotherm test is used to assess transport of chemicals in the environment.

Subpart D of Part 796 describes testing guidelines for determining transformation processes. A standard method for carbon dioxide evolution is used to determine aerobic aquatic biodegradation. Anaerobic biodegradability of organic chemicals is determined using a screening mechanism. A test is described allowing determination of rates of hydrolysis at any pH of environmental concern at 25°C.

Methodologies are described for determining direct photolysis rate constants and half-lives of chemicals in water in sunlight. The test method is designed to determine the molar adsorptivity and reaction quantum yield of a test chemical in aqueous solution. Combining these parameters with solar irradiance data will determine environmentally relevant rate constants and half-lives in aqueous solutions as a function of latitude and season of the year in the United States. A first-tier screening level test method is described to estimate the maximum direct photolysis rate constant and minimum half-life of chemicals in the atmosphere in sunlight as a function of latitude and season of the year in the United States.

In addition to codifying the testing guidelines originated by EPA for possible use in establishing standards in Section 4 test rules, testing guidelines developed by the international Organization for Economic Cooperation and Development (OECD) were also recently codified in the Code of Federal Regulations, included in 40 CFR Part 796.

Subpart B now also contains OECD testing procedures to determine absorption in aqueous solution (ultraviolet/visible spectra); boiling point/boiling range; dissociation constants in water; and particle size distribution/fiber length and diameter distributions.

Subpart C now also contains OECD testing procedures for ready biodegradability (modified AFNOR test, closed bottle test, modified MITI test [I], modified OECD screening test, and modified Sturm test); a simulation test for aerobic sewage treatment (a coupled units test); testing for inherent biodegradability (modified SCAS test, modified Zahn-Wellens test); testing for inherent biodegradability in soil; and testing for complex formation ability in water.

40 CFR Part 797 contains environmental effects testing guidelines which describe test procedures and methods for determining aquatic effects and terrestrial effects of chemical substances and mixtures to determine levels of toxicity.

Subpart B contains testing guidelines for aquatic effects. Regulatory sections within the Subpart contain testing procedures using freshwater and marine algae to determine acute toxicity and effects on primary production, survival, and growth. The freshwater aquatic plant *Lemna gibba* G3 is designated for testing to develop data on phytotoxicity of chemicals. A daphnid acute toxicity test exposes *Daphnia magna* or *D. pulex* to a chemical in static and flow-through systems, and a chronic toxicity test uses a renewal or flow-through system, exposing daphnids to a chemical from shortly after birth until well into adulthood to assess effects on survival, reproduction, maturation, fecundity, and growth.

Fish acute toxicity tests can be used to develop mortality data to calculate 24-, 48-, 72-, and 96-hour LC_{50}. Fish bioconcentration tests describe a procedure for continuous exposure of fathead minnows (*Pimephales promelas*) to a test substance in a flow-through system for up to 28 days; parameters include uptake rate constant (k_1), depuration rate constant (k_2), and steady-state bioconcentration factor, BCF (k_1/k_2). A fish early life stage toxicity test describes conditions and procedures for continuous exposure of several representative species to a chemical substance during egg, fry, and early juvenile life stages; species covered are fathead minnow, sheepshead minnow, brook trout, rainbow trout, Atlantic silverside, and tidewater silverside. A procedure is described to develop test data on acute toxicity (96-hour EC_{50}) for Eastern oysters (*Crassostrea virginica* [Gmelin]) and on oyster bioconcentration through continuous exposure of Eastern oysters to a test substance in a flow-through system. Mysid and penaeid shrimp acute toxicity testing develops data on concentration-response curves and 48- and 96-hour LC_{50} values; chronic mysid toxicity tests are described to develop a MATC (Maximum Acceptable Toxicant Concentration) and to quantify effects on specific chronic parameters.

Part 797 Subpart C contains guidelines for testing for terrestrial effects. Tests are designed to develop data on avian acute and chronic oral/dietary toxicity and data on reproductive effects, using bobwhite and mallard. Seed germination, root elongation, and early seedling growth toxicity testing is described, using seed of commercially important terrestrial plants to develop data on phytotoxicity. Plant uptake and translocation tests use commercially important terrestrial plants to develop data on the quantity of chemical substances incorporated in plant tissues and the potential for entry into food chains with resultant indirect human exposure.

40 CFR Part 798 describes health effects testing guidelines. Test methods described in this Part are used in determining human health effects of toxic chemicals released in the environment. The Part covers general toxicity testing, subchronic exposure, chronic exposure, specific organ or tissue toxicity, genetic toxicity, neurotoxicity, and special studies.

17.1.4 TSCA Section 4 Testing Process

An example of the operation of this Section 4 testing process can be found in the rulemaking procedure undertaken by EPA for the testing of certain chlorinated benzenes (51 FR 11756, 7 April 1986). For that procedure, a final test rule designated test standards for manufacturers and processors of various types of chlorinated benzenes. They were to be tested for: soil adsorption coefficient; environmental effects testing to include acute and chronic toxicity to mysid shrimp; and environmental effects testing to include 96-hour LC_{50} for the fathead minnow, 96-hour EC_{50} for *Gammarus,* and acute toxicity to silversides. Data from the required studies were to be submitted within certain time periods. Ordinarily, EPA has done the test rules in a two-phase process, first designating testing procedures, then studying industry-submitted study plans and schedules for testing, and finally designating appropriate testing guidelines which delimit testing standards. Pursuant to a court decision in 1984 (*NRDC v. EPA*, 595 F. Supp. 1255 [S.D.N.Y. 1984]), which upheld this process but imposed a "reasonable time frame" period on it, EPA has begun expediting the test rule procedures into a single-phase rulemaking process for most Section 4 test rules (51 FR 23706, 30 June 1986). Thus, when testing procedures are designated, now ordinarily the required testing guidelines will also be designated, as was done for these chlorinated benzenes.

Under TSCA Section 4(e), EPA is authorized to promulgate regulations under Section 4 requiring testing of chemicals in order to develop data relevant to determining the risks such chemicals may present to health and the environment. Section 4(e) established the Interagency Testing Committee (ITC) to recommend to EPA those chemicals to be given priority consideration in proposing Section 4 test rules. This list of priority chemicals, limited to 50 substances and mixtures at any one time, is updated at least every six months.

When the ITC formulates this priority list, it also can recommend to EPA which specific health or environmental effects testing should be done, in a general sense. For example, in its eighteenth report (51 FR 18368, 19 May 1986), the recommendation of tributyl phosphate was accompanied by a request for testing for certain ecological effects: chronic effects on aquatic and terrestrial plants; chronic effects on daphnids and/ or other aquatic invertebrates; and acute and chronic effects on benthic organisms and soil invertebrates, if these were found to be persistent under anaerobic conditions.

17.2 Clean Water Act (CWA)

Section 304(a)(1) of the Clean Water Act (CWA; 33 U.S.C. Section 1251 *et seq.*) requires EPA to publish and periodically update ambient water quality criteria. These criteria are to encompass the state-of-the-art scientific knowledge of the identifiable effects of pollutants on public health and welfare, aquatic life, and recreation. Such criteria have been issued periodically (e.g., "Blue Book" 1972; "Red Book" 1976). In 1980 (45 FR 79318, 28 November 1980) and 1984 (49 FR 5831, 15 February 1984), EPA published 65 individual ambient water quality criteria for toxic pollutants (termed "priority pollutants") listed under CWA Section 307(a)(1). New priority pollutants have since been added to this list. A number of these criteria, and national guidelines for criteria development, are currently being updated (50 FR 30784, 29 July 1985; 51 FR 8361, 11 March 1986; 51 FR 16205, 1 May 1986). Under CWA Sections 303 and 401, and regulations in 40 CFR Part 131, states are given primary responsibility for developing water quality standards and effluent limits to meet those standards; they can use the EPA-developed criteria as guidance in setting their water quality standards and effluent limits. EPA reviews the state standards and limits, and develops revised or additional standards or limits, as needed, to meet the requirements of the Clean Water Act.

For the standards for allowing effluent permits for the CWA Section 307 priority or toxic pollutants, consideration must be given to: toxicity; persistence/degradability; usual or potential presence of affected organisms in any waters; the nature and extent of the effect of the toxic pollutant on such organisms; and the extent to which effective control is being or may be achieved under other regulatory authority. The water quality criteria published by EPA for each of the priority/toxic pollutants may be used as the bases for setting standards for technology-based effluent limitations, which are translated into industry-specific National Pollutant Discharge Elimination System (NPDES) discharge permits. Where necessary, additional requirements can be imposed on discharge permits to assure attainment and maintenance of water quality standards and protect designated water uses, including an adequate margin of safety.

CWA Section 304(h) requires EPA to publish guidelines establishing test procedures for the chemical analysis of pollutants. Test procedures have previously been approved for about 115 different parameters. For any given parameter, the guideline regulations generally approved several different analytical methods. Alternate test procedures developed and proposed by effluent dischargers may be approved as equivalents by EPA (40 CFR Part 136).

In 1984, the guidelines establishing test procedures for the analysis of pollutants were amended by adding the priority toxic organic pollutant parameters and approved alternate test procedures (49 FR 43234, 26 October 1984). EPA divided priority toxic organic pollutants into 12 categories, based on their physical and chemical properties and chemical structures. A gas chromatography or high-pressure liquid chromatography test procedure was then developed for each category, with the expectation that the pollutants within each category could be measured by a single procedure. These procedures are to be routinely used where the pollutants are known to have a high probability of occurrence, or for qualitative identifications of unknown materials.

EPA is now also recommending the use of biological techniques as a complement to chemical-specific analyses to assess effluent discharges and express permit limitations. EPA has issued a national policy statement on the development of water quality-based permit limitations for toxic pollutants (49 FR 9016, 9 March 1984). In addition to enforcing specific numerical criteria for toxic pollutants, EPA (and the states in their role of granting NPDES permits) will also use biological techniques and available data on chemical effects to assess toxicity impacts and human health hazards based on the general standard of "no toxic materials in toxic amounts." EPA has stated that:

> In many cases, all potentially toxic pollutants cannot be identified by chemical methods. In such situations, it is more feasible to examine the whole effluent toxicity and instream impacts using biological methods rather than attempt to identify all toxic pollutants, determine the effects of each pollutant individually, and then attempt to assess their collective effect (49 FR 9017, 9 March 1984).

Data developed from an evaluation of effluent toxicity can then be analyzed in conjunction with chemical and ecological data, to aid in developing regulatory requirements to protect aquatic life. CWA Section 308 authorizes EPA to require of the owner/operator any information reasonably required to determine NPDES permit limits and to determine compliance with standards or permit limits; biological methods are specifically mentioned.

The principal advantages of using biological techniques for testing to assess water quality impacts are that: the effects of complex discharges of many known and unknown constituents can be measured only by biological analyses; bioavailability of pollutants after discharge is best measured by toxicity testing; and pollutants for which there are inadequate chemical analytical methods or criteria can thus be addressed (49 FR 9018, 9 March 1984).

The chemical, physical, and biological testing to be conducted by individual dischargers is determined on a case-by-case basis. Among the factors to be considered are: the degree of impact; the complexity and variability of the discharge; the water body type and hydrology; the potential for human health impact; the amount of existing data; the level of certainty desired in the water quality assessment; other sources of pollutants; the ecology of the receiving water; and the designated use for the receiving water. Where biological data are used to develop a water quality assessment or where the potential for water quality standards violations exist, biological monitoring may be required to insure continuing compliance with standards.

EPA has issued a Technical Support Document for water quality-based toxics control. Issues which EPA sees as currently involved in the water quality-based toxics control process are: correlation of effluent toxicity measurements to actual instream impact; inherent toxicity test method variability; on-site versus off-site toxicity testing; flow-through versus static and static-renewal toxicity testing; variability of effluent, exposure, and species sensitivity; acute/chronic toxicity ratio; and behavior of toxicants and toxicity after discharge, including persistence, additivity, antagonism, and synergism.

17.3 Clean Air Act (CAA)

Under Section 112 of the Clean Air Act (CAA; 42 U.S.C. Section 7401 *et seq.*), EPA is directed to prepare and update a list of hazardous air pollutants that it intends to regulate and to set national emission standards for hazardous air pollutants (NESHAPs), for existing and new stationary sources, that meet the regulatory goal of protection of public health. Hazardous air pollutants are defined as those that pose a localized risk of severe harm to human health. Hazardous air pollutants are those that "cause or contribute to air pollution that may reasonably be anticipated to result in an increase in mortality or an increase in serious irreversible, or incapacitating reversible, illness" (CAA Section 112 [a][1]).

Section 112 specifies two regulatory endpoints for setting NESHAPs: they must be emission or work practice standards, and they must protect public health "with an ample margin of safety." By 1980, EPA had listed seven pollutants: asbestos, benzene, beryllium, inorganic arsenic, mercury, radionuclides, and vinyl chloride; coke oven emissions were added in 1984. By 1984, standards had been proposed or finalized for most, though recently several proposed standards have been withdrawn or modified. There has been extensive litigation over the process of listing and then regulating pollutants under Section 112. The basic arguments revolve around the process and timing of listing pollutants and the weighing of benefits and costs in assessing risks.

The standards derived for the NESHAPs are organized around considerations of human health effects, rather than ecotoxicological effects. Early regulatory concern under the Clean Air Act centered on other pollutants which are termed "priority pollutants" (NO_2, CO, Pb, SOx, O_3, and particulates). These pollutants are subject to primary and secondary National Ambient Air Quality Standards (NAAQS), which could encompass ecotoxicological considerations. The primary standards (numerical emission amounts allowable) are designed specifically to protect public health and standards are designated for each pollutant after examination of human health effects. The secondary standards are to be designed to protect the public welfare, though primary consideration has historically been given only to building and crop damage, as opposed to more general areas of ecological concern. The criteria documents used in the setting of primary and secondary NAAQS are currently under review by EPA.

In contrast, ecological damage could be a major consideration in setting standards for protection from stratospheric ozone depletion under Section 157 of the CAA. EPA has initiated a research effort to assess issues related to protection of the stratosphere. International protocols limiting production of chlorofluorocarbons have recently been signed, in part because of the potential for ecological damage resulting from increased ultraviolet radiation as the stratospheric ozone layer becomes depleted.

17.4 Resource Conservation and Recovery Act (RCRA)

Under Sections 1108(a)(3) and 4004(a) of the Resource Conservation and Recovery Act (RCRA; 42 U.S.C. 6901 *et seq.*), EPA promulgated criteria for classification of solid waste disposal facilities and practices, codified as regulations in 40 CFR Part 257. These criteria include environmental performance standards that are used for determining the regulatory endpoint of which solid waste disposal facilities and practices pose a reasonable probability of adverse effects on human health or the environment. Those facilities that violate the criteria are termed "open dumps." Current enforcement is through the States or by suits brought in court by affected parties.

In 1984, the Hazardous and Solid Waste Amendments (HSWA) were passed. These amendments required EPA to submit a report to Congress addressing whether the current criteria in 40 CFR Part 257 are adequate to protect human health and the environment and recommending, if necessary, additional authorities to enforce the criteria. Additionally, EPA is required to revise the criteria for facilities that may receive household hazardous waste or small quantity generator hazardous waste, including provisions for groundwater monitoring, location criteria, and corrective actions (51 FR 20671, 6 June 1986).

Primary emphasis before the HSWA were enacted in 1984 was on the development of a "cradle-to-grave" manifest or tracking system for hazardous wastes. The 1984 Amendments imposed substantial new responsibilities, focusing on the land disposal of hazardous wastes. HSWA Section 201 establishes a statutory presumption against land disposal, which includes any placement of hazardous waste in a landfill, surface impoundment, waste pile, injection well, land treatment facility, salt dome formation, salt bed formation, or underground mine or cave. The presumption against disposal is rebuttable if EPA determines, for a particular waste, that one or more methods for land disposal are protective of human health and the environment. This can be demonstrated by a petitioner showing "to a reasonable degree of certainty that there will be no migration of hazardous constituents from the disposal unit or injection zone for as long as the waste remains hazardous" (RCRA Sections 3004[d][1], [e][1], [g][5]). There is a second method of rebuttal when EPA sets "levels or methods of treatment, if any, which substantially diminish the toxicity of the waste or substantially reduce the likelihood of migration of hazardous constituents from the waste so that short-term and long-term threats to human health and the environment are minimized" (RCRA Section 3004[m]). If EPA fails to set treatment standards or grant case-by-case exemptions, hazardous wastes will be subject to partial land disposal restrictions or banned from land disposal.

A schedule for land disposal restrictions, codified as 40 CFR Part 268, was made final in 1986, for a specified group of wastes known as the California list (i.e., liquid hazardous waste containing PCB's, and liquid and solid hazardous wastes containing halogenated organics, certain metals, free cyanides, or corrosives) (5 FR 19300, 28 May 1986). This rule established treatment standards and effective dates of land disposal restrictions.

Early in 1986, EPA proposed a framework for a regulatory program to implement the Congressionally mandated land disposal prohibitions (51 FR 1602, 14 January 1986). Included in these proposed regulations were procedures by which EPA will evaluate petitions demonstrating that continued land disposal is protective of human health and the environment; evaluate the best demonstrated achievable technology for minimizing toxicity or migration; and establish "screening levels" to identify the maximum concentration of a hazardous waste constituent below which there is no regulatory concern for land disposal and which is protective of human health and the environment. Screening levels are calculated based upon toxicological effects levels for these constituents and their fate and transport in air, groundwater, and surface water. The levels are expressed as a back calculation of maximum concentration levels in extracts. EPA will calculate screening levels from use of constituent fate and transport models to assess attenuative processes occurring during transport, based on a generic scenario with Monte Carlo simulations to accommodate variations, with the 90% level of the Monte Carlo probability distribution as the appropriate regulatory level. Concentrations at the point of human exposure should not exceed a designated human health effect level.

Because the standard for protection relates both to human health and the environment, EPA also proposes to evaluate effects such as toxicity to aquatic life in surface water in establishing regulatory thresholds (screening levels) (51 FR 1629, 14 January 1986). Aquatic life criteria for freshwater and saltwater organisms follow the ambient water quality criteria guidelines issued for the Clean Water Act (see discussion above). If a hazardous waste constituent does not have an already established water quality criteria guideline, EPA will identify, based on other available data, the level at which the hazardous constituent may cause harm to the environment.

For determining environmental effects when considering a petition for continued land disposal of hazardous wastes, consideration will be made of harmful effects on aquatic biota, wildlife, estuaries, vegetation, and protected lands. This can be demonstrated solely on the basis of fate and transport analyses showing concentration levels at or below established levels protective of the environment at the point of exposure. Included for analysis, at a minimum, are: endangered or threatened species and their habitats, wetlands, wilderness areas, wildlife refuges, coastal areas, and other areas of potential economic or ecological significance. A petitioner must demonstrate either no exposure or no-significant-effects level exposure. Established water quality criteria must not be exceeded, or, where no criteria exist, qualitative structure-activity relationships can be used to determine the significance of exposure. EPA is currently seeking more outside input on establishing final environmental effects testing standards.

17.5 Comprehensive Environmental Response, Compensation, and Liability Act (CERCLA) or Superfund

The Comprehensive Environmental Response, Compensation, and Liability Act (CERCLA or Superfund; 42 U.S.C. 9601 *et seq.*) establishes federal authority to deal with releases or threats of releases of hazardous substances from vessels and facilities into the environment. The list of covered hazardous substances contains approximately 700 names currently and EPA may designate additions to the list.

CERCLA requires a person in charge of a vessel or facility to notify the National Response Center immediately when there is a release of a designated hazardous substance in an amount equal to or greater than the reportable quantity (RQ) for that substance. CERCLA Section 102(b) establishes RQs for releases of designated hazardous substances at one pound unless other RQs are assigned under the Clean Water Act or are designated as an adjusted RQ by EPA. Government personnel will assess each release on a case-by-case basis under the National Contingency Plan.

EPA has adjusted RQs of 340 hazardous substances (50 FR 13456, 4 April 1985) and proposed adjustments of RQs for 105 more substances (50 FR 13514, 4 April 1985). The strategy used was an evaluation of intrinsic physical, chemical, and toxicological properties of each designated hazardous substance. The primary criteria are aquatic toxicity, mammalian toxicity (oral, dermal, and inhalation), ignitability, reactivity, and chronic toxicity and potential carcinogenicity. After the primary criteria RQs are determined, substances are further evaluated for their susceptibility to certain extrinsic degradation processes. These secondary criteria processes are biodegradation, hydrolysis, and photolysis (BHP). The BHP criteria are used, where appropriate, to raise RQ values one level from that suggested by the primary criteria analysis.

17.6 Safe Drinking Water Act (SDWA)

The Safe Drinking Water Act (SDWA; 42 U.S.C. 300f *et seq.*) directed EPA to promulgate National Revised Primary Drinking Water Regulations for organic, inorganic, microbial, and radionuclide contaminants in drinking water. Detailed assessments are to be made of experiences since interim regulations were proposed: occurrence frequency and human exposure potential; human health concerns and basic toxicology; water treatment technologies and costs; analytical chemistry and monitoring methods; and implementation options (48 FR 45502, 5 October 1983).

The primary drinking water regulations are to specify for each contaminant either maximum contaminant levels (MCLs) or treatment techniques, if it is not economically or technically feasible to ascertain the level of a contaminant in drinking water. The MCLs are the enforceable standards; recommended maximum contaminant levels (RMCLs) must also be specified, which are non-enforceable health goals for public water systems. RMCLs are to be set at a level at which no known or anticipated adverse health effects occur and which allows an adequate margin of safety. MCLs must be set as close to RMCLs as feasible.

In addition to the primary regulations, EPA is required to set secondary regulations that are to protect the public welfare, applying to contaminants which adversely affect the odor or appearance of water.

EPA has promulgated RMCLs and has proposed MCLs and primary regulations for some volatile synthetic organic compounds, synthetic organic compounds, inorganic chemicals, and microorganisms most commonly found in drinking water drawn from groundwater sources (50 FR 46880, 46902, 46936, 13 November 1985).

The SDWA also mandates regulation of underground injection of fluids through wells, accomplished through the Underground Injection Control program, in which EPA-approved programs are administered by the States.

EPA establishes Pesticide Guidance Levels for pesticides that have been found in drinking water and may do so for those that are likely to be found. These advisory levels are intended to help state and local authorities in their responses to drinking water contamination, but they are not mandatory federal standards. Such formal standards are established when MCLs and RMCLs are set for pesticide ingredients.

17.7 Federal Insecticide, Fungicide, and Rodenticide Act (FIFRA)

Under the Federal Insecticide, Fungicide, and Rodenticide Act (FIFRA; 7 U.S.C. 136 *et seq.*), all pesticides that are sold or distributed in commerce must be registered. For such registration, data must be available to allow EPA to evaluate a pesticide's risks and benefits. Products are registered by EPA only if there is sufficient information to make the positive risk–benefit determinations required by FIFRA. Regulations codified as 40 CFR Part 158 identify the types of data that EPA requires to make these determinations. Part 158 covers the full range of data requirements pertaining to the registration, reregistration, or experimental use of each pesticide product under FIFRA. Hereafter, use of the term "registration" will apply to new registrations as well as reregistrations accomplished under FIFRA Section 3(g) (see discussion below).

17.7.1 Registration Requirements

Since 1975, EPA has established basic requirements for registration of pesticide products in 40 CFR Part 162, Subpart A. Subsequently, guidelines were issued which described the kinds of data that must be submitted to satisfy the requirements of the registration regulations. The guidelines included sections detailing what data were required and when, standards for conducting acceptable tests, guidance on evaluation and reporting of data, and examples of acceptable protocols.

In 1981, EPA decided that it was impracticable and unnecessary to include in a regulation most of the detailed technical and scientific information contained in the guidelines. This was primarily because it was recognized that there may be several acceptable or even preferable protocols and provisions in addition to those in the guidelines; additionally, the diversity of pesticides and rapid changes in state-of-the-art chemical testing precluded making specified procedures universally applicable. Therefore, EPA decided to reorganize the guidelines and limit regulations to a presentation of data requirements and dates for fulfillment. Information on standards for conducting acceptable tests, guidance on evaluation and reporting of data, and examples of protocols are now not specified as requirements under FIFRA regulations but are available, as advisory information only, in Pesticide Assessment Guidelines from EPA.

40 CFR Section 158.130 designates environmental fate data requirements for registration of pesticides. This section sets forth the data required to demonstrate the fate of pesticides in the environment through degradation, metabolism, mobility, dissipation, and accumulation.

Degradation studies performed in the laboratory generally require data on hydrolysis and photodegradation in water. Data are normally not required on photodegradation on soil if the use the permit is being sought for involves application to soils solely by injection of the product into the soil or by incorporation of the product into the soil upon application. Data are required on a case-by-case basis on photodegradation in air depending on the product use pattern and other pertinent factors.

Laboratory data are required on aerobic soil metabolism for most permits. Anaerobic soil metabolism data are not required if an anaerobic aquatic metabolism study has been conducted. Anaerobic aquatic data are required for aquatic food crop and nonfood, and forestry uses, for a normal or an experimental use permit.

Data are required on leaching and adsorption/desorption for most conditional use and normal permits. Lab and field data on volatility are required on a case-by-case basis depending on product use patterns and other pertinent factors.

Field soil dissipation data are required for permits for terrestrial food crop and nonfood, forestry, and domestic outdoor uses of pesticides. Aquatic sediment dissipation data are required for aquatic food crop and nonfood uses. Combination and tank mixes data on dissipation are required on a case-by-case basis depending on product use patterns and other pertinent factors. Long-term soil dissipation data are required if pesticide residues do not readily dissipate in soil.

A confined accumulation study on rotational crops is required when it is reasonably foreseeable that any food or feed crop may be subsequently planted on the site of pesticide application. A field accumulation study on rotational crops is required if significant pesticide residue is likely to be present in the soil at the time of crop planting, as evidenced by residue data obtained from a confined accumulation study. Data are required on accumulation in irrigated crops if it is reasonably foreseeable that water at a treated site may be used for irrigation purposes. Data are required on accumulation in fish if significant concentrations of the active ingredient and/or its principal degradation products are likely to occur in aquatic environments and may accumulate in aquatic organisms. Data are required on accumulation in aquatic nontarget organisms if significant concentrations of the active ingredient and/or its principal degradation products are likely to occur in aquatic environments and may accumulate in aquatic organisms, unless a tolerance or an action level for fish has been granted.

In 1985, EPA promulgated new regulations governing Special Reviews of pesticides. The Special Review process determines whether the use of a pesticide poses unreasonable adverse effects to humans or the environment. (This review process has until recently been termed "Rebuttable Presumption Against Registration.") The Special Review is triggered when one or more of specified criteria are met or exceeded. The criteria are designed to assure that the determination of whether or not a pesticide poses an unreasonable risk will be based both on the toxic effects associated with the pesticide and the actual or projected exposure of humans and other nontarget organisms to the pesticide (50 FR 49003, 27 November 1985).

The criteria for initiation of a Special Review are contained in 40 CFR Section 154.7. EPA can call for a Special Review when, based on a validated test or other significant evidence, it is found that use of the subject pesticide may pose a risk of serious injury to humans or domestic animals. Other findings of risk which can trigger a Special Review are oncogenic, heritable genetic, teratogenic, fetotoxic, reproductive, or chronic or delayed toxic effects on humans. Environmental damage which can cause a Special Review involves a pesticide use which may result in residues in nontarget organisms at or above toxic levels or a use which produces adverse reproductive effects. Included in protection are endangered or threatened species. Additionally, pesticide use which threatens habitat destruction or adverse modification can be reviewed in a Special Review, as can any other use which poses a risk to humans or the environment which is not offset by social, economic, or environmental benefits.

FIFRA Section 3(g) requires the reassessment and *reregistration* of previously registered pesticide products that have not been evaluated according to the standards in FIFRA Section 3(c)(5). There are about 600 distinct active ingredients in the 48,000 pesticide products currently registered by EPA. Under the Registration Standards program, the scientific data base underlying each active ingredient is thoroughly reviewed and missing essential scientific studies are identified. The results of the review of the designated active ingredient are contained in a Registration Standard, which states EPA's regulatory positions regarding the products containing an active ingredient and the rationale for each position; also included are requirements for submission of additional data needed to complete the assessment, and label warnings or other restrictions required to protect health and the environment (51 FR 5246, 12 February 1986).

EPA encourages the public to provide information relevant to the review of individual active ingredients for which Registration Standards are scheduled. EPA lists pesticides under consideration for review each fiscal year. Particularly requested is information on human toxicology, residue chemistry, product chemistry, environmental fate, human exposure, and ecological effects. For FY86, 42 chemicals were scheduled for development of Registration Standards. After issuance of a Registration Standard, or when a significantly different use pattern for an already registered pesticide or a new chemical is registered, EPA prepares a Pesticide Fact Sheet, which summarizes information on chemical use patterns and formulations, scientific findings, EPA's regulatory position or rationale, and major data gaps.

17.7.2 Pesticides in Groundwater

EPA believes it is preferable to prevent the contamination of groundwater rather than remove pollutants from it (49 FR 42865, 24 October 1984). As part of the assessment of the safety of a pesticide, EPA evaluates potential for groundwater contamination based on data developed from the environmental fate requirements contained in 40 CFR Section 158.130 (see above), including hydrolysis, photodegradation, soil metabolism, adsorption/desorption, and dissipation under field conditions. Additionally, data on vapor pressure and solubility in water are required in Section 158.120 (product chemistry). These data, together with information on pesticide use patterns, are used in conjunction with predictive models to determine the likelihood of contamination from pesticide use. EPA can then, if warranted, impose restrictions on use through product labeling, such as limiting area of use, use rate, frequency of application, and formulation type. Groundwater monitoring may be required as a condition of registration. If risks outweigh possible benefits, registration can be denied, cancelled, or suspended.

EPA considered adoption of a separate groundwater criterion for initiation of a Special Review (see discussion above). It was concluded that the mere detection of a pesticide in groundwater was not a sufficient basis for a Special Review (50 FR 49008; 27 November 1985). If a pesticide is found in groundwater, EPA evaluates known contamination, the likelihood of more extensive contamination in the future, and available data concerning environmental fate and toxicity when deciding whether to initiate a Special Review. Health advisory levels may be set by EPA when pesticides are found, or thought likely to be present, in groundwater as a result of existing pesticide use.

Acknowledgments. This publication is ERC-167 of the Ecosystems Research Center, Cornell University, and was supported by the U.S. Environmental Protection Agency Cooperative Agreement Number CR812685–01. Additional funding was provided by Cornell University.

The work and conclusions published herein represent the views of the author and do not necessarily represent the opinions, policies, or recommendations of the Environmental Protection Agency.

Chapter 18

Environmental Decision Making in the Presence of Uncertainty

Mark A. Harwell[1] and Christine C. Harwell[1]

The previous chapter was an overview of the regulatory framework under which EPA effects environmental protection and management. When legislation is written in Congress, it contains both broad statements as to its general purpose and specific sections describing what activities are to be regulated and, sometimes, how such regulation is to occur. This language, along with the recorded legislative history (e.g., House and Senate reports, conference committee reports), indicates to the government agency charged with implementing the legislation the directions to follow in formulating regulations and enforcement mechanisms.

18.1 Regulatory and Ecological Endpoints

Regulations translate both the original general purposes of the legislation and the specific directions given by Congress into regulatory agency actions and requirements. When the legislation indicates, either through its general purpose language or through specific and more detailed sections of the legislative act, that there is a type of impact that is key to evaluating the actions being regulated, this is termed a *regulatory endpoint,* defined to be a regulatory decision-making norm which translates fundamental legislative purposes into regulatory action.

Making particular regulatory decisions concerning anthropogenic impacts on the environment requires translation of the generic regulatory endpoints into ecologically meaningful terms. This involves understanding not only the direct language used in regulation but also the intent of the regulatory community, by looking beyond the specific legal language

---

[1] Center for Environmental Research, Cornell University, Ithaca, New York 14853

to discern the essence of regulation intended with respect to ecological issues. The next step is the elaboration of this combination of language and intent into concepts of ecological relevance. For example, one generic regulatory endpoint widely used in water regulation is the phrase *balanced indigenous population*. But such a concept has no clear, intrinsic ecological meaning, and could be interpreted in a variety of ways. By understanding the intent and history of the regulatory community in using this phrase, ecologists can then translate it into its viable ecological meaning. For example, in regulation of the release into marine and estuarine ecosystems of less-than-secondarily treated municipal effluents under Clean Water Act §301(h), balanced indigenous population is much closer to the ecological concept of a natural, complete ecological community, especially within the benthos, representative of the vicinity of the potentially disturbed area, than it is to the idea of a single population maintained in a balanced or constant state, as the phrase might seem to imply (Harwell 1984). On the other hand, the same phrase, balanced indigenous population, is used with a different connotation in §316 of the Clean Water Act, pertaining to thermal pollution of freshwater systems by power plants; there it largely refers to population-level effects on specific species, especially on the important fish species in the water column. This example of differing connotations for the same regulatory endpoint applied in differing contexts highlights the necessity for appropriate linkage between the language and concepts of the regulatory arena and the language and concepts of the ecological community. There is no inherent impediment to bridging that language gap, but it must be done explicitly, or misinterpretation and misapplication of ecological principles will ensue.

The second major part of making particular regulatory decisions requires evaluation of the prospects for adverse effects from specific stresses under specific environmental conditions; this evaluation must rely on our full understanding of stress ecology. We focus attention here on the particular stress of toxic chemical inputs into ecosystems; however, these same ideas will largely apply to most other types of anthropogenic disturbances to the environment. The current understanding of how different ecosystems would respond to a variety of chemical inputs is certainly less than perfect. Many aspects of predicting ecosystem responses to stress are uncertain and will likely never be fully resolved. The present discussion will examine the origins and nature of uncertainties and will identify the current and prospective limits to ecological predictability.

18.2 Effects of Chemicals on Ecosystems

The effects of chemicals on ecosystems involve a large number of factors associated with the exposure of biota to a chemical and with the biotic responses to such exposures. Each of these factors incorporates physical or biological mechanisms; these mechanisms, in turn, vary in the degree to which they are understood, the availability of extensive data sets, and the sources and levels of uncertainty. These mechanisms are discussed elsewhere in this book and in other ERC reports (e.g., Levin et al. 1984); a brief recap of these ideas follows.

18.2.1 Exposure Regime

The definition of the regime by which the biotic constituents of an ecosystem are exposed to an introduced chemical is the initial step required for evaluating ecological effects. Understanding the exposure regime involves several facets, including:

1. the *fate and transport* of the chemical in the environment, i.e., how the chemical transits the ecosystem; the initial consideration is of the quantity, concentration, and distribution of the original chemical input into the ecosystem;
2. the *frequency and duration of chemical inputs* and other aspects related to acute versus chronic exposures; for example, there can be significant differences between a single pulse of a chemical into a stream versus a continual release of effluents;
3. *chemical speciation* or other transformations that occur as the environment interacts with the chemical; i.e., once the chemical is released into the open environment, it may undergo reactions with other constituents of the chemical and physical environment, thereby being transformed; as an example, mercury released as a metal can be converted into methyl mercury, with considerably different properties for fate, transport, bioavailability, and biotic effects;
4. *bioavailability* of the chemical, i.e., its mobility in being taken up by living organisms from an abiotic state; a chemical must somehow pass through into the biota for effects to take place;
5. *feedbacks* between biological effects and those environmentally induced transformations; for instance, toxicity to microorganisms could change the rates of degradation of the chemical;
6. *interactions* with other anthropogenic chemicals or stresses in the ecosystem; as an example, the presence of acidic precipitation could affect the physicochemical conditions in the environment sufficiently to alter the fate and transport of an anthropogenic metal;
7. *partitioning* of the chemical into components of the ecosystem; this is of considerable importance in establishing a differential exposure of a chemical to different parts of the ecosystem, rather than having merely an average exposure uniformly experienced by all biotic components; for example, sorption of a metal by sediments often result in much higher exposures for benthic infauna than for pelagic species with little contact with the sediments;
8. *remobilization* of the chemical from temporary sinks to become again

bioavailable for other components; this can be a critical issue in the continued chronic exposure; for example, the PCBs that are mostly bound into Hudson River sediments are being gradually released into the water column, thereby providing a chronic dose to fish and other important species downstream of the major zone of contamination; and

9. *loss* from the ecosystem through irreversible partitioning, chemical degradation, or export across the ecosystem boundaries; rates of this loss may have a major role in allowing ecosystem recovery, as discussed below.

18.2.2 Direct Biological Effects

There are several types of direct biological effects, including mortality of adults, mortality of other life-stages, and disruption of reproductive cycles. Each of these can result in direct changes in the populations of chemically affected species. On the other hand, direct mortality of *individuals* may not necessarily lead to *population* reductions, as differential toxicity of the chemical to individuals in the same population can allow compensation to occur, where less affected individuals have an enhanced survival rate because of lowered competition with individuals that are affected by the chemical. Other effects may not result in mortality directly, but could cause sublethal physiological or behavioral effects; however, these may produce *indirect* biological effects (see below) and consequently may cause population-level impacts. Thus, population changes can occur even though no individuals were directly killed by the chemical; conversely, lack of population changes can occur even though direct mortality of individuals does occur. This issue is discussed later when considering extrapolation.

Key issues associated with *direct* biological effects include:

1. the *differential sensitivity of target organisms* to chemical effects, including both within-species variance and across-species differences; within-species variances can be genetically determined, opening the possibility for evolutionary adaptation to continued chronic exposure to the chemicals, particularly for short-lived species; across-species variances opens the possibilities for changes in inter-species relationships, discussed below;

2. the *relative importance of the directly affected species* with respect to its ecological, aesthetic, economic, or other value within the ecosystem and/or vis-à-vis human interest; for example, direct biological effects on the primary producers of the ecosystem would be intrinsically important to the entire ecosystem, and direct mortality of an endangered species would have considerable importance to humans, even though its role in the ecosystem itself might be minimal;

3. influences of the *physicochemical environment* on the direct dose–response relationships; an aquatic organism's toxicity to a particular chemical in the environment may be enhanced when water temperatures are elevated in the summer; or another example, the humidity and moisture availability may determine the status of a plant's stomata, and if these are opened, the plant has an increased exposure compared to when the stomata are shut; and

4. *interactions* with other chemical stresses on the target organism, including the potential for additive, synergistic, or antagonistic responses; simultaneous exposure to acid precipitation and ozone apparently results in far greater damage to many plants than exposure to each independently.

18.2.3 Indirect Biological Effects

The mechanisms for these types of effects essentially involve effects on the biotic–biotic interactions in the ecosystem. These indirect effects are the major mechanisms by which the ecosystem can become a vector for propagating effects to populations not directly sensitive to the chemical perturbation. Many species–species interactions are important to the structure and/or functioning of the ecosystem, and, consequently, are particularly important with respect to adverse effects from toxic chemicals. Examples include:

1. *trophic effects,* i.e., in response to disruptions in food resources; this can involve changes through such relationships as plant–herbivore, predator–prey, host–parasite, and plant–pest interactions;

2. effects from *disruptions in habitat;* for example, toxic chemical effects near the smelters of Copper Hill, Tennessee, resulted in pervasive mortality to the dominant plant community, so that the soil was laid bare and subjected to extensive erosion, causing further damage and impeding ecosystem recovery;

3. changes in *competitive interactions,* such as competitive release following reductions in a species that shares a particular resource; this mechanism is one means for ecosystem-level compensation to population-level effects;

4. changes in *tissue concentrations* as the chemical passes through food webs, such as when it becomes concentrated in the tissues of organisms higher in the trophic structure; a well-known example is DDT concentrating in raptors; biomagnification is especially important for human exposure issues, where environmental toxicology (i.e., the environment as a vector for toxic exposure to humans) is the focus rather than ecotoxicology (i.e., the effects of a chemical on the ecosystem); and

5. impacts on populations that are parasites, symbionts, pollinators, or play other roles in *bi-specific interactions.*

As with direct biological effects, indirect effects may be substantially affected by the presence of multiple chemical inputs and by interactions with the physicochemical environment. Additionally, indirect effects mediated by biotic interactions, as well as indirect effects associated with ecosystem-level processes, may experience substantial time delays between the exposure, direct or primary biological response, and the eventual indirect effect.

18.2.4 Ecosystem-Level Effects

Ecotoxicological impacts at the ecosystem level integrate the direct and indirect biological responses to the chemical. This combination of mechanisms can lead to synergisms and other interactions, and can result in qualitatively new types of responses to chemical stress. Particularly important factors include:

1. direct or indirect effects on *critical species,* that is, on species that play crucial roles in the structure and/or functioning of ecological communities and ecosystems; for example, elimination of a keystone species that controls the populations of many other species through the interspecific interactions listed above can lead to substantial changes in the rest of the structure and functioning of the ecosystem;
2. changes in *community structure* caused by changes in the relative abundances of the constituent species and changes in inter-specific connections; for example, current research at the ERC on air pollution effects on forest dynamics suggests that such compositional alterations may occur over periods of many decades or centuries, as multi-species, complex competitive interactions that help determine ecosystem structure are altered by air pollution; and
3. effects on *ecosystem processes,* especially impacts on primary production, decomposition, and nutrient cycling; these process effects are often characterized by changes in photosynthesis rates of the dominant plants and/or by changes in the activities of the microbial communities that control decomposition and other nutrient cycling processes. Again the multi-chemical perturbation regime and the physicochemical state of the environment can greatly influence the nature of ecosystem-level responses.

18.2.5 Recovery Processes

The final element in evaluating chemical impacts on the environment is an assessment of the recovery mechanisms and rates following removal of the chemical input. Recovery here does not necessarily mean return to a pre-perturbed state or trajectory; indeed, establishment of a new base state may occur following normal or novel successional pathways. Interactions with the physicochemical environment, especially with other continued stresses, are of considerable importance to determining recovery. Recovery involves responses at a variety of hierarchical levels, including recovery of population levels, recovery of bi- or multi-species relationships, reestablishment of the trophic structure and the community composition, and redevelopment of ecosystem processes. Factors controlling recovery include:

1. the rates and effectiveness of *stress removal;* if a chemical is quickly lost from the ecosystem, either through chemical degradation, partitioning into a component from which there is little bioavailability, or export from the ecosystem, then the exposure of the ecosystem components after chemical inputs are removed decreases rapidly; by contrast, if the chemical is maintained within the environment in a bioavailable state, chronic, continued exposure will result, and recovery will be delayed;
2. the *maximum disturbance* experienced by biota and processes, that is, how extensive the ecosystem alterations are and which components are primarily affected; recovery from less-damaged states is typically much quicker and more complete than recovery from more extensively damaged states;
3. reestablishment of the *physical habitat,* such as soils or other substrates, and the *biological habitat,* such as the structural complexity of a forest canopy; since the habitat is intimately linked to the biotic constituents of the ecosystem, there is a close feedback here, as habitat reestablishment allows development of additional populations of biota, which further alters the habitat allowing other species in; and
4. sources of *reestablishment of populations,* such as proximity to unaffected populations, availability of propagules, mechanisms of propagation, and regeneration times; this is fundamentally tied to the spatial scale of the ecosystem damage, and recovery of a small patch of damaged ecosystem embedded within a larger expanse of relatively intact ecosystem will be much more rapid and effective than recovery of a very large damaged ecosystem, isolated from unaffected ecosystems of similar type.

Each of the above facets is potentially involved in the exposure and response of an ecosystem to a chemical or other disturbance. Consequently, *a priori* evaluations of ecosystem effects, if they are to reflect the full range of ecosystem responses, must address each of these issues independently *and* must address all in combination. That in itself is obviously a complex task, not yet accomplished for any ecosystem. However, there is a complicating overlay onto this paradigm: specifically, the spatial, temporal, and structural heterogeneity of natural ecosystems. That is, each response must be considered with respect to: the normal temporal variation experienced by an ecosystem in the absence of disturbance; the normal variation of the ecosystem across space; and the variety of response times and other characteristics of different ecosystems.

These are not trivial issues; for example, recognizing the impacts of a chemical on a species that naturally experiences large population fluctuations requires an extensive data base, with the incumbent power to distinguish a weak signal from substantial noise. As an example of a biological factor that affects the signal : noise response, low signal : noise ratios will occur for populations of species that are primarily controlled by density-independent mechanisms and consequently experience large normal fluctuations; high signal : noise ratios will occur for long-lived species that have considerable control over their own environment but are sensitive to the particular chemical stress. We learn from human epidemiology that the direct detection of population-level effects typically requires strong, direct causal relationships and massive data bases; consider in the case of the health effects of smoking the resources required to demonstrate dose–response relationships sufficiently for policy decisions to be made.

The situation for ecotoxicology is even more difficult, as it must contend not only with primary reliance on extrapolations, but also with great heterogeneity and diversity in the unit of interest (i.e., ecosystems rather than humans), with the potential for a tremendous number of mechanisms for indirect effects, and with the simultaneous assault on ecosystems by many different kinds of anthropogenic perturbations. Since we do not know all of the relationships discussed above individually, let alone in combination, and since we have to contend with a multitude of ecosystems and chemicals, there are a tremendous number of sources of uncertainty in predicting ecological responses to chemical disturbances. In short, it is clear that uncertainties are inevitable in any ecological evaluation. Nevertheless, the situation is not as hopeless as it might first appear, and we can make reasonable environmental decisions in the presence of uncertainties. We will first examine the nature of uncertainties in somewhat more detail.

18.3 Sources of Ecological Uncertainties

In general, uncertainties can be categorized as relating to: *insufficient data* and understanding about ecological systems and processes; the necessity to *extrapolate* from relationships established for particular ecosystems, particular perturbations, and particular conditions to the ecosystem and chemical of interest; the plethora of *potential indirect effects* in ecosystems, which can lead to unanticipated consequences; and inherent *environmental stochasticity*. We will consider each of these in turn.

18.3.1 Insufficient Data

The first category is intrinsic to complex systems in general. Consider the rather long list of steps involved in the translation of a chemical and its associated characteristics to population- and ecosystem-level effects, as discussed above. There are many additional factors associated with each of the listed items, so that a finer level of detail than presented above would make even more apparent the genuine complexity associated with ecosystem exposure–response relationships. In essence, each of those factors should be well understood before the whole ecosystem response can be well understood. This is a formidable task for even a single ecosystem and a single chemical because of the multiplicity of organisms and chemical processes involved. In most ecosystems, perhaps all, a thorough understanding of the biota present, their interactions and roles in the ecosystem, their responses to environmental conditions, their mutual feedbacks with geochemical processes, their redundancies and population dynamics, and many other relationships, is lacking. The obvious gaps in our understanding are further highlighted when one considers the enormous diversity of ecosystems on Earth; the internal diversity within ecosystems with respect to scales of processes, structure, time constants, and so on; the diversity of anthropogenic chemicals available or in development and their variety of chemical states and characteristics; and the host of other simultaneous anthropogenic perturbations on ecological systems. Fully characterizing an ecosystem and a chemical input would be extremely resource demanding, a prospect not possible for virtually any single ecosystem, let alone for all ecosystems and all combinations of chemicals of human interest.

18.3.2 Extrapolation Issues

This leads directly to the second category of uncertainties, the necessity for extrapolation from a limited information base to make predictions of specific ecosystem responses to specific chemical stresses. This need for extrapolation is more pervasive than is generally appreciated: extrapolation is required for data and understanding gained from *laboratory data* on single-species toxicity (the primary data source); use of laboratory developed *artificial microcosms* (gnotobiotics); use of *pieces of real ecosystems,* in the form of cores taken for maintenance in the laboratory as well as in the form of field enclosures; use of ecosystem mathematical and computer simulation *models;* and reliance on the few experiments that have been conducted on *intact ecosystems,* such as the watershed experiments at Hubbard Brook and Coweeta or the lake manipulations at Experimental Lakes Area in Canada. Even with extensive data from each of these types of information sources, including experiments involving toxic stresses to intact natural ecosystems, evaluation of the specific effects of a particular mix of chemicals on a particular ecosystem under a particular exposure regime must rely substantially on extrapolation. Only *a posteriori* explanation of what actually happened, such as after the release of Kepone in the James River, does not fundamentally rely on extrapolation. Yet extrapolation by necessity includes elements of the ecosystem, chemical, and/or physicochemical environment that were not a part of the empirical experiences; consequently, uncertainties are inevitable.

Extrapolation From Laboratory Bioassays

Many sources of uncertainty exist for each type of information source; examples are listed in Table 18.1, and a few are discussed here. Prediction based upon *laboratory testing* of the response of individuals to toxics has long been depended upon to set standards of acceptable release of chemicals into the environment. Single-species, acute toxicity tests are simple, replicable, relatively inexpensive, and provide statistical distributions of responses. However, these tests do not necessarily represent the toxicity of *individuals in the environment,* because the tests do not include all of the conditions that interact with dose–response relationships, such as the environmental conditions of temperature, moisture, nutrient availability, and other stresses the individual may experience in the environment, e.g., from other chemicals or other anthropogenic activities (see Levin et al. 1984). Differences in responses to acute (i.e., brief but intense) exposures to toxic chemicals, the usual aim of laboratory testing, versus chronic or repeated exposures, the usual condition in the environment, are considerable and potentially quite important. Reasonable simulation of real exposure regimes in laboratory tests is typically lacking, and the ecological issues of fate and transport, partitioning and bioavailability, chemical transformation and bioconcentration, are simply missed in these tests.

Further, laboratory tests do not necessarily represent the toxicity of *populations* in the environment. For example, testing of individuals in one life stage may not represent the toxicity of the total population to the chemical because of differential sensitivity during other parts of the species' life cycle. There may well be compensatory mechanisms by which individual-level toxicity does not correspond to population-level effects; one example is reduction in density-dependent factors limiting a population when significant numbers of the population die as a result of toxicity effects, allowing other members of the population to have an enhanced probability of survival. Another example is that toxicity to large numbers of the population may reduce the population to a point so low that even though the remaining individuals do not die from toxic reactions, the population further declines because of an insufficient breeding size. Interactions of the chemicals with the environment may also change the results in the environment compared with a laboratory testing situation. Consequently, chemicals characterized as being relatively safe based on con-

Table 18.1. Uncertainty inherent in different information sources

Information source	Examples of sources of uncertainty from errors in extrapolations
Laboratory tests on single species	disregard of physical environmental conditions on dose–response relationships
	selection of particular life stages to represent entire life-cycle
	selection of very few species to represent the large number of species in the actual environment
	lack of interactions with other species
	translation of individual-level effects into population-level effects, such as not including compensatory mechanisms, behavioral responses, etc.
Artificial microcosms	creation of artificial relationships among species
	exclusion of important trophic and other relationships
	effects of scale differences between microcosm and actual environment
	most of sources listed for laboratory tests
Cores/enclosures	effects of scale differences between core/enclosure microcosm and actual ecosystems
	edge effects
	spatial heterogeneity of natural ecosystem
	reproducibility of test conditions
	elimination of natural physical features
Models	inexact or incorrect mathematical formulations
	effects of aggregation
	insufficient data for model
	exceeding bounds for which model is applicable
	sensitivity of results to changes in parameters or to changes in model construct
Intact ecosystem manipulations	representativeness of specific ecosystem to general ecosystem type
	representativeness of stress of tested chemical to stress of chemical not tested
	effects of specific climate and other physical environmental conditions on ecosystem response
	reproducibility
	measurement of relevant information to define ecosystem dose–response relationships
	duration of experiment versus time of ecosystem response and recovery

trolled laboratory testing may exhibit enhanced adverse effects when released into ecological systems. The converse also can occur, i.e., chemicals that exhibit toxic effects in the laboratory can be rendered nontoxic in the field because of environmental mediation.

Yet another limitation of laboratory testing is the selection of particular individuals and particular species for testing, selections that may not represent the large number of species and individuals in the real environment. There may be a directional bias here, in that highly sensitive species with respect to toxic stress may also be highly sensitive to manipulation; as a result, the most sensitive species in the environment may not in practice be amenable to collection and maintenance in the laboratory. Further, the tested species may not reflect the toxicity of the species with greatest ecological, economic, aesthetic, or other importance. Here the issue is how adequate the species is as a surrogate for the remainder of the ecosystem. It is typical of laboratory testing to focus more on issues of precision and reproducibility of results than on accuracy and representativeness of results. This approach is easier to document, and experiments are easier to design, but in using this approach, we tend to develop a false sense of security in testing schemes and may well be misled in drawing ecosystem conclusions from laboratory testing.

Uncertainties associated with some of these limitations may be reduced by increased research; for example, conducting more tests on more species and more life-stages may allow a more complete selection of surrogates. Similarly, use of structure–activity relationships to characterize more adequately the variety of chemicals of concern (since all chemicals cannot be tested under all conditions) can enhance the ability to extrapolate from laboratory testing. However, many of the sources of uncertainties listed for laboratory tests are an inherent, irreducible aspect of the testing approach; examples include the lack of consideration of interspecific interactions and the potential for population-level responses to differ from individual-level responses.

Extrapolation from Microcosm Testing

The use of *artificially designed microcosms* can begin to include some of the factors missing from single-species, acute testing, especially with respect to effects on bi-specific interactions. However, these types of experiments are still limited by the artificiality of the interactions with respect to the nature of inter-specific relationships, and especially to the limited number of such interactions that can be designed into these types of systems. By having a limited number of relationships, the overall complexity of the complete matrix of bi-specific interactions is markedly smaller than in the real environment, even in a part of the real ecosystem that is the same size as the microcosm. In addition, by having only a relatively small system in the laboratory, many issues of scale become important, issues such as exclusion of spatial heterogeneity, importance of edge or wall effects, and effects of isolation from external inputs. Artificial microcosms, however, can be reasonably replicable, and they at least begin to address some facets of multi-species considerations, but in general emphasis is misdirected to precision rather than accuracy, that is, to reproducibility across experiments rather than to reliable characterization of real-world effects.

These problems are reduced significantly for microcosms developed from *intact cores of real ecosystems* established under laboratory conditions, as opposed to totally artificial constructs, since many of the intrinsic relationships of real ecosystems can be maintained without even having to identify or specify those relationships. Even more faithful representation of the real ecosystem can occur in enclosure microcosms, i.e., where parts of actual ecosystems are isolated to control inputs and outputs to the system. Examples include the use of large plastic enclosures in marine ecosystems. But even for these relatively natural microcosms, there are limitations to applicability to the real world. An important factor is again the issue of scale, in which the limited size of the core or enclosure mandates exclusion of certain components and processes that occur in the real world. Careful design can minimize these effects, but they cannot be eliminated. For example, the marine microcosms established at the Marine Ecosystem Research Laboratory at the University of Rhode Island reflect very well the dynamics of the chemistry and smaller populations of biota seen in the Narragansett Bay ecosystem, but especially large organisms, such as certain fish species, cannot be included in the microcosms, and the lack of their inter-relationships within the ecosystem might substantially influence the predicted responses to chemical or other stress. Similarly, natural spatial heterogeneity of real ecosystems may result in the collection of a core to be established as a microcosm which poorly reflects the structure of the overall real ecosystem; this problem can be minimized by judicious selection of the core sampled, but can never be eliminated for heterogeneous systems. Edge effects can also become important here, and the very fact of isolating the piece of the ecosystem, while useful for quantifying inputs and outputs, may well alter natural physical features, thereby affecting the ecosystem's structure (e.g., by eliminating recruitment) and functioning (e.g., by preventing exchange of gases with the atmosphere).

Extrapolation from Models

Mathematical and computer simulation models can be designed to include multiple scales of the ecosystem, and, thus, do not necessarily have the limitations of scale found in microcosms, although in practice most models are not so well designed. Model limitations often relate to the formulation of the model itself, such as how multiple parameters characterizing the environment are aggregated into a limited number of parameters to make the model more tractable, or how the equations are written, for example, with linearizing relationships in order for the mathematics to be solvable, where nonlinear relationships would be much more accurate.

Rarely can models *a priori* predict ecosystem "surprises" because the controlling mechanisms are not built into the model.

Additionally, models are typically limited by the availability of extensive, reliable, relevant data bases; the most precise and accurate simulation model that cannot be calibrated to the real world because the data requirements cannot be met is quite limited in its applicability. This limitation has often been reached in development and application of ecosystem models for use on environmental problems; for example, many of the large models developed under the International Biosphere Program a decade ago are still largely unused because of unavailability of data. These models are in effect under-aggregated, requiring far more environmental data collection efforts than reasonably can be expected; consequently, ecosystem-level models have unfortunately somewhat lost favor as applied tools. It is rare for ecological models to be examined for their full sensitivities to changes in parameters or changes in model construct. Further, it is rare for model error analyses to be performed, i.e., evaluating how error terms in input variables translate into error terms for the resultant output variables produced by the model; yet such error analysis is necessary in order for the uncertainties of the model to be characterized and understood. More effective utilization of ecosystem models can occur if such sensitivity analyses and error analyses are conducted.

Extrapolation from Intact-Ecosystem Experiments

Finally, extrapolation from experiments on whole, intact ecosystems, which would seem to require the least amount of extrapolation, nevertheless are replete with this source of uncertainty. The first issue is the appropriateness of the ecosystem as a representative of ecosystems of its general type. Specific conditions may make the ecosystem less than an ideal surrogate, such as particular soil or substrate, slope or aspect, precipitation patterns, historical disturbance by humans or rare, extreme events. Additionally, different types of ecosystems respond differently to the same disturbance; ideally, a classification scheme could be developed that would identify canonical ecosystem types, each category of which included all ecosystems that responded similarly to a given stress and followed similar recovery pathways; however, such a canonical scheme is yet to be developed. Differences in the particular experimental conditions compared to the anthropogenic stress of concern constitute another important factor, such as disturbance by different chemicals and exposures under different conditions. We often feel that reliable ecological predictions could be made if only we had data on a real ecosystem exposed to a real chemical stress, but there are vanishingly few examples of intact ecosystems subjected to chemical disturbance under controlled and well-monitored conditions; consequently, there are very wide gaps in our information base of whole ecosystem responses. Extrapolation from long-term ecological studies to real regulatory situations is very limited, both now and in the foreseeable future. On the other hand, in those cases where real-world responses to anthropogenic stresses can be causally related, they can have a marked influence on societal decision making, much more so than a plethora of theoretical extrapolations.

18.3.3 Indirect Effects

The third category of uncertainty, that is, the tremendous number of potential indirect effects, is the primary factor in the discussions of "surprise" in stress ecology. This relates in part to the first issue, the extreme complexity of the ecosystem-chemical interaction system. It is important to note that it is not merely the large numbers of organisms and chemicals involved. There are, after all, extremely large numbers of molecules in a gas; each molecule of a particular type is precisely identical in its characteristics, so that the total range of responses can be characterized adequately, and often precisely, by statistical techniques; indeed, often these are the only way to characterize such systems. But individual organisms of the same species are not identical in their characteristics, sensitivities, histories, and many other qualities. Because of this, responses to chemical exposure may vary substantially within a single species. Variation in sensitivities across species is even greater, and roles of species in communities and ecosystems vary tremendously. Similarly, all species–species interactions are not the same, and therefore all routes of indirect effects are not of equal potential importance. Treating the very large numbers of potential multi-specific interactions as being assignable to one or a few homogeneous groups with uniform probabilities of occurrence will inevitably result in inaccurate predictions and unanticipated results. Again, *a posteriori* analyses of documented indirect effects may identify the causal interactions of critical importance, but precisely predicting these in advance is not possible. We can better understand the relative *likelihood* of types of indirect effects to occur, perhaps through examination of many different case histories. And a detailing of as many potential indirect effects as possible can be useful, although distinguishing viable pathways from indirect effects versus the plethora of possible pathways is very difficult. Clearly, uncertainties here will always remain.

18.3.4 Environmental Stochasticity

Finally, uncertainties are associated with variations in environmental conditions; examples include the timing and intensity of precipitation events; the presence of conditions conducive to fire; the development of transient eddies in ocean currents; occurrences of hurricanes and other rare events; and so on. Strictly speaking, such phenomena may be deterministic, and if there were adequate information, each variation could in principle be predictable. However, the reality is that we will never be able reliably to forecast these events very far into the future. For example, at one level, we may be reasonably certain about climatic conditions likely to occur at distant times, such as predicting that maximum daily temperatures in February in, say, Ithaca, New York, are likely to be near freezing and certainly will be less than 30°C, whereas in Honolulu they are certain not to be near freezing. Similarly, we can be reasonably assured that average annual temperatures in the Northern Hemisphere will be within 1°C or so of recent past values. On the other hand, it is not possible to predict today what the actual weather conditions are to be on 13 February 2001 in Ithaca, nor is it possible to predict where and when a hurricane will strike land next year. Such phenomena must be treated as if they are stochastic, with concomitant best estimates enveloped within a band of uncertainty. The width of that band relative to the best estimate value will vary substantially, depending on the nature of the variable of interest, the spatial and temporal scale over which a single (e.g., average) value is applied, the distance into the future for which predictions are being made, and other factors. Again, it must be emphasized that uncertainties in this category are inevitable.

Each of the four identified categories of uncertainties (i.e., insufficient data, need for extrapolation, potential for indirect effects, and environmental stochasticity) has this characteristic of a core uncertainty that is essentially irreducible. That is not to say that nothing can be done to minimize uncertainties; indeed, there are compelling reasons to conduct the types of research needed to narrow uncertainties as much as is feasible. The first three categories can be substantially reduced by concerted ecological studies, particularly research that addresses the mechanisms by which ecosystems function and respond to stress. By conducting experimental research on a variety of ecosystems chosen to be representative of classes of ecosystems, by coupling mechanistic laboratory studies with field experiments, and by actively using sophisticated computer simulation models for sensitivity analyses and hypothesis testing, ecological science can make a quantum advance in its predictive capabilities. We believe such coupled approaches can provide a synergism that is essential to advancing our understanding of stress responses in ecosystems.

18.4 Environmental Decision Making

18.4.1 Acceptability of Risks

There are inherent constraints against eliminating uncertainties entirely that must be recognized. However, as suggested previously, recognition of inevitable uncertainties in predicting how ecosystems will respond to stress, chemical or otherwise, does not result in an incapability of making appropriate environmental management decisions.

Environmental decision making occurs now in the presence of uncertainties, and will always continue to do so; we will always feel the need for more information concerning chemicals, their effects on biota, the propagation of effects through ecosystems, and the prospects for recovery once the chemical is gone. In essence, our capabilities to predict ecosystem responses to a particular stress constitute a translucent crystal ball, sufficiently clear so that the nature of stress responses can be discerned, sufficiently cloudy so that precise predictions are not possible, and surprises are inevitable. The task of ecological research is to enhance the transparency of the crystal ball, i.e., to improve our methodologies for risk assessment; the task of decision making is to make optimal decisions based on the existing clarity of the crystal ball, i.e., to optimize our risk management. There is risk in making incorrect decisions because the projections are too obscure. Depending on the acceptability of particular levels of risk by society, decisions can be made at different points in the process of improving the clarity of the crystal ball and with differing degrees of added safety. This is the central issue in the current controversy over acid precipitation regulatory responses: whether sufficiently clear understanding exists to mandate action now versus whether the crystal ball is still too obscured for reasonably accurate projections to be made vis-à-vis the economic costs of making an overly stringent regulatory decision.

That environmental decision making is at least partially a function of societal values is demonstrated by comparing environmental decisions in the United States with decisions in other countries. For example, the intrinsic willingness to accept considerably greater environmental risks from nuclear energy facilities in the Soviet Union than in the United States resulted in much greater regulatory control and specificity in the U.S. over its nuclear industry, with significant public involvement in defining acceptability of risk and with extensive safeguards designed into systems. Consequently, the redundancies and containment systems at Three Mile Island were responsible for minimal human impacts from a major reactor accident there. By contrast, the much lower regulatory oversight in the U.S.S.R. reflected governmental (if not societal) acceptance of much higher risks to the population from adverse environmental activities, and the lack of redundancies and containment systems in the Soviet Union's nuclear industry has placed a substantial population of humans at serious health hazard from the Chernobyl accident.

This is not to imply that there are still not legitimate concerns about risks from the nuclear industry in the United States, but it is clear from this example how societal values fundamentally affect environmental decision making in the presence of uncertainties. Additionally, even within a single society, there are tremendous differences in the acceptability levels for different risks; for example, the risk to the U.S. population, as measured by deaths per hundred million persons, from poisoning by contaminated analgesic capsules is extremely low (on the order of 10^{-8} per year), yet we have a considerable regulatory system for reducing that risk even further, with extensive requirements of testing, quality assurance, packaging control. In contrast, the populational risk for death associated with cigarette smoking is extremely large (on the order of 10^{-4} per year), constituting the greatest single source of adult mortality, yet the regulatory efforts seem to be innately ineffective and, furthermore, are made futile by societal subsidies to the tobacco industry.

18.4.2 Ecological Information to Improve Decision Making

How these differential perceptions of acceptability of risk interact with the environmental decision-making process is beyond the scope of this paper and outside the arena of ecological analyses; the point here is to recognize that decisions occur in a larger societal framework. Given that a societally determined definition of environmental risk is specified, however, there are ways to improve the use of ecological information in making appropriate decisions in the presence of uncertainties.

Optimal Use of Existing Information

The first factor is to use the full suite of the best available ecological data and understanding. In the large majority of cases, using imperfect or uncertain ecological information is preferred to ignoring ecological considerations totally. Our understanding of the types of responses likely to occur from a disturbance is sufficiently developed so that the crystal ball is neither totally obscure nor substantially distorting. For the most part, the overall picture of an environmental response to an anthropogenic action can be portrayed with a greater assurance of accuracy the more extreme the perturbation is. For example, the recent studies of the global environmental consequences of a large-scale nuclear war constitute to a large degree an exercise in stress ecology, but address particularly intense and extensive perturbations to the global environment (see Harwell and Hutchinson 1985). Consequently, these studies focus on direct effects on the primary producers and other major supporting bases of ecosystems, recognizing that major disruption in these components is sufficient to understand the qualitatively important aspects of the world after nuclear war; details of effects on specific animal populations in this case are extremely uncertain, but are also largely unimportant. In the case of the effects of a toxic chemical released locally into the environment, however, we do not have the relative luxury of being able to ignore many population-level responses (these may in fact be of central importance to the environmental decision-making process), so an altered standard for the clarity of the crystal ball may be required.

We can optimize this by taking advantage of the full gamut of ecological information accrued from different experiences involving different systems and stresses. Thus, ecotoxicological regulation should be continually improving in its capability and reliability. But recall that this reliance on other systems and other stresses is basically an issue of extrapolation. Since we must rely fundamentally on extrapolations, the best way to minimize the introduced error from those extrapolations is to extrapolate from a variety of *types* of information sources. For example, using laboratory experimentation in concert with field enclosure experiments greatly enhances the predictive capabilities compared with either technique singly. Often ecology is too ad hoc for effective application to specific environmental decisions, yet *selective* reliance on the wealth of ecological experiences can considerably improve our efficacious use of ecological information in making decisions.

Research to Improve Predictive Capabilities

Improvements in the ecosystems prediction crystal ball should be continually sought through research programs, designed both to reduce and to characterize remaining uncertainties. Reductions in uncertainties should be a constant goal of research programs, but we have already discussed the reasons why many components of uncertainty are for all practical purposes irreducible. An important consideration, however, is that when uncertainties are uncertain, we are in a much weaker position to evaluate the risk of incorrect decision-making. Having well-defined uncertainties, even when irreducible, can substantially assist decision making, in that comparisons of the levels of uncertainty that remains, the levels of risk that are acceptable, and the relative risks that are assigned to a variety of decision alternatives, can form the basis for reasoned decisions to be made. Consequently, quantification of existing (reducible plus irreducible) uncertainties can be an integral part of the environmental decision-making process.

Selecting Reasonable Criteria for Decision Making

By drawing on the best available ecological expertise, the probability of protecting against unacceptable adverse responses is substantially improved, even though there may not be sufficient certainty about a scientific projection for, say, a 95% confidence level to be assigned. The point here is that expecting to achieve that 95% capability is both unreasonable, considering the previous discussion on sources of uncertainties, and unnecessary. We establish apparent, though unnecessary, obstacles to environmental decision making when we impose high confidence levels of scientific proof as a criterion for the ecological information being incorpo-

rated into decisions. For example, it may never be possible to demonstrate with a 95% confidence level that an epidemiological response has
occurred from a particular chemical; thus, awaiting this level of scientific
proof before regulating against the chemical is the same action as deciding
never to regulate it. A much more reasonable approach is to draw on
several lines of ecological reasoning and data to establish the best estimate of chemical effects, and use this as the basis for regulating. This is
the approach used in regulating most of our pharmaceuticals; consider,
for example, the Delaney standard for carcinogenicity as sufficient to
preclude use of a drug, even though no epidemiological evidence exists
that the chemical would actually lead to human deaths from cancer. The
obvious reason for acceptance of this standard is that there is a considerable time-lag in such cancer development, while the consequences are
unacceptable and cannot be mitigated. Precisely the same reasoning applies to ecotoxicology, as opposed to human toxicology, and mandates
greater reliance on the judgment of ecologists.

Margins of Safety Through Use of Protection Factors

For optimal environmental protection, it is usually better to err on the
side of predicting there would be adverse effects from a human activity
than predicting minimal effects. Thus, an important objective in environmental forecasting is protecting against false-negative predictions more
than protecting against false-positive predictions. This approach is appro-
priately embedded in the imposition of safety factors in setting acceptable
standards, such as in assigning acceptable levels of exposure to radiation.
In this example the available empirical data are used to assign a level
below which direct adverse effects are considered unlikely; a safety factor
of several-fold is then applied, resulting in the regulated level being as
much as ten times lower than the level of expected consequences. Different safety factors are applied to different situations; for example, safety
factors for exposure of the general public to radiation are much higher
(and, consequently, the standards are much lower) than are the safety
factors applied to occupational exposures, the assumption being that persons deliberately working in a hazardous environment assume a greater
risk than the population that assumes the risk involuntarily. Protecting
more against false negatives (i.e., against ecological surprises) can also
become an integral part of designing methodologies for environmental
predictions.

Reliance on Relative Assessments

Often there is less of a need for absolute predictions than for *relative
predictions*. Many environmental decisions are not simply yes or no
choices of particular activities, but involve selection among management

alternatives. For example, regulation of industry for water pollution effects may involve selection of the least adverse mix of effluents or the optimal medium for disposal, rather than just elimination of all effluents. This comparative component of decision making can rely on assessing relative effects, rather than precisely projecting specific effects. The advantage here is that often the reliability of ecological assessment methodologies for relative assignment of effects among alternatives considerably exceeds the reliability for specific absolute predictions. For example, sensitivity analyses of complex ecosystem models often show considerable sensitivity in the absolute predictions of the model to specific parameter values, but the rank order of relative effects from differing management alternatives is far less sensitive to input parameters. This effect should be used to advantage in developing greater confidence in the projections used as the basis for environmental decisions. On the other hand, excessive reliance on relative assessments could lead the least adverse effect to be considered an acceptable level of effect, even though the absolute magnitude of the effect may be unacceptable.

Timely Input of Science to Decision Making

Ecological information is likely to be more useful the *earlier* it becomes incorporated into the decision-making process. Relying on the best ecological understanding in initially developing the framework for regulating an anthropogenic activity can be much more effective than waiting until decisions are in the latter stages of implementation. Thus, ecological insights directly incorporated into legislation or regulations can accommodate the inevitable presence of uncertainties and can place these in proper perspective compared to societal issues such as risk and economic costs or benefits. Ecological information can be especially useful at the stage of developing guidelines for implementing the regulations at the local or regional level, such as in preparing guidance to regional offices of EPA for making decisions on §403(c) regulation of offshore oil exploration (cf. Kelly et al. 1987), as it is at this stage that the translation of regulatory endpoints into ecologically meaningful terms can most effectively interact with establishment of a decision making paradigm. On the other hand, delay in inclusion of ecological considerations until the very end of environmental decision making can be very unrewarding, such as when adversarial forces become operative in litigation and individual "experts" are called upon to make unequivocal judgments about phenomena that are intrinsically uncertain.

Tiered Approach

Ecological information can often be applied in a tiered approach, in which early screening identifies potentially higher risk actions. Increased testing and other information-gathering can then be required selectively, alleviating undue burdens on activities of low intrinsic risk. This approach is particularly appropriate for chemical regulation, since there is a tremendous number of chemicals under development or production, the large majority of which have little potential for causing adverse ecological effects. Early elimination of these from the set of chemicals of concern allows limited resources to be focused most effectively on those issues of greatest potential damage to the environment. In establishing this tiered scheme, trade-offs must be made between the desire to reduce the size of the list of chemicals for which more extensive (and expensive) testing and other attention are required and the desire to reduce the incidence of surprise, i.e., of false-negative evaluations.

Efficacious Use of Sensitivity Analyses

Ecological models can be used much more effectively than currently for characterizing the sensitivity of responses to different levels of chemical stress (a part of the comparative approach discussed above), and the sensitivity to different environmental parameters and relationships, thereby better identifying the nature of predictive uncertainties and better defining the research programs needed to reduce those uncertainties. The current research at Cornell University's Ecosystems Research Center highlights the efficacy of this approach, in which a very large, complex model of forested ecosystems is being extensively explored for its response to changes in parameter values and model formulation (Weinstein et al. in prep.). Information from these studies can be used to aggregate the model further, thereby reducing the data requirements without undue loss of resolution or predictive capability of the model; to identify those important aspects of the model for which increased data and understanding are required, thereby specifying research programs; and to characterize the nature of uncertainties in model predictions, thereby enhancing the decision-making process.

Mitigation of Effects

Environmental decision making should explicitly incorporate considerations of mitigation of effects; that is, attention should be given to the possibilities and methods for mitigating adverse effects in those situations where the initial projections were incorrect and substantial adverse responses occur. If there is a considerable capability for mitigation, then the level of acceptable risk should be higher than the situation where mitigation is difficult or impossible. As an example of this consideration, there is an enhanced concern for protection against adverse effects on marine ecosystems since it would be extremely difficult, if not impossible, to mitigate against an adverse effect once it had occurred; consequently, there are international agreements against disposal of high-level radioactive wastes in the oceans, since if containment were breached, there would be no way to recover the released radionuclides. In contrast, there is no such agreement against disposal within a geologic formation, for which local containment could presumably be retained even if the waste containers were to be breached. Similar concerns for mitigation exist for contamination of underground sources of drinking water.

Adaptive Approaches to Decision Making

Finally, the basic reliance on an adaptive approach to environmental risk management is essential. We should not expect our crystal ball to be accurate in its predictions forever, nor for us to be saddled irrevocably with an environmental decision made at a single point in time and based on limited information. It makes so much more sense, once a decision has been implemented, to have a system explicitly relying on feedbacks from monitoring of the environment, with continual reevaluation of the effects on the ecosystem and the prospects for adverse reactions. As new information is gained through actual exposure of the actual ecosystem to the actual stress of concern (i.e., information not inherently limited by the extrapolation issues), our understanding of dose–response relationships increases tremendously, and we can actively use the real situation for its empirical value. This aspect of adaptive management of the environment fundamentally relies on coupling of research, regulation, and monitoring.

This is rarely done currently, and there are few opportunities for it to be done under current regulatory schemes. One exception is the regulation of releases of municipal effluents into certain marine and estuarine ecosystems, regulated under §301(h) of the Clean Water Act. This section imposed control on this human activity based largely on the projected impacts on the ecosystem, rather than on technology considerations (such as the concepts of imposing best available or best practicable technology at the end-of-the-pipe regulation, the primary basis for water regulation in the United States). Furthermore, §301(h) imposes a monitoring requirement on the municipalities that are allowed these types of effluents, monitoring both of the immediate, near-field ecosystem, and also of parts of the ecosystem far from the effluents and, presumably, unaffected by them. If these regulatory and monitoring requirements were carefully married into an overall adaptive management scheme, then the most effective environmental management, both from an ecological and an economic basis, would ensue. This is the ultimate way to develop a safe and effective scheme for environmental management and protection.

Acknowledgments. This publication is ERC-168 of the Ecosystems Research Center, Cornell University, and was supported by the U.S. Environmental Protection Agency Cooperative Agreement Number CR812685–01. Additional funding was provided by Cornell University.

The work and conclusions published herein represent the views of the authors and do not necessarily represent the opinions, policies, or recommendations of the Environmental Protection Agency.

References

Harwell CC (1984) Regulatory Framework of Clean Water Act Section 301(h). ERC-29, Ithaca, NY: Ecosystems Research Center

Harwell MA, Hutchinson TC, with Cropper WP Jr, Harwell CC, Grover HD (1985) *Environmental Consequences of Nuclear War, Volume II. Ecological and Agricultural Effects.* Chichester, UK: Wiley

Kelly JR, Duke TW, Harwell MA, Harwell CC (1987) An ecosystem perspective on potential impacts of drilling fluid discharges on seagrasses. Environ Manage 11(4):537–562

Levin SA, Kimball KD, McDowell WH, Kimball SH (1984) New perspectives in ecotoxicology. Environ Manage 8(5):375–442

Weinstein DA, Harwell MA, Buttel L (in prep) A Sensitivity Analysis of a Forest Simulator: Methodology and Application. ERC-147, Ithaca, NY: Ecosystems Research Center

Index